AN INTRODUCTION TO

THE MATHEMATICS OF MEDICINE AND BIOLOGY

An Introduction to

THE MATHEMATICS OF MEDICINE AND BIOLOGY

J. G. DEFARES
Physiology Laboratory, University of Leyden, The Netherlands

and

I. N. SNEDDON
Department of Mathematics, University of Glasgow, Scotland

Second edition, prepared with the assistance of

M. E. WISE
J. A. Cohen Institute for Radiopathology and Radiation Protection, at the Physiology Laboratory, University of Leyden, The Netherlands

1973

NORTH-HOLLAND PUBLISHING COMPANY
AMSTERDAM · LONDON

Library of Congress Catalog Card Number 73-75530
ISBN North-Holland 0 7204 4131 5
ISBN Year Book Medical Publishers 0 8151 2392 2

PUBLISHERS:

NORTH-HOLLAND PUBLISHING COMPANY – AMSTERDAM
NORTH-HOLLAND PUBLISHING COMPANY, LTD. – LONDON

SOLE DISTRIBUTION IN THE WESTERN HEMISPHERE:

THE YEAR BOOK MEDICAL PUBLISHERS, INC.
CHICAGO, U.S.A.

First edition 1960

PRINTED IN THE NETHERLANDS

PREFACE TO THE FIRST EDITION

This book is written to help research workers in the biological and medical sciences to acquire the mathematical techniques which are being used increasingly in their own fields. An inspection of recent numbers of medical and biological journals reveals that workers in these fields must be familiar with mathematical (and not just statistical) procedures. The two authors approached the problem from different directions. One is a physiologist who has discovered the need to know these techniques himself and the other is an applied mathematician who has had to teach mathematics to students of biology.

It is assumed that, in general, the reader of such a book will be a graduate well launched on his chosen career who has ceased the study of mathematics many years previously. For that reason the first two chapters develop many ideas which will be familiar already to anyone who has studied mathematics at a high school. Although it is the needs of such readers which have been primarily in our minds we hope that the book will be of use for work in courses on mathematics for biologists. The basic mathematical tools are developed but always against a biological background. Indeed about a third of the book deals with the applications of mathematics to problems in medicine and biology.

It may not be inappropriate to offer a word of advice to the reader who has grown unfamiliar with methods of working in mathematics. He must acquire the habit of working with a paper and pencil while he is reading the book. The only way to master even the simplest ideas in mathematics is to try to reproduce the arguments stated and solve the problems set for the reader.

In studying mathematics progress should not be measured by the number of pages read per hour. Real progress may be made even if the pace is sometimes less than half a page per hour. The first three chapters perhaps constitute the most difficult part of the book. Especially Chapter 3, which discusses the concept of limits, requires slow and careful study, since the limit concept constitutes the foundation of calculus. Once Chapter 4 has been reached the reader will encounter few conceptual obstacles. It would be useful perhaps to point out that the important practical problem of plotting experimental data on semi-log paper is extensively discussed in Chapter 10.

The biological problems have been chosen to illustrate points in the mathematical theory. For that reason the theories discussed have not always been those which are most acceptable to biologists. For example, it is well known that Blair's theory of nervous excitation is an over-simplification of the physical situation and is inferior to the theory put forward by Monnier and Rashevsky; for all that, it is Blair's theory which is presented here because of its greater simplicity and the exercise it provides in setting up and solving a mathematical problem.* In writing the book we have tried to lay emphasis on how to formulate a given biological problem in mathematical terms, that is, how to set up the basic mathematical equations governing the model which describes the biological situation.

There are already many excellent books on statistics and statistical methods for workers in the medical and biological sciences. For that reason, we have not included any of the theory of statistics in this book. On the other hand, we hope to have included all the basic tools (such as theory of the exponential function, beta and gamma functions) which are necessary for the study of the elementary theory of mathematical statistics.

We shall take this opportunity to thank Professor J. W. Duyff for his stimulating interest and support and Mr. G. J. van Hees and Dr. M. E. Wise for their valuable suggestions. We are indebted also to Miss M. E. Gildart for her assistance in the preparation of the manuscript and to Miss C. Campbell for her assistance in compiling the index.

February 1960

J. G. DEFARES
I. N. SNEDDON

* In the first edition. In the second edition we have given the Hodgkin–Huxley equations.

PREFACE TO THE SECOND EDITION

We have replaced about a third of the physiological applications by new ones. This does not mean that those deleted are no longer interesting or important, but simply that our subject like others is affected by the "information explosion". There are many recent examples to choose from where the mathematics is both essential in the biomedical problem and well within the scope of the book. We felt we should include an inevitably small proportion of them.

There are a few mathematical additions, which were called for because of these recent applications, but there are no changes in the more advanced mathematical sections. We have expanded our outline of probability distributions, because these are so often involved directly in a biomathematical model or interpretation. This of course is separate from their role in analysing data subject to random fluctuations, i.e. in statistical analysis. There are more good books than ever on how to do this (and some less good ones), and so we do not give statistical methods. However the reader will find plenty of observed numerical data.

As in the first edition, these data occur in examples chosen mainly to show how to apply the mathematical theory. The mathematics may not be based on the most up to date model, but it should provide a good introduction to the biological or medical problem.

In some sections we have tried to provide more than this. In particular we have deliberately included two important biomedical problems in which there is widespread disagreement about the way to describe the data mathematically, the correct theoretical model, and the correct interpretation. We have done this because the choice of practicable (bio-)mathematical descriptions or models has increased explosively too, with the number and ease of use of computers. But with this progress it has also become easier to produce doubtful results and to accept a too complex model uncritically when the computer can cope in any case with all the numerical calculations. So often the problem of choosing a good model falls in a terrain that belongs neither to the biologist nor to the mathematician and they do not understand each other enough to occupy it. (The third author is an applied mathematician who has become a medical biometrician, and he is particularly concerned with this.) We hope this book may help towards bridging the gap.

The three "further reading" sections are new. They include, as an experiment, brief information on about 60 recent articles and the mathematics used in them. This is in addition to the references for applied sections, which are now listed together at the end of Chapters 6, 7, 8 and 10, and to various (mainly unchanged) references given in the text.

We are indebted to Mrs. H. A. C. Cornelissen, and also to Miss Brenda Marsh, for helping to prepare the new parts, to Dr. Cecily Tanner and Prof. P. R. J. Burch for some useful information, and the latter and the British Journal of Psychiatry for permission to reproduce a figure.

We are sorry to report that Professor Duyff, the head of the physiology laboratory in Leiden, died in 1969. He supported our subject very actively when it was less appreciated than now. We reproduce his foreword to the first edition: we think this is as valid now, except for some small details (e.g. statistical design in medicine seems not to have grown in importance), as 13 years ago.

Leiden and Glasgow, 1973

J. G. DEFARES
I. N. SNEDDON
M. E. WISE

FOREWORD TO THE FIRST EDITION

Physiology is rapidly developing into an exact, quantitative science borrowing its methods largely from experimental physics and physical chemistry; it is becoming the ground where biologists — including medical men, — and physicists—including physical chemists—meet. Progress in physiology is largely dependent on a close cooperation of biologists and physicists; a cooperation which is possible only if they understand each other's language.

For a problem to be solved it is first of all necessary that it be correctly stated, i.e. stated in terms of mathematics. The physicist is used to doing so; the average physiologist is not: thus it happens only too often that he is well aware of the existence of a problem, but unable to formulate it correctly, and thus to devise the experiment best suited to attack it; the physicist, on the other hand, is not acquainted with the problem, but could easily point the way to its solution. For a successful cooperation it is not necessary that the physicist becomes a physiologist also, nor that the physiologist be a full-fledged physicist; he should, however, be conversant with the methods of investigation used in physics — and, thus, with the mathematical operations involved, just as the physicist should at least have an elementary knowledge of biology. Here, the physiologist is at a disadvantage: while it is easy for the physicist to master the main facts and doctrines of biology, the biologist attempting to follow the arguments of the physicist encounters serious difficulties because of his lack of training in mathematics.

Clinical medicine, in its turn, is fast shedding empiricism, and develops into a branch of science firmly rooted in physiology; in consequence, mathematics also invades clinical science. In these circumstances it is not surprising that several textbooks of mathematics for physiologists, clinical scientists, and biologists in general — including a few excellent ones — have been published in the last twenty years. Even so, the present book fills a distinct need.

The biologist (the term taken in its broad sense) confronted with the task of teaching himself some mathematics is apt to lose interest and to give up unless he is, again and again, made aware of the fact that the mathematical methods and operations with which he is becoming acquainted can be, and have been, applied to problems within his own field of interest or investigation. In the present text well-chosen examples, taken from current literature, are used to illustrate the important part played in present-day physiology, pharmacology, and clinical medicine, by mathematical procedures. Roughly one-third of the total number of pages is given up to these examples, which are inserted at the appropriate places.

An extensive treatment of statistical methods and procedures in physiological and clinical investigation is not included in the text, but their mathematical basis comes under consideration in a separate section. Not so very long ago, the field of statistics was *terra incognita* to most research workers in the biological sciences; at present, most investigators are aware of the necessity of employing statistical methods to assess the significance of experimental data. Too many of them, however, go by a few rules of thumb, without asking themselves whether a fixed, indiscriminately used "standard" procedure is applicable to the material under consideration, and do not realize that the application of statistical methods to physiological, pharmacological, or clinical research should not be limited to an attempt to make some sense out of the jumble of data collected in the course of a badly organized experiment, but that statistical considerations should largely govern the design of the trials themselves. A sound mathematical basis, as given here, should be useful if only to convince the reader that, in many cases, it will be better to assure himself of the services of a professional statistician than to go and flounder in treacherously deep waters.

J. W. DUYFF

CONTENTS

CHAPTER 8

FUNCTIONS OF MORE THAN ONE VARIABLE

CHAPTER 9

DIFFERENTIAL EQUATIONS

APPENDIX

DETERMINANTS

CHAPTER 1

ALGEBRAIC PRELIMINARIES

Introduction

This book is written on the assumption that most readers of it will be biologists and medical practitioners who ceased the active study of mathematics several years before. The first chapter is written with a view to refreshing the knowledge of elementary algebra gained previously. Any reader who finds that the ideas contained in it are unfamiliar is advised to study these topics more closely in an introductory textbook on algebra, and, more important still, to work through a number of easy exercises to cultivate skill in performing the manipulations of elementary algebra. An excellent introduction of this kind (requiring no previous knowledge beyond that of arithmetic) is "Teach Yourself Algebra" by P. Abbott (English Universities Press, London, 1942). It covers in a more leisurely fashion the material covered in this chapter and some of the topics of Chapter 2. All that the present chapter does is to present briefly some fundamental aspects of "school" algebra to provide a background for the study of subsequent chapters.

1. The Number System of Mathematics *

The theory of the differential and integral calculus, which is what we shall mainly be concerned with in this volume, is based on the idea of a real number, and on operations with numbers. In this section we shall discuss briefly the system of real numbers starting with the simplest kind of number — the positive integers. In this respect we follow the well-known dictum of Leopold Kronecker that "God made the integers, all the rest is work of man".

We shall suppose that we know what the positive integers (or whole numbers) 1, 2, 3, . . . are. We have had such a wide experience of

* The ideas of this section are difficult (as are all fundamental ideas). The reader should read this section only with the aim of providing himself with some background of the foundations of mathematics. If he finds it difficult, he will find that he can pass straight to § 2, without having mastered *all* its detail.

manipulating these numbers that it is difficult for us to realise that even the idea of a positive integer is not immediately intuitive. The concept of the number 2 as that property shared by all "pairs" of objects no matter how diverse, is quite a sophisticated one which could only have made its appearance at a reasonably high level of civilisation. Indeed, a satisfactory answer to the question "what is a whole number?" has only been given within the last century – by Frege in his *Grundlagen der Arithmetik*, published in 1884. We shall not try here to come to a satisfactory definition of a positive integer but just regard it as an obvious generalization from experience.

Two operations, addition and multiplication, are defined for the positive integers, so that, if m and n are positive integers, so are $m + n$ and mn (or as it is more usually written in elementary mathematics $m \times n$). These operations on the whole numbers have well-known properties, e.g. $m + n = n + m$ and $mn = nm$ and so on, expressing the well-known fact that the result of a series of additions (or multiplications) is independent of the order in which the additions (or multiplications) are carried out.

However, subtraction and division are not always possible within the system of positive integers: if m is less than n then $m - n$ is not a positive integer; in other words there exists no *positive* integer such that when n is added to it the answer is m. Similarly m/n is a positive integer only if n is a factor of m. We can state these simple facts another way by saying that we cannot always solve the equation $n + x = m$ (or $nx = m$) to obtain an x which is a positive integer. To remedy this, we extend our concept of number to include more than just the positive integers. The first stage in this extension is to introduce the idea of a zero, 0, having the property that, for any positive integer n, $n + 0 = 0 + n = n$. By means of this new element, we can then define a new number, $-n$, having the property that $n + (-n) = (-n) + n = 0$. In this way we can build up the set of numbers $0, \pm 1, \pm 2, \pm 3, \ldots, \pm n, \ldots$, called the set of *signed integers*. The signed integers so introduced enable us to subtract, that is, to solve the equation $m + x = n$ where m and n are positive integers. We observe, however, that we can do more than that: if m and n are signed integers, then we can now always find the solution of the equation $m + x = n$ with x turning out to be a signed integer.

This, however, still does not allow us always to carry out divisions.

We cannot always solve the equation $nx = m$ within the set of signed integers. To be able to do this, we have to extend still further the idea of number. We now include all numbers of the form m/n, where m and n are signed integers and $n \neq 0$. We define the number m/n as that number which, when multiplied by n is equal to m; for example, $\frac{1}{4}$ is that number which when multiplied by 4 is equal to 1. The set of all such numbers is called the set of *rational numbers**. Within this set the operations of addition and multiplication may be performed, the sum and product of two rational numbers m/n and p/q being defined by the equations

$$\frac{m}{n} + \frac{p}{q} = \frac{mq + np}{nq}, \qquad \frac{m}{n} \times \frac{p}{q} = \frac{mp}{nq}.$$

These numbers are of course themselves both rational numbers and since $n \neq 0$, $q \neq 0$, $nq \neq 0$.

The operation of addition acting on rational numbers has the following properties. If a, b, c are rational numbers, then

$$(a + b) + c = a + (b + c) \tag{1}$$

expressing the fact that when three rational numbers are added together the manner in which they are grouped is not significant. Equation (1) symbolizes what is known as *the associative law of addition.* Similarly

$$a + b = b + a \tag{2}$$

which states that the order in which two rational numbers are added is irrelevant. This is known as *the commutative law of addition.* The third law of addition of rational numbers is that there exists a unique number 0 of the set such that, for all a in the set of rational numbers,

$$a + 0 = 0 + a = a. \tag{3}$$

Finally, to each a of the set of rational numbers there exists a unique member $-a$ of the set such that

$$a + (-a) = (-a) + a = 0. \tag{4}$$

A similar set of laws holds for the operation of multiplication. If a, b, c are rational numbers, we have the associative law of multi-

* Such numbers are called "rational", not because there is anything rational about their definition, but because they are defined as ratios.

plication

$$(ab)c = a(bc) \tag{5}$$

and the commutative law

$$ab = ba. \tag{6}$$

There also exists a unique element 1 of the set, such that, for all a in the set

$$a \cdot 1 = 1 \cdot a = a. \tag{7}$$

Finally, to each a of the set (except 0) there corresponds a unique element called its inverse and denoted by * a^{-1} such that

$$aa^{-1} = a^{-1}a = 1. \tag{8}$$

It is easily shown that if $a = m/n$ then

$$a^{-1} = \frac{n}{m}. \tag{9}$$

Situations in which the operations of addition and multiplication are involved simultaneously are governed by the two laws

$$a(b + c) = ab + ac \tag{10}$$

$$(b + c)a = ba + ca \tag{11}$$

which are known as the *distributive* laws of addition and multiplication.

Any system of symbols $a, b, c, \ldots$ which have the properties symbolized by equations (1) to (11) is called a *field*. For this reason we talk of the field of rational numbers.

The property (8) allows the operation of division by any rational number other than zero with a result which is itself an element of the field of rational numbers. This means that within the field of rational numbers we can solve any linear equation $ax + b = 0$ where a and b are rational numbers. It will be observed that this is more general than the problem of solving the same equation with a and b signed integers.

However, if we consider a very simple equation of second degree, namely $x^2 = 2$, it is easily proved that we cannot find a rational number to satisfy it. In other words, there is no rational number which, when multiplied by itself gives 2 as a result. We can, however,

* The reason for this notation, rather than the $1/a$ which is used in elementary mathematics is given in section 2 below.

find rational numbers whose squares are successively nearer and nearer to 2. For example, the numbers 1, 1.4, 1.41, 1.414, 1.4142, . . ., (which may be written as rational numbers 1, $\frac{14}{10}$, $\frac{141}{100}$, $\frac{1414}{1000}$, $\frac{14142}{10000}$, . . .) have squares 1, 1.96, 1.9881, 1.999396, 1.99996164, . . ., which are successively nearer and nearer to 2. The number whose square is 2 (and this is the number which we write as $\sqrt{2}$) may be regarded as the "limiting" number of this "sequence" of rational numbers. We can think of the sequence as "tending" to a "limit" which is a new kind of number. Any number which can be considered as the limit of a sequence of rational numbers in this way is called a *real number*. The numbers 3, π, $2+\sqrt{3}$, $\sqrt{5}$, etc. are all real numbers. The rational numbers themselves can be expressed as limits of sequences of this kind and so are themselves real numbers. A real number which is not rational is called *irrational*. It can be shown that the operations of addition and multiplication can be performed within the system of real numbers and that they possess all the properties (1) to (11) above. In other words, the real numbers form a field, called the *real field*. We can introduce the idea of order in the real field. We say that a is greater than b (or b is less than a) if the real number $a - b$ is positive and we write $a > b$ (or $b < a$). The symbols $\geqq$, $\leqq$ mean "greater than or equal to" and "less than or equal to" respectively.

The real numbers are obtained from the rational numbers by allowing a limiting process to take place. The special advantage of the real numbers over the rational numbers is in fact that they enable limiting processes of this kind to take place with the results remaining in the same field. Now the subject of the calculus rests essentially on limiting processes so that the natural field of numbers to use in the calculus is the real field. However, even quite simple things cannot be done with real numbers alone. For example, we cannot solve every quadratic equation whose coefficients are real numbers. If a, b and c are real numbers, then the equation

$$ax^2 + bx + c = 0$$

has real solutions

$$x = \frac{-b \pm \sqrt{(b^2 - 4ac)}}{2a}$$

only if $b^2 \geqq 4ac$ since, if $b^2 < 4ac$ we have to take the square root of a negative number and there is no real number whose square is negative. Hence, if we want to be able to solve every quadratic equation, we must again extend our number system.

In order to be able to solve every quadratic equation we need a number whose square is negative, that is, we need a solution of the equation $x^2 = -1$. We define a number i by the equation $i^2 = -1$ and we assume that in every other way, i behaves like a real number. The fact that i alone is sufficient to solve every quadratic equation is then easily seen. In the case in which $b^2 < 4ac$ the equation $ax^2 + bx + c = 0$ has solutions

$$x = -\frac{b}{2a} \pm i\frac{\sqrt{(4ac - b^2)}}{2a}.$$

A *complex number* is any number of the form $a + ib$, with a and b real. The sum and product of complex numbers are defined consistently with the manipulation of i as if it were real. Thus,

$$(a + ib) + (c + id) = (a + c) + i(b + d), \tag{12}$$

$$(a + ib)(c + id) = (ac - bd) + i(bc + ad). \tag{13}$$

Zero in the set of complex numbers is $0 + i0$ and unity is $1 + i0$. The negative of $a + ib$ is $-a - ib$. The *reciprocal* (or *inverse*) of any complex number other than zero, say $a + ib$ where a and b are not both zero, is defined as

$$\frac{1}{a + ib} = \frac{a - ib}{(a + ib)(a - ib)} = \frac{a - ib}{a^2 + b^2}. \tag{14}$$

With these definitions it is easy to see that the set of complex numbers satisfy the properties embodied in equations (1) to (11), so that the complex numbers form a field. Any real number a can be written in the form $a + i0$, so that the field of real numbers forms part of the field of complex numbers.

If z denotes the complex number $a + ib$, where a and b are both real, we call a the real part of z and b the imaginary part of z. The importance of the real and imaginary parts of a complex number lies in the fact that if two complex numbers are equal, then their real parts are equal and their imaginary parts are equal. In other words, if $a + ib = c + id$, then $a = c$ and $b = d$. Thus from an equation in complex numbers we can deduce two equations in real numbers by taking the real and imaginary parts separately.

Having had to extend our number system to solve first-order equations and then quadratic equations we might imagine that a further extension would be needed to solve cubic equations (that is, equations involving terms like x^3) and so on. In fact, it can be shown that any equation of any finite degree in which the coefficients are either real or complex numbers, can be solved entirely in terms of complex numbers.

When we are dealing with real numbers we are often interested only in the numerical value of a number (i.e. not in its sign $\pm$). We denote the numerical value of a real number x by the symbol $|x|$. If x is positive $|x| = x$, but if x is negative, $|x| = -x$. $|x|$ is called the *modulus* of x. Suppose that x and y are positive real numbers, then $x + y$ will be positive and $|x + y| = x + y$. If, on the other hand we assume that x is positive and y is negative, then $|y| = -y$ and if $x > |y|$, $|x + y| = x + y$. But $|x| + |y| = x - y > x + y$, so that we obtain the result

$$|x + y| < |x| + |y|.$$

We can easily establish the same result if $x < |y|$. We have therefore shown that if x and y are any real numbers

$$|x + y| \leqq |x| + |y|. \tag{15}$$

A relation of this kind is called an *inequality*; if $\leqq$ is replaced by $<$ the inequality is said to be *strict*. In a similar way we can prove the results

$$\left|\frac{x}{y}\right| = \frac{|x|}{|y|}, \tag{16}$$

$$|xy| = |x| \cdot |y|. \tag{17}$$

PROBLEMS

It is essential that the reader should be thoroughly familiar with the manipulation of the laws governing real numbers. He should be capable of doing all the routine problems in Chapters 1—6, 10—12, 17 of Abbott's book. The following are suggested as additional typical problems.

1. The average weight in a class consisting of a boys and b girls is w. If the average weight of the boys is x and that of the girls is y, show that

$$y = \frac{a(w - x)}{b} + w.$$

2. If

$$\frac{a}{b} = \frac{c}{d},$$

prove that

$$(ab + cd)^2 = (a^2 + c^2)(b^2 + d^2).$$

3. If

$$x = \sqrt{\left(\frac{y - 1}{y + 1}\right)},$$

express y in terms of x.

4. Express the following statement in algebraic symbols and prove that it is true:-

The cube of the sum of two numbers added to twice the sum of their cubes is equal to three times the sum of the numbers multiplied by the sum of their squares.

5. The following are special cases of the same identity, obtained by giving a variable n different numerical values:-

$$3^2 = 2 \times 4 + 1$$
$$7^2 = 6 \times 8 + 1$$
$$13^2 = 12 \times 14 + 1.$$

Find the identity in n, and use it to find the square of 9 999.

2. Indices

In elementary mathematics we often encounter repeated products of a number with itself of the form

$$a \times a \times a \times \ldots \times a$$

in which there are n identical factors, so that it is convenient to denote them by a simple symbol. We write such a product in the form a^n, and say that it is the *n-th power* of the number a; for example, we write

$$a \times a = a^2,$$
$$a \times a \times a = a^3.$$

It is a simple matter to establish the rules of manipulation of such products. For instance $a^m \times a^n$ means the product of the number

$a \times a \times a \ldots$ to m factors and the number $a \times a \times a \ldots$ to n factors; when we write this product out we see that it is merely $a \times a \times a \ldots \times a$ to $(m + n)$ factors, i.e. it is a^{m+n}. We have therefore proved that, if m and n are positive integers

$$a^m \times a^n = a^{m+n}. \tag{1}$$

Similarly, if we write out the meaning of $a^m \div a^n$ where $m > n$, we find that

$$a^m \div a^n = a^{m-n} \qquad (m > n), \tag{2}$$

and if we write out what we mean by $(a^m)^n$ we find that

$$(a^m)^n = a^{m \times n}. \tag{3}$$

The definition of a^n which we have given is intelligible only if n is a positive integer; it is obviously meaningless to say that $a^{\frac{1}{2}}$ means a multiplied by itself a $\frac{1}{2}$-number of times! If, therefore, we wish to define a^n in the case where n is no longer a positive integer we must seek a new definition. In framing this definition we keep in mind the desirability of as many as possible of the three basic laws (1)—(3) remaining true. If we use the result (2) formally in the case in which $m = n$ we see that it yields simply that $a^0 = a^m \div a^m$, which we know to be equal to 1. Therefore if we take

$$a^0 = 1 \tag{4}$$

as our definition of a^n when $n = 0$, we see that this definition is consistent with the rule (2). It is also readily shown that this interpretation does not conflict with either of the rules (1) and (3).

In a similar way if we put $m = n = \frac{1}{2}$ formally in equation (1) we find that it takes the form

$$a^{\frac{1}{2}} \times a^{\frac{1}{2}} = a^{\frac{1}{2}+\frac{1}{2}} = a^1 = a$$

showing that $a^{\frac{1}{2}}$ may be interpreted as being that number which when multiplied by itself is equal to a; in other words $a^{\frac{1}{2}}$ may be defined to mean $\sqrt{a}$, the *square root* of a. Similarly, putting $m = n = p = \frac{1}{3}$ in the equation

$$a^m \times a^n \times a^p = a^{m+n+p} \tag{5}$$

we see that

$$a^{\frac{1}{3}} \times a^{\frac{1}{3}} \times a^{\frac{1}{3}} = a^1 = a$$

showing that $a^{\frac{1}{3}}$ may be interpreted as meaning $\sqrt[3]{a}$, the cube root of a. Further,

$$a^{\frac{2}{3}} = a^{\frac{1}{3}} \times a^{\frac{1}{3}} = \sqrt[3]{a} \times \sqrt[3]{a} = \sqrt[3]{(a^2)}.$$

By considering the general case of these examples we find that we incur no inconsistency in our scheme if we interpret $a^{m/n}$ to mean the n-th root of a^m.

By means of this definition we have managed to extend the definitions of a^n to cases in which n is a positive rational number. We shall now consider the case in which the index n is negative. If we let $m = -n$ in equation (1) and apply the rule formally we find that it suggests that the definition of a^{-n} should be such that $a^{-n} \times a^n = a^0 = 1$. This suggests that we take as our definition of a^{-n}

$$a^{-n} = \frac{1}{a^n}. \tag{6}$$

By these new definitions we have been able to give the expression a^n a meaning whenever n is a rational number.

EXAMPLE 1: *Simplify the expression*

$$\frac{(x^{-\frac{3}{4}} - 4x^{\frac{1}{4}})^{\frac{2}{3}}[(8x)^{\frac{2}{3}} + 4x^{-\frac{1}{3}}]^{-\frac{1}{2}}}{(8x)^{-\frac{1}{3}}}.$$

Writing $x^{\frac{1}{4}}$ as $x^{1-\frac{3}{4}} = x \cdot x^{-\frac{3}{4}}$ we see that $x^{-\frac{3}{4}} - 4x^{\frac{1}{4}} = x^{-\frac{3}{4}}(1 - 4x)$ so that $(x^{-\frac{3}{4}} - 4x^{\frac{1}{4}})^{\frac{2}{3}} = x^{-\frac{3}{4}\cdot\frac{2}{3}}(1 - 4x)^{\frac{2}{3}} = x^{-\frac{1}{2}}(1 - 4x)^{\frac{2}{3}}$.
Since $(8x)^{\frac{2}{3}} = (2^3)^{\frac{2}{3}}x^{1-\frac{1}{3}}$, we find that $[(8x)^{\frac{2}{3}} + 4x^{-\frac{1}{3}}]^{-\frac{1}{2}} = (x^{-\frac{1}{3}})^{-\frac{1}{2}} \cdot (4)^{-\frac{1}{2}}(1 + x)^{-\frac{1}{2}} = x^{\frac{1}{6}} \cdot \frac{1}{2}(1 + x)^{-\frac{1}{2}}$. Further

$$\frac{1}{(8x)^{-\frac{1}{3}}} = (8x)^{\frac{1}{3}} = 2x^{\frac{1}{3}}$$

so that the expression we are given is equal to

$$\begin{aligned} 2x^{\frac{1}{3}} \cdot x^{-\frac{1}{2}}(1 - 4x)^{\frac{2}{3}} \cdot \tfrac{1}{2}x^{\frac{1}{6}}(1 + x)^{-\frac{1}{2}} & \\ &= x^{\frac{1}{3}-\frac{1}{2}+\frac{1}{6}}(1 - 4x)^{\frac{2}{3}}(1 + x)^{-\frac{1}{2}} \\ &= (1 - 4x)^{\frac{2}{3}}(1 + x)^{-\frac{1}{2}}. \end{aligned}$$

PROBLEMS

1. Simplify

$$\frac{(x^{-\frac{1}{2}} - x^{\frac{1}{2}})^{-2} + (x^{-\frac{1}{2}} + x^{\frac{1}{2}})^{-2}}{(x^{-\frac{1}{2}} - x^{\frac{1}{2}})^{-2} - (x^{-\frac{1}{2}} + x^{\frac{1}{2}})^{-2}}.$$

2. Express in their simplest forms

(i) $(x-y)^{\frac{2}{3}} + (x+3y)(x-y)^{-\frac{1}{3}} - (x^2-3y^2)(x-y)^{-\frac{4}{3}}$;

(ii) $$1 - \frac{1+(1+3a^{-1})(1+2a^{-1}+3a^{-2})^{-\frac{1}{2}}}{1+a+(a^2+2a+3)^{\frac{1}{2}}}.$$

3. Without using tables, find the cube root of

$$\frac{6.4}{0.27}\left\{\frac{4^{-\frac{4}{3}}}{3^{-2}\cdot\sqrt[3]{(12.5\times 2^{-5})}}\right\}^{\frac{3}{2}}.$$

The index notation we have introduced above may be usefully employed in writing down very large or very small numbers. For instance the number

$$1\ 256\ 000\ 000$$

may be written in the form

$$1.256 \times 1\ 000\ 000\ 000$$

and 1 000 000 000 is the product of 9 tens, i.e. it is 10^9, so that the given number can be written in the form

$$1.256 \times 10^9.$$

A number written in this way is said to be written in *standard form.* One digit only is retained in the whole number part and this is multiplied by a power of 10 whose index is equal to the number of figures which follow the digit retained. Similarly a small number such as

$$0.000\ 017\ 58$$

can be written as

$$1.758 \div 100\ 000 = 1.758 \div 10^5 = 1.758 \times 10^{-5};$$

in this case the numerical part of the index is one greater than the number of zeros following the decimal point in the given number.

This notation is of great value in handling problems, such as growth problems, in which the numbers involved may be very large, or, in measurements of minute organisms where the linear dimensions may be exceedingly small.

3. Logarithms

We saw in the last section that, if n is a rational number, positive or negative, it is possible to assign a meaning to the symbol a^n if a

denotes a positive number. If, given a, it is possible to find an index n so that

$$x = a^n \tag{1}$$

we say that n is the *logarithm of x to the base a*, and we write

$$n = \log_a x. \tag{2}$$

For example

$$16 = 2^4,$$

so that

$$\log_2 16 = 4,$$

and

$$1000 = 10^3, \quad 0.001 = 10^{-3}$$

so that

$$\log_{10} 1000 = 3, \quad \log_{10} 0.001 = -3.$$

In this section we shall discuss the most frequently used properties of logarithms but we shall postpone until Chapter 6 the answer to the fundamental question of how to calculate $\log_a x$ when the numbers x and a are prescribed. For the moment we shall assume that it is always possible to find the logarithms of a positive number x to any positive base a, and indeed that tables of logarithms to the base 10 are available.

The rules for the manipulation of logarithms follow immediately from the laws for indices which we outlined in the previous section. If $\log_a x = m$, and $\log_a y = n$, then it follows from the definition given above that

$$x = a^m, \quad y = a^n$$

and from the index law expressed by equation (1) of the previous section we see that

$$xy = a^m \times a^n = a^{m+n}$$

showing that

$$\log_a (xy) = m + n = \log_a x + \log_a y; \tag{3}$$

i.e. *the logarithm of the product of two numbers is equal to the sum of the logarithms of these numbers.*

A similar rule can be derived for the logarithm of a quotient. With the same notation we see that

$$\frac{x}{y} = \frac{a^m}{a^n} = a^{m-n},$$

so that

$$\log_a \frac{x}{y} = m - n = \log_a x - \log_a y, \tag{4}$$

showing that *the logarithm of the quotient of two numbers is equal to the logarithm of the numerator minus the logarithm of the denominator.*

Similarly, if $\log_a x = m$,

$$x^n = (a^m)^n = a^{mn}$$

showing that

$$\log_a (x^n) = n \log_a x. \tag{5}$$

In other words, *the logarithm of the power of a number is equal to the logarithm of the number multiplied by the index of the power.* This includes the case in which n is fractional. For example,

$$\log_a \sqrt{x} = \tfrac{1}{2} \log_a x. \tag{6}$$

A problem which might well arise in practice is the one of determining the value of the logarithm of a number x to a base b when $\log_a x$ is known. Suppose that $\log_a x = m$, $\log_a b = n$, then from the definition of a logarithm we see that $x = a^m$, and $b = a^n$. From this second relation it follows that $a = b^{1/n}$, so that substituting in the first relation we have $x = (b^{1/n})^m = b^{m/n}$, proving that

$$\log_b x = \frac{m}{n} = \frac{\log_a x}{\log_a b}. \tag{7}$$

The rules of manipulation expressed by equations (3)—(7) are sufficient to enable the performance of arithmetic calculations to be carried out speedily. It is assumed that the reader is already familiar with the use of tables of logarithms to the base 10 and is capable of performing arithmetical calculations easily. A good account of the use of such tables is given in Chapter 16 of the book by P. Abbott cited in the Introduction.

EXAMPLE 2: *Without using tables determine the value of the expression*

$$\log_3 \tfrac{49}{2} + \log_3 \tfrac{27}{35} - \log_3 \tfrac{21}{10}.$$

We can write this expression in the form

$$\log_3 49 - \log_3 2 + \log_3 27 - \log_3 35 - \log_3 21 + \log_3 10$$
$$= 2\log_3 7 - \log_3 2 + 3 - \log_3 7 - \log_3 5 - \log_3 7 - 1 + \log_3 5 + \log_3 2$$
$$= 2.$$

Although the idea of a functional relationship is not introduced until Chapter 2 a single example of functional relationships involving a logarithm may be given here.

In physiology the relationship between the intensity E of excitation of an afferent (nervous) pathway and the intensity S of the stimulus may (for S well above threshold) be approximated by

$$E = \alpha \log S \tag{8}$$

which expresses the fact that the rate of increase of E with S decreases as S increases; α is a constant of proportionality. This relationship, which is of great generality in the nervous system, has been applied by H.D. Landahl to the problem of flicker and its fusion.

Consider a movie, which involves the projection of successive pictures at a rate of about 10 per second. When you are sitting at the rear you have a smooth continuous view, but when you are sitting in the front row the picture dances before your eyes. This is due to the fact that at the front of the cinema the light intensity is greatest, so that the flicker frequency (rate of successive pictures) is too low to produce flicker fusion, i.e. smoothness of "vision". If you cannot afford to sit in the royal box all you have to do is put on sun glasses to get a smooth view when sitting near the screen. This useful piece of information, viz. that the fusion frequency f^* increases with the intensity of light, S, is mathematically couched in the well-known Perry–Porter law, which states:

$$f^* = K \log S \tag{9}$$

where K is a constant.

Landahl ‡ was able to show theoretically that flicker fusion occurs when

$$\tfrac{1}{8} AET = h_0 \tag{10}$$

where E is the intensity of excitation, T is the period ($= 1/f$ where f

‡ H.D. Landahl, Bull. Math. Biophysics **19** (1957) 157.

is the frequency of flicker), h_0 is a threshold, and A is a constant of proportionality. The reader should make no attempt to "understand" this proposition which is here merely presented as a "truth" from which the empirical Perry–Porter law may be derived by substituting eq. (8) in eq. (10) and introducing the frequency $f = 1/T$ in eq. (10), thus giving for the fusion frequency f^*, the expression

$$f^* = \frac{A\alpha}{8h_0} \log S \tag{11}$$

which, since the quantity $A\alpha/8h_0$ is a constant, K, say, in any given situation represents the Perry–Porter's law.

After completing Chapter 6 the student will be able to follow the elegant theoretical derivation as presented by N. Rashevsky in his little book "Mathematical principles in biology and their applications" (Thomas, Springfield, 1961).

PROBLEMS

1. If $\log(a + b) = \log a + \log b$, find b in terms of a and prove that

$$\log a - \log b = \log (a - 1).$$

2. If $x = \log_{10} (p - qy) - \log_{10} p$, prove that

$$y = \frac{p}{q} (1 - 10^x).$$

3. Without using tables show that

$$\frac{\log 2\sqrt{3} + \frac{1}{2} \log 12 - \log 9}{\log 2 - \log \sqrt{3}} = 2$$

whatever base the logarithms are taken to.

4. Series

In many mathematical operations we fix our attention in succession on one number, then on another, then on a third and so on. If the first number is denoted by T_1, the second by T_2, and so on until we reach the n-th one which is denoted by T_n, we therefore have to consider the succession of numbers

$$T_1, T_2, T_3, \ldots, T_n. \tag{1}$$

(The three dots ... mean "and so on".) Such a set of numbers is called a *sequence* and the r-th member, denoted by T_r, is called the *r-th term* of the sequence. From the set of terms of such a sequence we can form a sum

$$T_1 + T_2 + T_3 + \ldots + T_n \tag{2}$$

which is called a *series*. The value

$$S_n = T_1 + T_2 + T_3 + \ldots + T_n \tag{3}$$

of this sum is called *the sum to n terms* of the series (2).

The simplest series occurring in mathematics is one in which each term differs from its predecessor by a constant. If the first term T_1 of such a series is denoted by a and the common difference by d, then we have

$$T_1 = a,\ T_2 = a + d,\ T_3 = a + 2d, \ldots, T_n = a + (n - 1)d$$

and the series takes the form

$$a + (a + d) + (a + 2d) + \ldots + \{a + (n - 1)d\}. \tag{4}$$

For this series

$$S_n = a + (a + d) + (a + 2d) + \ldots + \{a + (n - 1)d\} \tag{5}$$

and, since the order in which we perform additions is not important, we can write this result in the form

$$S_n = \{a + (n - 1)d\} + \{a + (n - 2)d\} + \{a + (n - 3)d\} + \ldots + a. \tag{6}$$

Adding corresponding terms in the sums (5) and (6) we find that

$$2S_n = \{2a + (n - 1)d\} + \{2a + (n - 1)d\} \\ + \{2a + (n - 1)d\} + \ldots \{2a + (n - 1)d\}.$$

Each term on the right hand side has the value $2a + (n - 1)d$ and there are n such terms so that the sum of the terms on the right is $n\{2a + (n - 1)d\}$. Hence for the series (4),

$$S_n = \tfrac{1}{2}n\{2a + (n - 1)d\}. \tag{7}$$

If we denote the n-th term of the series by l (for "last" term) then, since $l = a + (n - 1)d$, it follows that

$$S_n = \tfrac{1}{2}n(a + l). \tag{8}$$

As an example consider the sum of the first n terms of the series

$$2 + 5 + 8 + \ldots.$$

Here $a = 2$ and $d = 5 - 2 = 3$ so that the sum of the first n terms is given, by (7), as

$$\tfrac{1}{2}n\{4 + (n - 1)3\} = \tfrac{1}{2}n(3n + 1).$$

The simplest example of all is the sum

$$1 + 2 + 3 + \ldots + n$$

of the first n integers. Here the first term a is 1 and the last term l is n so that, by equation (8), we have that

$$1 + 2 + 3 + \ldots + n = \tfrac{1}{2}n(n + 1). \tag{9}$$

We now introduce a notation which greatly simplifies formal work in series. For the sum of the series

$$T_1 + T_2 + T_3 + \ldots + T_n$$

we write

$$\sum_{r=1}^{n} T_r.$$

The sigma sign means that we take the sum of all the terms like T_r starting with the one appropriate to $r = 1$, i.e. with T_1, and ending with the one appropriate to $r = n$, i.e. with T_n. For instance for the series on the left of (9) the typical term T_r is simply r so that we can rewrite equation (9) in the form

$$\sum_{r=1}^{n} r = \tfrac{1}{2}n(n + 1). \tag{10}$$

Similarly equation (7) could be put in the form

$$\sum_{r=1}^{n} \{a + (r - 1)d\} = \tfrac{1}{2}n\{2a + (n - 1)d\}. \tag{11}$$

We shall now consider the sum

$$\sum_{r=1}^{n} r^2 = 1^2 + 2^2 + 3^2 + \ldots + n^2.$$

It takes only a simple multiplication to show that for any r we have

$$(r + 1)^3 - r^3 = 3r^2 + 3r + 1.$$

Putting $r = 1, 2, 3, \ldots, n$ we find that

$$
\begin{aligned}
2^3 - 1^3 &= 3 \cdot 1^2 + 3 \cdot 1 + 1 \\
3^3 - 2^3 &= 3 \cdot 2^2 + 3 \cdot 2 + 1 \\
4^3 - 3^3 &= 3 \cdot 3^2 + 3 \cdot 3 + 1 \\
&\vdots \\
n^3 - (n-1)^3 &= 3(n-1)^2 + 3(n-1) + 1 \\
(n+1)^3 - n^3 &= 3n^2 + 3n + 1.
\end{aligned}
$$

Adding we find that

$$(n+1)^3 - 1 = 3\sum_{r=1}^{n} r^2 + 3\sum_{r=1}^{n} r + n.$$

Substituting from eauation (10) we find that

$$(n+1)^3 - 1 = 3\sum_{r=1}^{n} r^2 + 3 \cdot \tfrac{1}{2}n(n+1) + n$$

so that

$$
\begin{aligned}
3\sum_{r=1}^{n} r^2 &= (n^3 + 3n^2 + 3n + 1) - 1 - \tfrac{3}{2}n(n+1) - n \\
&= \tfrac{1}{2}\{2n^3 + 6n^2 + 6n - 3n(n+1) - 2n\}
\end{aligned}
$$

from which it follows that

$$\sum_{r=1}^{n} r^2 = \tfrac{1}{6}n(n+1)(2n+1). \tag{12}$$

Similarly starting from the identity

$$(r+1)^4 - r^4 = 4r^3 + 6r^2 + 4r + 1$$

we can show that

$$(n+1)^4 - 1 = 4\sum_{r=1}^{n} r^3 + 6\sum_{r=1}^{n} r^2 + 4\sum_{r=1}^{n} r + n.$$

Substituting from equations (10) and (12) and simplifying the result we can show that

$$\sum_{r=1}^{n} r^3 = \tfrac{1}{4}n^2(n+1)^2. \tag{13}$$

Comparing equations (10) and (13) we obtain the interesting result

$$1^3 + 2^3 + 3^3 + \ldots + n^3 = (1 + 2 + 3 + \ldots + n)^2. \tag{14}$$

EXAMPLE 3: *Find the sum to n terms of the series*

$$4^2 + 7^2 + 10^2 + \ldots.$$

For this series the r-th term is

$$(3r + 1)^2 = 9r^2 + 6r + 1$$

so that the sum to n terms of the series is

$$S_n = 9\sum_{r=1}^{n} r^2 + 6\sum_{r=1}^{n} r + n$$

and this is equal to

$$9 \cdot \tfrac{1}{6}n(n + 1)(2n + 1) + 6 \cdot \tfrac{1}{2}n(n + 1) + n$$

by equations (10) and (12). Hence

$$\begin{aligned} S_n &= \tfrac{1}{2}n\{3(2n^2 + 3n + 1) + 6(n + 1) + 2\} \\ &= \tfrac{1}{2}n(6n^2 + 15n + 11). \end{aligned}$$

We consider another type of series which occurs frequently in mathematics. The series (2) is said to be a *geometric series* if every term is a constant multiple of its predecessor. If we denote this constant multiple by x and take a as the first term of the series then a geometric series with n terms can be written as

$$a + ax + ax^2 + \ldots + ax^{n-1} = \sum_{r=1}^{n} ax^{r-1}. \tag{15}$$

If we denote the sum of this series by S_n then we have that

$$\begin{aligned} xS_n &= ax + ax^2 + ax^3 + \ldots + ax^n \\ &= S_n - a + ax^n \end{aligned}$$

from which it follows that

$$(1 - x)S_n = a(1 - x^n).$$

If $x \neq 1$ we can divide both sides of this equation by $(1 - x)$ to obtain the result

$$\sum_{r=1}^{n} ax^{r-1} = \frac{a(1 - x^n)}{1 - x}. \tag{16}$$

EXAMPLE 4: *Find the sum to n terms of the series*

$$\frac{1}{2} + \frac{1}{2^2} + \frac{1}{2^3} + \ldots.$$

This is a series of the type (15) with $a = \frac{1}{2}$, $x = \frac{1}{2}$ so that by equation (16)

$$\sum_{r=1}^{n} (\tfrac{1}{2})^r = 1 - (\tfrac{1}{2})^n.$$

It is interesting to note some numerical values for this series.

$$1 - \sum_{r=1}^{n} (\tfrac{1}{2})^r = (\tfrac{1}{2})^n \approx \begin{cases} 10^{-2} & \text{if } n = 7 \\ 10^{-3} & \text{if } n = 10 \\ 10^{-4} & \text{if } n = 13. \end{cases} \tag{17}$$

Sometimes the sums of quite complicated looking series can be found by putting the series into a suitable form. For instance, for the series

$$\tfrac{1}{2} + \tfrac{1}{6} + \tfrac{1}{12} + \tfrac{1}{20} + \ldots$$

we note that the r-th term

$$\frac{1}{r(r+1)}$$

can be written in the form

$$\frac{1}{r} - \frac{1}{r+1}$$

so that the sum to n terms of the series is

$$S_n = (1 - \tfrac{1}{2}) + (\tfrac{1}{2} - \tfrac{1}{3}) + (\tfrac{1}{3} - \tfrac{1}{4}) + \ldots + \left(\frac{1}{n-1} - \frac{1}{n}\right) + \left(\frac{1}{n} - \frac{1}{n+1}\right)$$

$$= 1 - \frac{1}{n+1}. \tag{18}$$

If we look at equations (17) and (18) we see that as n, the number of terms in the series, increases the difference

$$1 - S_n$$

decreases steadily. In other words by taking n sufficiently great we can make S_n as nearly equal to 1 as we please. Thus for the series $\sum_{r=1}^{n} (\frac{1}{2})^r$ we need only take n to be 20 for S_n to be 1 to within one part in a million. In such circumstances we say that "S_n tends to 1 as n tends to infinity" and write "$S_n \to 1$ as $n \to \infty$". In general, if for any series $\sum_{r=1}^{n} T_r$ with sum to n terms S_n, we can find a number S such that by taking n sufficiently large the numerical value of the difference $S - S_n$ can be made as small as we please we say that the series *converges to the sum* S or that S is the *sum to infinity* of the series. In symbols we write $S_n \to S$ as $n \to \infty$, or

$$\sum_{r=1}^{\infty} T_r = S. \tag{19}$$

It follows, for instance, from equations (17) and (18) that

$$\sum_{r=1}^{\infty} (\tfrac{1}{2})^r = 1$$

and

$$\sum_{r=1}^{\infty} \frac{1}{r(r+1)} = 1.$$

It should be emphasised that ∞ should not be thought of as a number, but merely as a symbol occurring in expressions of this kind. Let us consider the convergence of the geometric series (15). It follows from (16) that

$$\frac{a}{1-x} - \sum_{r=1}^{n} ax^{r-1} = \frac{ax^n}{1-x}. \tag{20}$$

Now if x is numerically greater than 1, x^n will get larger and larger as n increases so that the series does not converge if $|x| > 1$. On the other hand if x is numerically less than 1, the numerical value of x^n will diminish steadily as n increases (cf. equation (17) for the case $x = \frac{1}{2}$). Hence if $|x| < 1$ the series converges to the sum $a/(1-x)$ and we write

$$\sum_{r=1}^{\infty} ax^r = \frac{a}{1-x}, \quad \text{if} \quad |x| < 1. \tag{21}$$

EXAMPLE 5: *Find the vulgar fraction equivalent to the recurring decimal* $0.1\dot{2}\dot{3}$.

$0.1\dot{2}\dot{3}$ is a shorthand way of writing the decimal

$$0.1232323\ldots$$

which is equal to

$$\frac{1}{10} + \frac{23}{1\,000} + \frac{23}{100\,000} + \frac{23}{10\,000\,000}$$

$$= \frac{1}{10} + \frac{23}{1000}\left\{1 + \frac{1}{100} + \left(\frac{1}{100}\right)^2 + \ldots\right\}.$$

The infinite series in the bracket has for sum

$$\frac{1}{1 - \dfrac{1}{100}} = \frac{100}{99}$$

so that the repeating decimal is equivalent to

$$\frac{1}{10}+\frac{23}{990}=\frac{99+23}{990}=\frac{61}{495}.$$

PROBLEMS

1. Show that

$$1^3+3^3+5^3+\ldots+(2n-1)^3=\sum_{r=1}^{2n-1} r^3-8\sum_{r=1}^{n-1} r^3,$$

and hence find the sum of the cubes of the first n odd integers.

2. If S_n denotes the sum of the first n terms of the series

$$a+(a+d)x+(a+2d)x^2+(a+3d)x^3+\ldots,$$

show, by considering $(1-x)S_n$, that

$$S_n=\frac{a}{1-x}+\frac{x(1-x^n)d}{(1-x)^2}.$$

Deduce that if $|x|<1$ the series has a sum to infinity of amount

$$\frac{a+(d-a)x}{(1-x)^2}.$$

Find the sum to infinity of the series

$$1+2x+3x^2+4x^3+\ldots$$

if $|x|<1$.

3. The second term of a geometric series is 3, and the fifth term is $\frac{81}{64}$. Show that the series has a sum to infinity, and find by how much the sum exceeds the sum of the first ten terms, giving the result to three places of decimals.

4. If the r-th term, T_r, of a series is given by a formula of the type

$$T_r=u_r-u_{r+1},$$

for all values of r, show that the sum of the first n terms of the series is u_1-u_{n+1}.

Show that

$$\frac{1}{2}\left\{\frac{1}{2r-1}-\frac{1}{2r+1}\right\}=\frac{1}{4r^2-1}.$$

Hence find the sum of the first n terms of the series

$$\tfrac{1}{3} + \tfrac{1}{15} + \tfrac{1}{35} + \tfrac{1}{63} + \dots$$

and deduce the value of the sum to infinity.

5. If the r-th term, T_r, of a series is given by a formula of the type

$$T_r = u_{r+2} - 2u_{r+1} + u_r,$$

show that the sum of the first n terms of the series is $u_1 - u_2 + u_{n+2} - u_{n+1}$. Deduce that, if u_n becomes very small as n becomes large, the sum to infinity of the series is $u_1 - u_2$.

Prove that

$$\frac{1}{2}\left\{\frac{1}{r+2} - \frac{2}{r+1} + \frac{1}{r}\right\} = \frac{1}{r(r+1)(r+2)}$$

and deduce the value of the sum to infinity of the series

$$\frac{1}{1\cdot 2\cdot 3} + \frac{1}{2\cdot 3\cdot 4} + \frac{1}{3\cdot 4\cdot 5} + \dots.$$

5. Permutations and Combinations

In the theory of probability it is sometimes necessary to calculate the number of ways in which a set of r objects can be chosen from a larger group consisting of n objects. For example, suppose we are given four letters a, b, c, d and we wish to determine the number of ways that three letters may be chosen from these four, attention being paid to the order in which the letters occur. There are obviously four ways of choosing the first letter. After the choice of the first letter has been made, there are three letters remaining from which to choose, so that for each choice of the first letter there are three ways of choosing the second. For example, if we choose a as the first letter, then the corresponding first pairs of letters are ab, ac, ad. Since there are three such possibilities for each choice of the first letter, it follows that there are in all 4×3 possible ways of choosing the first two letters. After the first two letters have been chosen there are only two letters left, so that for each pair of first two letters, there are only two ways of choosing the third letter. Since there were 4×3 ways of choosing the first pair it follows that the number of ways of choosing three letters is $4 \times 3 \times 2 = 24$.

This fact can also be seen by listing the combinations of letters which are possible. These are readily seen to be:-

abc	*abd*	*acb*	*acd*	*adb*	*adc*
bac	*bad*	*bca*	*bcd*	*bda*	*bdc*
cab	*cad*	*cba*	*cbd*	*cda*	*cdb*
dab	*dac*	*dba*	*dbc*	*dca*	*dcb*

and the total number is found to be 24. Each group of letters such as *abc*, when the order in which the letters occur is important, is called a *permutation* and what we have calculated is that there are twenty four different permutations possible from four letters taken three at a time.

There may be some problems where the order in which the letters occur in the group is not important, i.e., the permutations *abc, acb, bac, bca, cab, cba* contain the same letters, *a*, *b* and *c* though in different orders. If we look at the complete set of twenty four permutations listed above, we see that when we do not pay attention to the order in which the letters occur we obtain the four possibilities

abc *abd* *acd* *bcd*

since the permutations *abc, cba, bca, acb, cab* are all produced by the same combination of letters *a*, *b*, *c*, etc. Each such group is called a *combination* of letters, so we have shown that it is possible to choose four different combinations of three letters from a set of four letters. It should be emphasised that a permutation is altered if a single member, i.e., letter, of the group is altered or if the order in which the members occur is altered; a combination is altered if a single member of the group is changed, but the different permutations that are possible within the group do not change the combination.

We shall now generalise these results to obtain the number of possible permutations of n objects taken r at a time, which is denoted by the symbol ${}^nP_r$, and the number of possible combinations of n objects taken r at a time, which is denoted by ${}^nC_r$.

If we are given n distinct objects to choose from, then the number of ways in which we can choose the first object is n. Once the choice of this object has been made there are $n - 1$ objects left and therefore $n - 1$ ways in which the second object can be chosen. Similarly, the third object can be chosen in $n - 2$ ways and so on, until we reach the r-th object which, since $r - 1$ objects have been removed from the given set n, can be chosen in $n - (r - 1)$, or $n - r + 1$ ways. The total number of ways, ${}^nP_r$, of selecting the r objects is the product of the number of ways of selecting each object* so that

$$^nP_r = n(n-1)\cdots(n-r+1). \tag{1}$$

By multiplying the right hand side of this equation by $(n-r)(n-r-1)\cdots 3\cdot 2\cdot 1$ and dividing by the same quantity, we see that this result may be put in the form

$$^nP_r = \frac{n(n-1)\cdots 3\cdot 2\cdot 1}{(n-r)(n-r-1)\cdots 3\cdot 2\cdot 1}. \tag{2}$$

The numerator of this expression is the product of the first n integers. This product arises so frequently that we introduce a symbol to denote it. If n is a positive integer, we use the symbol $n!$ (called *factorial* n), to denote the product

$$n! = n(n-1)(n-2)\cdots 3\cdot 2\cdot 1. \tag{3}$$

With this notation equation (2) can be written in the form

$$^nP_r = \frac{n!}{(n-r)!}. \tag{4}$$

For example, the number of permutations of 4 things taken 3 at a time is

$$^4P_3 = \frac{4!}{1!} = 4! = 24$$

as found above by direct calculation.

If we write out the expression for $(n+1)!$ and separate off the first factor it is readily shown that

$$(n+1)! = (n+1)\cdot n!. \tag{5}$$

Putting $n = 0$ in this equation and using the fact that $1! = 1$ we find that *formally* $0! = 1$. This equation has no meaning since in the definition of $n!$ it is assumed that n is a positive integer, not zero. This result can be used formally in calculations since, in cases in which it is used, the method of calculation could be modified to produce a result identical with that obtained by assuming that $0! = 1$.

As a particular case of equation (4), we find that the number of permutations of n objects taken n at a time is

* This assumption that, if one selection can be made in any one of p ways, and if, after one of these p has been chosen, a second selection can be made in any one of q ways, then the two selections can be made together, and in that order, in pq ways, is called the fundamental principle of choice.

$$^nP_n = n! \tag{6}$$

a result which is readily established from first principles.

To determine the number of combinations, ^nC_r of n objects taken r at a time we make use of equations (4) and (6). By equation (6) we see that each combination containing r distinct objects can produce $r!$ permutations. If there are ^nC_r such combinations possible from n distinct objects then the total number of permutations obtainable from n objects taken r at a time is $^nC_r \times r!$ But this, by definition, is the number ^nP_r, so that we have the relation

$$^nP_r = {}^nC_r r!\,. \tag{7}$$

Substituting the value for ^nP_r given by equation (4) and dividing both sides of equation (7) by $r!$ we obtain the relation

$$^nC_r = \frac{n!}{r!(n-r)!}\,. \tag{8}$$

For example, the number of combinations of 4 objects taken 3 at a time is

$$^4C_3 = \frac{4!}{3!\,1!} = \frac{4\cdot 3!}{3!\,1!} = 4,$$

as we saw by a direct method above.

It follows immediately from the definition that

$$^nC_{n-r} = {}^nC_r. \tag{9}$$

The values of ^nC_r for integral values of n and r up to 12 are given in Table 1.

TABLE 1

Values of ^nC_r for small values of n and r

n \ r	0	1	2	3	4	5	6	7	8	9	10	11	12
1	1	1	—	—	—	—	—	—	—	—	—	—	—
2	1	2	1	—	—	—	—	—	—	—	—	—	—
3	1	3	3	1	—	—	—	—	—	—	—	—	—
4	1	4	6	4	1	—	—	—	—	—	—	—	—
5	1	5	10	10	5	1	—	—	—	—	—	—	—
6	1	6	15	20	15	6	1	—	—	—	—	—	—
7	1	7	21	35	35	21	7	1	—	—	—	—	—
8	1	8	28	56	70	56	28	8	1	—	—	—	—
9	1	9	36	84	126	126	84	36	9	1	—	—	—
10	1	10	45	120	210	252	210	120	45	10	1	—	—
11	1	11	55	165	330	462	462	330	165	55	11	1	—
12	1	12	66	220	495	792	924	792	495	220	66	12	1

EXAMPLE 6: *Prove that* ${}^nC_r + {}^nC_{r-1} = {}^{n+1}C_r$.

From equation (8) we have

$$
\begin{aligned}
{}^nC_r + {}^nC_{r-1} &= \frac{n!}{r!(n-r)!} + \frac{n!}{(r-1)!(n-r+1)!} \\
&= \frac{n!(n+1-r+r)}{r!(n+1-r)!} \\
&= \frac{(n+1)!}{r!(n+1-r)!}
\end{aligned}
$$

and this is equal to ${}^{n+1}C_r$.

In discussing the number of permutations of n objects (equation (6) above) we assumed that the objects are all different. We shall now consider how this result is affected when some of the objects being permuted are alike. Suppose, for instance, that we have n letters, p of which are a's, q are b's and that the remaining $n - p - q$ letters are all different. Let N be the total number of permutations of all these letters. Consider any one of these permutations, and in it replace the p letters a by different symbols $a_1, a_2, \ldots, a_p$, these letters being different from b and all the other letters.

By mutually arranging the p new letters we obtain $p!$ permutations, so that when replacement is made we have $N(p!)$ permutations.

In any one of these $N(p!)$ permutations we now replace the q letters b by different symbols $b_1, b_2, \ldots b_q$ all different and different from the a's and the other letters. By arranging these q new symbols among themselves we obtain $q!$ permutations. Since this holds for each of $N(p!)$ permutations, we have after the double replacement $N(p!)(q!)$ permutations. But the n letters are now all different and hence, by equation (6), there are $n!$ permutations of them. Hence

$$Np!\,q! = n!$$

so that

$$N = \frac{n!}{p!q!}\,. \tag{10}$$

This result can be extended as follows:

The number of permutations of n objects of which p are alike of one kind, q alike of another, r alike of a third kind, and so on, is

$$\frac{n!}{p!q!r!\ldots}. \tag{11}$$

EXAMPLE 7: *How many permutations are there of all the letters in the word "cataract"?*

In how many of these are the three a's separate?

In the word "cataract" there are 8 letters, made up of 3 a's, 2 c's, 2 t's and an r. Hence in the notation of (11) we have $n = 8$, $p = 3$, $q = r = 2$ so that the total number of permutations is

$$\frac{8!}{3!2!2!} = \frac{8 \cdot 7 \cdot 6 \cdot 5 \cdot 4 \cdot 3 \cdot 2 \cdot 1}{3 \cdot 2 \cdot 1 \cdot 2 \cdot 1 \cdot 2 \cdot 1} = 8 \cdot 7 \cdot 6 \cdot 5 = 1680.$$

For the second part, first of all arrange the separators — in this case the five consonants of which 2 are c's and 2 are t's. The total number of ways of arranging these consonants is

$$\frac{5!}{2!2!}.$$

To each of these arrangements of consonants there are 6 vacant spaces in which to place the three a's. The number of ways of placing the vowels in any consonant arrangement is therefore

$${}^6C_3 = \frac{6!}{3!3!}.$$

The number of permutations in which the three a's are separate is therefore

$$\frac{5!}{2!2!} \cdot \frac{6!}{3!3!} = \frac{5 \cdot 4 \cdot 3 \cdot 2 \cdot 1 \cdot 6 \cdot 5 \cdot 4 \cdot 3 \cdot 2 \cdot 1}{2 \cdot 1 \cdot 2 \cdot 1 \cdot 3 \cdot 2 \cdot 1 \cdot 3 \cdot 2 \cdot 1} = 5 \cdot 6 \cdot 5 \cdot 4 = 600.$$

PROBLEMS

1. Calculate the number of ways in which a president, vice-president, secretary and four committee members may be chosen from a club of 40 members.

2. Of 12 different balls, 3 are white. In how many ways can a group of 4 balls be selected to include at least one white ball?

3. There are twelve men who have to sit six on each side of a long table. Three wish to sit on one side and four on the other. If the order in which they are seated is not important show that there are

$$\frac{(6!)^2}{4!(3!)^2 2!}$$

ways in which they may be seated, but that if order is important there are

$$\frac{(6!)^2 5!}{3!2!}$$

ways.

4. In how many ways can two birds sit on six trees?

5. In how many ways can n people be arranged at a round table?

6. In how many ways can n different beads be arranged on a string to form a necklace?

6. The Binomial Theorem

By multiplying the two factors together we find that

$$(1+x)^2=(1+x)(1+x)=1+2x+x^2 \tag{1}$$

$$(1+x)^3=(1+x)(1+x)^2=(1+x)(1+2x+x^2)=1+3x+3x^2+x^3 \tag{2}$$

$$\begin{aligned}(1+x)^4&=(1+x)(1+x)^3=(1+x)(1+3x+3x^2+x^3)\\&=1+4x+6x^2+4x^3+x^4\end{aligned} \tag{3}$$

$$\begin{aligned}(1+x)^5&=(1+x)(1+x)^4=(1+x)(1+4x+6x^2+4x^3+x^4)\\&=1+5x+10x^2+10x^3+5x^4+x^5\end{aligned} \tag{4}$$

and so on. If we look at the values of ${}^nC_r$ in Table 1 we see that these equations can be written in the forms

$$(1 + x)^2 = 1 + {}^2C_1x + x^2$$
$$(1 + x)^3 = 1 + {}^3C_1x + {}^3C_2x^2 + x^3$$
$$(1 + x)^4 = 1 + {}^4C_1x + {}^4C_2x^2 + {}^4C_3x^3 + x^4$$
$$(1 + x)^5 = 1 + {}^5C_1x + {}^5C_2x^2 + {}^5C_3x^3 + {}^5C_4x^4 + x^5.$$

This suggests that, if n is a positive integer,

$$\begin{aligned}(1 + x)^n &= 1 + {}^nC_1x + {}^nC_2x^2 + \dots + {}^nC_rx^r + \dots + {}^nC_{n-1}x^{n-1} + x^n\\ &= \sum_{r=0}^{n} {}^nC_rx^r.\end{aligned} \tag{5}$$

To prove the result (5) we use the *method of induction*, that is we show that if the theorem is true for $n = m$ it is true for $n = m + 1$ and by observing that it is true for $n = 2$ it follows that it is true for

$n = 3, 4, 5, \ldots$. If (5) is true for $n = m$ we have

$$(1 + x)^m = \sum_{r=0}^{m} {}^mC_r x^r$$

and

$$\begin{aligned}(1+x)^{m+1} &= (1+x)(1+x)^m \\ &= (1+x)(1+{}^mC_1 x+\ldots+{}^mC_r x^r+\ldots+{}^mC_{m-1}x^{m-1}+x^m) \\ &= 1+({}^mC_1+1)x+\ldots+({}^mC_r+{}^mC_{r-1})x^r+\ldots+({}^mC_{m-1}+1)x^m+x^{m+1}.\end{aligned}$$

Since ${}^mC_1 + 1 = m + 1 = {}^{m+1}C_1$ and ${}^mC_r + {}^mC_{r-1} = {}^{m+1}C_r$ (by example 6 above) we find that

$$(1 + x)^{m+1} = 1 + {}^{m+1}C_1 x + \ldots + {}^{m+1}C_r x^r + \ldots + {}^{m+1}C_m x^m + x^{m+1}$$

showing that the result is true for $n = m + 1$. Comparison of equations (1) and (5) shows that the result (5) is true for $n = 2$. Hence the result (5) is true for all positive values of n.*

A more general form of the result can be obtained from equation (5) We may write

$$(a + b)^n = a^n \left(1 + \frac{b}{a}\right)^n$$

so it follows from equation (5) that

$$(a + b)^n = a^n \sum_{r=0}^{n} {}^nC_r \left(\frac{b}{a}\right)^r$$

which gives

$$(a + b)^n = \sum_{r=0}^{n} {}^nC_r a^{n-r} b^r, \tag{6}$$

for every positive integer n.

Equation (6) is the binomial theorem for a positive integer n. It should be noted that:

(i) The number of terms in the expansion of $(a + b)^n$ is $n + 1$.

(ii) The coefficients from left to right are the same as those from right to left (since ${}^nC_{n-r} = {}^nC_r$).

(iii) The $(r + 1)^{\text{th}}$ term of the expansion is ${}^nC_r a^{n-r} b^r$.

* Since the result is true for $n = 2$, it is true for $n = 3$; if it is true for $n = 3$, it is true for $n = 4$, and so on.

EXAMPLE 8: *Prove that*

(i) $\sum_{r=0}^{n} {}^nC_r = 2^n$;

(ii) $\sum_{r=0}^{n} (-1)^r\ {}^nC_r = 0$;

(iii) $\sum_{r=0}^{m} {}^nC_r\ {}^nC_{m-r} = {}^{2n}C_m$.

(i) Put $x = 1$ in equation (5) to obtain the required result.
(ii) Put $x = -1$ in equation (5) to obtain the required result.
(iii) If we write

$$(1+x)^n(1+x)^n = (1+x)^{2n},$$

the coefficient of x^m in the expansion of the right hand side is ${}^{2n}C_m$. But the coefficient of x^m in the expansion of the left hand is

$${}^nC_m + {}^nC_{m-1}\,{}^nC_1 + {}^nC_{m-2}\,{}^nC_2 + \ldots + {}^nC_1\,{}^nC_{m-1} + {}^nC_m$$

and this is equal to

$$\sum_{r=0}^{m} {}^nC_r\,{}^nC_{m-r}$$

so that the result follows.

EXAMPLE 9: *Find the coefficient of x^2 in the expansion of*

$$\left(x^3 - \frac{2}{x^2}\right)^9.$$

The $(r+1)^{\text{th}}$ term in this expansion is

$${}^9C_r(x^3)^{9-r}\left(-\frac{2}{x^2}\right)^r$$

i.e.

$$(-2)^r \frac{9!}{r!(9-r)!} x^{27-5r}.$$

To find the coefficient of x^2 we must put $27 - 5r = 2$, i.e. $r = 5$. The coefficient of x^2 in the expansion is therefore

$$(-2)^5 \times \frac{9!}{5!4!} = -32 \times 126 = -4032.$$

We shall now consider the expansion of $(1 + x)^n$ in ascending powers of x in the case where n is not a positive integer but any rational number, positive or negative. We introduce the symbol

$$\binom{n}{r} = \frac{n(n-1)\cdots(n-r+1)}{r!} \tag{7}$$

where n is any rational number and r is a positive integer. When n is a positive integer it is obvious that

$$\binom{n}{r} = {}^nC_r$$

but when n is not a positive integer, the symbol ${}^nC_r$ is not defined. What equation (5) states is therefore that if n is a positive integer

$$\sum_{r=0}^{n} \binom{n}{r} x^r = (1 + x)^n.$$

When n is not a positive integer the series

$$1 + nx + \frac{n(n-1)}{2!}x^2 + \ldots + \binom{n}{r}x^r + \ldots \tag{8}$$

is an infinite series and the question naturally arises as to whether or not the series is convergent.

It can in fact be shown that, if $|x| < 1$, the infinite series (8) converges and has as its sum to infinity the positive value of $(1 + x)^n$. If $|x| > 1$ the series (8) diverges. This is the BINOMIAL THEOREM for a rational index n. It is stated here without proof, since a rigorous proof requires fairly advanced mathematical ideas. The proof for the cases $n = -1$ and $n = -2$,

$$(1+x)^{-1} = 1 - x + x^2 - x^3 + \ldots + (-1)^r x^r + \ldots, \qquad |x| < 1, \tag{9}$$

$$(1+x)^{-2} = 1 - 2x + 3x^2 - 4x^3 + \ldots + (-1)^r (r+1) x^r + \ldots, \quad |x| < 1, \tag{10}$$

can be readily constructed by the methods of Section 4. (Cf. equation (16) and Problem 2).

Other commonly occurring series are

$$(1+x)^{\frac{1}{2}} = 1 + \tfrac{1}{2}x - \tfrac{1}{8}x^2 + \tfrac{1}{16}x^3 - \tfrac{5}{128}x^4 + \tfrac{7}{256}x^5 - \ldots, \quad |x| < 1, \tag{11}$$

$$(1+x)^{-\frac{1}{2}} = 1 - \tfrac{1}{2}x + \tfrac{3}{8}x^2 - \tfrac{5}{16}x^3 + \tfrac{35}{128}x^4 - \tfrac{63}{256}x^5 + \ldots, \quad |x| < 1, \tag{12}$$

$$(1+x)^{\frac{1}{3}} = 1 + \tfrac{1}{3}x - \tfrac{1}{9}x^2 + \tfrac{5}{81}x^3 - \tfrac{10}{243}x^4 + \tfrac{22}{729}x^5 + \ldots, \quad |x| < 1. \tag{13}$$

The equations (11) and (13) can be used to find square roots and cube roots respectively. For example, suppose we wish to find the square root and the cube root of 37. We note that $37 = 36 + 1$ so that

$$(37)^{\frac{1}{2}} = 6(1 + \tfrac{1}{36})^{\frac{1}{2}}$$

so that, from equation (11),

$$\begin{aligned}(37)^{\frac{1}{2}} &= 6\{1 + \tfrac{1}{2} \cdot \tfrac{1}{36} - \tfrac{1}{8}(\tfrac{1}{36})^2 + \tfrac{1}{16}(\tfrac{1}{36})^3\} \\ &= 6\{1 + \tfrac{1}{72} - \tfrac{1}{10368} + \tfrac{1}{746496}\} \\ &= 6\{1 + 0.01389 - 0.00009 + 0.00000\} \\ &= 6.0828.\end{aligned}$$

To evaluate the cube root of 37 we note that $27 \times 37 - 999$ so that

$$(37)^{\frac{1}{3}} = (\tfrac{999}{1000})^{\frac{1}{3}} \times \tfrac{10}{3} = \tfrac{10}{3}(1 - \tfrac{1}{1000})^{\frac{1}{3}}.$$

Making use of the expansion (13) we therefore have

$$\begin{aligned}(37)^{\frac{1}{3}} &= \frac{10}{3}\left\{1 - \frac{1}{3000} - \frac{1}{9\,000\,000}\right\} \\ &= 3.33333 - 0.00111 \\ &= 3.33222.\end{aligned}$$

For application to problems in calculus an important result is contained in:

EXAMPLE 10: *Show that if h/x is small*

$$\frac{(x + h)^n - x^n}{h} = nx^{n-1} + hS_n(x)$$

where

$$S_n(x) = \tfrac{1}{2}n(n - 1)x^{n-2} + \tfrac{1}{6}n(n - 1)(n - 2)x^{n-3}h + \ldots.$$

From the binomial theorem we have that, for $|h/x| < 1$,

$$\begin{aligned}(x+h)^n &= x^n\left(1 + \frac{h}{x}\right)^n \\ &= x^n\left\{1 + n\left(\frac{h}{x}\right) + \tfrac{1}{2}n(n-1)\left(\frac{h}{x}\right)^2 + \tfrac{1}{6}n(n-1)(n-2)\left(\frac{h}{x}\right)^3 + \ldots\right\} \\ &= x^n + nhx^{n-1} + \tfrac{1}{2}n(n-1)h^2x^{n-2} + \tfrac{1}{6}n(n-1)(n-2)h^3x^{n-3} + \ldots\end{aligned}$$

and the result follows by simple algebraic manipulations.

We can use the binomial theorem to obtain quite complicated expansions. The method is illustrated by

EXAMPLE 11: *Show that, if* $|x| < 1$,

$$\frac{1}{2 + 3x + x^2} = \tfrac{1}{2} - \tfrac{3}{4}x + \tfrac{7}{8}x^2 - \tfrac{15}{16}x^3 + \ldots$$

and find the coefficient of x^r *in the expansion.*

We note that $2 + 3x + x^2 = (2 + x)(1 + x)$ and that

$$\begin{aligned}\frac{1}{(1 + x)(2 + x)} &= \frac{1}{1 + x} - \frac{1}{2 + x} \\ &= \frac{1}{1 + x} - \frac{1}{2(1 + \frac{1}{2}x)}.\end{aligned} \tag{14}$$

If $|x| < 1$ both of these expressions can be expanded by means of equation (9) to give

$$\begin{aligned}&\{1 - x + x^2 - x^3 + \ldots + (-1)^r x^r + \ldots\} \\ &- \frac{1}{2}\left\{1 - \tfrac{1}{2}x + \tfrac{1}{4}x^2 - \tfrac{1}{8}x^3 + \ldots + \frac{(-1)^r}{2^r}x^r + \ldots\right\} \\ &= \tfrac{1}{2} - \tfrac{3}{4}x + \tfrac{7}{8}x^2 - \tfrac{15}{16}x^3 + \ldots + (-1)^r\frac{2^{r+1} - 1}{2^{r+1}}x^r + \ldots.\end{aligned}$$

PROBLEMS

1. Find the term independent of x in the expansion of

$$\left(\frac{3}{x^3} - \frac{x}{3}\right)^8.$$

2. Find the first three non-zero terms in the expansion of

$$\frac{(3x - 1)^2}{(2x - 1)^3}$$

in ascending powers of x, and prove that the coefficient of x^r is

$$2^{r-3}(r - 8)(1 - r).$$

State the range of values of x for which the expansion is valid.

3. Show that

$$\tfrac{1}{2}\sqrt{5} = 1 + \frac{1}{10} + \frac{1 \cdot 3}{10 \cdot 20} + \frac{1 \cdot 3 \cdot 5}{10 \cdot 20 \cdot 30} + \cdots;$$

$$\sqrt[3]{4} = 1 + \frac{1}{4} + \frac{1 \cdot 4}{4 \cdot 8} + \frac{1 \cdot 4 \cdot 7}{4 \cdot 8 \cdot 12} + \cdots.$$

4. Prove that $100 = 2^7(1 - \frac{7}{32})$ and hence calculate $(100)^{\frac{1}{7}}$ to three decimal places.

5. Write down in its simplest form the coefficient of x^r in the expansion of

$$\frac{a + bx + cx^2}{(1 - x)^3}$$

and determine a, b, c so that this expansion reduces to

$$\sum_{r=1}^{\infty} r^2 x^r.$$

Show that

$$\sum_{r=1}^{\infty} \frac{r^2}{2^r} = 6.$$

7. Approximations and Errors

If x is a small proper fraction, the terms of the sequence

$$1,\ x,\ x^2,\ x^3, \ldots$$

are in rapidly decreasing order of magnitude. If a quantity y can be expressed as a series of ascending powers of x by a formula of the type

$$y = a_0 + a_1 x + a_2 x^2 + a_3 x^3 + \ldots$$

and if x is small, we say that $a_0 + a_1 x$ is a *first approximation* to the value of y and that $a_0 + a_1 x + a_2 x^2$ is a *second approximation* to the value of y.

For instance it follows from the binomial theorem that, for small values of x,

$$(1 + x)^n \simeq 1 + nx. \tag{1}$$

(The symbol $\simeq$ should be read "is approximately equal to".) Equation (1) gives a first approximation to $(1 + x)^n$ with an error of order $\frac{1}{2}n(n - 1)x^2$. A second approximation is given by

$$(1 + x)^n \simeq 1 + nx + \tfrac{1}{2}n(n - 1)x^2. \tag{2}$$

A particular case of (2) which is of great value in approximate calculations is that corresponding to $n = \frac{1}{2}$:-

$$(1 + x)^{\frac{1}{2}} \simeq 1 + \tfrac{1}{2}x - \tfrac{1}{8}x^2 \tag{3}$$

which can be used for finding approximate values of the square roots of numbers nearly equal to 1. Thus we get the approximation

$$\sqrt{0.92} \simeq 1 - 0.04 - 0.0008 = 0.9592$$

(obtained by putting $x = -0.08$).

Similarly the formula

$$(1 + x)^{\frac{1}{3}} \simeq 1 + \tfrac{1}{3}x - \tfrac{1}{9}x^2 \tag{4}$$

may be used to obtain the approximate values of the cube roots of numbers nearly equal to 1.

The method can be readily adapted to meet more complicated situations as illustrated by

EXAMPLE 12: *Find approximations to* $y = (a + bx + cx^2)^n$ *when* x *is* (i) *small,* (ii) *large,* (iii) *nearly* 1.

(i) We can write

$$y = a^n \left(1 + \frac{b}{a}x + \frac{c}{a}x^2\right)^n$$

$$\simeq a^n \left\{1 + n\left(\frac{b}{a}x + \frac{c}{a}x^2\right) + \tfrac{1}{2}n(n-1)\left(\frac{b}{a}x + \frac{c}{a}x^2\right)^2\right\}.$$

If we retain only terms up to order x^2 we have

$$y \simeq a^n \left\{1 + \frac{nb}{a}x + \frac{nc}{a}x^2 + \tfrac{1}{2}n(n-1)\frac{b^2x^2}{a^2}\right\}$$

that is

$$y \simeq a^n + na^{n-1}bx + [na^{n-1}c + \tfrac{1}{2}n(n-1)a^{n-2}b^2]x^2 \tag{5}$$

for small values of x.

(ii) If x is large, x^{-1} will be small so that writing

$$a + bx + cx^2 = cx^2\left(1 + \frac{b}{c}\cdot\frac{1}{x} + \frac{a}{c}\cdot\frac{1}{x^2}\right)$$

we find that

$$y = c^n x^{2n}\left(1 + \frac{b}{c}\cdot\frac{1}{x} + \frac{a}{c}\cdot\frac{1}{x^2}\right)^n$$
$$\simeq c^n x^{2n}\left\{1 + n\left(\frac{b}{c}\cdot\frac{1}{x} + \frac{a}{c}\cdot\frac{1}{x^2}\right) + \tfrac{1}{2}n(n-1)\cdot\frac{b^2}{c^2x^2}\right\}$$

so that

$$y \simeq x^{2n}\left\{c^n + \frac{nbc^{n-1}}{x} + \frac{2nac^{n-1} + n(n-1)b^2c^{n-2}}{2x^2}\right\} \tag{6}$$

for large values of x.

(iii) If x is nearly equal to 1 we can write $x = 1 + \xi$ where ξ is small and then

$$a + bx + cx^2 = a + b(1+\xi) + c(1 + 2\xi + \xi^2)$$
$$= \alpha + \beta\xi + c\xi^2$$

where $\alpha = a + b + c$, $\beta = b + 2c$, and it follows from (5) that

$$y \simeq \alpha^n + n\alpha^{n-1}\beta\xi + [n\alpha^{n-1}c + \tfrac{1}{2}n(n-1)\alpha^{n-2}\beta^2]\xi^2$$

giving the approximation

$$y \simeq (a+b+c)^n + n(a+b+c)^{n-1}(b+2c)(x-1)$$
$$+ [n(a+b+c)^{n-1}c + \tfrac{1}{2}n(n-1)(a+b+c)^{n-2}(b+2c)^2](x-1)^2 \tag{7}$$

for $x \simeq 1$.

We can readily extend the method to cover formulae involving more than two quantities x and y. The simplest case is that in which

$$w = (1+x)^m(1+y)^n(1+z)^p. \tag{8}$$

For small values of x, y and z

$$(1+x)^m \simeq 1 + mx,\quad (1+y)^n \simeq 1 + ny,\quad (1+z)^p \simeq 1 + pz,$$

so that

$$w \simeq (1+mx)(1+ny)(1+pz).$$

Multiplying these three factors together and retaining only the first order terms we find the approximate expression

$$w \simeq 1 + mx + ny + pz, \tag{9}$$

when x, y and z are small.

In the experimental sciences we are often faced with a situation in which the quantity in which we are interested, w say, cannot be measured directly but is derived by means of a formula from measured values of other quantities $x, y, z, \ldots$ say. We can often estimate the experimental errors in the values of $x, y, z, \ldots$ and we wish to estimate the error in the value of w derived from these measurements. The general problem requires for its solution the theory of functions of several variables and is postponed until Section 9 of Chapter 8. Here we shall consider only the simple case in which

$$w = kx^m y^n z^p \tag{10}$$

where k, m, n and p are constants.

We shall illustrate the procedure by first considering a special case. Suppose that

$$w = 6.58 \frac{x^{\frac{1}{2}} y^{\frac{3}{2}}}{z^{\frac{1}{4}}} \tag{11}$$

and that the observed values of x, y, z are in error by $+1\%$, $+2\%$ and -2% respectively. We wish to estimate the percentage error in the value of w calculated from the formula (11).

Suppose that the correct values of x, y and z are a, b, c respectively, then the values actually measured will be

$$x = a(1 + \tfrac{1}{100}), \quad y = b(1 + \tfrac{2}{100}), \quad z = c(1 - \tfrac{2}{100})$$

so that the calculated value of w is

$$w = 6.58 \frac{a^{\frac{1}{2}} b^{\frac{3}{2}}}{c^{\frac{1}{4}}} (1 + \tfrac{1}{100})^{\frac{1}{2}} (1 + \tfrac{2}{100})^{\frac{3}{2}} (1 - \tfrac{2}{100})^{-\frac{1}{4}}$$

and, by equation (9), this is approximately equal to

$$6.58 \frac{a^{\frac{1}{2}} b^{\frac{3}{2}}}{c^{\frac{1}{4}}} \left\{1 + \frac{\frac{1}{2} + \frac{3}{2} \cdot 2 + (-\frac{1}{4})(-2)}{100}\right\} = 6.58 \frac{a^{\frac{1}{2}} b^{\frac{3}{2}}}{c^{\frac{1}{4}}} (1 + \tfrac{4}{100}).$$

Now the correct value of w is $6.58 a^{\frac{1}{2}} b^{\frac{3}{2}} c^{-\frac{1}{4}}$ so that the error in the calculated value is approximately $+4\%$.

Let us return now to the general case (10). If the percentage errors in the measurements of x, y and z are ξ, η and ζ respectively then we may write

$$x = a\left(1 + \frac{\xi}{100}\right), \quad y = b\left(1 + \frac{\eta}{100}\right), \quad z = c\left(1 + \frac{\zeta}{100}\right)$$

where a, b and c are the correct values. The calculated value of w is

$$ka^m b^n c^p \left(1 + \frac{\xi}{100}\right)^m \left(1 + \frac{\eta}{100}\right)^n \left(1 + \frac{\zeta}{100}\right)^p$$
$$\simeq ka^m b^n c^p \left(1 + \frac{m\xi + n\eta + p\zeta}{100}\right)$$

by equation (9). Since the correct value of w is $ka^m b^n c^p$ it follows that the percentage error $\varepsilon(w)$ in the calculated value of w is approximately

$$\varepsilon(w) = m\xi + n\eta + p\zeta. \tag{12}$$

In this calculation we have assumed that we have knowledge of the actual values of the errors made. A more usual situation is to know the limits of error in each measurement, so that the error in x lies between $\pm\xi\%$ etc. It is readily shown that the error in w will lie within the range $\pm\varepsilon$ where

$$\varepsilon = |m\xi| + |n\eta| + |p\zeta|. \tag{13}$$

EXAMPLE 13: *The acceleration due to gravity, g, may be determined by measuring the length l and the period T of a simple pendulum and using the formula*

$$g = \frac{4\pi^2 l}{T^2}.$$

If errors of $\pm 1\%$ can be made in the measurements of T and l, find the probable error in g.

This is governed by equation (13) with $m = -2$, $n = 1$, $p = 0$ and $\xi = \eta = 1$. This gives $\varepsilon = 2 + 1$ so that the probable error in the calculated value of g is $\pm 3\%$.

PROBLEMS

1. If x is so small that x^3 and higher powers of x can be neglected show that

$$(1 - \tfrac{3}{2}x)^5 (2 + 3x)^6 = 64 + 96x - 720x^2.$$

2. Show that, when x is small,

$$\frac{(4 + 3x)^{\frac{2}{3}}}{(2 + x)^{\frac{1}{3}}} \simeq 2 + \tfrac{2}{3}x - \tfrac{13}{72}x^2,$$

and that, when x is large,

$$\frac{(4 + 3x)^{\frac{2}{3}}}{(2 + x)^{\frac{1}{3}}} \simeq (9x)^{\frac{1}{3}} \left\{1 + \frac{2}{9x} + \frac{8}{81x^2}\right\}.$$

3. Neglecting powers of x higher than the second prove that

$$\frac{(2 + x^2)^3\sqrt{(2 - 3x^2)}}{\sqrt{(2 + x)^3}} \simeq 4 - 3x + \tfrac{39}{8}x^2 + \ldots,$$

when x is small.

4. The volume, V, of a sphere is given in terms of its radius, r, by the formula

$$V = \tfrac{4}{3}\pi r^3.$$

How accurate must the measurement of the radius be if the error in the volume must be less than 1%?

5. The coefficient of viscosity, η, of a fluid can be calculated by measuring the total volume Q of fluid flowing per second through a tube of length l and internal radius a under pressure p and using the formula

$$Q = \frac{\pi p a^4}{8l\eta}.$$

If errors of $\pm 1\%$ can be made in the measurement of the lengths a and l, $\pm 2\%$ in the measurement of Q and $\pm \frac{1}{2}\%$ in the measurement of p, how accurate is the estimate of the value of η?

6. According to the special theory of relativity the mass m of a body moving with velocity v is given in terms of the rest mass m_0 and the velocity of light c (3×10^{10} cm/sec) by the formula

$$m = m_0(1 - v^2/c^2)^{-\frac{1}{2}}.$$

Deduce that

$$m = m_0\left(1 + \frac{v^2}{2c^2} + \ldots\right).$$

If E is the kinetic energy of a body of mass M, equal to $m - m_0$, moving with velocity v, show that, for $v \ll c$, $E = Mc^2$.

8. Partial Fractions

We have already seen (in example 11 on p. 34) the value of being able to write an expression of the type

$$\frac{1}{(1 + x)(2 + x)} \tag{1}$$

as the difference of two simple terms

$$\frac{1}{1+x} - \frac{1}{2+x}. \tag{2}$$

These are called *partial fractions* for the original expression. The process of finding partial fractions, i.e. of writing a complicated expression of the type (1) as the sum of simpler expressions of the type (2), is an important one with many applications to calculus.

We shall consider here only the simpler aspects of the theory. An expression

$$\frac{px^{n-1} + qx^{n-2} + \ldots + r}{(a_1x + b_1)(a_2x + b_2) \cdots (a_nx + b_n)} \tag{3}$$

in which the denominator is the product of n simple linear factors can be written as a sum of n simple terms:-

$$\frac{A_1}{a_1x + b_1} + \frac{A_2}{a_2x + b_2} + \ldots + \frac{A_n}{a_nx + b_n}. \tag{4}$$

The coefficients $A_1, A_2, \ldots, A_n$ can be found by giving x particular values or by multiplying both sides of the identical relation

$$\frac{px^{n-1} + qx^{n-2} + \ldots + r}{(a_1x + b) \cdots (a_nx + b_n)} = \frac{A_1}{a_1x + b_1} + \ldots + \frac{A_n}{a_nx + b_n}$$

by $(a_1x + b) \cdots (a_nx + b_n)$ and equating coefficients of corresponding powers of x. We illustrate the procedure by

EXAMPLE 14: *Find partial fractions for*

$$\frac{2x^2 + 4x}{2x^3 + x^2 - 2x - 1}.$$

We note that the denominator has factors $(2x + 1)(x + 1)(x - 1)$ so that we may write

$$\frac{2x^2 + 4x}{2x^3 + x^2 - 2x - 1} = \frac{A}{2x + 1} + \frac{B}{x + 1} + \frac{C}{x - 1}.$$

Multiplying both sides of this identity by $(2x + 1)(x + 1)(x - 1)$ we find that

$$2x^2 + 4x = A(x + 1)(x - 1) + B(2x + 1)(x - 1) + C(2x + 1)(x + 1). \tag{5}$$

We may now proceed in either of two ways:

(i) If the equation holds for all values it holds for:-

(*a*) $x = 1$ implying that $2 + 4 = C(3)(2)$, i.e. that $C = 1$.

(*b*) $x = -1$ implying that $2 - 4 = B(-1)(-2)$, i.e. that $B = -1$.

(*c*) $x = -\frac{1}{2}$ implying that $\frac{1}{2} - 2 = A(\frac{1}{2})(-\frac{3}{2})$, i.e. that $A = 2$.

(ii) The right hand side of equation (5) can be written in the form

$$A(x^2 - 1) + B(2x^2 - x - 1) + C(2x^2 + 3x + 1)$$
$$= (A + 2B + 2C)x^2 + (3C - B)x - (A + B - C).$$

If this has to be identical with $2x^2 + 4x$ we must have

$$\begin{aligned} A + 2B + 2C &= 2 \\ 3C - B &= 4 \\ A + B - C &= 0 \end{aligned}$$

and these equations have solution $A = 2, B = -1, C = 1$ in agreement with (i). This method is probably the easier one to use. We therefore have that

$$\frac{2x^2 + 4x}{2x^3 + x^2 - 2x - 1} = \frac{2}{2x + 1} - \frac{1}{x + 1} + \frac{1}{x - 1}.$$

If the denominator of the expression to be resolved into partial fractions contains one or more repeated linear factors, i.e. factors of the type $(a_1x + b_1)^r$ where r is an integer $(r > 1)$, then this factor gives rise to partial fractions of the form

$$\frac{A_1}{a_1x + b_1} + \frac{A_2}{(a_1x + b_1)^2} + \ldots + \frac{A_r}{(a_1x + b_1)^r}.$$

The remaining factors can be dealt with as previously.

Example 15: *Find partial fractions for*

$$\frac{3x^2 + 4x - 1}{(x + 2)^2(2x + 1)}.$$

We write

$$\frac{3x^2 + 4x - 1}{(x + 2)^2(2x + 1)} = \frac{A}{x + 2} + \frac{B}{(x + 2)^2} + \frac{C}{2x + 1}.$$

Multiplying both sides of this relation by $(x + 2)^2(2x + 1)$ we find that A, B, C must be such that

$$3x^2 + 4x - 1 = A(2x + 1)(x + 2) + B(2x + 1) + C(x + 2)^2.$$

Putting $x = -2$ we find that

$$12 - 8 - 1 = B(-3)$$
$$B = -1.$$

Substituting this value we obtain the relation

$$3x^2 + 6x = A(2x + 1)(x + 2) + C(x + 2)^2$$

which is equivalent to

$$3x = A(2x + 1) + C(x + 2).$$

Putting $x = -2$ we find that $A = 2$ and putting $x = \frac{1}{2}$ we find that $C = -1$. Hence

$$\frac{3x^2 + 4x - 1}{(x + 2)^2(2x + 1)} = \frac{2}{x + 2} - \frac{1}{(x + 2)^2} - \frac{1}{2x + 1}.$$

If the denominator of the expression we wish to resolve into partial fractions has a factor of the form $(ax^2 + bx + c)^r$ where r is a positive integer, then the partial fractions expression must contain terms of the form

$$\frac{A_1x + B_1}{ax^2 + bx + c} + \frac{A_2x + B_2}{(ax^2 + bx + c)^2} + \ldots + \frac{A_rx + B_r}{(ax^2 + bx + c)^r}$$

corresponding to this type of fraction. The remaining factors are dealt with as before. To illustrate the procedure we give:-

EXAMPLE 16: *Find partial fractions for*

$$\frac{4}{(x^2 + 1)^2(x - 1)^2}.$$

We write

$$\frac{4}{(x^2 + 1)^2(x - 1)^2} = \frac{Ax + B}{x^2 + 1} + \frac{Cx + D}{(x^2 + 1)^2} + \frac{E}{x - 1} + \frac{F}{(x - 1)^2}.$$

Multiplying both sides of the relation by $(x^2 + 1)^2(x - 1)^2$ we find that

$$4 = (Ax + B)(x^2 + 1)(x - 1)^2 + (Cx + D)(x - 1)^2$$
$$+ E(x - 1)(x^2 + 1)^2 + F(x^2 + 1)^2.$$

If we put $x = 1$ we find that $F = 1$. Substituting this value into the equation and dividing throughout by $(x - 1)$ we find that

$$-(x^2+3)(x+1) = (Ax+B)(x^2+1)(x-1) + (Cx+D)(x-1) + E(x^2+1)^2.$$

If we put $x = 1$ we find that $E = -2$. Substituting this value into the equation and dividing throughout by $(x - 1)$ we find that

$$\begin{aligned} 2x^3 + x^2 + 4x + 1 &= (Ax + B)(x^2 + 1) + (Cx + D) \\ &= Ax^3 + Bx^2 + (A + C)x + (B + D). \end{aligned}$$

Equating coefficients we find that $A = 2$, $B = 1$, $A + C = 4$ and so $C = 2$, $B + D = 1$ and so $D = 0$.

Hence

$$\frac{4}{(x^2 + 1)^2(x - 1)^2} = \frac{2x + 1}{x^2 + 1} + \frac{2x}{(x^2 + 1)^2} - \frac{2}{x - 1} + \frac{1}{(x - 1)^2}.$$

The technique of finding partial fractions plays an important part in the solution of a certain class of what are called differential equations, of which the first example is encountered in § 10 of Chapter 6. Another application is found in integration (§ 4 of Chapter 7). Perhaps the most important application of partial fractions occurs in calculating what are known as inverse Laplace transforms (§ 16 of Chapter 7).

PROBLEMS

1. Show that*

$$\frac{x^3 + 3x^2 + 4x + 3}{x^2 + 3x + 2} = x + \frac{1}{x + 1} + \frac{1}{x + 2}.$$

2. Find partial fractions for

$$\frac{2x^3 - 4x^2 + 2x - 2}{(x^2 + 1)(x - 1)^2}.$$

3. Find partial fractions for

$$\frac{36 - 31x + 6x^2}{(3 + 2x)(5 - 2x)^2},$$

* It will be observed that the methods developed above can be used only if the degree of the numerator is less than that of the denominator. Hence we begin by "dividing out" x.

and hence show that, for small values of x, it is approximately equal to

$$0.4800 - 0.3493x + 0.2126x^2.$$

4. Prove that

(i) $$\frac{2x}{(x^2+1)(x+1)^2} = \frac{1}{x^2+1} - \frac{1}{(x+1)^2},$$

(ii) $$\frac{x^3+2x-1}{(x^2+1)^2(x+1)} = \frac{x}{x^2+1} + \frac{x}{(x^2+1)^2} - \frac{1}{x+1}.$$

CHAPTER 2

FUNCTIONS OF A SINGLE VARIABLE

1. The Graphical Representation of Data

In the biological sciences we frequently encounter situations in which it is desirable to analyse the relationships between various entities which are varying simultaneously. In such instances the data can be analysed numerically only if they are expressed in some way by numbers. This, of course, is not always possible; in the social sciences, for instance in the analysis of Gallup polls, it is not always possible to express the facts completely by numbers — differing reasons for supporting a political party are not easily expressed numerically! In this book we shall not be concerned with data of that kind, but with experimental facts which can be expressed by a series of numbers. Our problem will be to attempt to discover the significance of the figures made available to us.

It is only rarely that we encounter a scientist who can see the general trend of a variation by quoting a series of numbers. Most scientific workers require the relevant numbers to be presented pictorially to them before they grasp their significance.

The simplest form of graphical representation is one with which most people are familiar from their reading of newspapers and government publications. This works on the principle that the most vivid way of presenting a series of numbers is by representing them by drawings which are suggestive both of the size of the number and of the nature of the quantity to which that number refers. For instance, it is common practice to represent one million pounds sterling by a drawing of a coin; in this representation two million pounds would be shown as two coins, and so on. This method of providing the general reader with a picture of a complex numerical situation is often employed in newspapers discussing economic problems. Any method of this kind which depends on the idea of giving each item to be discussed a group of pictorial units, all the units being of the same size, is called an *ideograph*.

A slightly more sophisticated version of a simple ideograph is shown in Fig. 1 which illustrates graphically the effect of androgenic steroids on cortisone-inhibited growth rate*. In this diagram the percentage increase in weight is represented graphically by the length of a rectangle in each case, and the relative merits of the various treatments leap instantly to the eye.

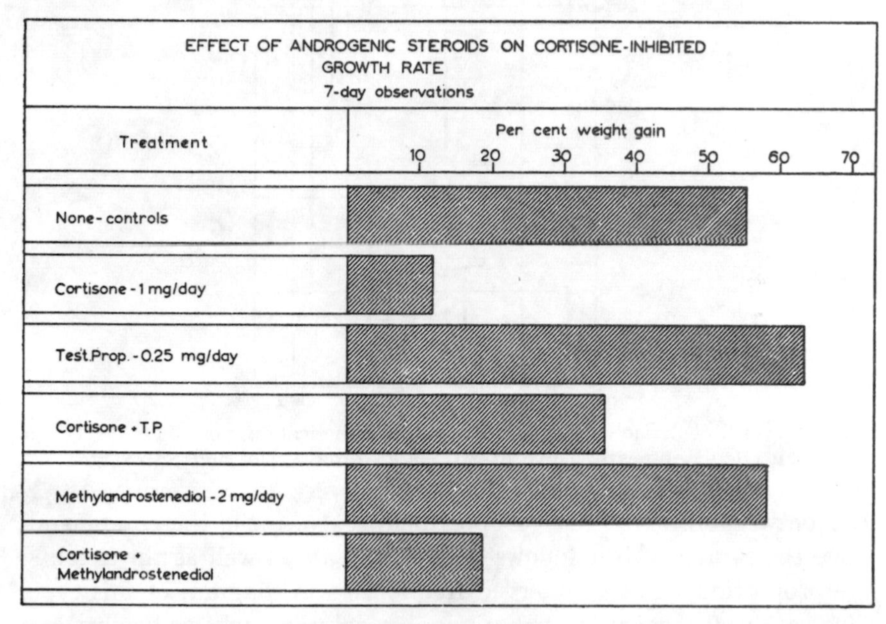

Fig. 1. The effect of testosterone propionate in moderate dosage on the growth increment in seven-day tests in young male rats whose normal growth rate is inhibited with cortisone.

Closely related to "bar diagrams" of this kind are the diagrams which occur in the discussion of frequency distributions. In an investigation† of the numbers and contraction-values of individual motor-units examined in some muscles of the limb, Eccles and Sherrington presented some of their results in the form shown in Fig. 2. This diagram represents the frequency of distribution of nerve fibres of different diameters in nerve of *gast. med.*. The lengths of the fibres are marked along the horizontal axis and above each stretch corresponding to a range of length μ a column is erected whose

* R. Gaunt, "Chemical Control of Growth in Animals", in "Dynamics of Growth Processes". (Princeton University Press, 1954.) Fig. 7, p. 199.

† Eccles and Sherrington, Proc. Roy. Soc. B **106** (1930) 326–357.

height is proportional to the percentage of fibres of that length occurring in the nerve. For example, since 15% of the fibres had lengths between 12μ and 13μ, the column raised on the part of the horizontal

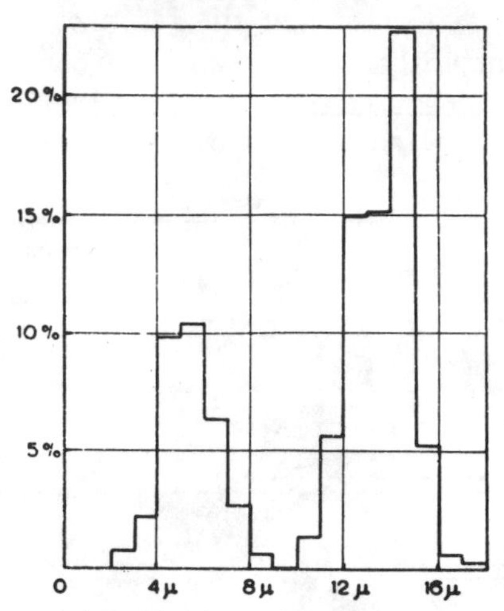

Fig. 2. Nerve fibre diameters in μ plotted against numerical distribution in percentage of total efferent fibres in nerve of *gast. med.* (after Eccles and Sherrington).

line between 12 and 13μ is proportional to 15. Since the columns all have the same width it follows that the areas, as well as the heights, are proportional to the observed frequencies. A diagram of this type, illustrating frequency distributions by columns whose heights are proportional to the observed frequencies, is called a *histogram*. The use of histograms is widespread in the representation of biological and medical data, but their proper analysis belongs to the study of statistics, not mathematics. In mathematics, diagrams such as ideograms and histograms have their uses in helping us to appreciate the general structure of the numerical data with which we are confronted but they are of little value in the precise analysis of the data. We shall now consider some more useful ways of representing numerical data graphically.

2. The Drawing of Graphs

In discussing the best methods of representing numerical data graphically we shall confine ourselves to the simplest case. This is when only two quantities are involved, and the variation of one depends on that of the other. Probably the most widely known example of this type is the variation of the temperature of a hospital patient with time. Readings of the temperature and the time are recorded as numbers, and are then plotted on a chart.

In an experiment to determine the variation with time of the temperature of a certain substance the figures shown in Table 2 were obtained. The time shown in the top row of the table is the time

TABLE 2

Actual time	3.16	3.17	3.18	3.19	3.20	3.21	3.22	3.23	3.27	3.31	3.35	3.39
Time (min)	1	2	3	4	5	6	7	8	12	16	20	24
Temperature (°C)	23.6	25.5	25.6	25.7	25.6	25.6	25.6	25.6	25.5	25.5	25.4	25.2

actually observed on a watch. The time in the second row is the time reckoned from 3.15 p.m. as zero, that is, it is the number of minutes which have elapsed from 3.15 p.m. To put these results in pictorial form we associate each reading of the temperature with a pair of numbers, and then represent that pair of numbers by a point plotted on a sheet of paper. For instance the temperature reading at 3.23 p.m.

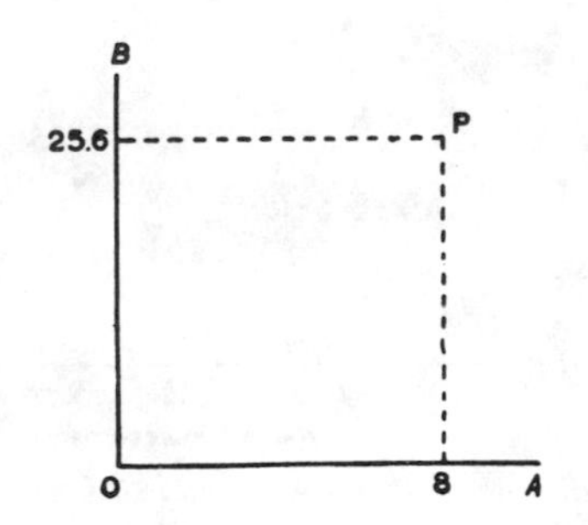

Fig. 3. Plotting the point (8, 25.6).

may be associated with the pair of numbers (8, 25.6) and this ordered pair can be represented by a point in the following way: through a fixed point O draw two fixed lines OA, OB at right angles to one another (cf. Fig. 3). Along OA mark off a distance proportional to

8 min and long OB mark off a distance proportional to 25.6°C.* From the end points of these segments draw lines parallel to OA and OB as shown in Fig. 3. The point P at which these lines intersect may be taken to represent the ordered pair of numbers (8, 25.6). Since the number corresponding to the time is marked along the line OA that line is known as the *time-axis*; similarly the line OB is called the *temperature-axis*. If the point O is so chosen that it represents the ordered pair (0, 0), then it is called the *origin* of the coordinates of time and temperature. It will be noted that in Fig. 4 representing the data of Table 2 the origin of coordinates is not shown; in order to show the variation of temperature clearly the temperature-axis is marked off from 23°C to 26°C.

Proceeding in this way we can associate a point in the plane OAB with each reading of the thermometer listed in Table 2. By joining up the points so obtained by a smooth curve we get a rough idea of

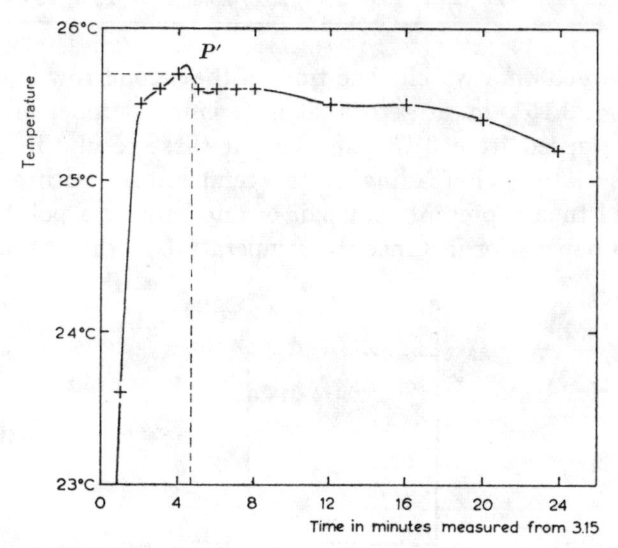

Fig. 4. Variation of temperature with time in a certain experiment.

how the temperature varies. The smooth curve is called the *graph* of the relation between time and temperature. If we assume that the temperature varies in a continuous fashion then we would not go far wrong if we connected the points we have plotted with a smooth

* It will be observed that the scales used along OA and OB are not necessarily the same.

curve of the type shown in Fig. 4. By means of such a curve we can at least estimate the probable value of the temperature at a time when its exact experimental value has not been recorded. We do this by a process which is rather the reverse of plotting the points of the curve. Suppose, for instance, that we wish to know the value of the temperature at the time 3 h 19 min 45 sec on the afternoon of the experiment. With the choice of time zero which we have made, this corresponds to a time of 4.75. We then mark off on OA a distance corresponding to a value of 4.75 for the time, erect a perpendicular cutting the curve we have drawn at the point P'. Through this point we draw a line parallel to the time-axis, and find that this line cuts the temperature-axis at a point corresponding to a value of 25.68°C for the temperature, or, to the same degree of accuracy as the other value of Table 2, 25.7°C. The guide lines are shown dotted in Fig. 4. In this way the curve or graph we have drawn enables us to have information readily available not only about the times at which measurements were actually made but also about the times in between. We can also see certain significant results at a glance. From the curve we have drawn the following information about the experiment emerges:-

(a) in the first minute the temperature of the body rose rapidly;
(b) the rate of increase of the temperature fell sharply after the first minute, the maximum temperature of 25.8°C being reached just about 30 sec after the fourth minute;
(c) the temperature decreased from the maximum very slowly with time — falling only half a degree centigrade in twenty minutes.

Where our reasoning may have gone wrong is in the assumption that a smooth curve can be drawn through the points whose accuracy is guaranteed by the experimental results. Here we are guided by our physical intuition which tells us that, in the circumstances of the experiment, violent variations in temperature within small time intervals are improbable. It is true that it might be possible to draw more than one smooth curve through a given set of points but it is not very likely that two such different curves would lead to estimates of the temperature in violent disagreement with one another and, in experiments of this kind, we would not hope to give more than a reasonable estimate of the temperature at any intermediate time.

It should be emphasized that exact values can be determined only by direct experimental measurement — all that a graph can do is to give reasonable estimates at times when it was either impossible or inconvenient to record the temperature experimentally.

In certain problems in the social sciences estimates of this kind are not desired, the main point in constructing a graph being to

TABLE 3
The number of males under 21 years of age received in Borstal detention in England and Wales from 1927 to 1935

Year	1927	1928	1929	1930	1931	1932	1933	1934	1935
Number	568	635	679	725	873	1011	854	793	686

obtain a pictorial image of the situation. Significant fluctuations in production or in certain types of population are often more easily detected when the available data are presented in a graphical way

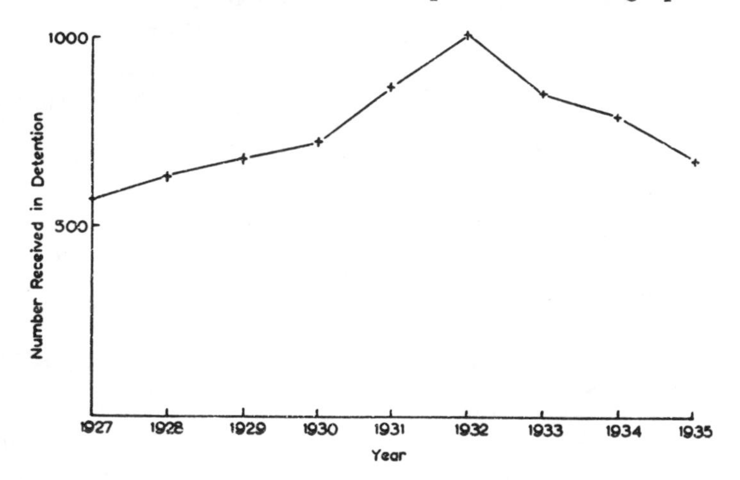

Fig. 5. Variation of the number of males under 21 received in Borstal detention in England and Wales from 1927 to 1935.

than when they appear merely as numerical entries in a table. Fig. 5 is a graph of the values displayed in Table 3.* Once the points corresponding to the figures of Table 3 have been plotted, it is not really

* The data of Table 3 are taken from "Report of the Commissioners of Prisons and the Directors of Convict Prisons for the Year 1935" (Great Britain, Cmd. 5430), p. 20.

feasible to join them with a smooth curve, for there is no reason to believe that the process of receiving young men into Borstal detention is a continuous process. Indeed, if the annual statistics were broken down to show the month-by-month variation, it might emerge that there were great differences between the summer and winter months. If, however, we join the points by straight lines (as is done in Fig. 5) we obtain a graph, called a *polygonal line,* which enables to follow the fluctuations of Borstal populations throughout the years 1927—1935.

The two cases we have considered have differed quite markedly in their character but they have one thing in common — we should hardly expect the quantities whose variation we are studying to be related to the time by any firm "law of nature". In the first case the temperature could change because of chance effects such as fluctuations in the temperature of the room in which the experiment was carried out; in the second case the size of the Borstal population was likely to be effected by economic and human factors which could not readily be predicted, or, if they could be predicted, not in any sense be measured accurately.

We shall now consider two cases in which we might reasonably expect that the quantities we are discussing are connected by some "law of nature". In both cases the relevant graphs assume particularly simple forms which, later, will enable to discuss the situation theoretically.

In the first instance we consider the stretching of a long thin wire, suspended from one end and with weights applied to the other end. In an experiment of this kind with a wire of length 466 cm the set of experimental values recorded in Table 4 was obtained. If we

TABLE 4

The elongation of a thin wire by the application of a load

Load (in kg)	0	4	8	12	16	20
Elongation (in mm)	0.00	0.65	1.25	1.90	2.50	3.10

plot these points we get the crosses shown in Fig. 6. A simple curve passing through (or very close to) each of these points is the straight line through the origin shown in the figure.* With the aid of Fig. 6

* It is obvious that this line must pass through the origin since a zero load will produce zero elongation.

we can now read off values of the elongation for values of the applied load intermediate to those given in Table 4. Suppose, for instance, that we wish to know what elongation will be produced by a load of 13 kg. We go along the horizontal (load) axis a distance representing 13 kg and through it draw a line parallel to the vertical (elongation-) axis intersecting the straight line we have drawn in a point, P say.

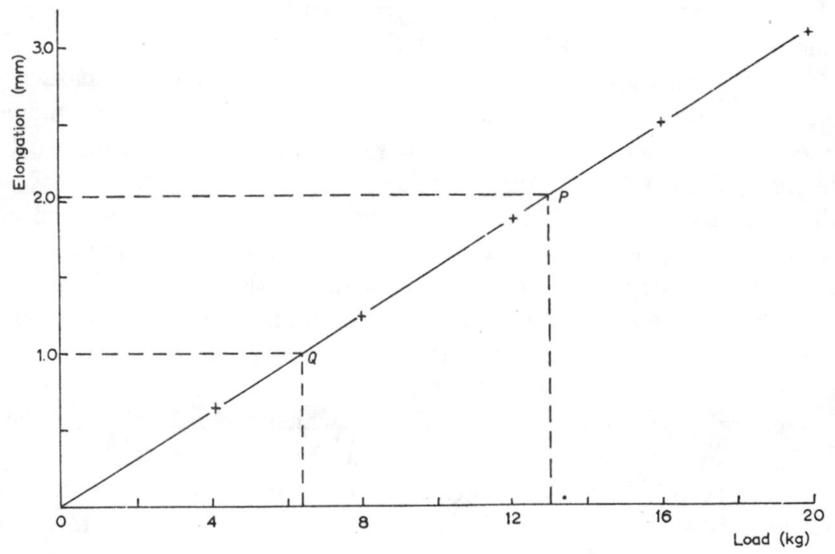

Fig. 6. The elongation of a long thin wire by the application of a load.

Through P we draw a line parallel to the load-axis; the point at which this line cuts the elongation-axis is the point corresponding to the desired elongation. If we read off the value in this case we find it to be approximately 2.03 mm. On the other hand, we might want to know what load we should apply if we wish to produce an elongation of say 1 mm. To do this we mark off the point corresponding to 1 mm on the vertical axis and through that point draw a line parallel to the horizontal axis which cuts the straight line in the point Q. Dropping a perpendicular from Q to the horizontal axis we get the value of the load corresponding to an elongation of 1 mm; it turns out to be approximately 6.33 kg.

As a second example of the kind of situation which is liable to arise in the discussion of physical experiments we shall consider the oscillations of a small but heavy sphere attached to the end of a piece of light string whose length is very great in comparison with the diameter of the "bob". Such a mechanical system is known as a *simple pendulum*. The experiment we shall consider consists in letting the bob swing through a small distance on either side of the vertical and measuring the time it takes to go from one extreme position to the

TABLE 5

The oscillations of a simple pendulum

Length of the pendulum (in cm)	25	50	75	100	125	150
Period (in sec)	0.995	1.405	1.720	2.000	2.230	2.430
(Period)^2	0.99	1.98	2.97	4.00	4.97	5.93

other and back again. The time for such a complete swing is called the *period* of the pendulum. The object of the experiment is to try to

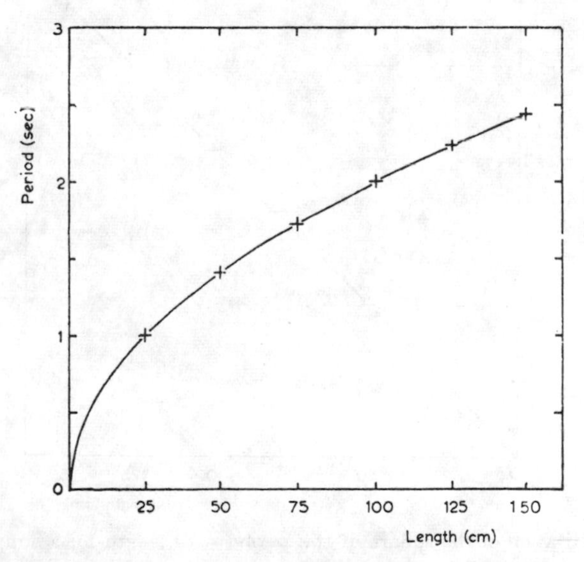

Fig. 7. The variation of period with length for a simple pendulum.

determine how variations in the length of the pendulum — that is, in the distance between the bob and the point of suspension — affect the period of the pendulum. In one such experiment the set of values shown in Table 5 was obtained. The physical quantities measured in the course of the experiment appear in the first two rows of the table. The entries in the third row are not measured quantities — their origin will be explained a little later. If we plot the quantities contained in the first two rows of this table and join them with a smooth curve we get a curve of the form shown in Fig. 7. Just as in the previous case, we can use this curve to determine the period of a pendulum of length other than those experimentally determined. Thus, a pendulum of length 80 cm is seen to have a period of 1.80 sec. We see also that if we want a pendulum with a period of 1.5 sec we must ensure that it is of length 56 cm approximately. Since the curve joining the experimental points is not a straight line, it is not easy to draw it very accurately so we do not have as much confidence in these predictions as we had in the case of the experiment of stretching the thin wire. A straight line is more easily drawn and values can be read from it with more confidence.

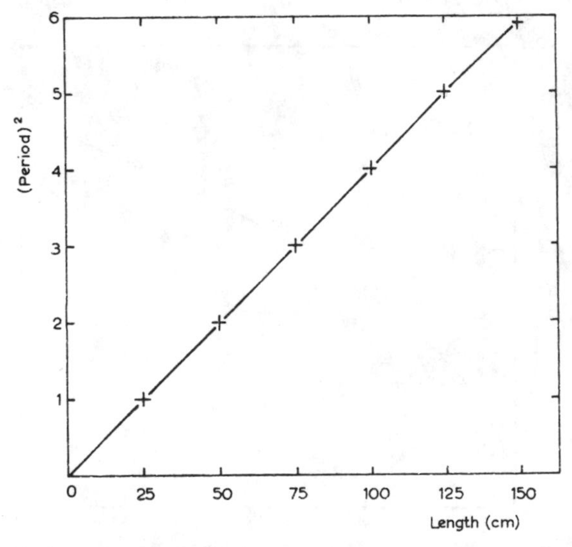

Fig. 8. The variation of the square of the period with length for a simple pendulum.

If we square the period in each instance we get the third row of Table 5. Plotting the values of (period)^2 against the length of the pendulum we get the crosses shown in Fig. 8. It is seen from this diagram that the observed points lie very close to a straight line through the origin. By means of this straight line we may read off values of the kind we might require in the solution of a particular problem. For instance, if we want to know the length of a pendulum whose period is 1.5 sec we form the square of 1.5, that is 2.25, and read off from the straight line the value of the length corresponding to that value of the $(\text{period})^2$. In this way we again reach the approximate value of 56 cm for the length of the pendulum. The inverse problem can be tackled similarly: if we wish to know the period of a pendulum of length 80 cm we take from the straight line the corresponding value of the $(\text{period})^2$. It turns out to be 3.25 approximately. The desired value of the period is therefore the square root of 3.25 which gives a period of approximately 1.8 sec.

A graph of an entirely different shape is obtained by considering the variation of the volume of a gas with pressure if its temperature is kept constant. We consider a fixed mass of gas whose volume can be measured easily and which can be subjected to varying pressures. Keeping the gas at a steady temperature we proceed to measure the volumes which are produced by the application of known pressures. For example, in a certain experiment the values recorded in Table 6 were obtained, the third row of figures being obtained from the second as a result of a simple division.

If we plot a graph connecting the pressure with the volumes we

TABLE 6
Variation of volume with pressure in a gas at a fixed temperature

Pressure (lb/in^2)	150	125	100	75	50	25
Volume (cu.ft)	1.00	1.20	1.50	2.00	3.00	6.00
1 ÷ Volume	1.000	0.833	0.667	0.500	0.333	0.167

get the curve shown in Fig. 9, the experimentally determined points being denoted by crosses. If, on the other hand, we plot against the pressure, not the volume but its reciprocal (that is, 1/volume) we get the crosses shown in Fig. 10. It looks from the position of these points as if the graph of the pressure against the reciprocal of the volume is a straight line. This is a much more manageable curve than the one we obtained by plotting the values of the pressure and the volume (Fig. 9).

The process we have been considering, of finding intermediate numerical values from those which we were originally given, is called *interpolation*. This is a process which may be carried out purely

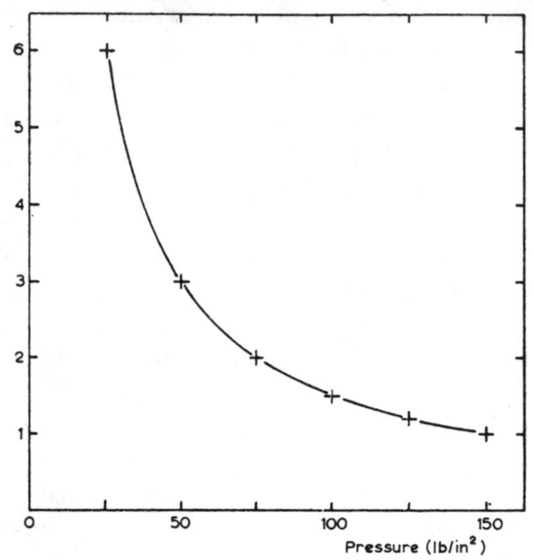

Fig. 9. The variation of volume with pressure in a gas at a fixed temperature.

numerically with the aid of what is known as the Calculus of Observations — that is, it can be carried out solely by operations with

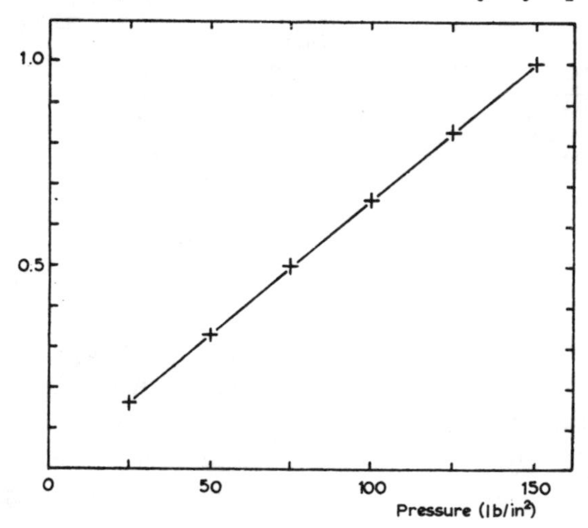

Fig. 10. The variation of the reciprocal of the volume with pressure in a gas at a fixed temperature.

numbers alone, without the use of graphs — so that what we are doing ought more strictly to be called *graphical interpolation.*

It will have been observed that, in the graphical representation of the data provided by Table 6, we did not, in the second case — Fig. 10 above — make the straight line we had drawn pass through the origin of coordinates. That was because we had no direct knowledge of the behaviour of the gas at very low pressures. It might be considered reasonable to *assume* that the gas would behave under low pressures in much the same way as it does at the pressures recorded in the experiment and that we should therefore continue the straight line until it passes through the origin. Such a process of continuing the graph beyond the region for which reliable information is available is called *extrapolation.* The dangers in extrapolating data are obvious. If, for instance, on a Monday a patient's temperature at 8 a.m., 12 noon, 4 p.m. and 8 p.m. was 99.6°F, 99.2°F, 98.8°F and 98.4°F and the readings were then ceased, it would be inadvisable to assume that this temperature at 8 a.m. on the following Friday morning would be 90.0°F, though that is the value which a graphical extrapolation would suggest!

PROBLEMS

1. When the depth of water in a reservoir is h metres the area of the water surface in square metres is 100 S. The following table gives a number of simultaneous values of h and S:

h	0	2	4.5	6	7.3	9.5	12	13.4
S	13.3	19.6	31.5	40.0	48.3	58.5	63.8	65.2

Plot the curve, and determine the surface area when the depth is 2.5, 5.0, 7.5, 10.0, 12.5 metres.

2. The following table gives the atmospheric temperature every two hours from 4 a.m. until midnight on a certain day:-

4 a.m.	6 a.m.	8 a.m.	10 a.m.	Noon	2 p.m.	4 p.m.	6 p.m.	8 p.m.	10 p.m.	Mid-night
39°	38.7°	42°	50°	58°	66°	70.5°	69°	57°	45°	39.5°

Taking 1 inch on the axis of x to represent 2 hours and 1 inch on the axis of y to represent 10°, draw a graph showing the temperature throughout the period of twenty hours.

From your graph —

(i) find the temperature at 1 p.m., and also the times at which the temperature was 60°;

(ii) estimate the maximum temperature and the time at which it was reached;

(iii) determine the period during which the temperature appeared to rise at a uniform rate and what this uniform rate was;

(iv) determine the period during which the temperature was decreasing most rapidly.

3. The height above the ground of a ball thrown up into the air after the lapse of 1, 2 . . . seconds is given in the table below:-

t (seconds)	1	2	$2\frac{1}{2}$	3	4	5
y (feet)	72	112	120	120	96	40

Draw a graph connecting y and t, using a scale of

$$1 \text{ in} = 1 \text{ sec for } t, \text{ and } 1 \text{ in} = 10 \text{ ft for } y.$$

Determine from your curve the times at which the ball will be at a height of 80 feet.

3. The Idea of a Functional Relationship

We shall now begin to develop a few general ideas on the basis of the examples we discussed in the last section. The fundamental link between the graphical and the purely numerical approach to the kind of problems we have been considering is provided by the method of coordinate, or analytical, geometry. In this approach each point in a plane can be uniquely specified by a pair of real numbers, (x, y) say. The interpretation of these numbers is precisely that which we have given in the rather special examples we considered in the last section. The number x can be thought of as the distance of the point in question from a fixed y-axis, which is denoted by OY in Fig. 11. In a similar way the y can be thought of as the distance of the point from a fixed x-axis, OX, which is perpendicular to the axis OY. The converse of this statement is also true:- each pair of numbers (x, y) determines a unique point in the plane. The numbers x and y are known as the *coordinates* of the point; if it is desirable to distinguish between them x is called the *abscissa* of the point and

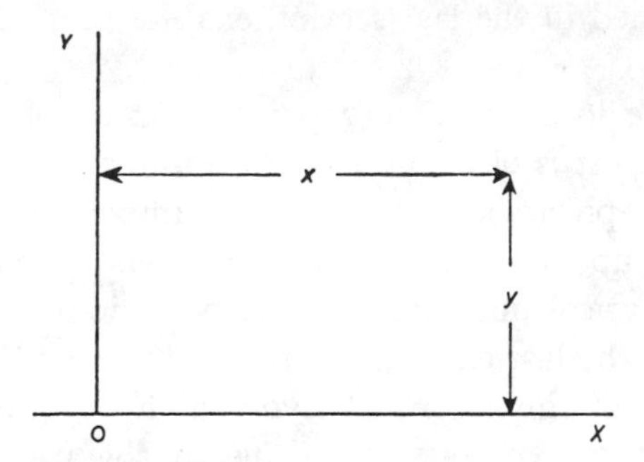

Fig. 11. The coordinates of a point.

y the *ordinate*. The point O is called the *origin of coordinates*; it corresponds to the number pair (0,0). The lines OX, OY are known as the *coordinate axes*.

Not all the numbers with which we deal in this way need be positive. If the point P lies to the right of OY its x-coordinate will be reckoned to be positive; if it lies to the left its x-coordinate will be reckoned to be negative. Similarly, if P lies above OX its y-coordinate is reckoned to be positive while if P lies below OX it has a negative y-coordinate. These relations are illustrated graphically in Fig. 12.

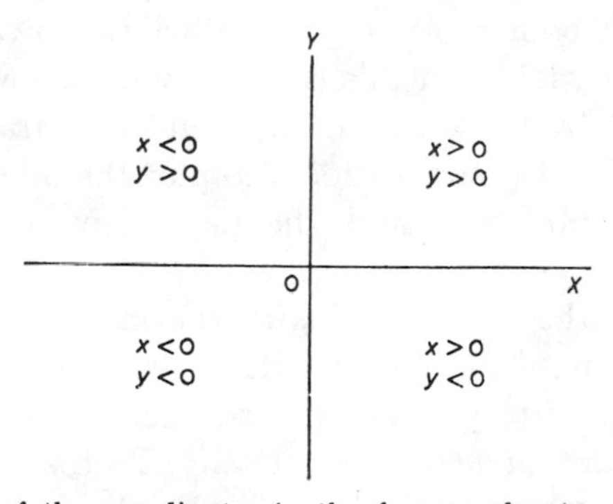

Fig. 12. Signs of the coordinates in the four quadrants of the xy-plane.

The drawing of a graph is equivalent to assigning to every number x a number y in such a way that the associated point given by the number pair (x, y) lies on the curve we have drawn. The graph is, in fact, a prescription enabling us to find the value of y which corresponds to a given value of x; for, if a value of x is given, then the procedure outlined in the last section enables us to determine y from the graph.

In the unique location of any point on a graph by means of its coordinates (x, y) it is obvious that two numbers are involved. Every point on the graph associates the two numbers and the totality of points making up the graph expresses some kind of relationship between the physical quantities whose behaviour is being discussed. For instance, in the last example of the last section, both the pressure and the volume of the gas were involved. Every point of the graph in Fig. 9 therefore expresses the value of the pressure which must be associated with a prescribed value of the volume at a certain stage of the experiment. During the progress of the experiment the volume of the gas and its pressure both changed, and the curve we drew expressed the relationship between the two quantities as they both changed. Both the pressure and the volume of the gas are *variables* that is to say their values are not fixed but change throughout the course of the observations. The two quantities, though not fixed, are nevertheless not independent. Our method of exhibiting their variations graphically shows that if we make a change in the pressure we can then determine the corresponding change in the volume of the gas. The variable (in this case the pressure) which changes by amounts which may be chosen arbitrarily is called the *independent variable,* while the other variable (in this case the volume) whose changes are then determined by those of the independent variable is called the *dependent variable.* It is conventional to plot the independent variable along the horizontal axis and the dependent variable along the vertical axis.

In the case of the experiment with the mass of gas, we could, of course, have arranged matters in such a way that the volume could be changed by arbitrary amounts and the pressure changes would then have been determined automatically. In that case, then, it is a matter of choice which variable we make the independent variable and which the dependent one. In the case of the figures relating to admission to Borstal institutions, it is obviously best to regard the time as the independent variable. Indeed time is invariably taken as the independent variable in any problem in which it occurs as one of the variables.

When the relationship between two variables is such that to each value of x there corresponds a single value of y, we say that y is a *single-valued function* of x. If to each value of x there are several values of y, we say that y is a *many-valued function* of x. For example, the volume of a gas at constant temperature is a single-valued function of its pressure; the period of a simple pendulum is a single-valued function of its length; the elongation of a thin wire is a single-valued function of the load applied to its end. As an example of a many-valued function of x, we may cite $\sqrt{x}$, the square root of x; to every value of x there correspond two values of y. For instance, if $x = 4$ then y is either $+2$ or -2.

Instead of always writing "y is a function of x" we put briefly $y = f(x)$.

From what we have said it might appear that a functional relationship was some kind of property of graphs. Graphical representation certainly is an aid to our thinking about functions but any kind of prescription for giving the value y when the value of x is known is a functional relationship. For instance the data might be presented in graphical form, as the record of a recording barometer, in numerical form, or by a formula. As an example of the latter way of expressing a functional relation we have the formula

$$S = P(1 + nr) \tag{1}$$

expressing the sum S accruing from a principal P invested for n years at simple interest of rate r. If we regard the principal and the rate of interest as fixed, then the equation (1) expresses the sum S as a function of the single variable n, since it provides a means of assigning a value to S to any arbitrary value of n.

In mathematical analysis we are mainly concerned with functional relationships which can be expressed by equations of the type

$$y = f(x) \tag{2}$$

and we adopt the convention that $f(a)$ should denote the value of the function when the independent variable x takes the value a. For instance, if

$$f(x) = 3x^3 + 2x^2 + 4x + 5$$

then

$$f(1) = 3 \times 1^3 + 2 \times 1^2 + 4 \times 1 + 5 = 14$$
$$f(2) = 3 \times 2^3 + 2 \times 2^2 + 4 \times 2 + 5 = 45$$
$$f(0) = 3 \times 0^3 + 2 \times 0^2 + 4 \times 0 + 5 = 5$$

and so on.

We shall now discuss some of the simpler functional relationships and the graphs which correspond to them. This will provide us with a simple but effective tool in the interpretation of biological data.

(i) *Proportionality*

If the number y is related to the number x through the equation

$$y = kx \tag{3}$$

we say that y is proportional to x, and that k is the constant of proportionality. For simplicity, let us suppose that $k = 2$ so that

$$y = 2x. \tag{4}$$

If we let x take integral values from 0 to 10 and calculate the corresponding values of y we get the following table of values:-

x	0	1	2	3	4	5	6	7	8	9	10
y	0	2	4	6	8	10	12	14	16	18	20

If we plot these points on graph paper we see that the points all lie on a straight line which passes through the origin (cf. Fig. 13). The obvious way of characterising a line through the origin — i.e. of distinguishing it from a second line through the origin — is to state what its slope is. We all have an intuitive idea of what we mean by the slope of a line and the mathematical procedure is merely to formalise that concept into the definition of the *gradient* of a line. We define the gradient of a line through the origin by means of the formula

$$\text{gradient} = \frac{\text{the } y\text{-coordinate of any point on the line}}{\text{the } x\text{-coordinate of that point}}. \tag{5}$$

The point with $x = 5$, $y = 10$ lies on the straight line we have drawn in Fig. 13 so that that line has gradient $10/5 = 2$. We can therefore assert that the graph of the functional relation $y = 2x$ is a straight line passing through the origin and with gradient 2.

By a precisely similar argument it can be shown that the graph of the functional relation $y = kx$, where k is a pure number, is a straight line through the origin of gradient k.

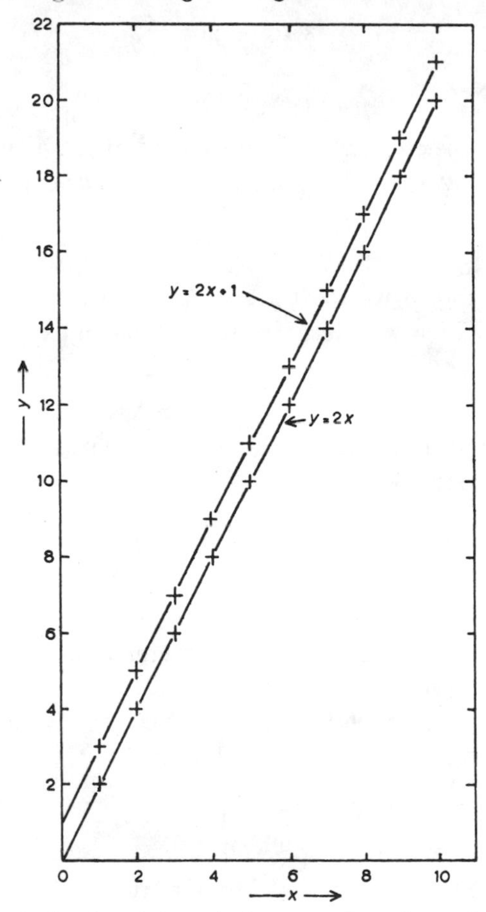

Fig. 13. Graphs of the relations $y = 2x$ and $y = 2x + 1$.

Conversely if we plot the values of a variable y against the corresponding values of a variable x and find that these values lie on a straight line through the origin of gradient k, we can assert that y is proportional to x with constant of proportionality k; that is, the relation between y and x is expressed by equation (3).

For example, we see from Fig. 6 that the elongation e of a thin wire is related to the applied load W by an equation of the form

$$e = kW.$$

To find the constant k, we note from the table of values (Table 4 above) that when $W = 20$ kg, $e = 3.10$ mm so that

$$3.10 = 20k,$$

that is,

$$k = 0.155.$$

If, therefore, the elongation e is measured in mm and the load W is measured in kg then the relation between them is expressed by the equation

$$e = 0.155W.$$

Similarly from Fig. 8 we see that if T is the period of a pendulum of length l, then $T^2 = kl$ where, from the table of experimental values, we have

$$4.00 = k \cdot 100$$

showing that $k = 0.04$ and so the relation between T and l is expressed by the equation

$$T^2 = 0.04l$$

which is equivalent to

$$T = 0.2\sqrt{l},$$

provided T is measured in sec. and l in cm. Finally, from the extrapolation of the curve of Fig. 10 we see that for the mass of gas under consideration in the experiment described the volume v and the pressure p are connected by the relation

$$\frac{1}{v} = kp$$

and from the table of values we find that

$$0.667 = k \cdot 100,$$

showing that, in the units used, $k = 0.00667$ and hence that the relation between p and v can be written in the form

$$\frac{1}{v} = 0.00667p$$

which is equivalent to

$$v = \frac{150}{p}$$

it being understood that if p is measured in lb/in², v is measured in cu.ft.

(ii) *The general linear relation*

Let us now consider the functional relationship characterised by the equation

$$y = kx + c. \tag{6}$$

We shall begin by considering one or two special cases. Suppose, for instance, that

$$y = 2x + 1,$$

then we have the table of values:-

x	0	1	2	3	4	5	6	7	8	9	10
y	1	3	5	7	9	11	13	15	17	19	21

If we plot these values on the same diagram (Fig. 13) as the graph of the relation $y = 2x$, we find that the points obtained all lie on a straight line which is parallel to the line with equation $y = 2x$. In terms of our definition of the term gradient, this means that the graph of $y = 2x + 1$ is a straight line with the same gradient as the line with equation $y = 2x$. The line we have drawn cuts the line OY at a distance of one unit from the origin O.

In the general case of the relation $y = kx + c$ if we substitute the value $x = 0$ we find that the corresponding value of y is c, whatever the value of k. The graph of the relation (6) therefore cuts the OY axis at a distance c from the origin. By considering further special cases we can also show that, in every case, the graph of the relation (6) is a straight line which is parallel to the line with equation $y = kx$. The functional relationship (6) may therefore be represented graphically by a straight line which cuts the OY axis at a distance c from the origin, and which is parallel to the straight line through the origin of gradient k. We say that k is the *gradient*, and c the *intercept* on the y-axis of the line with equation (6).

In interpreting the results of experiments or observations we are often confronted with the converse problem, that is, we know that the relationship between x and y, when expressed graphically, is linear, and we wish to express that relationship in the form of an algebraic equation. Suppose that the observed points are $A, B, C, D, E, \ldots$ (cf. Fig. 14) and that a straight line can be drawn through them.

Suppose that this line (produced if necessary) cuts the line OY in the point M, and that OM is c units in length. Then, obviously, when $x = 0$, $y = c$. Therefore it only remains to determine the gradient of

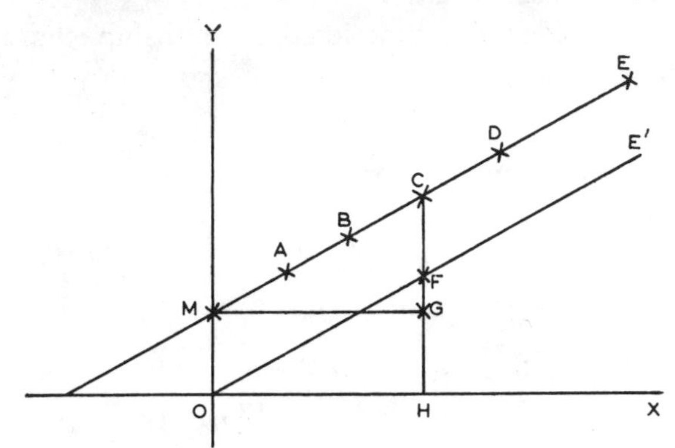

Fig. 14. The derivation of the equation of a straight line.

the line. If through one of the points, C, we draw the line CGH parallel to OY such that it cuts OX in H and the point G is such that MG is parallel to OX, then from the definition of the gradient of the parallel line OE', which meets HC at F, we see that

$$k = \frac{HF}{OH}$$

and it is a matter of elementary geometry to show that $OH = MG$ and $HF = GC$. We therefore have

$$k = \frac{GC}{MG}.$$

Now if (x_C, y_C) are the coordinates of the point C, then $OH = x_C$, $HC = y_C$ and $GC = HC - HG = HC - OM = y_C - c$, so that

$$k = \frac{y_C - c}{x_C}. \tag{7}$$

EXAMPLE 1: *Find the equation of the straight line joining the points* (x_1, y_1) *and* (x_2, y_2).

Suppose that the relation between x and y is of the form (6). Then, since the two given points lie on this line, their coordinates must satisfy this relation. That is, we must have

$$y_2 = kx_2 + c,$$

and (8)

$$y_1 = kx_1 + c.$$

Subtracting these two equations we find that

$$y_2 - y_1 = k(x_2 - x_1)$$

from which it follows that

$$k = \frac{y_2 - y_1}{x_2 - x_1}. \tag{9}$$

Now from equation (8) we have

$$c = y_2 - kx_2.$$

Substituting for the constant k into this equation from equation (9) we find that

$$c = \frac{x_2 y_1 - x_1 y_2}{x_2 - x_1}$$

so that the equation of the line is

$$y = \frac{y_2 - y_1}{x_2 - x_1} x + \frac{x_2 y_1 - x_1 y_2}{x_2 - x_1}. \tag{10}$$

(iii) *The parabolic relation*

We shall now consider the graph corresponding to the algebraic relation

$$y = kx^2, \tag{11}$$

which we shall call the parabolic relation. To begin with we shall plot the curves corresponding to certain particular values of k. To do this we make use of the following table:-

x	0	+1	+2	+3	+4	+5
x^2	0	1	4	9	16	25
$2x^2$	0	2	8	18	32	50
$3x^2$	0	3	12	27	48	75
$4x^2$	0	4	16	36	64	100

The curves corresponding to the relations $y = x^2$, $y = 2x^2$, $y = 3x^2$ and $y = 4x^2$, the points of which were plotted from the above table of values, are shown in Fig. 15. It will be noted that each of the curves has the shape of a "U", but that the U is 'flatter' if the value of the constant k is small. Such a curve is called a *parabola*, with vertex at the origin.

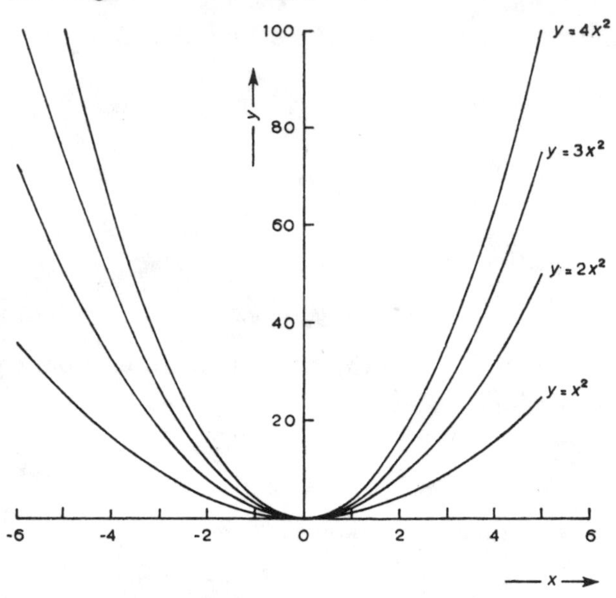

Fig. 15. Curves corresponding to the relation $y = kx^2$ with $k = 1, 2, 3, 4$.

If it is suspected that the relationship between two quantities is of a parabolic nature we try to fit our observed curve to a curve with equation $y = kx^2$ and determine k as follows: if the point (x_C, y_C) lies on the parabola then

$$y_C = kx_C^2,$$

from which it follows that

$$k = \frac{y_C}{x_C^2}.$$

For example, the shape of the curve in Fig. 7 suggests that there is a parabolic relation between l, the length of a pendulum, and T, the period of the pendulum, of the form $l = kT^2$. When $l = 100$ cm, $T = 2$ sec, so that $100 = k(2)^2$ and, therefore, $k = 25$. The relation between the period and the length of the pendulum then comes out to be

$$l = 25\,T^2.$$

In point of fact the best way of handling a relationship of this kind is to make it linear by taking x^2, not x, as the independent variable. The constant k can then be determined by the method of case (ii). This procedure has already been illustrated under subsection (i) above.

(iv) *The hyperbolic relation*

Finally, we shall investigate the form of the graphs of relations of the type

$$y = \frac{k}{x}. \tag{12}$$

In books on elementary algebra this relationship is usually expressed

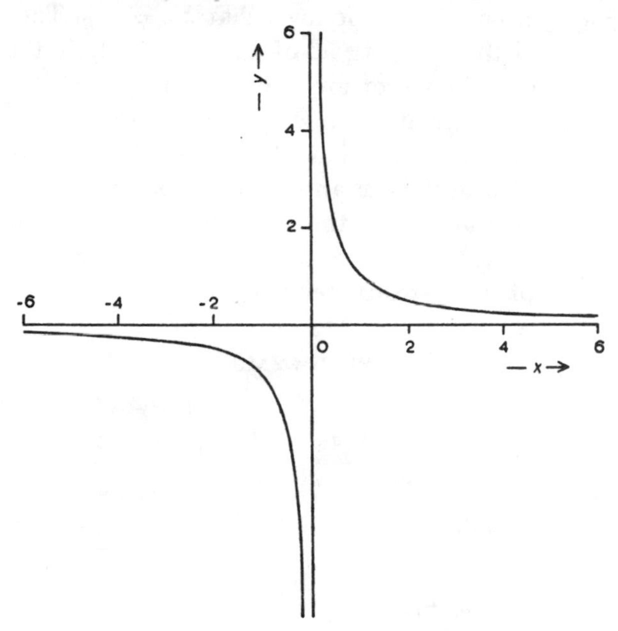

Fig. 16. Graph of the relation $y = 1/x$.

by saying that y is inversely proportional to x. The relationship itself is often called "inverse proportionality".

The typical curve of this set is

$$y = \frac{1}{x}$$

for which a suitable table of values would be:-

x	± 0.5	± 1.0	± 1.5	± 2.0	± 2.5	± 3.0	± 3.5	± 4.0	± 4.5	± 5.0
y	± 2.00	± 1.00	± 0.67	± 0.50	± 0.40	± 0.33	± 0.29	± 0.25	± 0.22	± 0.20

The curve obtained from these points is shown in Fig. 16. A curve of this shape is called a *rectangular hyperbola*. One of its important properties is that as x tends to zero the value of y increases without bound, and that as x increases steadily, y tends more and more closely to zero.

To obtain the value of k for a curve of this kind we select a typical point (x_C, y_C) lying on it, and measure its coordinates x_C and y_C as accurately as possible. Since the point lies on the curve we must have $y_C = k/x_C$ from which it follows that $k = x_C y_C$. The parameter k of any curve with an equation of the type (12) is therefore the product of the coordinates of any point lying on it.

In any practical case it is much more satisfactory to plot the points, and, having observed that the shape of the curve is hyperbolic, to re-plot $1/y$ against x and obtain a straight line. This was the procedure we adopted in the last section when discussing the relation between the pressure and the volume of a given mass of gas which is kept at a steady temperature.

PROBLEMS

1. If

$$f(x) = 5x^4 + 4x^3 + 3x^2 + 2x + 1$$

find $f(0)$, $f(1)$, $f(-1)$, $f(2)$, $f(-2)$, $f(3)$ and $f(-3)$.

2. Find the equations of the lines joining the following pairs of points:

(i) (1, 2) and (3, 4);
(ii) (−1, −1) and (2, 1);
(iii) (0, 1) and (2, 0).

3. The maximum amount of a certain substance, w grams, that dissolved in 100 grams of water at temperature t degrees centigrade was found to be as shown in the following table:-

t ...	5	15	25	35	45	55
w ...	75	82	90	99	106	115.

Choosing suitable scales, plot w against t and draw what you consider to be the best-fitting straight line graph.

Assuming that for this range of temperature the relationship between w and t is of the form $w = at + b$, where a and b are constants, use your graph to find the values of a and b.
From your graph read off the value of w when t is 52 and the value of t when w is 100.

4. The resistance to the motion of a train, R kg per 100 kg of the weight, is given by the law $R = a + bv^2$, where v is the speed of the train in km/hour and a and b are constants. If $R = 5$ when $v = 40$ and $R = 7$ when $v = 56$, find a and b.

5. It is believed that the values of x and y in the following table conform to a law of the type

$$y = \frac{a}{1 + bx}.$$

Show by plotting y as a function of xy that this is the case, and find approximately the values of a and b.

x	6	7	8	9	10	11	12
y	6.77	6.23	5.77	5.38	5.04	4.73	4.46

6. How would you plot the curve $y = ax^2 + b$ on graph paper so that the resulting graph is a straight line? Give a similar method for $y = ax^3 + b$.

Apply your methods to determine whether $y = ax^2 + b$ or $y = ax^3 + b$ is the more suitable law for describing the following approximate data:-

x	1	2	3	4	5
y	6	14	23	38	57

By drawing the "best" straight line through the data plotted in the correct way, find values for a and b.

4. Hyperbolic Curves in Physiology

(*a*) *The strength-duration curve*

Weiss' law, well-known in the physiology of muscle and nerve, is based on the finding that the intensity of an electric current required to just excite the tissue (threshold-strength) depends on the duration of current flow. More precisely, the longer the *duration* of current flow the weaker will be the (least) current required to excite.* This relation may be represented by the experimental curve shown in Fig. 17. This relationship between current strength and duration (at threshold-level) was found to be a hyperbola and could be represented by the following *empirical formula*

$$i = \frac{a}{t} + b \tag{1}$$

where i = current strength, t = time, and a and b are constants.

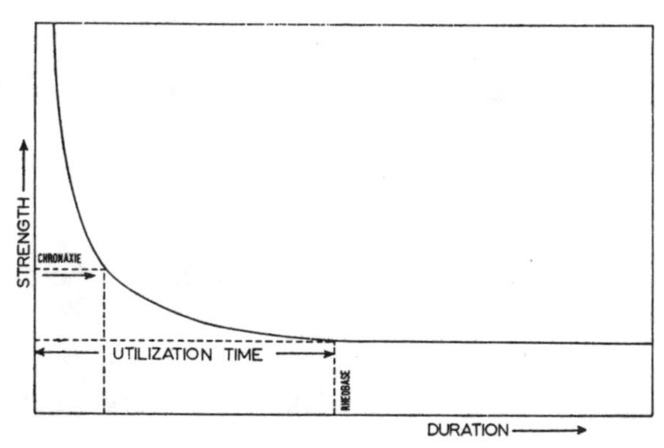

Fig. 17. The strength-duration curve. The curve relates the least strength of a constant current and the least time during which it must flow in order to reach threshold, i.e. to excite the tissue.
[From Howell's textbook of physiology, 15th ed. (1947).]

It should be emphasised that this formula is purely empirical, that is, that it is not derived from theoretical considerations, but is just a convenient way of expressing the experimental data.

From (1), it may be noted that when $t \to \infty$, $i \to b$. Obviously, the current of strength b is the smallest current that can possibly excite the tissue, whatever the duration of the current flow. This critical strength of current is called the *rheobase*.

Theoretically, the rheobase is attained at $t = \infty$. Practically, however, the rheobase is attained within a finite time. The "shortest

* M. Cramer, Erregungsgesetze des Nerven, Handb. norm. path. Physiol. **9** (1929) 244–284, **18** (1932) 241–246; H. Schaefer, "Allgemeine Electrophysiologie" (Vienna, Deutick, 1940); H. Davis and A. Forbes, "Chronaxie", Physiol. Reviews **16** (1936) 407.

time" within which this rheobasic current is attained is called the *utilization time*. Lapique introduced, for practical purposes, the concept *chronaxie*, which is defined as the minimum time required to just excite a tissue with a current twice the rheobasic strength. The concepts rheobase and chronaxie are clarified in Fig. 17. The concept of chronaxie plays an important role in comparative physiology and neurology.

Tasaki *et al.** have obtained strength-duration curves obeying Weiss' law from isolated single nerve fibres. Their result shown in Fig. 18 shows that the curves are influenced by temperature. The two

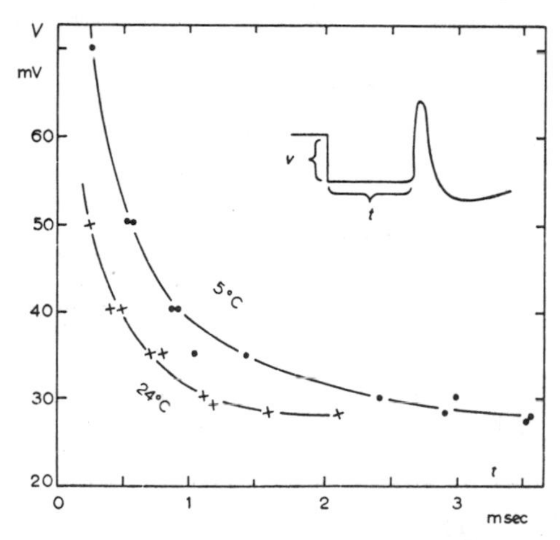

Fig. 18. Strength-duration curves of a nerve fibre taken at different temperatures. [From I. Tasaki and M. Fujita, J. Neurophysiol. **11** (1948) 311.]

curves in this figure are taken from one and the same preparation at two different temperatures. It can be shown that the value of a is approximately doubled by a fall of 10°C, whereas the rheobase b is not appreciably changed by temperature changes. Note that here the applied voltage is plotted against time, but with R constant, current and voltage are proportional.

(*b*) *Minimal times to register temperature jumps*

When the external temperature jumps required to attain a fixed

* I. Tasaki, Nervous Transmission (Charles C. Thomas, Springfield, 1953); I. Tasaki and M. Fujita, J. Neurophysiol. **11** (1948) 311.

temperature change of e.g. 2.5°C at a skin depth of e.g. 0.5 mm are plotted against the minimal times, just as in the case discussed above, a strength-duration curve is again obtained*. This curve, which is shown in Fig. 19 is found to be a hyperbola which may be represented by the equation

$$T = \frac{a}{t} + b \tag{2}$$

where T = change of external temperature, t = time, and a and b are constants.

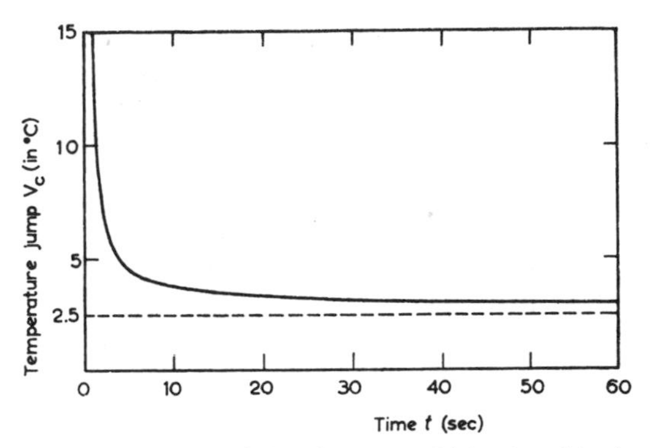

Fig. 19. Graphical representation of the times at which at a skin depth of 0.5 mm a temperature change of 2.5° is obtained, as a function of the external temperature jump V_e. [From H. Hensel, Ergebnisse der Physiol. **47** (1952) 166.]

(c) *Speed of shortening of muscle*

When a muscle contracts against a load (weights) the speed of shortening decreases with increasing weights. This relationship between speed of shortening (during isotonic contraction) in cm/sec and load (in grams) was found to be hyperbolic.† It is interesting to note, that although Fenn and Marsh established empirically the hyperbolic nature of this curve, A. V. Hill by thermodynamic reasoning derived a hyperbolic equation relating the shortening speed and load. Lack of space prohibits further discussion of this chapter of muscle physiology.

* H. Hensel, "Physiologie der Thermoreception", Ergebnisse der Physiol. **47** (1952) 166–348; V. Petrow and I. Jakowlew, Fiziol. Z. **28** (1940) 343.

† W. O. Fenn and J. Marsh, J. of Physiol. **85** (1935) 277; K. Gassner and H. Reichel, Z. Biol. **10** (1952) 7; A. V. Hill, Proc. Roy. Soc. B **126** (1938) 136; H. Reichel, Muskelelastizität, Ergebn. der Physiol. **47** (1952) 469–554.

(*d*) *The oxygen dissociation curve of blood*

Oxygen enters into a loose chemical combination with the haemoglobin molecule to form oxyhaemoglobin. The relationship between the percentage saturation of haemoglobin with oxygen — i.e., the pro-

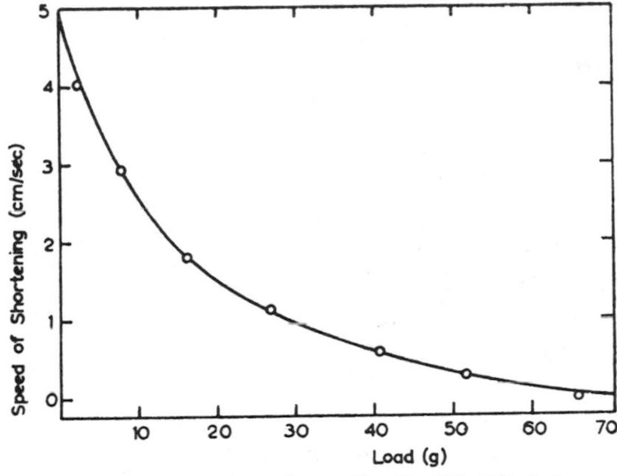

Fig. 20. Speed of shortening as a function of load. [Modified from A. V. Hill, Proc. Roy. Soc. Lond. B **126** (1938) 136.]

portion of oxyhaemoglobin to reduced haemoglobin — and the partial pressure of oxygen, when expressed graphically is called the oxygen dissociation curve of haemoglobin. The oxygen dissociation curve of a solution of haemoglobin in distilled water is found to be a rectangular hyperbola. Hüfner has derived (from first principles) a theoretical formula that agreed with the hyperbola found empirically in the case of a haemoglobin solution. Hüfner's derivation, however, is known to be incorrect, and has been superseded by Adair's theory which accounts satisfactorily for the known facts.

Adair's "Intermediate compound theory"

This theory, advanced by G. S. Adair* in 1925, followed rather naturally from the fact that the 4 iron atoms in the Hb-molecule are each capable of holding one O_2 molecule. This fact was deduced from the accurately determined molecular weight of haemoglobin (66 800) together with the known iron-oxygen ratio. The rather tedious derivation of equation (3) below may be skipped, if the reader is only interested in the hyperbola application. The assumption is

* G. S. Adair, J. Biol. Chem. **63** (1925) 529.

made that the following stages occur:

(1) $Hb_4 + O_2 \rightleftarrows Hb_4O_2$ and by the law of mass action $K_1 = \dfrac{[Hb_4O_2]}{[Hb_4][O_2]}$, hence $[Hb_4O_2] = K_1\ [Hb_4][O_2]$. ($Hb_4$, since Hb contains 4 iron atoms capable of holding O_2.)

$$(2) \qquad Hb_4O_2 + O_2 \rightleftarrows Hb_4O_4;\ K_2 = \frac{[Hb_4O_4]}{[Hb_4O_2][O_2]},$$

therefore $[Hb_4O_4] = K_2K_1[Hb_4][O_2]^2$

$$(3) \qquad Hb_4O_4 + O_2 \rightleftarrows Hb_4O_6;\ K_3 = \frac{[Hb_4O_6]}{[Hb_4O_4][O_2]},$$

therefore $[Hb_4O_6] = K_3K_2K_1\ [Hb_4][O_2]^3$

$$(4) \qquad Hb_4O_6 + O_2 \rightleftarrows Hb_4O_8;\ K_4 = \frac{[Hb_4O_8]}{[Hb_4O_6][O_2]},$$

therefore $[Hb_4O_8] = K_4K_3K_2K_1\ [Hb_4][O_2]^4$
(K_1, K_2, K_3, K_4 are the respective equilibrium constants of the intermediate reactions.)*

The concentration of combined O_2 is obviously

$$[Hb_4]\{K_1[O_2] + 2K_1K_2[O_2]^2 + 3K_1K_2K_3[O_2]^3 + 4K_1K_2K_3K_4[O_2]^4\}$$

while the total concentration of haemoglobin is:

$$[Hb_4]\{1 + K_1[O_2] + K_1K_2[O_2]^2 + K_1K_2K_3[O_2]^3 + K_1K_2K_3K_4[O_2]^4\}.$$

Since four oxygen molecules combine with haemoglobin when complete oxygenation has occurred:

$$\text{percentage oxyhaemoglobin} = \frac{\frac{1}{4}(\text{concentration of combined } O_2)}{\text{Total Hb-concentration}} \times 100.$$

If the symbol Y denotes p.c. oxyhaemoglobin/100 we obtain

$$Y = \frac{K_1[O_2]+2K_1K_2[O_2]^2+3K_1K_2K_3[O_2]^3+4K_1K_2K_3K_4[O_2]^4}{4\{1+K_1[O_2]+K_1K_2[O_2]^2+K_1K_2K_3[O_2]^3+K_1K_2K_3K_4[O_2]^4\}}. \quad (3)$$

A similar expression is obtained if we replace O_2 by O_2 pressure p.

Now Adair, Ferry and Green, Forbes and Roughton and others†

* For the sake of simplicity we omit a discussion of activities and activity coefficients.

† W. H. Forbes and F. J. W. Roughton, J. Physiol. **71** (1931) 229; Paul W. and F. J. W. Roughton, J. Physiol. **113** (1951) 23; F. J. W. Roughton, Physiol. Rev. **15** (1935) 278; Ferry and A. A. Green, J. Biol. Chem. **81** (1929) 175.

were able to obtain a good correspondence between their data and a theoretical curve based on formula (3) using suitably selected values for the unknown equilibrium constants K_1, K_2, K_3 and K_4.

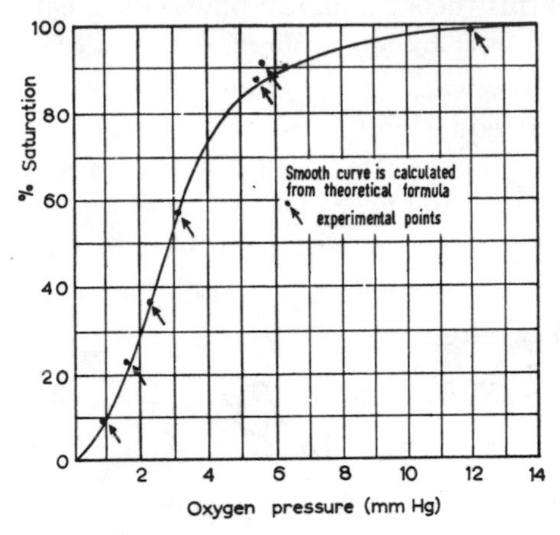

Fig. 21. From Forbes and Roughton, J. Physiol. **71** (1931) 229.

Fig. 21 shows "the goodness of the fit" obtained by Forbes and Roughton. There was one weak point, however. With four unknown constants at one's disposal, it is easy to choose several widely varying sets of values for K_1, K_2, K_3, K_4, all of which give sound agreement between calculated and experimental results.

For example the values of K_2 and K_4 in one set (e.g. I in Table 7) may differ markedly from those of K_2 and K_4 in another set (II), without giving rise to an appreciable difference in the calculated results. Table 7 serves to illustrate this point.

TABLE 7

Oxyhaemoglobin (percentage)

	Set I	Set II
pO_2 in mm	$K_1 = 0.4$ $K_2 = 0.15$ $K_3 = 0.06$ $K_4 = 4.5$	$K_1 = 0.4$ $K_2 = 0.075$ $K_3 = 0.06$ $K_4 = 9$
1	10.2	9.3
2.5	37.8	36.7
3.75	63.6	64.2
5	79.2	81.2
10	95.9	96.8

[from W. H. Forbes and F. J. W. Roughton, J. of Phys. **71** (1931)]

Although the satisfactory agreement between theoretical and experimental results together with the very sound theoretical basis involving only one single assumption, i.e. the existence of intermediates, makes this theory a highly plausible one, it remains a matter of considerable interest to determine the values of K_1, K_2, K_3 and K_4 experimentally. How can this be done? Now, when p is taken very small, the general equation (3) reduces to

$$Y = \frac{K_1 p}{4(1 + K_1 p)} \tag{4}$$

since p^2, p^3, p^4 become negligibly small.

When p is very large

$$Y = \frac{3K_1K_2K_3p^3 + 4K_1K_2K_3K_4p^4}{4K_1K_2K_3p^3 + 4K_1K_2K_3K_4p^4} = \frac{3 + 4K_4p}{4 + 4K_4p} \tag{5}$$

since now the p and p^2 terms are negligible.

Thus, by accurate determinations at the extreme ends of the dissociation curve, it should be possible to obtain K_1 and K_4 independently. The number of arbitrary constants will thus be reduced to two and a far more specific test for the validity of equation (3) will be obtained.

In a later study Paul and Roughton made independent determinations of K_1 and K_4 using equations (4) and (5). We shall give a brief account of their method. Firstly, the validity of equation (4) had to be tested. Equation (4) yields a rectangular hyperbola which is not a convenient curve for our purposes. Re-arranged, we can obtain a linear relationship, which gives a more specific test. We obtain:

$$\frac{1}{Y} = \frac{4}{K_1 p} + 4$$

so that if $1/Y$ is plotted against $1/p$ a straight line should result, cutting the ordinate axis at $1/Y = 4$.

Fig. 22 shows their results plotted in this manner. Considering the difficulty of the determinations, these results appear to be highly satisfactory. Since $4/K_1$ is equal to the slope of the straight line, K_1 can easily be determined. (In a later paper it was stated that at pH $= 9.1$ two systematic errors were introduced making the value of K_1 $\pm 10\%$ too low.)

K_4 has been determined experimentally by a rather indirect method. For this, the reader is referred to the original papers.*

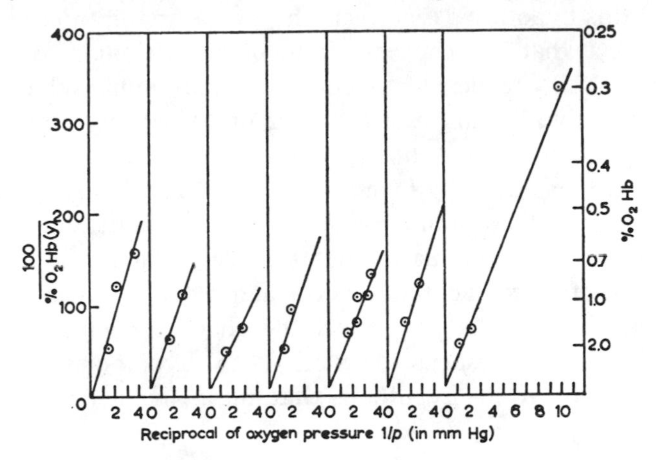

Fig. 22. From Paul and Roughton, J. Physiol. (1951).

(*e*) *The chemical transmission of nerve impulses*

The concentration-action curves of acetylcholine and adrenaline, i.e. the curves relating to the responses of some effector organs (e.g. muscles) and the concentration of applied or injected acetylcholine or adrenaline, have been shown to be rectangular hyperbolas. Clark†, for example, studied the effect of acetylcholine on the frog heart, and obtained the empirical equation

$$R = \frac{x}{k + k'x}$$

where R is the response, x the concentration, k and k' are parameters. (by "inverting" the above equation and writing

$$y = \frac{1}{R} = \frac{k + k'x}{x} = k' + \frac{k}{x}$$

the familiar expression of the hyperbola equation is recognised.)

Clark also tested the experimental curves for parabolic and exponential equations, but found consistent deviations when using these equations.

* See, for instance, F. J. W. Roughton, J. Physiol. **126** (1954) 359–383.
† A. J. Clark, J. Physiol. **61** (1926) 530; **64** (1927) 123.

The hypothesis has been advanced that autonomic nerves stimulate the effectors by the liberation of acetylcholine (or adrenaline in some cases). If this hypothesis is correct, then, as a first approximation, we might expect that the concentration, of acetylcholine for example, will vary with the frequency of stimulation. This implies that we might expect the curve relating the response of the effector organ to the frequency of stimulation through the nerve to be also hyperbolic. Rosenblueth* has actually found this to be the case, for example, the curve connecting the isotonic contraction of the nictitating membrane and the frequency of stimulation of its nerve was found to be a hyperbola. The use of an exponential equation here also gave rise to systematic deviations. The similarity between the concentration action curves and the frequency-response curves (both hyperbolas) lends considerable support to the hypothesis that autonomic nerve fibres act by releasing acetylcholine or adrenaline.**

5. The Law of Allometric Growth

The first quantitative analysis of differential growth on a general basis was made by Huxley† in 1924. Huxley made use of the formula

$$y = bx^k \tag{1}$$

to describe the relation of the growth of a part or organ to that of the whole organism and to suggest that this might express a general law of differential growth. The variables x and y denote weights or lengths of parts of an organism and b and k are constants, b being known as the *initial growth index* and k the *equilibrium constant*: for instance, in one of the cases considered by Huxley y denoted the mean weight of the large chela of *Uca pugnax*, and x denoted the mean weight of the rest of the body after the removal of the large chela. The equation (1) is called by Huxley the *simple allometry* formula. The relation is described as *isometry* in the special case of $k = 1$, which gives simple proportion between x and y with $y = 0$ when $x = 0$. If the equilibrium constant k is negative the relation is described as *enantiometry*. We shall now consider certain aspects of this relationship.

* A. Rosenblueth, "The transmission of nerve impulses at neuro-effector junctions and peripheral synapses", (John Wiley and Sons, New York, 1950.)

** For criticism see G. L. Brown and J. C. Eccles, J. Physiol. **82** (1934) 211 and J. C. Eccles, Ergebn. Physiol. 88 (1936) 1.

† J. S. Huxley, Nature **114** (1924) 895. For a complete survey of the work of Huxley and others see J. S. Huxley, "Problems of Relative Growth" (Methuen, London, 1932).

The simple allometry formula (1) can be put into a dimensionless form in a particularly simple way. If y takes the value y_0 when x takes the value x_0 then, since these values must satisfy the formula (1), we must have

$$y_0 = bx_0^k$$

from which it follows that

$$b = \frac{y_0}{x_0^k}.$$

Substituting this value for the constant b into equation (1) we find that this equation may be written in the form

$$\frac{y}{y_0} = \left(\frac{x}{x_0}\right)^k. \tag{2}$$

Since (y/y_0) and (x/x_0) are pure numbers, equation (2) is the most suitable form through which to represent graphically the law of allometric growth.

The form of the relation (2) for various values of the equilibrium

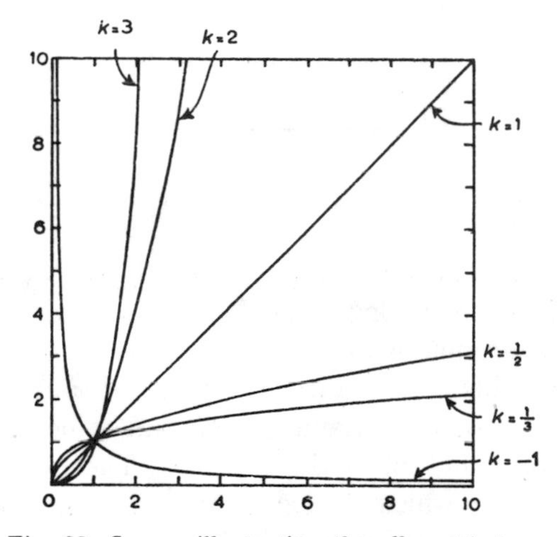

Fig. 23. Curves illustrating the allometric law.

constant k is shown in Fig. 23. Since, in biological problems, the ratios (x/x_0) and (y/y_0) will always be positive, the curves are shown only for positive values of the variables.

It is the converse problem which is of most interest to biologists. That is, if we are given tables of values of two quantities x and y we want to know if they are related by the simple allometry formula and, if so, what the appropriate values of the constants b and k are. We solve this problem by making use of the properties of logarithms. If we take the logarithm of both sides of equation (1) we see that this equation is equivalent to the equation

$$\log y = \log b + k \log x \tag{3}$$

so that if we write

$$\log y = \eta, \quad \log x = \xi, \quad \log b = \beta, \tag{4}$$

we have

$$\eta = \beta + k\xi. \tag{5}$$

It follows from this equation that the graph of η against ξ is a straight line (cf. Fig. 24) such that β is the intercept (OB) on the η-axis and

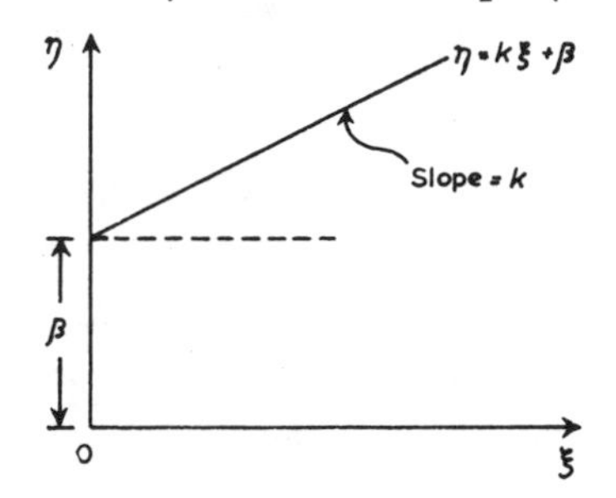

Fig. 24. Determination of the constants in an allometric relation.

k is the slope of the line. In terms of the original variables it follows that the relation between $\log y$ and $\log x$ is a linear relation, so that if we wish to determine whether the relation between x and y is of the type (1) we should plot the values not of x and y but of $\log x$ and $\log y$; if the points so obtained are found to lie on a straight line then the relation is in fact of the type (1). The relation (3) is valid whatever the base to which the logarithms are taken, but in practical applications it is simplest, because of the simplicity of operation and the availability of tables, to work to the base 10.

Once it has been verified by this graphical procedure that the variation of y with x is of the form (1) it remains to determine the numerical values of the constants b and k. This can be done by considering the coordinates (ξ_1, η_1), (ξ_2, η_2) of two points on the line of Fig. 24. Since these points lie on the line with equation (5) it follows from equation (9) of the last section that

$$k = \frac{\eta_2 - \eta_1}{\xi_2 - \xi_1}, \quad \beta = \frac{\xi_2 \eta_1 - \xi_1 \eta_2}{\xi_2 - \xi_1}. \tag{6}$$

We can transform these results back to the original variables x and y by making use of the equations (4). If (x_1, y_1) and (x_2, y_2) are two points on the curve whose equation is (1), then putting $\xi_1 = \log x_1$, $\xi_2 = \log x_2$, $\eta_1 = \log y_1$, $\eta_2 = \log y_2$, $\beta = \log b$ in equations (6) we have

$$k = \frac{\log y_2 - \log y_1}{\log x_2 - \log x_1}, \quad \log b = \frac{\log x_2 \log y_1 - \log x_1 \log y_2}{\log x_2 - \log x_1}. \tag{7}$$

If the plotted points do not actually all lie on a straight line, then it is possible to find "the best possible" line through them; there is a numerical procedure for obtaining the best fit* but we shall assume here that it is done by eye.

We shall illustrate the method outlined above by:-

EXAMPLE 2: *The quantities x and y have the values given in the table*:-

x	1.60	1.92	2.15	3.27	4.06	5.10
y	1.11	1.38	1.58	2.61	3.39	4.46

Show that x and y obey a simple allometric law and determine the value of the initial growth index and the equilibrium constant.

From these values we form the subsidiary table:-

$\xi = \log x$	0.2041	0.2833	0.3324	0.5145	0.6085	0.7076
$\eta = \log y$	0.0453	0.1399	0.1987	0.4166	0.5302	0.6493

If we plot these points we get the crosses shown in Fig. 25 and it is obvious that these points all lie on the straight line shown. The variables x and y are therefore connected by a simple allometric formula. From an inspection of Fig. 25 we see that $\log y = -0.2$ when $\log x = 0$, and that the gradient of the line is 1.2, so that

$$\log y + 0.2 = 1.2 \log x$$

that is,

$$\log y = (-1 + 0.8) + 1.2 \log x.$$

* See B. C. Brookes and W. F. L. Dick, "Introduction to Statistical Method", (Heinemann, London, 1951) pp. 184—188.

In the form (1) this is

$$y = 0.6312\, x^{1.20}.$$

Once we have verified that the points lie on a line we can calculate the constants k and b from (7) instead of reading them straight from

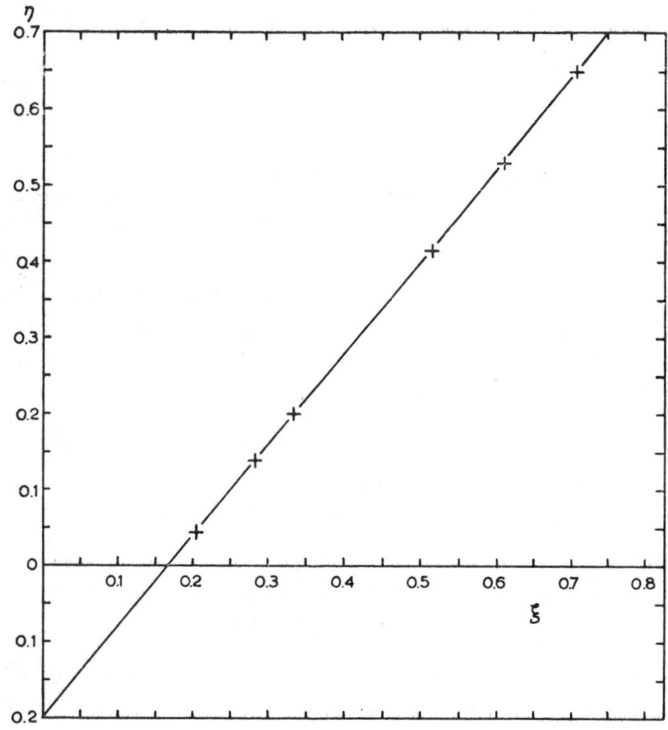

Fig. 25. Variation of $\xi = \log x$ with $\eta = \log y$ in example 2.

the graph. If we take $x_1 = 1.60$, $y_1 = 1.11$, $x_2 = 5.10$, $y_2 = 4.46$ in these equations we find that

$$k = \frac{0.6493 - 0.0453}{0.7076 - 0.2041} = \frac{0.6050}{0.5035} = 1.2016$$

$$\log b = \frac{0.7076 \times 0.0453 - 0.2041 \times 0.6493}{0.7076 - 0.2041} = 0.1995$$

which gives $b = 0.6317$. Hence we have

$$y = 0.6317\, x^{1.2016}$$

for the law connecting x and y.

A mathematical relation of a similar kind arises from a consideration of Fechner's law, which states that a relative increment in the stimulus energy produces a *constant* increment in the apparent magnitude (e.g. loudness). This law is in fact incorrect, and this is not due to any failure to produce a reliable scale for subjective loudness. Several of such scales have been developed, and in all of them a tone of 100 decibels sounds about 40 times as loud as the same tone of 50 decibels, whilst from Fechner's law it should sound only twice as loud.

A more adequate mathematical description of the relationship between stimulus ϕ (e.g. sound pressure) and sensation ψ (e.g. loudness) is given by "a power law".

$$\psi = K\phi^{\beta}$$

where β is the exponent of the power function and K is a constant. In log-log coordinates we obtain a straight line with slope β, i.e.

$$\log \psi = \beta \log \phi + \log K.$$

Although it remains to be proven it is most likely that the nonlinearity in the coupling between stimulus and sensation is effected in the sense organ rather than in the central nervous system*.

PROBLEMS

1. The corresponding values of two variables x and y, as obtained by experiment, are given in the following table:

x	1	2	3	4	5
y	30.4	10.7	7.1	4.4	3.2

Show that, allowing for experimental error, the relation between x and y is of the form $y = bx^k$, and deduce approximate values for b and k.

* Ciba Foundation Symposium. Touch, Heat and Pain, Churchill, 1966. S.S. Stevens, Neural Events and the Psychophysical Law, Science **170** (1970) 1043–1050.

2. The mean values of the pre-ocular length and total length in *Planaria gonocephala* are given by the following table*:

Total length (mm)	1.5	3.0	5.0	18.0
Pre-ocular length (mm)	0.23	0.43	0.50	1.125

Show that the two quantities are related by an allometric law and find approximate values for the parameters involved.

3. The values of the cranium length and face length in the baboon *Papio porcarius* at different sizes are**:-

Mean cranium length (mm)	78.5	100.25	108.9	114.7	118.25	122.0
Mean face length (mm)	31.0	64.6	94.8	131.0	140.8	144.25

Verify that the relation between these two lengths is of the type (3) and determine approximate values for the relevant constants.

4. The variation of abdomen breadth with carapace length in *Carcinus Maenas* is given by the table†:-

Mean carapace length (mm)	3.09	3.80	4.22	4.76	5.19
Mean abdomen length (mm)	0.578	0.680	0.823	0.979	1.019

Find the allometric equation connecting the two quantities.

5. The index of cortical folding and the weight of hemisphere are given for several species by the table†† :-

Species	Weight of hemisphere	Index of folding
Man	571.0	2.81
Orang	79.7	2.56
Ateles ater	64.5	1.73
Macaca	31.2	1.87
Canis fam.	25.0	1.71
Felis dom.	14.0	1.49

Show that the relation between the two quantities can be expressed approximately by an allometric equation, and determine approximate values of the constants.

* M. Abeloos, C. R. Soc. Biol. **98** (1928) 917.

** S. Zuckermann, Proc. Zool. Soc. (1926) 843.

† J. S. Huxley and O. W. Richards, Journ. Mar. Biol. Assn. **17** (1931) 1001.

†† This data has been collected from various authors by G. v. Bonin, J. Gen. Psychol. **25** (1941) 273.

6. How can the relation $y = k\,10^{ax}$, where a, k are constants, be represented on ordinary graph paper so that the graph obtained is a straight line?

Figures for the population y of a certain town in year x are as follows:

x	1910	1920	1930	1940
y	180 000	270 000	400 000	600 000

Verify that these figures conform approximately to a law of the above type, determine values for a and k, and, assuming that the same law of growth is subsequently maintained, estimate the year in which the population will be 800 000.

6. Polynomials and Rational Functions

Next in order of difficulty to the functions we have already considered are those which can be made up from combinations of simple forms of the type ax^n where a is a constant and n is a positive integer. The simplest extension of this idea is to the construction of functions which are sums of such terms, namely functions of the type

$$f(x) = a_0 + a_1 x + a_2 x^2 + \ldots + a_n x^n \tag{1}$$

where $a_0, a_1, a_2, \ldots a_n$ are constants and a_n is supposed to be non-zero. Such a function is called a *polynomial* of degree n.

The graphs of polynomials are easily drawn by constructing a table of values, plotting the points, and joining them by a smooth curve. A case which arises so often that it is worth considering in more detail is that of the quadratic polynomial

$$y = px^2 + qx + r. \tag{2}$$

We can rearrange the right hand side of equation (2) to give

$$p\left(x^2 + \frac{q}{p}x\right) + r = p\left[x^2 + 2\left(\frac{q}{2p}\right)x + \left(\frac{q}{2p}\right)^2\right] + r - \frac{q^2}{4p}$$
$$= p(x-a)^2 + b,$$

where

$$a = -\frac{q}{2p}, \quad b = r - \frac{q^2}{4p}. \tag{3}$$

The equation (2) can then be written in the form

$$y - b = p(x - a)^2. \tag{4}$$

Suppose that P is the point (a, b) (cf. Fig. 20) and that through the point P we draw two axes — an ξ-axis and an η-axis — parallel

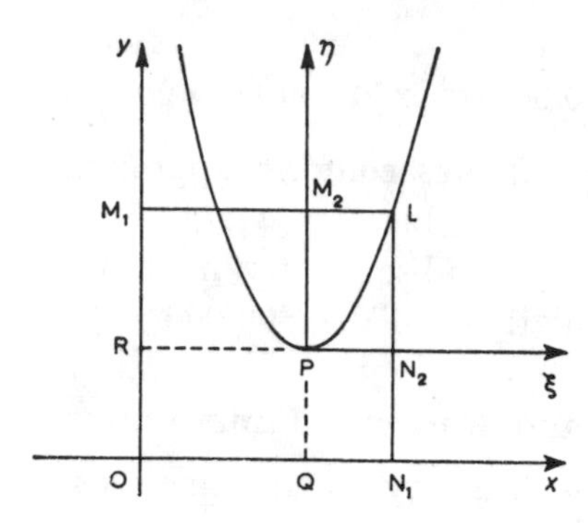

Fig. 26. Graph of the function $y = px^2 + qx + r$.

to the x- and y-axes. If L is a point whose coordinates are (x, y) in one system and (ξ, η) in the other, then, in the notation of Fig. 26, $ON_1 = x$, $OM_1 = y$, $PN_2 = \xi$, $PM_2 = \eta$. Since in the xy-system P has coordinates (a, b) it follows that $OQ = a$, $OR = b$. Now the obvious geometrical relations

$$M_2L = ON_1 - OQ, \quad N_2L = OM_1 - OR,$$

are equivalent to the algebraic relations

$$\xi = x - a, \quad \eta = y - b. \tag{5}$$

Substituting from these equations into equation (4) we see that, referred to our new axes, the equation (2) takes the form

$$\eta = p\xi^2 \tag{6}$$

showing that the curve is a parabola (cf. Fig. 15) with vertex at the point P. Drawing this parabola we get the curve of Fig. 26. The point P is called a minimum turning value of the graph. At this point y has a smaller value than at any point in the vicinity. This curve has only one such point.

EXAMPLE 3: *Sketch the curve whose equation is* $y = 4x(2 - x)$. We can write this equation in the form

$$y = -4x^2 + 8x = -4(x^2 - 2x + 1) + 4$$

which is equivalent to

$$y - 4 = -4(x - 1)^2,$$

so that if we transform to parallel axes through the point (1, 4) we see that the equation of the curve becomes $\eta = -4\xi^2$ and hence

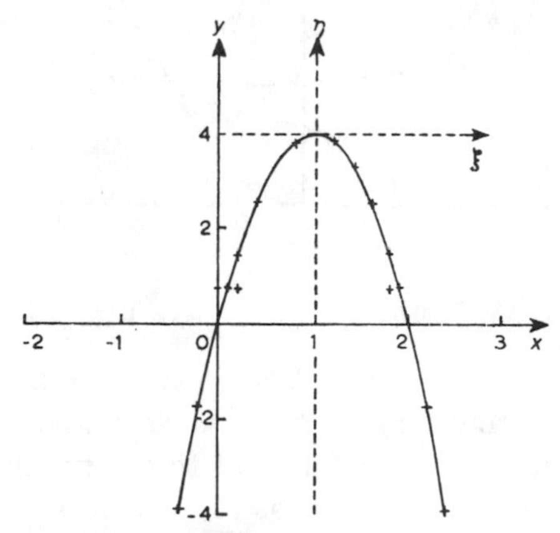

Fig. 27. Graph of the function $y = 4x(2 - x)$.

the curve has the shape shown in Fig. 27. From this curve we see that $y = 4$ is the greatest value achieved by the function, and also that $y = 0$ when $x = 0$ and $x = 2$, a fact which is obvious from the original form of the functional relationship.

An important idea we must now consider is that of the tangent at a point to a parabola. Apart from the circle, there is no general geometrical procedure for drawing the tangent to a curve at any point on it, so we must find some other means of defining the tangent. Suppose that, for simplicity, we consider the parabola whose equation is $y = kx^2$ and the point P on it. Let $A, B, C, D, \ldots$ be a series of points on the parabola which are in increasing proximity to P (cf. Fig. 28). We can imagine that the lines AP, BP, CP, $DP \ldots$ are successive positions of a line rotating in a counterclockwise direction about the point P. Eventually we would reach a position PT of the line which meets the curve at the one point P. The straight line PT so obtained is called the *tangent to the parabola* at P. Another way we could look at the line PT is by regarding it as a line which does meet

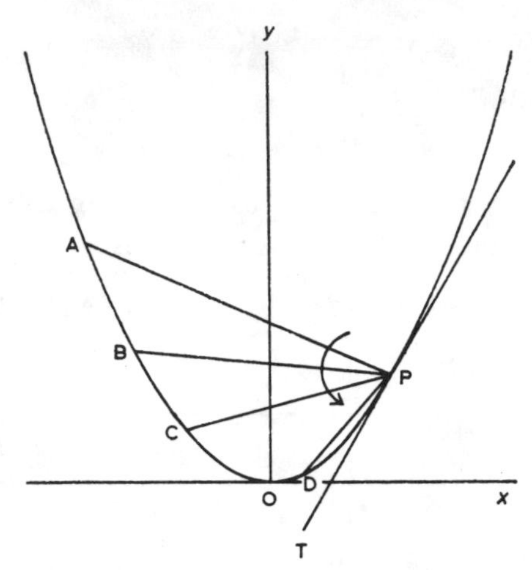

Fig. 28. Tangent at a point to a parabola.

the parabola in two points but that these two points coincide in P — i.e. PT cuts the parabola in *two coincident points*.

To derive the equation of the tangent to the parabola at the point P with coordinates (x_1, kx_1^2) we first of all consider the equation of the line PQ joining P to another point (x_2, kx_2^2) of the parabola. If we denote the ordinates of P and Q by y_1 and y_2 respectively, then it is a matter of elementary algebra to show that

$$\frac{y_2 - y_1}{x_2 - x_1} = k(x_1 + x_2), \quad \frac{x_2 y_1 - x_1 y_2}{x_2 - x_1} = -kx_1 x_2.$$

Substituting these expressions into equation (10) of Section 3 above we see that the equation of the line PQ is

$$y = k(x_1 + x_2)x - kx_1 x_2. \tag{7}$$

As Q gets nearer to P the chord PQ gets nearer and nearer to the line with equation

$$y = 2kx_1 x - kx_1^2, \tag{8}$$

obtained by putting $x_2 = x_1$ in equation (7). Hence, by the above reasoning equation (8) is the equation of the tangent PT to the parabola at the point P.

The graphs of more complicated polynomials of the type (1) can be treated in a similar way. The procedure of curve drawing can be simplified considerably by use of the methods of the differential calculus, so we shall postpone a fuller discussion until these techniques are available to us. (See Chapter 4 below.) We shall merely illustrate the elementary procedure by considering

EXAMPLE 4: *Draw the graph of the function*

$$y = x^3 - 3x^2 - 4x + 12.$$

We first of all note that this polynomial can be factorized to give

$$y = (x + 2)(x - 2)(x - 3).$$

It follows immediately from this form that $y = 0$ when x is -2, 2 or 3. If $x < -2$ then all three factors will be negative so that $y < 0$, and if $x > 3$ all three factors will be positive so that $y > 0$. If $-2 < x < 2$ the first factor will be positive and the remaining two factors will each be negative so that $y > 0$; similarly if $2 < x < 3$ the first two factors will be positive and the third will be negative giving $y < 0$.

For small values of x, x^2 and x^3 are numerically much smaller than x so that for small values of x, y is approximately equal to $12 - 4x$. On the other hand if x is large, x^3 is numerically greater than x^2 or x so that for large values of x, y is approximately equal to x^3; so if x is large and positive, y is large and positive and if x is large and negative, y is large and negative.

Even this limited amount of calculation enables us to draw a rough sketch of the graph of the function of the kind shown in Fig. 29. A more accurate curve can of course be obtained by calculating the coordinates of a number of points and plotting them, but for many purposes in mathematics a rough sketch of this kind is adequate.

Next in order of complexity are the *rational functions*, so called because they are ratios of polynomials, i.e. they are of the form

$$y = \frac{P(x)}{Q(x)} \tag{9}$$

where $P(x)$ and $Q(x)$ are polynomials in x.

We have already considered the simplest rational function, which is not a simple polynomial, namely

$$y = \frac{1}{x}.$$

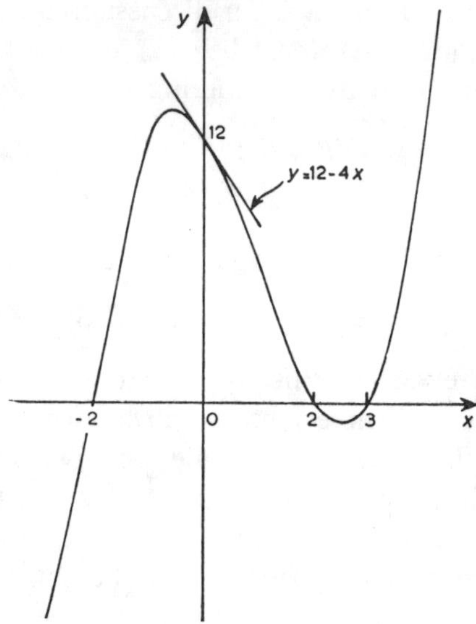

Fig. 29. Sketch of the graph of the function $y = x^3 - 3x^2 - 4x + 12$.

We saw (cf. Fig. 16) that as x became larger, y became smaller and that the larger x became the more nearly did the points of the curve approach the line $y = 0$. The line $y = 0$ is called an *asymptote* of this hyperbola, an asymptote being defined as a straight line to which the curve approaches more and more closely as x or y or both become numerically larger and larger. Since the equation of the hyperbola may be written $x = 1/y$ it follows that the line $x = 0$ is also an asymptote as is obvious from Fig. 16.

It is obvious that we obtain the vertical asymptotes, if any, of the graph of the function (9) by finding the values of x which make $Q(x) = 0$. To illustrate the procedure in the case of rational functions we consider:-

EXAMPLE 5: *Sketch the graph of the function*

$$y = \frac{x-1}{x-2}.$$

We first of all note that $y = 0$ when $x = 1$.

To find the vertical asymptote we have the equation $x - 2 = 0$. If we put $x = 2 + \varepsilon$ where ε is small, then we have

$$y = \frac{1+\varepsilon}{\varepsilon} \simeq \frac{1}{\varepsilon}$$

so that if ε is small and positive, y will be large and positive, whereas if ε is small and negative, y will be large and negative.

The behaviour of the function for small values of x is obtained by rearranging the denominator to obtain

$$y = \frac{1-x}{2(1-\frac{1}{2}x)} = \tfrac{1}{2}(1-x)(1-\tfrac{1}{2}x)^{-1}$$

and using the binomial theorem to give the approximation

$$y = \tfrac{1}{2}(1-x)(1+\tfrac{1}{2}x) \simeq \tfrac{1}{2}(1-\tfrac{1}{2}x).$$

Similarly, if x is large we have

$$y = \left(1-\frac{1}{x}\right)\left(1-\frac{2}{x}\right)^{-1} \simeq \left(1-\frac{1}{x}\right)\left(1+\frac{2}{x}\right) \simeq 1+\frac{1}{x}$$

showing that the line $y = 1$ is an asymptote. The curve approaches the asymptote from above when x is large and positive, and from below when x is large and negative.

Making use of these facts we construct a rough sketch of the function as in Fig. 30.

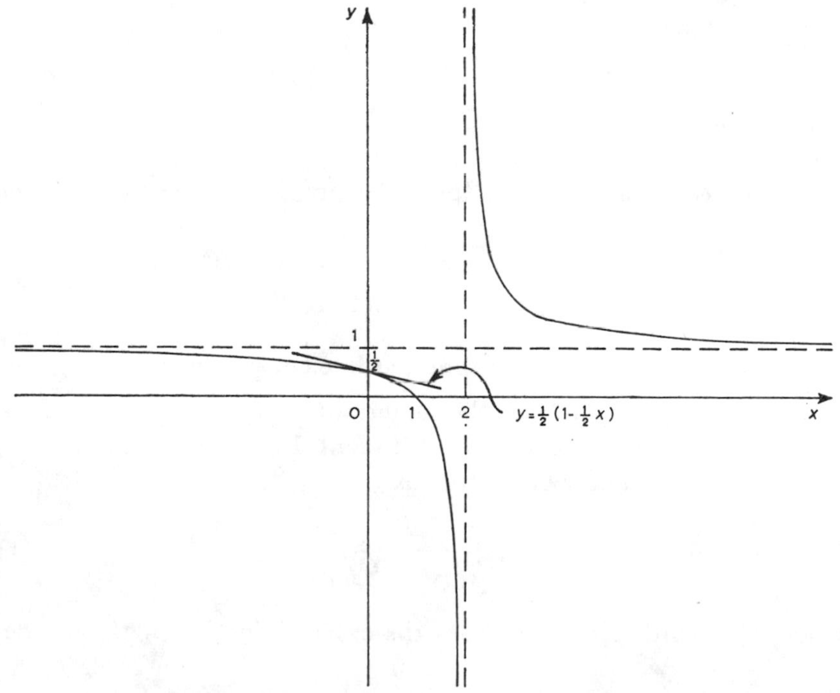

Fig. 30. Sketch of the graph of the function $y = (x-1)/(x-2)$.

In a great many mathematical problems it is useful to be able to draw a rough graph of a function in this way, and the reader is recommended to acquire some facility in the procedure.

PROBLEMS

1. Sketch the parabola whose equation is

$$y = x^2 + 4x + 3.$$

Show that there are two points on the parabola which have ordinate 8 and derive the equations of the tangents to the parabola at these points.

2. Sketch the graphs of the polynomials

(i) $y = x^3 - x^2 + x - 1$;
(ii) $y = x^4 - 5x^2 + 4$;
(iii) $y = x^3 - x^2$.

3. Sketch the graphs of the rational functions

(i) $y = \dfrac{(x+1)(x-2)}{x}$.

(ii) $y = \dfrac{x^2 - 10x + 9}{x^2 + 2x + 1}$;

(iii) $y = \dfrac{x}{(x+1)(x+2)}$.

4. On the same diagram, plot the graphs of $y = 2x^2 - 5$ and $y = 3x + 4$.

From the two graphs find the roots of the equation

$$2x^2 - 3x - 9 = 0.$$

7. A Simple Method to Determine Cardiac Output in Man

The cardiac output may be determined by the socalled Fick principle familiar to every medical student. Using CO_2 as a "tracer", the Fick equation may be written as

$$Q = \frac{F_e}{C_{\bar{v}CO_2} - C_{aCO_2}} \tag{1}$$

where Q is cardiac output, F_e is the exhaled (net) CO_2 flow, C denotes concentration (of CO_2) and subscripts $\bar{v}$ and a denote mixed venous blood (in the pulmonary artery) and arterial blood respectively. Equation (1) is simply a so-called equation of continuity, which is a balance sheet based upon the axiom that no matter (in this case CO_2 molecules) can be "created" or "lost".

By writing eq. (1) as

$$\left(C_{\bar{v}_{CO_2}} - C_{a_{CO_2}}\right) Q = C_{\bar{v}_{CO_2}} \cdot Q - C_{a_{CO_2}} \cdot Q = F_e \tag{2}$$

we see that it must hold, since $C_{\bar{v}_{CO_2}} \cdot Q$ is the CO_2-flow entering the lungs on the venous side and $C_{a_{CO_2}} \cdot Q$ equals the CO_2-flow leaving the lungs on the arterial side.

In the *steady state* their difference must equal the CO_2-flow (l/min, say), escaping through the airways.

Measuring F_e and $C_{a_{CO_2}}$ presents no difficulties.

The main problem is to measure $C_{\bar{v}_{CO_2}}$, the (total) CO_2 concentration in the *mixed venous blood* entering the lungs.

This can only be done directly by using heart-catheterization. Indirectly we might, at least "theoretically", use the lungs as a tonometer. If you hold your breath the CO_2 tension in your lungs will gradually rise towards an asymptotic value, the partial pressure of CO_2 in the mixed venous blood (cf. Chapter 6, §11).

Due to "premature" recirculation of the blood this equilibrium during breath holding is not attained so that alveolar (end-expiratory) P_{CO_2} (P = partial pressure) obtained at the end of breath holding cannot be used as an index of mixed venous P_{CO_2}.

If we can determine P_{CO_2} in the blood then we can by using a (standard) CO_2 absorption curve of blood, calculate the CO_2 *concentration* required in the Fick equation. (This curve is the main subject of Chapter 10, §17).

The simplest way to determine $P_{\bar{v}_{CO_2}}$, the partial pressure of CO_2 in the mixed venous blood, is to use some form of mathematical extrapolation. A very simple approach has been developed by Wise and Defares. The interesting feature of this method of extrapolation is that it requires almost no theoretical assumptions, i.e.

is not based on a (simplified) model. For example, an assumption such as "simple exponential" rise of CO_2 during breath holding, is *not* required.

The subject starts rebreathing in a small bag. As a result the P_{CO_2} in the bag at the end of each expiration rises in a stepwise manner with the number of breaths. Due to recirculation of the blood only the bag P_{CO_2} values obtained within the first 20 seconds can be utilized for the mathematical extrapolation to obtain the asymptotic value, $P_{\bar{v}CO_2}$.

Let us denote $P_{\bar{v}CO_2}$ by q, and end-expiratory bag P_{CO_2} by p, and let n be the number of breaths.

Using the experimental values a p_n versus p_{n-1} plot is made (i.e. p_3 is plotted against p_2, etc.). When the points fall on a straight line, we may obviously write,

$$p_n = ap_{n-1} + b \tag{3}$$

where a and b are constants whose numerical values are found from this plot. It is easily seen that when n is large $p_n \simeq p_{n-1} \simeq q$ ($\simeq$ means almost equal to), so that (3) becomes, for large n (long times)

$$q = aq + b \tag{4}$$

or

$$q = \frac{b}{1 - a} \tag{5}$$

In words, the mixed venous P_{CO_2} equals $b/(1-a)$.

Deviations of the first few points from the straight line drawn through the remaining points may be ignored, as can be shown mathematically. More complete mathematical treatments have been presented by Defares and Wise*. The above method of computing cardiac output has been found to agree with other methods (dye-dilution method, etc.), and has been used in the physiological testing of astronauts (Lovelace foundation, Albuquerque) and other NASA projects.

* J.G. Defares in Physicomathematical Aspects of Biology, ed. N. Rashevsky, Academic Press, New York, 1962, pp. 265–286. See also M.E. Wise and J.G. Defares, Clinical Science 43 (1972) 303.

In Leyden and Houston (Dr. D. Cardus) the method is computer controlled and is being used to measure arterial "elasticity" (compliance) and other cardiovascular parameters.

After reading Chapter 9, or earlier, the student will appreciate the assertion that the above linear equation (3) is in fact a first order linear difference equation, which is the "discrete" counterpart of a first order linear differential equation. In general the process is of second or higher order with the rapid components died out after the first two or three initial breaths, so that in most cases the empirical equation (3) can be used for extrapolation.

8. Trigonometrical Functions

At the root of the discussion of trigonometric functions is the definition of *angle* which we employ in mathematical analysis. In elementary mathematics we have an intuitive concept of what we mean by the angle $\angle POQ$ between two lines OP and OQ (cf. Fig. 31).

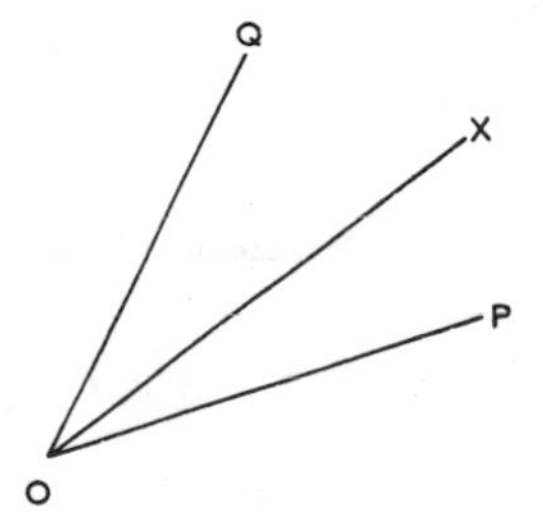

Fig. 31. The concept of an angle.

We consider a variable line OX which starts by coinciding with OP and then rotates about O until it coincides with OQ, and we say that in the course of this rotation OX describes the *angle* POQ. The angle is taken to be positive if, in describing it, OX has revolved in an anti-clockwise direction and negative if OX has revolved in a clockwise direction. In elementary mathematics and in practical applications of trigonometry (to surveying etc.) we measure angles in the *sexagesimal system* in which the basic geometrical unit, the right angle, is divided into 90 equal parts called *degrees*. The degree is subdivided into 60 equal parts called *minutes*, and the minute into 60 equal parts called *seconds*. The abbreviation $^\circ$ is used for degrees, $'$ for minutes and $''$ for seconds so that an angle of 30 degrees 15 minutes 25 seconds is written as $30°15'25''$. In the sexagesimal system the line OX in one complete revolution would describe an angle of 360°. Such a system is of course purely arbitrary.

Although the sexagesimal system is used exclusively in trigonometrical calculations, we employ another system in theoretical discussions. This rests on the geometrical result that if PQ and pq are two arcs of concentric circles with centre O and radii R and r respectively such

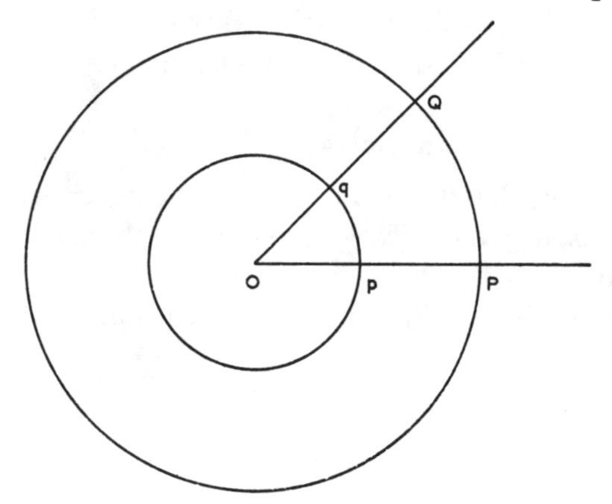

Fig. 32. The definition of an angle.

that PQ and pq subtend the same angle POQ at the common centre of the circles (cf. Fig. 32) then

$$\frac{\text{arc } PQ}{R} = \frac{\text{arc } pq}{r}.$$

This may be stated in the form: the ratio (arc PQ)$/R$ is independent of R if the angle POQ is fixed. We may therefore take the ratio (arc PQ)$/R$ to be the measure of the angle POQ. This ratio is called the *circular measure*, or *radian measure*, of the angle POQ. The same convention of sign is adopted as in the case of sexagesimal measure. The unit of circular measure is the angle for which the arc PQ is equal in length to R; this angle is called 1 *radian*.

It is a simple matter to find the relation between the sexagesimal measure of an angle and its radian measure. We recall that π is defined to be the ratio of the circumference of a circle to its diameter so that by the above definition

$$1 \text{ revolution} = \frac{\text{circumference}}{\text{radius}} = 2\pi \text{ radians}$$

while, in the sexagesimal system, we take 1 revolution to be 360°. Thus $360° = 2\pi$ radians so that $1° = \pi/180$ radians, and 1 radian $= 180°/\pi$ and this is approximately $57°17'44''$.

It follows at once from the definition of circular measure that if an arc PQ subtends an angle of x radians at the centre of a circle of radius R, then the length, s, of the arc PQ is given by

$$s = Rx. \tag{1}$$

In the rest of this book it will be assumed that *all angles are measured in radians* unless the contrary is stated explicitly.

In the remarks above we have assumed that the angles with which we are dealing are less than one complete revolution, but it will be recalled from elementary geometry that given two arms OP and OQ there is not one angle POQ but a whole set of such angles. If α is any one of these angles and x is the simple angle defined above, then α will be equal to x plus an integral multiple (n, say) of complete revolutions. Thus we may write (in radian measure)

$$\alpha = x + 2n\pi. \tag{2}$$

We shall now define the circular or trigonometrical functions. Let x be any given angle. Suppose that OA and OB are two lines at right angles, and that LOL' is a straight line through their point of intersection O such that $\angle AOL = x$ (cf. Fig. 33). If we choose any point

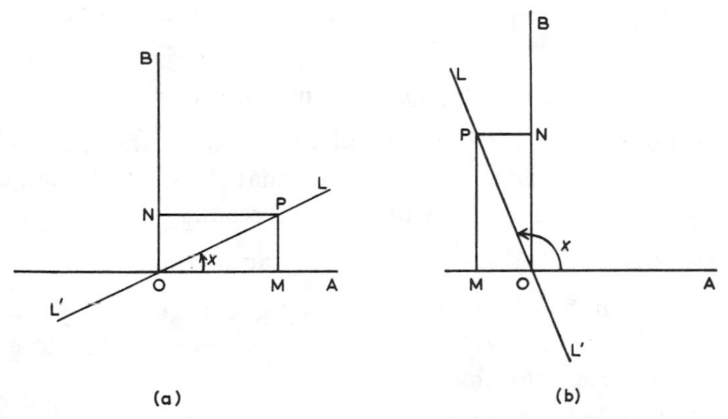

Fig. 33. Definition of the trigonometrical ratios.

P on LOL' and let M and N be the feet of the perpendiculars from P to OA and OB, then it is a matter of elementary geometry to show that each of the three ratios of measures of steps

$$\text{(i)} \quad \frac{OM}{OP}, \quad \text{(ii)} \quad \frac{ON}{OP}, \quad \text{(iii)} \quad \frac{ON}{OM} \tag{3}$$

is independent of the particular point P chosen and will be dependent only on the angle x. We define these ratios to be

(i) $\cos x$, (ii) $\sin x$, (iii) $\tan x$

respectively. Their reciprocals are called

(i) $\sec x$, (ii) $\operatorname{cosec} x$, (iii) $\cot x$

respectively. It is an immediate consequence of these definitions that

$$\tan x = \frac{\sin x}{\cos x}. \tag{4}$$

In defining the trigonometrical functions by the ratios (3) we adopt the convention that OM is reckoned to be positive if M lies to the right of O, negative if it lies to the left; similarly ON is taken to be positive if N lies above O, negative if it lies below it. Distances along OL are always taken to be positive. Thus in Fig. 33(a) both OM and ON are positive, while in Fig. 33(b) OM is negative and ON is positive. With this convention of signs and the sign convention we have adopted for angles it follows that

$$\sin(-x) = -\sin x, \quad \cos(-x) = \cos x. \tag{5}$$

It is obvious from equation (2) that since x and $x + 2n\pi$ are geometrically indistinguishable,

$$\cos(x+2n\pi) = \cos x, \ \sin(x+2n\pi) = \sin x, \ \tan(x+2n\pi) = \tan x \tag{6}$$

for any integer n (positive or negative). Now in the limit when x is zero, $OM = OP$ and $ON = 0$ so that $\cos 0 = 1$, $\sin 0 = 0$, $\tan 0 = 0$, and, from (6), we obtain the results

$$\cos 2n\pi = 1, \quad \sin 2n\pi = 0, \quad \tan 2n\pi = 0. \tag{7}$$

Similarly when in the limit when x is a right angle (i.e. $\frac{1}{2}\pi$ radians) we have $OM = 0$ and $ON = OP$ so that $\cos \frac{1}{2}\pi = 0$, $\sin \frac{1}{2}\pi = 1$ and it follows from (6) that

$$\cos(2n + \tfrac{1}{2})\pi = 0, \quad \sin(2n + \tfrac{1}{2})\pi = 1. \tag{8}$$

Suppose that x is an acute angle and that we construct the line OL'' by making the angle $AOL'' = \frac{1}{2}\pi + x$. Then on OL'' we can take a point P' such that $OP' = OP$ as shown in Fig. 34 and draw perpendiculars $P'M'$ and $P'N'$ as shown. It is a simple matter to show that the triangles OPM, $OP'N'$ are congruent, so ON' is equal in length to OM and, since both are positive, we may write $ON' = OM$; similarly $OM' = P'N' = PM = ON$ in length but OM' is negative

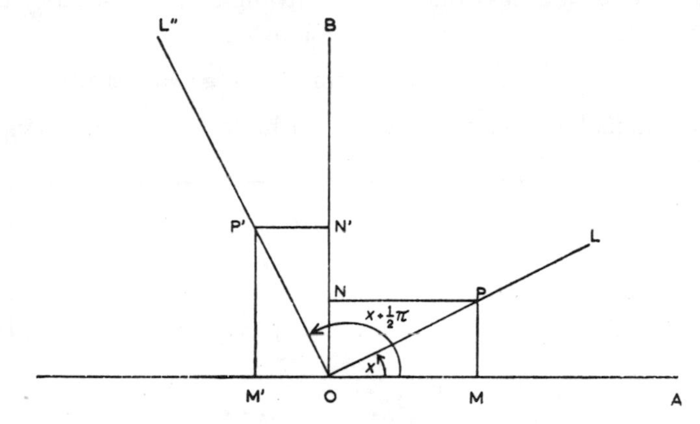

Fig. 34. Relation of trigonometrical ratios of $\frac{1}{2}\pi + x$ to those of x.

while ON is positive so that we have $OM' = -ON$. From the definitions of the trigonometric functions we have the following results:

$$\cos\,(x + \tfrac{1}{2}\pi) = \frac{OM'}{OP'} = -\frac{ON}{OP} = -\sin x$$

$$\sin\,(x + \tfrac{1}{2}\pi) = \frac{ON'}{OP'} = \frac{OM}{OP} = \cos x.$$

We have demonstrated these results in the case in which the angle x is acute, but the proof can readily be extended to show that, for a general angle x,

$$\cos\,(x + \tfrac{1}{2}\pi) = -\sin x, \quad \sin\,(x + \tfrac{1}{2}\pi) = \cos x. \tag{9}$$

From these results and equation (4) it follows that

$$\tan\,(x + \tfrac{1}{2}\pi) = -\cot x. \tag{9a}$$

From the first of equations (9) we can write

$$\cos\,(x + \pi) = \cos\,\{(x + \tfrac{1}{2}\pi) + \tfrac{1}{2}\pi\} = -\sin(x + \tfrac{1}{2}\pi) = -\cos x;$$

$$\cos(x + \tfrac{3}{2}\pi) = \cos\{(x + \pi) + \tfrac{1}{2}\pi\} = -\sin(x + \pi) = \sin x.$$

Similar results hold for the sine function and we obtain the pairs of equations

$$\cos(x + \pi) = -\cos x, \quad \sin(x + \pi) = -\sin x, \tag{10}$$

$$\cos(x + \tfrac{3}{2}\pi) = \sin x, \quad \sin(x + \tfrac{3}{2}\pi) = -\cos x. \tag{11}$$

When x is acute (i.e. less than one right angle in magnitude) then in Fig. 34 $\angle\, OPM = \frac{1}{2}\pi - x$ and it follows that

$$\cos(\tfrac{1}{2}\pi - x) = \sin x, \quad \sin(\tfrac{1}{2}\pi - x) = \cos x. \tag{12}$$

A few particular values are useful to have in the memory. If the

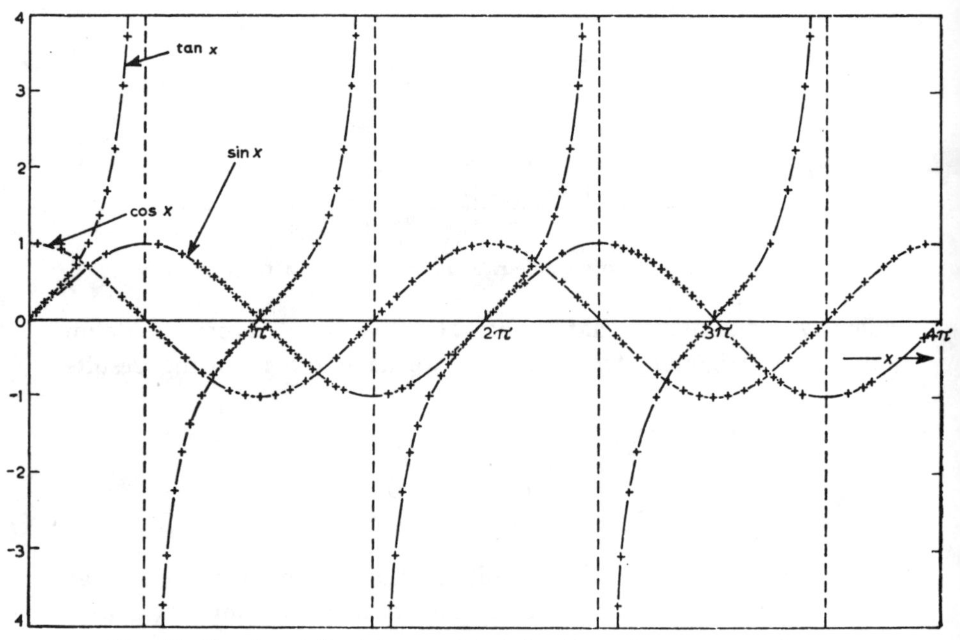

Fig. 35. Graphs of the trigonometrical functions $\sin x$, $\cos x$, and $\tan x$.

angle x is $\frac{1}{4}\pi$ (i.e. 45°), $OM = ON$ and $OP = \sqrt{2} \cdot OM$ so that

$$\cos \tfrac{1}{4}\pi = \frac{1}{\sqrt{2}}, \quad \sin \tfrac{1}{4}\pi = \frac{1}{\sqrt{2}}, \quad \tan \tfrac{1}{4}\pi = 1. \tag{13}$$

Similarly if x is $\frac{1}{6}\pi$ (i.e. 30°), $OP = 2 \cdot ON$, $OM = \sqrt{3} \cdot ON$ and we find that

$$\cos \tfrac{1}{6}\pi = \frac{\sqrt{3}}{2}, \qquad \sin \tfrac{1}{6}\pi = \tfrac{1}{2}, \qquad \tan \tfrac{1}{6}\pi = \frac{1}{\sqrt{3}} \tag{14}$$

and it follows from (12) that

$$\cos \tfrac{1}{3}\pi = \tfrac{1}{2}, \qquad \sin \tfrac{1}{3}\pi = \frac{\sqrt{3}}{2}, \qquad \tan \tfrac{1}{3}\pi = \sqrt{3}. \tag{15}$$

By means of these results and tables of trigonometrical functions it is a simple matter to draw the graphs of $\cos x$, $\sin x$, and $\tan x$. These are shown in Fig. 35 for values of x in the range 0 to 4π. The graphs for wider ranges can easily be constructed since the curve repeats its shape over each interval of 2π (cf. equations (6) above). We say that such a function is *periodic* with *period* 2π.

We shall now derive some general properties of trigonometrical functions. If we assume that x is an acute angle, then it follows from Fig. 33(a) that, since, by Pythagoras' theorem,

$$OP^2 = OM^2 + MP^2,$$

we have

$$\left(\frac{OM}{OP}\right)^2 + \left(\frac{MP}{OP}\right)^2 = 1$$

from which we deduce immediately that

$$\cos^2 x + \sin^2 x = 1. \tag{16}$$

We proved this theorem with reference to a diagram in which x is acute but it is easily seen that the result is true for all angles x. If we divide both sides of equation (16) by $\cos^2 x$ we find that

$$1 + \left(\frac{\sin x}{\cos x}\right)^2 = \left(\frac{1}{\cos x}\right)^2$$

which may be written in the form

$$1 + \tan^2 x = \sec^2 x. \tag{17}$$

Similarly if we divide both sides of equation (16) by $\sin x$ we obtain the result

$$1 + \cot^2 x = \operatorname{cosec}^2 x. \tag{18}$$

Equations (16), (17) and (18) are the three forms of the *fundamental identity* of trigonometry.

Suppose now that x and ξ are two acute angles such that $x + \xi$

is also acute (cf. Fig. 36). Then if we take a pair of perpendicular lines OA, OB intersecting at O as axes of reference, we can draw lines OL and OU such that $\angle AOL = x$, $\angle LOU = \xi$. We choose a point W on OU such that $OW = 1$ unit of length, and draw a line WP cutting OL in P at right angles so that OPW is a right angle. Through W and P we drop perpendiculars WW' and PP' onto OA, and through P we drop a perpendicular PQ to WW'. Then it is easily seen that $\angle AOU = x + \xi$ and $\angle W'WP = x$. In the right-angled triangle OWP, $OW = 1$ so that $WP = \sin \xi$ and $OP = \cos \xi$. Similarly in the right-angled triangle WPQ, $WP = \sin \xi$ and $\angle PWQ = x$ so

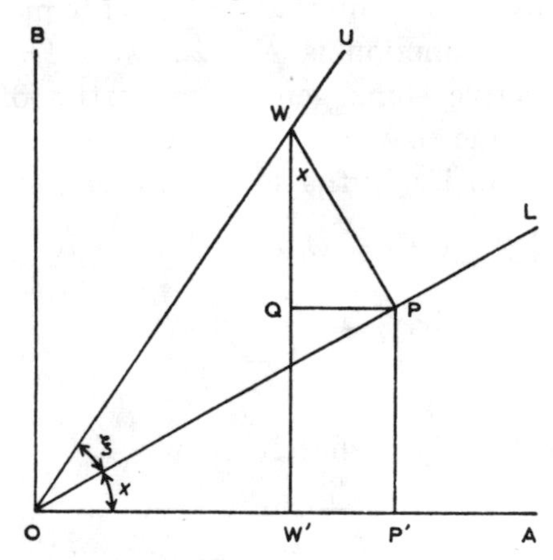

Fig. 36. The addition theorems for the trigonometrical functions.

that $PQ = \sin x \sin \xi$, and in the triangle OPP', $OP = \cos \xi$ and so $OP' = \cos x \cos \xi$. Finally in the right-angled triangle OWW', $OW = 1$ and $\angle W'OW = x + \xi$ so that $OW' = \cos (x + \xi)$. Now it is obvious from the figure that

$$OW' = OP' - W'P' = OP' - PQ,$$

from which it follows that

$$\cos (x + \xi) = \cos x \cos \xi - \sin x \sin \xi. \tag{19}$$

We have proved this result for a very special case but it can easily be proved quite generally.* If we replace ξ by $-\xi$ and make use of equations (5) we find that

* For complete proofs see T. M. MacRobert and W. Arthur, "Trigonometry" (Methuen & Co., London, 1937) Part I, Chapter VI and Section 8 of Chapter IX.

$$\cos(x - \xi) = \cos x \cos \xi + \sin x \sin \xi. \tag{20}$$

Furthermore if in equation (19) we replace ξ by $\frac{1}{2}\pi + \xi$ we find that

$$\cos(\tfrac{1}{2}\pi + x + \xi) = \cos x \cos(\tfrac{1}{2}\pi + \xi) - \sin x \sin(\tfrac{1}{2}\pi + \xi)$$

and making use of the results (9) we see that this reduces to the formula

$$\sin(x + \xi) = \sin x \cos \xi + \cos x \sin \xi. \tag{21}$$

Similarly we can show that

$$\sin(x - \xi) = \sin x \cos \xi - \cos x \sin \xi. \tag{22}$$

From equations (4), (19) and (21) we have that

$$\tan(x + \xi) = \frac{\sin(x + \xi)}{\cos(x + \xi)}$$

$$= \frac{\sin x \cos \xi + \cos x \sin \xi}{\cos x \cos \xi - \sin x \sin \xi} = \frac{\dfrac{\sin \xi}{\cos \xi} + \dfrac{\sin x}{\cos x}}{1 - \dfrac{\sin x}{\cos x}\,\dfrac{\sin \xi}{\cos \xi}}$$

showing that

$$\tan(x + \xi) = \frac{\tan x + \tan \xi}{1 - \tan x \tan \xi}. \tag{23}$$

Replacing ξ by $-\xi$ and making use of (9) we obtain the result

$$\tan(x - \xi) = \frac{\tan x - \tan \xi}{1 + \tan x \tan \xi}. \tag{24}$$

If we put $\xi = x$ in equations (19), (21) and (23) we obtain the results

$$\left.\begin{aligned} \cos 2x &= \cos^2 x - \sin^2 x, \\ \sin 2x &= 2 \sin x \cos x, \\ \tan 2x &= \frac{2 \tan x}{1 - \tan^2 x}. \end{aligned}\right\} \tag{25}$$

EXAMPLE 6: *Show that*

$$a \cos \theta + b \sin \theta = R \cos(\theta - \phi)$$

where

$$R = \sqrt{(a^2 + b^2)} \text{ and } \tan \phi = b/a.$$

We can write

$$a \cos \theta + b \sin \theta = \sqrt{(a^2 + b^2)} \left\{ \cos \theta \frac{a}{\sqrt{(a^2 + b^2)}} + \sin \theta \frac{b}{\sqrt{(a^2 + b^2)}} \right\}. \quad (26)$$

Now if we have a right-angled triangle OMP in which the angle OMP is the right angle and $OM = a$, $MP = b$ it follows from Pythagoras' theorem that $OP = \sqrt{(a^2 + b^2)}$. If we denote the angle MOP by ϕ it follows from the definitions (3) that

$$\cos \phi = \frac{a}{\sqrt{(a^2 + b^2)}}, \quad \sin \phi = \frac{b}{\sqrt{(a^2 + b^2)}}, \quad \tan \phi = \frac{b}{a}. \quad (27)$$

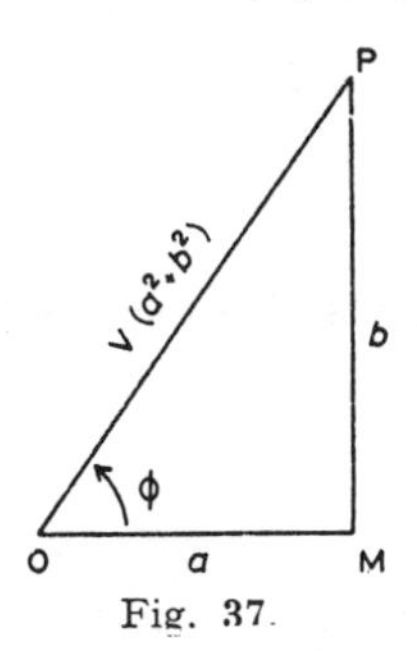

Fig. 37.

If we put $R = \sqrt{(a^2 + b^2)}$ it follows from equations (26) and (27) that

$$a \cos \theta + b \sin \theta = R(\cos \theta \cos \phi + \sin \theta \sin \phi)$$

and hence, from equation (20), that

$$a \cos \theta + b \sin \theta = R \cos (\theta - \phi). \quad (28)$$

PROBLEMS

1. Show that

$$\cos x \cos \xi = \tfrac{1}{2}\{\cos (x - \xi) + \cos (x + \xi)\},$$
$$\sin x \sin \xi = \tfrac{1}{2}\{\cos (x - \xi) - \cos (x + \xi)\},$$
$$\sin x \cos \xi = \tfrac{1}{2}\{\sin (x + \xi) + \sin (x - \xi)\}.$$

Deduce that

$$\cos^2 x = \tfrac{1}{2}(1 + \cos 2x), \quad \sin^2 x = \tfrac{1}{2}(1 - \cos 2x).$$

2. Prove that

(i) $\cos 3x = 4 \cos^3 x - 3 \cos x$,

(ii) $\sin 3x = 3 \sin x - 4 \sin^3 x$,

and hence that

(iii) $\cos^3 x = \frac{1}{4}(\cos 3x + 3 \cos x),$

(iv) $\sin^3 x = \frac{1}{4}(3 \sin x - \sin 3x).$

Deduce from (iii) that

$$\cos^6 x = \tfrac{1}{32}(10 + 15 \cos 2x + 6 \cos 4x + \cos 6x).$$

3. Prove that

$$1 - \tan^4 \tfrac{1}{2}x = \frac{8 \cos x}{\cos 2x + 4 \cos x + 3}.$$

4. Prove that

$$\cos^4(x + \tfrac{1}{4}\pi) + \cos^4(x - \tfrac{1}{4}\pi) = \tfrac{1}{4}(3 - \cos 4x).$$

9. Inverse Trigonometrical Functions

If we write $x = \sin y$, then to any given value of x which is numerically less than, or equal to, unity, there is an infinite number of real values of y. This is obvious from the second of equations (5) of the last section, or from Fig. 35. We say that $y = \text{Sin}^{-1} x$, i.e. we use the symbol* $\text{Sin}^{-1} x$ to denote *any* angle whose sine is x. Similarly $\text{Cos}^{-1} x$ means *any* angle whose cosine is x, and $\text{Tan}^{-1} x$ means *any* angle whose tangent is x. For example any one of the angles $\frac{1}{6}\pi, \frac{5}{6}\pi, \frac{13}{6}\pi, \ldots$ can be taken as a value of $\text{Sin}^{-1} \frac{1}{2}$, and any one of the angles $\frac{1}{4}\pi, \frac{5}{4}\pi, \frac{9}{4}\pi, \ldots$ can be taken as a value of $\text{Tan}^{-1} 1$.

When a function $f(x)$, such as $\text{Sin}^{-1} x$, has more than one real value for a given value of x, the numerically least value is called the *principal value* of $f(x)$ corresponding to the given value of x; if there are two numerically equal least values, the positive one is taken as the principal value. For example, if $y = \text{Sin}^{-1} \frac{1}{2}$ then $y = -7\pi/6, \pi/6, 5\pi/6, \ldots$ so that the principal value is $\pi/6$. From now onwards we shall use the symbols $\sin^{-1} x$, $\cos^{-1} x$, $\tan^{-1} x$ to denote the approximate principal values. We shall sometimes also follow the continental practice of writing $\sin^{-1} x = \text{arc} \sin x$, $\cos^{-1} x = \text{arc} \cos x$, $\tan^{-1} x = \text{arc} \tan x$. If, however, we wish, in some particular context, to emphasize that the general value rather than the principal value is intended, we shall retain the use of capital letters and write $\text{Sin}^{-1} x$,

* The symbol $\text{Sin}^{-1} x$ is arrived at by treating "Sin" as though it were a number in the equation $x = \text{Sin}\, y$ and "solving" for y.

$\text{Cos}^{-1} x$, $\text{Tan}^{-1} x$.

If $x > 0$ the numerically least value of $\text{Sin}^{-1} x$ lies in the range $(0, \frac{1}{2}\pi)$ but if $x < 0$ the numerically least value lies in the range $(-\frac{1}{2}\pi, 0)$. Also if $x = 0$, the numerically least value of $\text{Sin}^{-1} x$ is zero. Hence we have the relation

$$-\tfrac{1}{2}\pi \leqq \sin^{-1} x \leqq \tfrac{1}{2}\pi \qquad (-1 \leqq x \leqq 1). \tag{1}$$

Similarly, if $x > 0$ the numerically least value of $\text{Cos}^{-1} x$ lies in $(0, \frac{1}{2}\pi)$ while if $x < 0$ it lies in the range $(\frac{1}{2}\pi, \pi)$. Also when x is 0 it has the value $\frac{1}{2}\pi$. Hence we have

$$0 \leqq \cos^{-1} x \leqq \pi \qquad (-1 \leqq x \leqq 1). \tag{2}$$

In the same way we can show that

$$-\tfrac{1}{2}\pi \leqq \tan^{-1} x \leqq \tfrac{1}{2}\pi. \tag{3}$$

The graphs of $\sin^{-1} x$, $\cos^{-1} x$, $\tan^{-1} x$ are readily deduced from

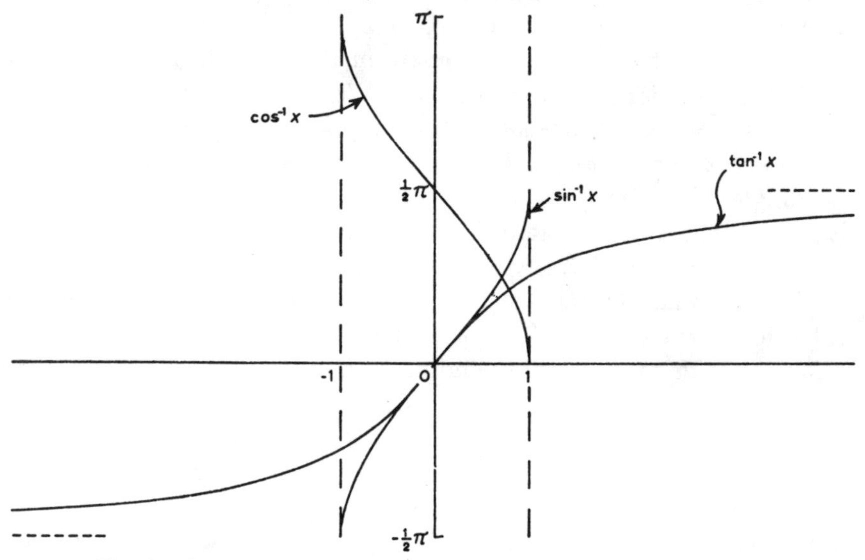

Fig. 38. Principal values of the inverse trigonometrical functions.

those of the trigonometrical functions (Fig. 35, p. 104). They are shown in Fig. 38.

Example 7: *Evaluate* $\cos(\sin^{-1} x)$.

If we let $\theta = \sin^{-1} x$ then $x = \sin\theta$ and

$$\cos(\sin^{-1} x) = \cos\theta = \pm\sqrt{(1 - \sin^2\theta)} = \pm\sqrt{(1 - x^2)}$$

so that

$$\cos(\sin^{-1} x) = \pm\sqrt{(1 - x^2)}.$$

PROBLEMS

1. If $2\sin^{-1} x = \sin^{-1} y$, express y in terms of x.

2. Show that

$$\tan^{-1} x + \tan^{-1} a = \tan^{-1}\left(\frac{x + a}{1 - ax}\right).$$

3. Prove that

(i) $\sin^{-1} x + \cos^{-1} x = \frac{1}{2}\pi.$

(ii) $\tan^{-1} x + \cot^{-1} x = \begin{cases} \frac{1}{2}\pi & \text{if } x \geqq 0 \\ -\frac{1}{2}\pi & \text{if } x < 0. \end{cases}$

4. Prove that $2\tan^{-1}\frac{1}{3} + \tan^{-1}\frac{1}{7} = \frac{1}{4}\pi.$

CHAPTER 3

LIMITS AND DERIVATIVES

In many biological and physical problems the processes we are investigating are dynamic (i.e. they depend essentially on a time variable) and we are interested in rates of change. For this reason, we now consider such processes and the ideas of 'limit' and 'derivative' to which such discussions give rise

1. Rates of Change

We begin by looking at a few typical elementary examples of situations in which we are interested in a rate of change. The first of these concerns the increase in weight of an organism with age (or, what is the same thing, the increase in the size of a population with time). Then we consider the velocity of a body moving in a straight line, and, finally, the gradient of the tangent to a curve. This last case may not, at first sight, appear "dynamic" but we can think of a curve as being traced out by a moving point.

(*a*) *Rate of growth*

We shall begin by considering a simple growth process. We shall suppose that, during the period under consideration the weight or size of a body, denoted by W obtains a law of the form

$$W = f(t) \tag{1}$$

where t denotes the time. We shall consider changes in W from a time t_1. The weight W at that time will be $f(t_1)$ and at a time $t_1 + h$ slightly later it will be $f(t_1 + h)$. Hence, during the time interval h, the change in weight will be $f(t_1 + h) - f(t_1)$ and the average rate of change of weight during that interval will be

$$r_h = \frac{f(t_1 + h) - f(t_1)}{h}. \tag{2}$$

To show how this quantity is calculated we have

EXAMPLE 1: *During a certain time the weight of a body is governed by the parabolic law*

$$W = W_1(t/t_1)^2$$

where W_1 and t_1 are constants. Find the average rate of change of weight in the neighbourhood of the time t_1.

When $t = t_1$, $W = W_1$ and in the neighbourhood of $t = t_1$ we have the following table of values:

h	$0.1t_1$	$0.2t_1$	$0.4t_1$	$0.6t_1$	$0.8t_1$	$1.0t_1$
$t = t_1 + h$	$1.1t_1$	$1.2t_1$	$1.4t_1$	$1.6t_1$	$1.8t_1$	$2.0t_1$
$W = f(t_1+h)$	$1.21W_1$	$1.44W_1$	$1.96W_1$	$2.56W_1$	$3.24W_1$	$4.00W_1$
r_h	$2.1\dfrac{W_1}{t_1}$	$2.2\dfrac{W_1}{t_1}$	$2.4\dfrac{W_1}{t_1}$	$2.6\dfrac{W_1}{t_1}$	$2.8\dfrac{W_1}{t_1}$	$3.0\dfrac{W_1}{t_1}$

in which the last row is calculated from the first and third by means of equation (2). From this table of values we see that the shorter the time interval h becomes the nearer does the average rate of change of weight r_h in the interval tend to the value $2W_1/t_1$, and this value might therefore be interpreted as the rate of change of W at the instant $t = t_1$. We can see this analytically from equation (2). If we write

$$f(t_1 + h) = W_1\left(\frac{t_1 + h}{t_1}\right)^2 = W_1\left(1 + \frac{h}{t_1}\right)^2 = W_1\left(1 + \frac{2h}{t_1} + \frac{h^2}{t_1^2}\right)$$

then it follows from equation (2) that

$$r_h = \frac{W_1}{h}\left\{1 + \frac{2h}{t_1} + \frac{h^2}{t_1^2} - 1\right\}.$$

If $h \neq 0$ we can divide the expression in the brace by h to obtain the result

$$r_h = \frac{2W_1}{t_1}\left\{1 + \frac{1}{2}\frac{h}{t_1}\right\}. \tag{3}$$

Now if h is very small in comparison with t_1 (but not actually zero, which would forbid us dividing by h) then $h/2t_1$ is very small in comparison with 1 and the average rate of change of weight r_h calculated over this very small interval is indistinguishable from $2W_1/t_1$. We express this fact by saying that "the limit of r_h as h tends to zero is $2W_1/t_1$" or that "r_h tends to $2W_1/t_1$ as h tends to zero", it being understood that h may become as small as we please but it must always remain different from zero, since if h were zero the ratio (2) would be of the form 0/0 which is not defined. The two sentences in quotation marks may be written symbolically as

$$\lim_{h \to 0} r_h = \frac{2W_1}{t_1} \tag{4}$$

and

$$r_h \to 2W_1/t_1 \text{ as } h \to 0$$

respectively.

The fact expressed by equations (4) and (2) can be stated differently thus: by taking h sufficiently small we can (for this particular form of $f(t)$) make the difference between $\{f(t_1 + h) - f(t_1)\}/h$ and $2W_1/t_1$ as small as we please.

Although we have considered the growth rate at a particular time $t = t_1$ for the particular law of growth $W = W_1(t/t_1)^2$, it is obvious that the process can be extended to deal with the general case of finding the rate of growth at an instant t when the law of growth is $W = f(t)$. The rate of growth is then

$$\lim_{h \to 0} \frac{f(t + h) - f(t)}{h}. \tag{5}$$

(b) Velocity of a point

As a second example we consider the motion of a body along a straight line. If the position, P, of the body is determined with reference to a fixed origin O the distance OP may be denoted by x. Since the position of the point P will vary with the time t, it follows that the distance x will be a function of t and we may write

$$x = f(t).$$

For instance at a time $t = t_1$ the distance of the body from O will be $x_1 = f(t_1)$. At a later instant $t_1 + h$ this distance will be $x_2 = f(t_1 + h)$. Hence in the time interval h the body will have gone a distance $x_2 - x_1 = f(t_1 + h) - f(t_1)$ and hence during this time its average velocity will be

$$v_h = \frac{f(t_1 + h) - f(t_1)}{h}. \tag{6}$$

EXAMPLE 2: *A body is moving along a straight line in such a way that its distance x from a given fixed point O is given for various* *times t by the table of values*

t(sec)	0.0	0.1	0.2	0.3	0.4	0.5	0.6	0.7	0.8
x(cm)	0.000	0.201	0.408	0.627	0.864	1.125	1.416	1.743	2.112

Find the average velocity over intervals near the beginning of the motion.

Since $x = 0$ when $t = 0$ it follows that if x_h is the value of x at a time h later the average value of the velocity during the interval h will be

$$v_h = \frac{x_h}{h}.$$

Substituting the observed values of x from the above table and performing the simple divisions we obtain the table

h(sec)	0.1	0.2	0.3	0.4	0.5	0.6	0.7	0.8
v_h(cm/sec)	2.01	2.04	2.09	2.16	2.25	2.36	2.49	2.64

from which it will be observed that the smaller the interval h becomes the more nearly does v_h have the value 2, so that we might conjecture that if we could measure a sufficiently small time interval h the value of v_h would be 2.00. In the notation of (a) above it would seem that $v_h \to 2$ as $h \to 0$ or, with the notation of equation (6)

$$\lim_{h \to 0} \frac{f(h) - f(0)}{h}$$

would seem to be 2.

It is obvious that, mathematically, the general situation here would seem to be identical with that of subsection (a) above: if the position of a particle at time t is given by the law $x = f(t)$ its average velocity between times t_1 and $t_1 + h$ is given by equation (6) and its velocity at the instant t_1 is given by

$$\lim_{h \to 0} \frac{f(t_1 + h) - f(t_1)}{h}.$$

To illustrate the general procedure we have:

EXAMPLE 3: *If a body, starting with velocity u at a point 0, moves with constant acceleration f in a straight line, its distance x from 0 at time t is given by the formula $x = ut + \frac{1}{2}ft^2$. Find the average velocity of the body during the interval from t_1 to $t_1 + h$ and deduce the "instantaneous velocity" at the instant t_1.*

In this case

$$f(t_1) = ut_1 + \tfrac{1}{2}ft_1^2$$

so that

$$f(t_1 + h) = u(t_1 + h) + \tfrac{1}{2}f(t_1 + h)^2 = ut_1 + uh + \tfrac{1}{2}ft_1^2 + fht_1 + \tfrac{1}{2}fh^2$$

and

$$\frac{f(t_1 + h) - f(t_1)}{h} = \frac{uh + ft_1h + \tfrac{1}{2}fh^2}{h}.$$

If h is not actually zero we may divide the numerator and denominator of the fraction by h to obtain the formula

$$v_h = u + ft_1 + \tfrac{1}{2}fh$$

for the average velocity of the body over the time interval from t_1 to $t_1 + h$. As the time difference h gets smaller and smaller this quantity will become more and more nearly equal to $u + ft_1$ and this is the value of the velocity of the body at time t_1. Another way of saying the same thing is to say that the difference between v_h and $u + ft_1$ can be made as small as we please by choosing h to be sufficiently small. We shall return to this point later (in Section 4 below).

(c) Tangent to a curve

We saw previously (in Section 6 of Chapter 2) how to calculate the gradient of a line joining two points on a parabola and how to deduce the gradient of the tangent to the parabola at one of the points. We shall now generalize that procedure to derive the gradient of the line joining two points on the curve whose equation is

$$y = f(x) \tag{8}$$

and the gradient of the tangent to the curve at a point on it.

If $P(x_1, y_1)$ and $Q(x_2, y_2)$ are two points on the curve with equation (8) then $y_1 = f(x_1)$ and $y_2 = f(x_2)$, since this is what we mean by saying that P and Q lie on the curve. The average gradient of the curve between these points P and Q may be defined to be the gradient of the line * PQ and this will, of course, vary with the positions of the points P and Q. The average gradient between the points P and Q of the curve with equation (8) is found, (by example 1 of Chapter 2 p. 68 above), to be

* Such a line is often called a *chord* of the curve.

$$m_{PQ} = \frac{y_2 - y_1}{x_2 - x_1} = \frac{f(x_2) - f(x_1)}{x_2 - x_1}. \tag{9}$$

We might write this result in another way. If we suppose that the abscissa of Q differs from that of P by h so that $x_2 = x_1 + h$, then equation (9) takes the form

$$m_{PQ} = \frac{f(x_1 + h) - f(x_1)}{h}. \tag{10}$$

To illustrate the procedure we consider:

EXAMPLE 4: *Consider the average gradient of the curve with equation*

$$y = x^3 - x^2 + 1$$

in the neighbourhood of the point P with coordinates (1, 1).

Here $x_1 = 1$ and $f(1) = 1 - 1 + 1 = 1$. Also we have the following table of values:

h	0.6	0.4	0.2	0.1	0.02	0.01	−0.01	−0.02	−0.1
x_1+h	1.6	1.4	1.2	1.1	1.02	1.01	0.99	0.98	0.9
$f(x_1+h)$	2.536	1.784	1.288	1.121	1.0208	1.0102	0.9902	0.9808	0.919
m_{PQ}	2.56	1.96	1.44	1.21	1.04	1.02	0.98	0.96	0.81

where, in accordance with equation (10), the last row is calculated by means of the equation

$$m_{PQ} = \frac{f(1 + h) - f(1)}{h}. \tag{11}$$

From the above table we see that the nearer the point Q tends to the point P (i.e. the nearer that h gets to zero) from either side, the nearer the average gradient of the curve between P and Q approaches the value 1. This is expressed by saying that as the point Q tends to the point P the gradient of PQ tends to the value 1; the chord PQ tends to a limiting position, a line of gradient 1 passing through P. We note that this limiting position of the line PQ is not itself a chord of the curve.

If we substitute the form for $f(1 + h)$ in equation (11) we see that provided h is not zero *

* The value $h = 0$ is excluded since in that case the right hand side of equation (4) would have the meaningless form 0/0.

$$m_{PQ} = \frac{(1+h)^3 - (1+h)^2 + 1 - 1}{h}$$

$$= \frac{1 + 3h + 3h^2 + h^3 - 1 - 2h - h^2 + 1 - 1}{h} = 1 + 2h + h^2$$

showing that, when h is very small, m_{PQ} is practically 1. We then say that "in the limit as h tends to zero, m_{PQ} tends to 1" and we write

$$\lim_{h \to 0} m_{PQ} = 1.$$

Although a particular curve and the particular point with abscissa 1 has been used, it is obvious that the procedure can be extended to any point which lies on the curve with equation (8). The average gradient of the curve between points P and Q with abscissae x_1 and $x_1 + h$ respectively is given by equation (10). The tangent to the curve at the point P is defined to be the limiting position of the chord PQ as the point Q approaches the point P along the curve, and we further define the gradient of this tangent to be the gradient of the curve at P. Thus the gradient of the curve at P is the value to which m_{PQ} tends as Q tends to P, or, in symbols, it is

$$\lim_{Q \to P} m_{PQ}$$

which by equation (10) is the same thing as saying that it is

$$\lim_{h \to 0} \frac{f(x_1 + h) - f(x_1)}{h}. \tag{12}$$

2. The Formal Definition of a Derivative

We have seen that in several situations, involving the discussion of rates of change, we are concerned with the evaluation of limits of the kind (12) of the last section. If this limiting process is carried out for each value of x for which the function $f(x)$ is defined, the quantity

$$\lim_{h \to 0} \frac{f(x + h) - f(x)}{h} \tag{1}$$

will be a function of x, which we shall call the *gradient function* of $f(x)$ and which we shall denote by $f'(x)$.

We can interpret in another way the limiting procedure which was set up in the last section for the solution of the geometrical problem of determining the gradient of a curve $y = f(x)$ at a point with abscissa x_1 on it. If x is changed from x_1 to some other value then the value of y will change also; if x is changed by a small amount δx to $x_1 + \delta x$, we may suppose that the corresponding increase in y is δy. The changes δx, δy, are called *increments* in x, y respectively. (It should be observed that δx does not mean δ multiplied by x but should be regarded as a single symbol meaning the increment in x). The ratio

$$\frac{\delta y}{\delta x}$$

is the average rate of change of y per unit change of x during a change in x of amount δx. We then *define* the instantaneous rate of change of y with respect to x, for the value $x = x_1$, to be

$$\lim_{\delta x \to 0} \frac{\delta y}{\delta x}.$$

This may be interpreted as the increase in the value of y which would result from a unit increase in x if y were increasing at this constant rate throughout the change in x. As in the case of finding the gradient function, the method used does not depend upon the particular choice of the value x_1 of x. We can therefore refer to the rate of change of y for any x and it is defined to be

$$\frac{\mathrm{d}y}{\mathrm{d}x} = \lim_{\delta x \to 0} \frac{\delta y}{\delta x}. \tag{2}$$

The quantity $\mathrm{d}y/\mathrm{d}x$ (which should be looked upon as a single symbol, not a quotient) is called the *derivative* or *differential coefficient* of y with respect to x. The process of finding the derivative is called *differentiation with respect to x*.

If $y = f(x)$ we see, by comparing (1) and (2), that

$$\frac{\mathrm{d}y}{\mathrm{d}x} = f'(x),$$

i.e. the derivative introduced in this way is identical with the gradient function defined previously.

Several equivalent notations are used to denote the same function:-

$$f'(x), \quad \frac{\mathrm{d}y}{\mathrm{d}x}, \quad Dy, \quad D_x y, \quad y'.$$

The symbols

$$\left(\frac{\mathrm{d}y}{\mathrm{d}x}\right)_{x=a}, \quad \left(\frac{\mathrm{d}y}{\mathrm{d}x}\right)_a, \quad f'(a)$$

denote that the derivative has to be evaluated at the particular value a of the variable x.

A simple geometrical interpretation of the derivative is illustrated in

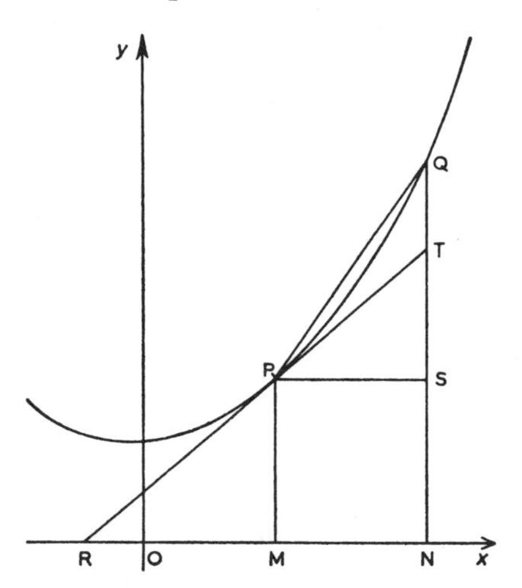

Fig. 39. Geometrical representation of a derivative.

Fig. 39. If P and Q are the points on the curve $y = f(x)$ with abscissae x and $x + \delta x$ respectively, PM and QN are perpendicular to the x-axis and PS is parallel to the x-axis, then $PS = \delta x$ and $SQ = \delta y$, the difference between the ordinates of P and Q. If the tangent to the curve at P meets QS in T, then the gradient of the line PT is $f'(x)$ and (considering the triangle PTS) the gradient of the line PT is ST/PS so that

$$f'(x) = \frac{ST}{PS} = \frac{ST}{\delta x}$$

from which it follows that $ST = f'(x)\delta x$. Hence in the diagram $PS = \delta x$, $SQ = \delta y$, $ST = f'(x)\delta x$.

In growth problems, if $W(t)$ is, say, the weight of an organ at time t then $W'(t) = \mathrm{d}W/\mathrm{d}t$ will be the *growth rate*. In the discussion of many growth problems it is convenient to introduce the quantity

$$\frac{1}{W}\,\frac{\mathrm{d}W}{\mathrm{d}t}$$

which is known as the *specific growth rate.*

Similarly if the distance, from a fixed point, of a body moving along a straight line is $x(t)$ at the instant t, then the velocity of the body at that instant is

$$v(t) = \frac{\mathrm{d}x}{\mathrm{d}t}.$$

3. Limits

In the preceding section we used the idea of a limit in an intuitive way. We shall now make it more precise and discuss the main theorems concerning the evaluation of limits. Before stating the formal definition let us consider a very simple example. Consider the ratio

$$\frac{(1+h)^2-1}{h}.$$

If h is not actually equal to zero we can write this ratio as $2+h$, and it is obvious that the smaller h gets the more nearly will this quantity approach the value 2. What we mean by saying that

$$\lim_{h\to 0}\frac{(1+h)^2-1}{h} = 2$$

is that the numerical value of the difference

$$\frac{(1+h)^2-1}{h} - 2$$

can be made as small as we please by choosing the numerical value of h to be sufficiently small. We now generalize this to define that what we mean by the statement

$$\lim_{h\to 0} F(h) = A \tag{1}$$

is that the numerical difference $|F(h) - A|$ can be made as small as we please by making $|h|$ sufficiently small.

More generally, if it is possible to make $|F(h) - A|$ as small as we please by making $|h - a|$ sufficiently small, we say that

$$\lim_{h\to a} F(h) = A. \tag{2}$$

To illustrate the procedure we consider

EXAMPLE 5: *Prove that* $\lim_{h \to 1} \dfrac{(h+1)^3 - 8}{h-1} = 12.$

Since

$$(h+1)^3 - 8 = h^3 + 3h^2 + 3h + 1 - 8$$

we find (by division) that, if $h \neq 1$,

$$\frac{(h+1)^3 - 8}{h-1} = \frac{h^3 + 3h^2 + 3h - 7}{h-1} = h^2 + 4h + 7$$

so that

$$\frac{(h+1)^3 - 8}{h-1} - 12 = h^2 - 1 + 4(h-1) = (h-1)\{4 + (h+1)\}.$$

If we write $F(h)$ for the function $\{(h+1)^3 - 8\}/(h-1)$ we see that

$$F(h) - 12 = (h-1)\{4 + h + 1\}$$

and if $h \simeq 1$, this reduces to

$$F(h) - 12 = 6(h-1).$$

Hence we can make $|F(h) - 12|$ as small as we please by suitably choosing $|h-1|$. For instance if we wish to make $|F(h) - 12| < 1/100$ we need only choose $|h-1|$ so that

$$|h-1| < \tfrac{1}{600}.$$

Hence, it follows from the definition (2) that

$$\lim_{h \to 1} F(h) = 12$$

as was required.

Another example may be simply constructed from the argument at the end of sub-section (*b*) of Section 1 above.

In our earlier discussion of series (Section 4 of Chapter 1) we encountered a limit of another sort — one in which the independent variable becomes increasingly large. To see the same ideas at work in a slightly different context, let us consider the function

$$f(x) = 5 + \frac{1}{x^3}.$$

For the successive values 1, 2, 5, 10, 100, 1000, . . . of the independent variable x, the function $f(x)$ takes the values 6.0, 5.125, 5.008, 5.001, 5.000 001, 5.000 000 001 ... and it is obvious that as x gets larger and larger, $f(x)$ approaches more closely to the value 5. Furthermore by choosing x to be sufficiently large we can make the quantity $|f(x) - 5|$ as small as we please. For instance we can make

$$|f(x) - 5| < 10^{-3n}$$

(for any given n) merely by taking $x > 10^n$. We then say that $f(x)$ tends to 5 as x tends to "infinity" or in symbols: $f(x) \to 5$ as $x \to \infty$, or

$$\lim_{x\to\infty} f(x) = 5.$$

More generally, if by making x sufficiently large we can make $|f(x)-A|$ as small as we please we say that

$$\lim_{x\to\infty} f(x) = A, \tag{3}$$

and obviously the simplest such result is

$$\lim_{x\to\infty} x^{-n} = 0, \tag{4}$$

for all positive real values of n.

Another result which we shall need (and which we shall assume) is that if $|x| < 1$,

$$\lim_{n\to\infty} x^n = 0. \tag{5}$$

It follows from the reasoning of Section 4, Chapter 1, that if

$$S_n = 1 + r + r^2 + \ldots + r^n, \quad |r| < 1$$

then

$$\lim_{n\to\infty} S_n = \frac{1}{1-r}.$$

In general we say that a series is *convergent* if the limit as $n \to \infty$ of the sum S_n of n terms is finite.

EXAMPLE 6: *Prove that* $\lim\limits_{x\to\infty} \dfrac{ax+b}{cx+d} = \dfrac{a}{c}$, $(c \neq 0)$.

If we divide the numerator and denominator of this expression by x we find that

$$f(x) = \frac{ax+b}{cx+d} = \frac{a+\dfrac{b}{x}}{c+\dfrac{d}{x}} = \frac{a\left(1+\dfrac{b}{ax}\right)}{c\left(1+\dfrac{d}{cx}\right)}.$$

Now if x is very large $|d/cx|$ will be very small if $c \neq 0$ and we can replace $(1 + d/cx)^{-1}$ by the first term in its binomial expansion. Retaining only terms of the first order in $1/x$ we find that

$$\begin{aligned} f(x) &= \frac{a}{c}\left(1+\frac{b}{ax}\right)\left(1+\frac{d}{cx}\right)^{-1} \\ &= \frac{a}{c}\left\{1+\left(\frac{b}{a}-\frac{d}{c}\right)\frac{1}{x}\right\} \end{aligned}$$

and this tends to a/c as $x \to \infty$.

Another notation we use is: if $f(x)$ can be made as large as we please for all x sufficiently close to a we say that $f(x) \to +\infty$ as $x \to a$ if the sign of $f(x)$ is positive and that $f(x) \to -\infty$ if the sign is negative. For example, $x^{-2} \to \infty$ as $x \to 0$.

Sometimes the value of a limit as $x \to a$ depends on whether $x \to a$ through values of x greater than a, or through values less than a. For example if $x \to a$ through values of $x > a$ then $(x - a)^{-1} \to +\infty$. We denote this by

$$\lim_{x\to a+} (x-a)^{-1} = +\infty.$$

A limit of this kind is known as a *limit from the right.* Similarly if $x \to a$ through values of $x < a$ then $(x - a)^{-1} \to -\infty$, a result which we write in the form

$$\lim_{x\to a-} (x-a)^{-1} = -\infty.$$

This kind of limit is known as a *limit from the left.* It should be noticed, in passing, that the right and left hand limits

$$\lim_{x\to a+} f(x), \quad \lim_{x\to a-} f(x)$$

need not coincide, but it is only when they do that we can say that

$$\lim_{x\to a} f(x)$$

exists. For instance, $\lim_{x\to 0+} x^{-1} = +\infty$ and $\lim_{x\to 0-} x^{-1} = -\infty$ so that $\lim_{x\to 0} x^{-1}$ does not exist.

There are a number of theorems on limits which are of great importance.

THEOREM 1: *If*

$$\lim_{x\to a} f(x) = A, \qquad \lim_{x\to a} g(x) = B,$$

then

$$\lim_{x\to a} [f(x) + g(x)] = A + B.$$

To prove this theorem we go back to our formal definition of a limit. Since $\lim_{x\to a} f(x) = A$, it follows that by taking $|x - a|$ sufficiently small we can make $|f(x) - A|$ as small as we please. Now we saw in Section 1 of Chapter 1 that

$$\begin{aligned} |f(x) + g(x) - (A + B)| &= |\{f(x) - A\} + \{g(x) - B\}| \\ &\leqq |f(x) - A| + |g(x) - B| \end{aligned}$$

from which it follows that by taking $|x - a|$ to be sufficiently small we can make the difference $|f(x) + g(x) - (A + B)|$ as small as we please. In other words

$$\lim_{x\to a} [f(x) + g(x)] = A + B$$

and the theorem is proved.

By proofs similar to this we can establish:

THEOREM 2: *If* $\lim_{x\to a} f(x) = A$ *and* c *is any constant then*
$$\lim_{x\to a} cf(x) = cA.$$

THEOREM 3: *If* $\lim_{x\to a} f(x) = A \neq 0$ *then* $\lim_{x\to a} 1/f(x) = 1/A$.

THEOREM 4: *If* $\lim_{x\to a} f(x) = A$, $\lim_{x\to a} g(x) = B$, *then*
$$\lim_{x\to a} f(x)g(x) = AB.$$

From these last two results we can deduce

THEOREM 5: *If* $\lim_{x\to a} f(x) = A$, *then* $\lim_{x\to a} [f(x)]^m = A^m$, *provided that if* $m < 0$, $A \neq 0$

and

THEOREM 6: *If* $\lim_{x\to a} f(x) = A$, $\lim_{x\to a} g(x) = B \neq 0$, *then*

$$\lim_{x\to a} \left\{\frac{f(x)}{g(x)}\right\} = \frac{A}{B}.$$

In these theorems a can be any real number (including zero) or it can be ∞ provided that $\lim_{x\to\infty}$ is interpreted in the sense described above.

Another device which is useful in the evaluation of limits is to change the variable. This is often used because limits in which the variable tends to zero are often easier to evaluate than those in which it tends to a non-zero value. If we write $x = a + h$ then it is obvious that as $x \to a$, $h \to 0$ so that we obtain the result

$$\lim_{x\to a} f(x) = \lim_{h\to 0} f(a+h). \tag{6}$$

To illustrate the use of this result we have

EXAMPLE 7: *Evaluate* $\lim_{x\to 2} \dfrac{x^2 - x - 2}{x\,(x-2)}$.

From equation (6) we have that

$$\begin{aligned}\lim_{x\to 2} \frac{x^2 - x - 2}{x(x-2)} &= \lim_{h\to 0} \frac{(2+h)^2 - (2+h) - 2}{(2+h)h} \\ &= \lim_{h\to 0} \frac{4 + 4h + h^2 - 2 - h - 2}{h(2+h)} \\ &= \lim_{h\to 0} \frac{h^2 + 3h}{h(h+2)} \\ &= \lim_{h\to 0} \frac{h+3}{h+2} \\ &= \tfrac{3}{2}.\end{aligned}$$

PROBLEMS

1. Evaluate the limit

$$\lim_{x\to 0} \frac{\sqrt{(1+x)} - 1}{x}.$$

2. Evaluate the limits

$$\lim_{x\to 1} \frac{2x^2 - x - 1}{x(x-1)}, \qquad \lim_{x\to -1} \frac{(x-1)^2 - 4}{x+1}.$$

3. Evaluate the limit

$$\lim_{x\to\infty} \frac{ax^2 + bx + c}{px^2 + qx + r}$$

where a, b, c, p, q and r are constants.

4. Some Simple Limits

We shall now consider certain simple limits involving algebraic and trigonometrical functions.

The basic one required in the differentiation of polynomials is easily derived from a result we have already established in Section 6 of Chapter 1. We saw, in example 10 there, that if n is a rational number and h is small (so that we may use the binomial theorem),

$$\frac{(x+h)^n - x^n}{h} = nx^{n-1} + hS$$

where S is a series of powers of h. As $h \to 0$ it is obvious that $hS \to 0$ so that we obtain the limit theorem

$$\lim_{h\to 0} \frac{(x+h)^n - x^n}{h} = nx^{n-1}. \tag{1}$$

The most important limit theorem relating to the trigonometrical functions is that

$$\lim_{x\to 0} \frac{\sin x}{x} = 1. \tag{2}$$

It will be noted that, when $x = 0$, $\sin x/x$ is the meaningless quantity $0/0$; but we are concerned with the behaviour of the function, not when $x = 0$ but when x *tends to* 0.

To prove the result (2) we consider a small angle AOB of circular measure $2x$ (cf. Fig. 40). With centre O and unit radius we draw a circular arc cutting the arms of the angle in P and Q and from P and Q we draw tangents PT and QT to the circle. Then by the symmetry of the figure, OT will bisect $\angle\, POQ$ and PQ will be at right angles to OT. The arc PMQ lies inside the triangle PQT and is everywhere convex to PT and TQ. Hence the arc PMQ will be less than $QT + TP = 2QT$ and greater than $PQ = 2NQ$. Hence

$$NQ < \text{arc } MQ < TQ.$$

Now in the triangle OQT, $OQ = 1$, $\angle\, OQT$ is a right angle and $\angle\, TOQ = x$, so that $NQ = \sin x$, arc $MQ = x$ and $TQ = \tan x$. Hence

$$\sin x < x < \tan x. \tag{3}$$

Therefore

$$1 < \frac{x}{\sin x} < \frac{1}{\cos x}.$$

Now as $x \to 0$, $\cos x \to 1$ so that $x/\sin x \to 1$ and the result (2) follows immediately. We have proved the result for the case in which $x \to 0$

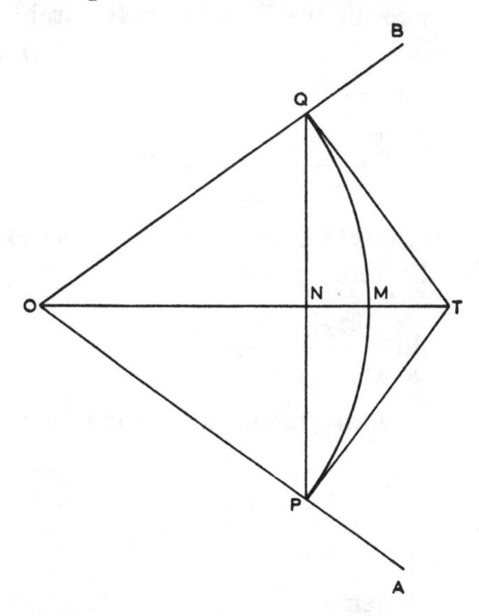

Fig. 40. Evaluation of $\lim_{x\to 0} \sin x/x$.

through positive values but it can readily be shown that the result is also true when $x \to 0$ through negative values.

When x is positive we can also write the inequality (3) in the form

$$\cos x < \frac{x}{\tan x} < 1$$

so that

$$\lim_{x\to 0} \frac{\tan x}{x} = 1. \tag{4}$$

A useful form of the limit theorems (2) and (4) is to state that, for very small values of x,

$$\sin x \simeq x, \quad \tan x \simeq x. \tag{5}$$

EXAMPLE 8: *Evaluate the limit* $\lim_{x\to 0} \dfrac{1 - \cos x}{x^2}$.

By equation (25) of Section 6 of Chapter 2,

$$1 - \cos x = 2 \sin^2 \tfrac{1}{2}x$$

so that

$$\frac{1 - \cos x}{x^2} = \frac{2 \sin^2 \frac{1}{2}x}{x^2} = \frac{1}{2}\left(\frac{\sin \frac{1}{2}x}{\frac{1}{2}x}\right)^2.$$

Now as $x \to 0$, the ratio $\sin \frac{1}{2}x/\frac{1}{2}x \to 1$ so that

$$\lim_{x\to 0} \frac{1 - \cos x}{x^2} = \frac{1}{2}.$$

PROBLEMS

1. If $f(x)$ is a polynomial of degree n in x:-

$$f(x) = p_n x^n + p_{n-1}x^{n-1} + \ldots + p_1 x + p_0$$

find

$$\lim_{x\to\infty} x^{-n} f(x), \qquad \lim_{x\to\infty} x^{-m} f(x), \qquad (m > n).$$

2. Evaluate $(n > 0)$

$$\lim_{x\to 0} \frac{(\sin x)^n}{x^n}, \qquad \lim_{x\to 0} \frac{(\sin x)^n}{\sin x^n}.$$

5. The Derivatives of Simple Algebraic Functions

In this section we shall find the derivatives of simple algebraic functions. Before doing this we shall, however, prove a quite general theorem:

THEOREM 7: *If* $y(x) = f(x) + g(x)$ *then* $y'(x) = f'(x) + g'(x)$.
By the definition of a derivative

$$\begin{aligned} y'(x) &= \lim_{h\to 0} \frac{y(x+h) - y(x)}{h} \\ &= \lim_{h\to 0} \left\{\frac{f(x+h) + g(x+h) - f(x) - g(x)}{h}\right\} \\ &= \lim_{h\to 0} \left\{\frac{f(x+h) - f(x)}{h} + \frac{g(x+h) - g(x)}{h}\right\}. \end{aligned}$$

Using Theorem 1 we see that the limit on the right hand side can be written as

$$\lim_{h\to 0} \frac{f(x+h)-f(x)}{h} + \lim_{h\to 0} \frac{g(x+h)-g(x)}{h}.$$

From the definition of a derivative these limits are $f'(x)$ and $g'(x)$ respectively. We have thus proved the theorem.

The theorem is readily extended to the sum of n functions: if

$$y(x) = f_1(x) + f_2(x) + \ldots + f_n(x)$$

then

$$y'(x) = f_1'(x) + f_2'(x) + \ldots + f_n'(x). \tag{1}$$

We shall begin by finding the derivative of a constant, c. If $f(x) = c$ for all values of x then $f(x+h) = c$ for all x and h. Fixing x we have

$$f(x+h) - f(x) = 0$$

so that

$$\lim_{h\to 0} \frac{f(x+h)-f(x)}{h} = 0$$

showing that

$$\frac{\mathrm{d}c}{\mathrm{d}x} = 0 \tag{2}$$

if c is a constant.

Next let us consider the simple function

$$f(x) = ax^n$$

where a is a constant. By the definition

$$f'(x) = \lim_{h\to 0} \frac{a(x+h)^n - ax^n}{h}$$
$$= a \lim_{h\to 0} \frac{(x+h)^n - x^n}{h}.$$

This limit was considered in the last section. By equation (1) of the last section we therefore have $f'(x) = nax^{n-1}$, a result which we write in the form

$$\frac{\mathrm{d}}{\mathrm{d}x}(ax^n) = nax^{n-1}. \tag{3}$$

It follows immediately from equations (1), (2) and (3) that

$$\frac{\mathrm{d}}{\mathrm{d}x}(a_0 + a_1x + \ldots + a_nx^n) = a_1 + 2a_2x + \ldots + na_nx^{n-1} \tag{4}$$

in which $a_0, a_1, \ldots, a_n$ are constants.

EXAMPLE 9: *Show that if n is a positive integer*

$$\frac{d}{dx}(ax+b)^n = na(ax+b)^{n-1}. \tag{5}$$

If we write $f(x) = (ax+b)^n$ it follows from the binomial theorem that

$$f(x) = a^n x^n + na^{n-1}x^{n-1}b + \ldots + {}_nC_r a^{n-r}x^{n-r}b^r + \ldots + naxb^{n-1} + b^n$$

and it follows from (4) that

$$f'(x) = na^n x^{n-1} + n(n-1)a^{n-1}x^{n-2}b + \ldots$$
$$+ (n-r)\,{}_nC_r a^{n-r}x^{n-r-1}b^r + \ldots + nab^{n-1}.$$

Now

$$(n-r)\,{}_nC_r = (n-r)\frac{n!}{r!(n-r)!} = \frac{n!}{r!(n-r-1)!} = \frac{n(n-1)!}{r!(n-r-1)!}$$

so that $(n-r)\,{}_nC_r = n\,{}_{n-1}C_r$ and it follows that

$$f'(x) = na\{a^{n-1}x^{n-1} + (n-1)a^{n-2}x^{n-2}b + \ldots$$
$$+ {}_{n-1}C_r a^{n-1-r}x^{n-1-r}b^r + \ldots + b^{n-1}\}.$$

By means of the binomial theorem we can identify the series in the braces on the right hand side as $(ax+b)^{n-1}$. Hence we have proved the result.

It should be observed that, since we have assumed the binomial theorem to be true for any rational n, the result (3) above holds not only for n a positive integer but for n equal to any rational number (positive or negative). For example if $n = \frac{1}{2}$ we get

$$\frac{d}{dx}(x^{\frac{1}{2}}) = \tfrac{1}{2}x^{-\frac{1}{2}}$$

which may be written in the form

$$\frac{d}{dx}(\sqrt{x}) = \frac{1}{2\sqrt{x}}. \tag{6}$$

Similarly if we put $n = -m$ in equation (3) we get the result

$$\frac{d}{dx}(x^{-m}) = -mx^{-m-1}$$

which is equivalent to

$$\frac{\mathrm{d}}{\mathrm{d}x}\left(\frac{1}{x^m}\right) = -\frac{m}{x^{m+1}}. \tag{7}$$

PROBLEMS

1. Find the derivatives of the functions

(i) $3x^4 + 4x^3 - 6x^2 + 7$;

(ii) $\dfrac{5}{2x^2}$;

(iii) $\sqrt{x} + \dfrac{1}{\sqrt{x}}$;

(iv) $\dfrac{x^6 + 3x^3 + x - 3}{x^4}$;

(v) $(2x + 1)^6$.

2. Write down the binomial series for $(1 + x)^n$ where n is any rational number (not a positive integer) and, assuming that the derivative of this series is the derivative of the function, show that

$$\frac{\mathrm{d}}{\mathrm{d}x}(1 + x)^n = n(1 + x)^{n-1}.$$

Hence find the derivatives of the functions

(i) $\sqrt{(1 + x)}$, (ii) $\dfrac{1}{(1 + x)^2}$, (iii) $\sqrt[3]{(1 + x)}$.

3. For an allometric growth law $W = kt^n$ show that the specific growth rate (cf. page 121 above) is n/t.

6. The Derivatives of the Trigonometrical Functions

By definition, the derivative of $\sin x$ is

$$\frac{\mathrm{d}}{\mathrm{d}x}(\sin x) = \lim_{h\to 0} \frac{\sin(x + h) - \sin x}{h}.$$

Now by

$$\sin(x + h) - \sin x = 2\cos(x + \tfrac{1}{2}h)\sin\tfrac{1}{2}h$$

we may write the limit on the right as

$$\lim_{h\to 0} \cos\left(x + \tfrac{1}{2}h\right) \cdot \lim_{h\to 0} \frac{\sin \frac{1}{2}h}{\frac{1}{2}h}.$$

The first of these limits is $\cos x$ and, by equation (2) of Section 4, the second is unity so that

$$\frac{d}{dx}(\sin x) = \cos x. \tag{1}$$

Similarly the derivative of $\cos x$ is

$$\frac{d}{dx}(\cos x) = \lim_{h\to 0} \frac{\cos(x+h) - \cos x}{h}$$

$$= -\lim_{h\to 0} \sin\left(x + \tfrac{1}{2}h\right) \cdot \lim_{h\to 0} \frac{\sin \frac{1}{2}h}{\frac{1}{2}h}$$

showing that

$$\frac{d}{dx}(\cos x) = -\sin x. \tag{2}$$

The derivative of $\tan x$ may be calculated by means of equation (4) of Section 4. We have

$$\frac{d}{dx}(\tan x) = \lim_{h\to 0} \frac{\tan(x+h) - \tan x}{h}$$

$$= \lim_{h\to 0} \frac{1}{h}\left\{\frac{\tan x + \tan h}{1 - \tan x \tan h} - \tan x\right\}$$

$$= (1 + \tan^2 x) \lim_{h\to 0} \frac{\tan h}{h} \cdot \lim_{h\to 0} \frac{1}{1 - \tan x \tan h}.$$

Using equation (4) of Section 4 and equation (17) of Section 6 of Chapter 2 we find that

$$\frac{d}{dx}(\tan x) = \sec^2 x. \tag{3}$$

Using the fact that

$$\frac{\sec(x+h) - \sec x}{h} = \frac{1}{h}\left\{\frac{1}{\cos(x+h)} - \frac{1}{\cos x}\right\}$$

$$= -\frac{1}{\cos(x+h)\cos x}\left\{\frac{\cos(x+h) - \cos x}{h}\right\},$$

if we let h tend to zero and make use of the limit which leads to equation (2), we find that

$$\frac{d}{dx}\sec x = \frac{\sin x}{\cos^2 x} = \sec x \cdot \tan x. \tag{4}$$

Similarly we can show that

$$\frac{d}{dx}\operatorname{cosec} x = -\frac{\cos x}{\sin^2 x} = -\operatorname{cosec} x \cdot \cot x \tag{5}$$

and that

$$\frac{d}{dx}\cot x = -\operatorname{cosec}^2 x. \tag{6}$$

EXAMPLE 10: *Find the derivative of* $\sin(ax + b)$.
From the definition of a derivative

$$\begin{aligned}\frac{d}{dx}\sin(ax+b) &= \lim_{h\to 0}\frac{1}{h}\{\sin(ax+ah+b) - \sin(ax+b)\}\\ &= \lim_{h\to 0}\frac{1}{h}\cdot 2\sin(\tfrac{1}{2}ah)\cos(ax+b+\tfrac{1}{2}ah)\\ &= \lim_{h\to 0}\frac{\sin(\frac{1}{2}ah)}{\frac{1}{2}ah}\cdot a\cdot\cos(ax+b+\tfrac{1}{2}ah)\\ &= a\cos(ax+b).\end{aligned}$$

PROBLEMS

1. Prove the results (5) and (6) above.
2. Show that

$$\frac{d}{dx}\cos(ax+b) = -a\sin(ax+b).$$

3. Show that

$$\frac{d}{dx}(x\sin x) = \sin x + x\cos x;$$

$$\frac{d}{dx}(x\cos x) = \cos x - x\sin x.$$

7. Higher Derivatives

If we are given a function of x, $f(x)$ say, and we differentiate it with respect to x then the derivative

$$\frac{\mathrm{d}}{\mathrm{d}x} f(x) = f'(x) \tag{1}$$

will also be a function of x so that it can, in turn, be differentiated to yield

$$\frac{\mathrm{d}}{\mathrm{d}x} f'(x) = f''(x). \tag{2}$$

The symbol $f''(x)$, denoting the derivative of $f'(x)$ is called the *second derivative* of $f(x)$ with respect to x. If we use the notation y to denote a function of x and $\mathrm{d}y/\mathrm{d}x$ to denote its derivative, then the second derivative will be

$$\frac{\mathrm{d}}{\mathrm{d}x}\left(\frac{\mathrm{d}y}{\mathrm{d}x}\right)$$

and this is denoted by

$$\frac{\mathrm{d}^2 y}{\mathrm{d}x^2}.$$

For example if $x(t)$ is the distance travelled along a straight line in a time t, the velocity at time t will be $x'(t)$ and the acceleration, being $v'(t)$, the rate of change of the velocity, will be equal to $x''(t)$.

Since the second derivative of $f(x)$ is a function of x we can form its derivative; this we denote by $f'''(x)$. We can proceed in this way forming, in general, derivatives of the n-th order, denoted by $f^{(n)}(x)$, the n-th derivative being formed from the $(n-1)$th according to the rule

$$f^{(n)}(x) = \frac{\mathrm{d}}{\mathrm{d}x} f^{(n-1)}(x). \tag{3}$$

The alternative notations for those higher derivatives are shown in Table 8.

For example, if

$$f(x) = x^n$$

then

$$f'(x) = nx^{n-1}$$

and

$$f''(x) = n\frac{\mathrm{d}}{\mathrm{d}x}(x^{n-1}) = n(n-1)x^{n-2}.$$

Further

$$f'''(x) = n(n-1)\frac{d}{dx}(x^{n-2}) = n(n-1)(n-2)x^{n-3}.$$

TABLE 8

Notations for the higher derivatives of a function $y = f(x)$

Function	y	y	y	$f(x)$
1st derivative	Dy	$\frac{dy}{dx}$	y'	$f'(x)$
2nd derivative	D^2y	$\frac{d^2y}{dx^2}$	y''	$f''(x)$
⋮	⋮	⋮	⋮	⋮
n^{th} derivative	D^ny	$\frac{d^ny}{dx^n}$	$y^{(n)}$	$f^{(n)}(x)$

Proceeding in this way we find that

$$f^{(r)}(x) = n(n-1)\cdots(n-r+1)x^{n-r}.$$

If n is a positive integer, then

$$f^{(n)}(x) = n(n-1)\cdots 2\cdot 1 = n!.$$

This function is a constant with respect to x so that if we differentiate again the result will be zero so that

$$f^{(n+1)}(x) = 0.$$

We can write these results in the form

$$\frac{d^r}{dx^r}(x^n) = \begin{cases} n(n-1)\cdots(n-r+1)x^{n-r} & \text{if } r < n \\ n! & \text{if } r = n \\ 0 & \text{if } r > n. \end{cases} \quad (4)$$

PROBLEMS

1. If $y = A\cos(ax + b)$ find the first and second derivatives of y with respect to x (by using example 9 and Problem 2 of Section 6) and show that, whatever the values of A and b,

$$\frac{d^2y}{dx^2} + a^2y = 0.$$

2. If $y = Ax^4 + Bx^2$, where A and B are constants, find the first

four derivatives of y with respect to x and show that, whatever the values of A and B,

$$x^2 \frac{\mathrm{d}^2 y}{\mathrm{d}x^2} - 5x \frac{\mathrm{d}y}{\mathrm{d}x} + 8y = 0.$$

3. A body moving along straight line has its distance x from a fixed point given by the formula $x = ut + \frac{1}{2}ft^2$, where u and f are constants. Show that the acceleration of the body is constant.

8. Continuous Functions

Many of the functions which we encounter in applications of mathematics have graphs which are continuous curves, i.e. they have no breaks in them. Such functions occupy a special place in mathematics.

A function $f(x)$ is said to be *continuous at the point* $x = a$, provided, firstly, that a definite finite value is attributed to the function at that point and secondly that $f(x)$ tends to $f(a)$ as x tends to a from above or below. In other words, $f(a)$ is a definite finite number and

$$\lim_{x \to a-} f(x) = \lim_{x \to a+} f(x) = f(a). \tag{1}$$

If a function is not continuous at $x = a$, it is said to be *discontinuous* at that point.

A function $f(x)$ is said to be continuous in an interval $a \leqq x \leqq b$ if it is continuous at every point of the interval. For example, the function $f(x) = \frac{2}{3}(x^3 - 3x)$, whose graph is shown in Fig. **44**(a), p. **140** is continuous for all finite values of x.

We can obtain a better understanding of continuity by studying some cases of discontinuous function. For instance the function

$$f(x) = \frac{x^2 - a^2}{x - a}$$

is discontinuous at the point $x = a$, because no definite value is attached to $f(x)$ at $x = a$; if we put $x = a$ in the formula for $f(x)$ we get the indeterminate form $0/0$. The function whose graph is shown in Fig. **41** is discontinuous at the point $x = a$ since $\lim_{x \to a-} f(x)$ is not equal to $\lim_{x \to a+} f(x)$, the former being positive and the latter negative. Such a function is said to have a *finite discontinuity* at the point $x = a$. It will be observed that this function is continuous in the intervals $0 < x < a$ and $x > a$. Another function of this type is shown in Fig. 45(b) the discontinuity here being at the point $x = 0$.

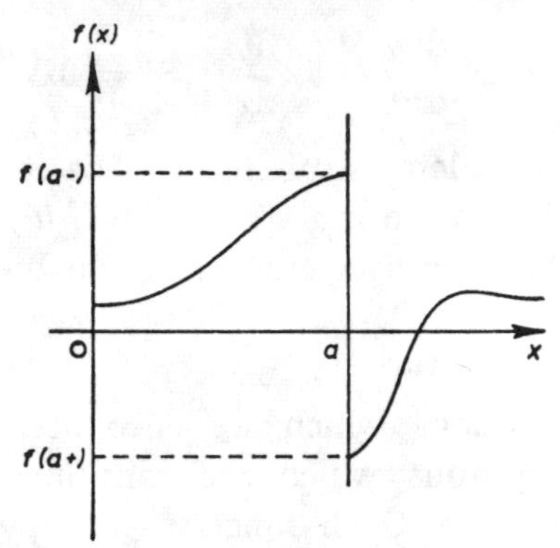

Fig. 41. The graph of a function with finite discontinuity at $x = a$.

A different type of discontinuity is shown in Figs. 42 and 43. In these cases no finite value is ascribed to the function $f(x)$ at the point $x = a$. In Fig. 42,

$$\lim_{x\to a-} f(x) = +\infty, \ \lim_{x\to a+} f(x) = +\infty$$

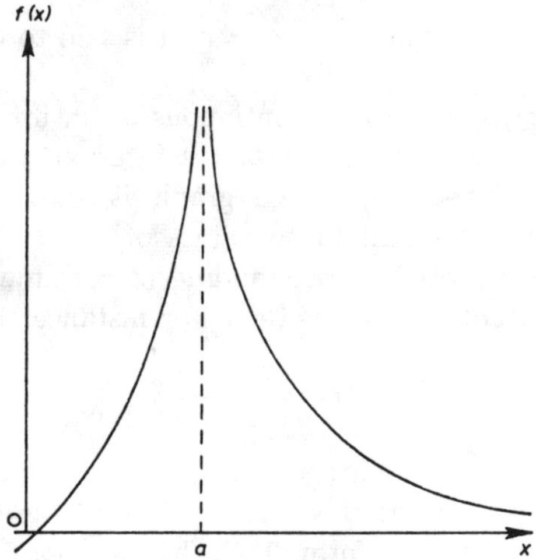

Fig. 42. The graph of a function with infinite discontinuity at $x = a$.

while in Fig. 43

$$\lim_{x\to a-} f(x) = -\infty, \ \lim_{x\to a+} f(x) = +\infty;$$

in both cases we say that $f(x)$ has an *infinite discontinuity* at the point $x = a$. The function $f(x) = x^{-2}$ has an infinite discontinuity at $x = 0$

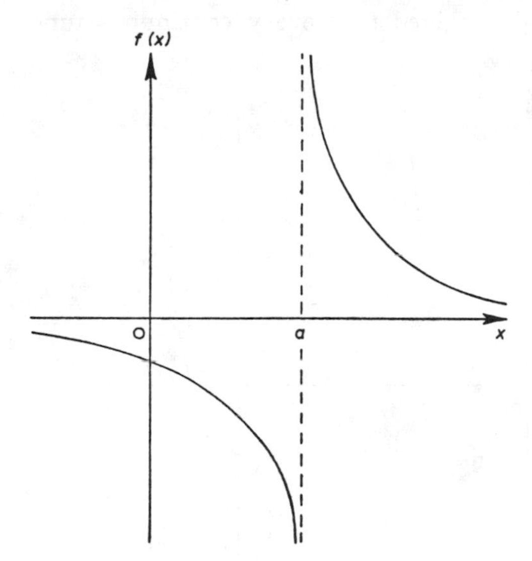

Fig. 43. The graph of a function with infinite discontinuity at $x = a$.

of the same type as the function in Fig. 42, while the function $f(x) = x^{-1}$ has an infinite discontinuity at $x = 0$ of the same type as the function in Fig. 43.

Before discussing the principal properties of continuous functions we return for a moment to subsection (c) of Section 1, and Section 2. We saw there that the gradient of the curve $y = f(x)$ at the point (x, y) is $f'(x)$, so that the gradient of the curve $y = f(x)$ is given by the ordinate of the corresponding point on the curve $y = f'(x)$. The gradient of this second curve is in turn given by the ordinate of the curve $y = f''(x)$. The graphs of the three functions $f(x)$, $f'(x)$, $f''(x)$ for a typical case (it is, in fact, $f(x) = \frac{2}{3}(x^3 - 3x)$) are shown in Fig. 44. In this case the function $f(x)$ is continuous for all values of x and so are its first and second derivatives. The general shape of the graph of $f'(x)$ can be arrived at simply from an inspection of that of $f(x)$. We see from Fig. 44(a) that between U and A (i.e. $-\infty < x < -1$) the gradient of the curve $y = f(x)$ is positive, at A (i.e. $x = -1$) it is zero, between A and B (i.e. $-1 < x < 1$) it is negative, at B $(x = 1)$ it is zero again, and between B and V it is positive. These facts are embodied in Fig. 44(b) where corresponding points are denoted by the same capital letter.

It might be imagined that every continuous function has a con-

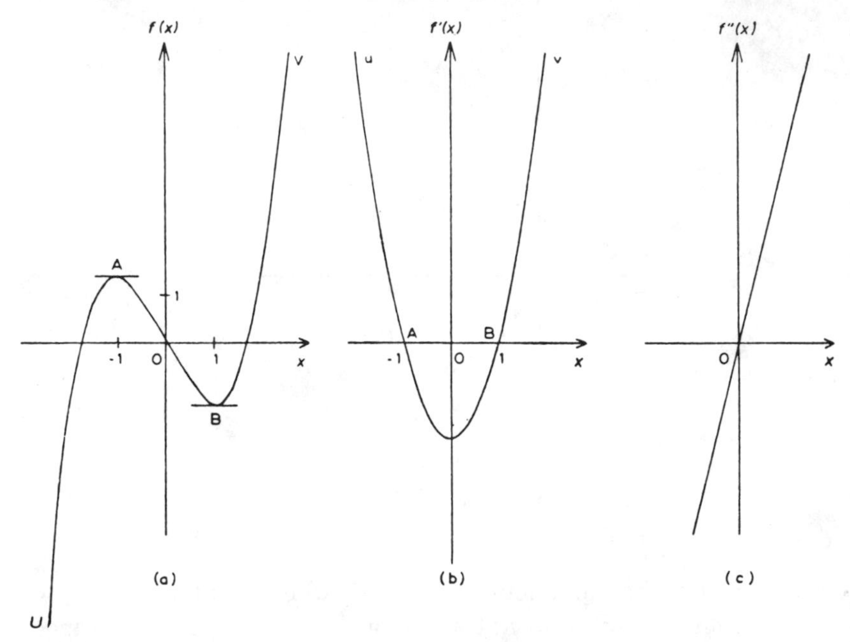

Fig. 44. The graph of the function $f(x) = \frac{2}{3}(x^3 - 3x)$ and of $f'(x)$ and $f''(x)$.

tinuous derivative. That this is not so is shown in Fig. 45, where a continuous function has a derivative with a finite discontinuity at the origin.

The main theorem relating to continuous functions is:

THEOREM 8: *If $f(x)$ is continuous in the interval $a \leqq x \leqq b$ and if in that interval $m \leqq f(x) \leqq M$ then given any number μ such that $m < \mu < M$, there exists at least one number ξ such that $a \leqq \xi \leqq b$ and $f(\xi) = \mu$.*

It is intuitively obvious from a simple diagram such as Fig. 46 that a function which is continuous in an interval $a \leqq x \leqq b$ takes every value between m and M at least once — i.e. there are no breaks in the graph of the function. An argument such as this, based on a geometrical intuition does not, of course, constitute a proper mathematical proof of the theorem, but a rigorous proof based purely on arithmetical concepts is too difficult for inclusion here. Theorem 8 is sometimes called the *fundamental theorem on continuous functions*.

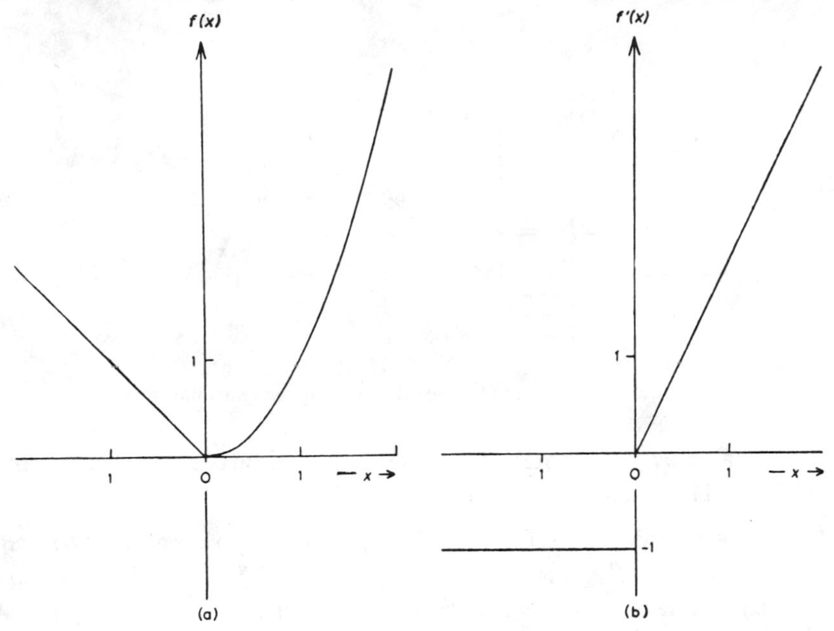

Fig. 45. A continuous function with discontinuous derivative.

An important corollary of this last theorem is:-

THEOREM 9: *If $f(x)$ is a continuous function in the interval $a < x < b$ and if $f(a) < 0$ and $f(b) > 0$ then there exists at least one number ξ with the properties: $a < \xi < b$ and $f(\xi) = 0$.*

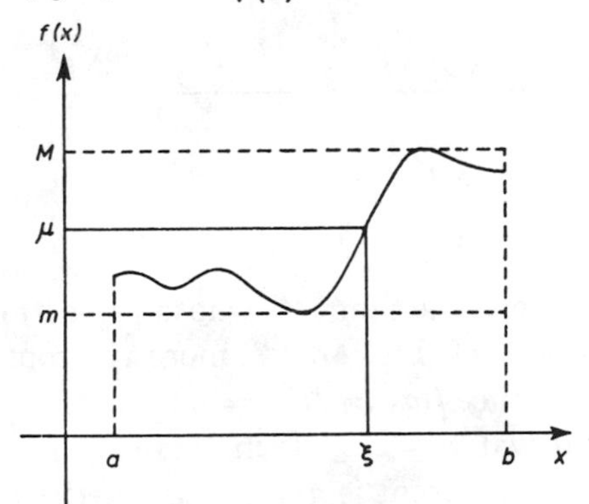

Fig. 46. The fundamental theorem on continuous functions.

This form of the fundamental theorem is illustrated graphically in Fig. 47.

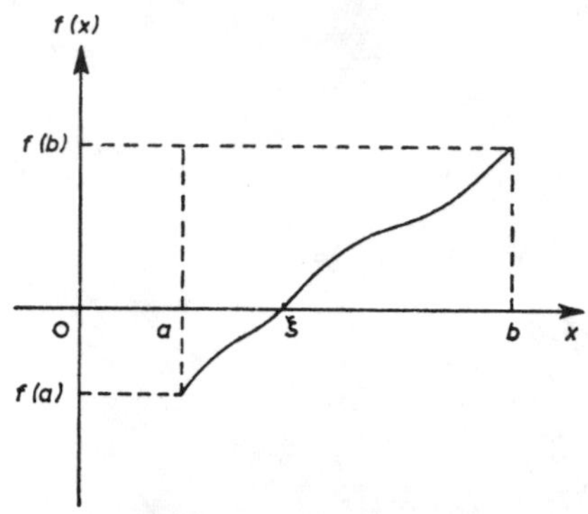

Fig. 47. Corollary to the fundamental theorem on continuous functions.

We shall now say something about the derivative of a continuous function. The basic result is:

THEOREM 10 (ROLLE'S THEOREM): *If $f(x)$ and $f'(x)$ are both continuous functions of x in the interval $a \leqq x \leqq b$ and if $f(a) = f(b) = 0$, then there exists at least one number ξ with the properties: $a < \xi < b$, $f'(\xi) = 0$.*

Geometrically, Rolle's theorem asserts that if a smooth curve joins two points A and B on the x-axis, the tangent to the curve is parallel

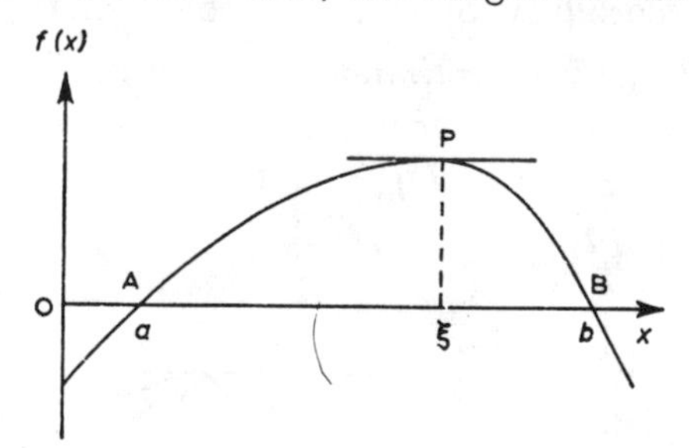

Fig. 48. Rolle's theorem.

to the x-axis at at least one point P whose projection onto the x-axis lies between A and B (cf. Fig. 48). A more rigorous proof would run something like this: Since $f(a) = f(b) = 0$ then it may be that $f(x) = 0$ for all x in the interval $a \leqq x \leqq b$; in this case $f'(x) = 0$ for all x in the interval and the theorem is trivial. If on the other hand $f(x)$ is not constant for all x in the interval then $f(x)$ increases (or decreases) at some point; to return to the value 0 at $x = b$ it must necessarily decrease (or increase). Thus $f'(x) > 0$ at some point and then $f'(x) < 0$. But we postulated that $f'(x)$ is continuous and therefore, by Theorem 9, there is at least one value of x, ξ say, such that $f'(\xi) = 0$.

As a consequence of this theorem we have:

THEOREM 11 (MEAN VALUE THEOREM): *If $f(x)$ and $f'(x)$ are both continuous functions of x in the interval $a \leqq x \leqq b$, then there exists at least one number ξ such that $a \leqq \xi \leqq b$ and*

$$f'(\xi) = \frac{f(b) - f(a)}{b - a}. \tag{2}$$

Geometrically this means that if A and B are any two points on the graph of a continuous function then there is at least one point P

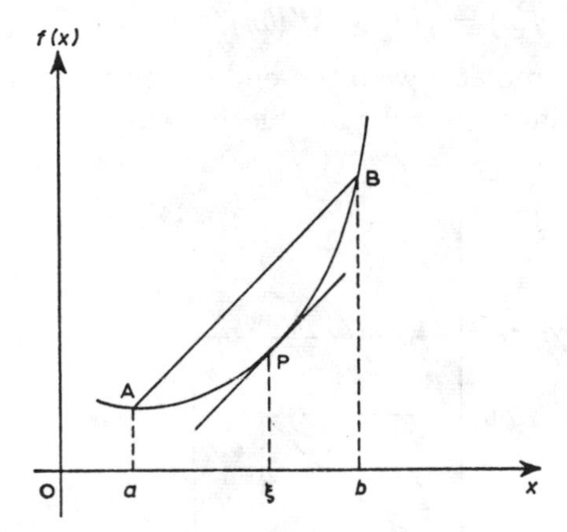

Fig. 49. The mean value theorem.

on the curve between A and B such that the tangent at P is parallel to the chord AB of the curve (cf. Fig. 49). The result can, however, be derived analytically from Rolle's theorem in the following way:

If, from $f(x)$, we construct a new function

$$\phi(x) = f(x) - \frac{f(b) - f(a)}{b - a} x + \frac{af(b) - bf(a)}{b - a}$$

then we can readily show that $\phi(a) = \phi(b) = 0$ and that

$$\phi'(x) = f'(x) - \frac{f(b) - f(a)}{b - a}. \tag{3}$$

If we apply Rolle's theorem to the function $\phi(x)$, it follows that there is a number ξ such that $a \leqq \xi \leqq b$ and $\phi'(\xi) = 0$. Putting $x = \xi$ in equation (3) we obtain equation (2).

We can write equation (2) in the form

$$f(b) = f(a) + (b - a)f'(\xi), \qquad (a \leqq \xi \leqq b). \quad (4)$$

Putting $b = a + h$, we can write $\xi = a + \theta h$, where $0 \leqq \theta \leqq 1$ so that equation (4) becomes

$$f(a + h) = f(a) + hf'(a + \theta h), \qquad (0 \leqq \theta \leqq 1). \quad (5)$$

Similarly putting $a = b - h$ we have

$$f(b - h) = f(b) - hf'(b - \theta h), \qquad (0 \leqq \theta \leqq 1). \quad (6)$$

Putting $a = 0$ in equation (5) we get the special case

$$f(h) = f(0) + hf'(\theta h), \qquad (0 \leqq \theta \leqq 1). \quad (7)$$

In biological problems we often encounter functions which increase steadily as the independent variable increases and these are given a

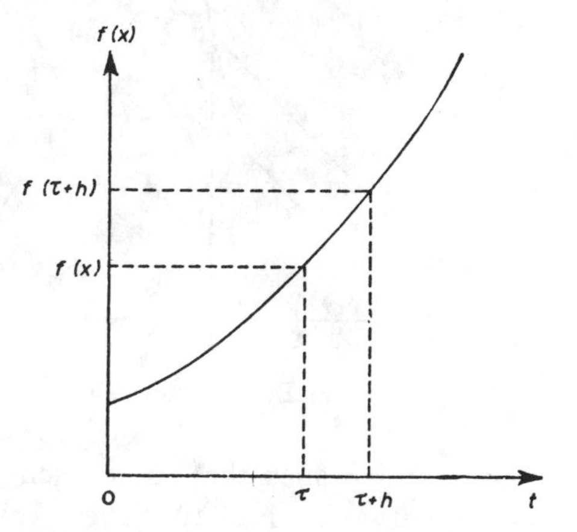

Fig. 50. A monotonic increasing function.

special name. The graph of such a function is shown in Fig. 50. If $f(t)$ is a function of t and if h is positive, then

$$f(\tau + h) > f(\tau), \quad (h > 0), \quad (8)$$

for every τ. A function which satisfies this condition for every pair of values τ and $\tau + h$ (with $h > 0$) in an interval is said to be *monotonically increasing*. For such a function, if h is positive, $f(\tau + h) - f(\tau)$

is positive, and the ratio $\{f(\tau + h) - f(\tau)\}/h$ is always positive; the ratio is still positive if h is negative since, then, $f(\tau + h) - f(\tau)$ is negative. As $h \to 0$ the ratio remains positive so that, for a monotonic increasing function

$$f'(t) > 0$$

for all values of t in the interval. This result is obvious geometrically.

A function which decreases steadily (such as the one in Fig. 51)

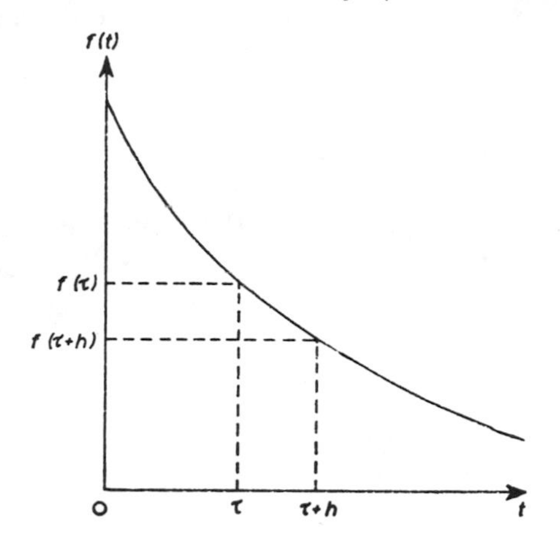

Fig. 51. A monotonic decreasing function.

is said to be *monotonically decreasing*. For such a function

$$f(\tau + h) < f(\tau), \quad (h > 0) \tag{9}$$

and

$$f'(t) < 0$$

for all values of t in the interval.

9. Derivatives in Fluid Transport

In applying the ideas of differentiation to biological situations a student must acquire facility in writing down rates as derivatives. To illustrate this we now consider the derivation of the differential equation of the "Starling hypothesis".

Starling's hypothesis of the fluid transport across the capillary wall states that the driving force is given by the difference between the hydrostatic and osmotic forces. Our problem is to obtain an expression for the hydrostatic pressure and the osmotic pressure as a function of the distance along the capillary. To obtain a simple equation, we must make some rather drastic simplifications, so that our model resembles a "capillary" containing, say, a sucrose solution rather than the actual situation in the body. However, it is relatively easy to introduce the "non-linearities" inherent to the physiological situation. We wish to stress that what follows is merely a *sketch* leading to the setting up of a differential equation describing the process. From the present view-point once the differential equation is obtained, the problem is solved.

We make the following assumptions:

1. The capillary wall is rigid and of constant diameter.

2. The distribution of pores and their dimensions are uniform.

3. Fluid flow along the capillary is "Poiseuille" (deviations from the Poiseuille law due to the presence of corpuscles, etc. are ignored).

4. Fluid flow across the wall of the capillary is proportional to the difference between the hydrostatic and osmotic pressures where, of course, it is implicit that these pressures are zero outside the capillary. This assumption is, however, by no means necessary.

In this first approximation only steady-state conditions are considered so that the time t is not involved.

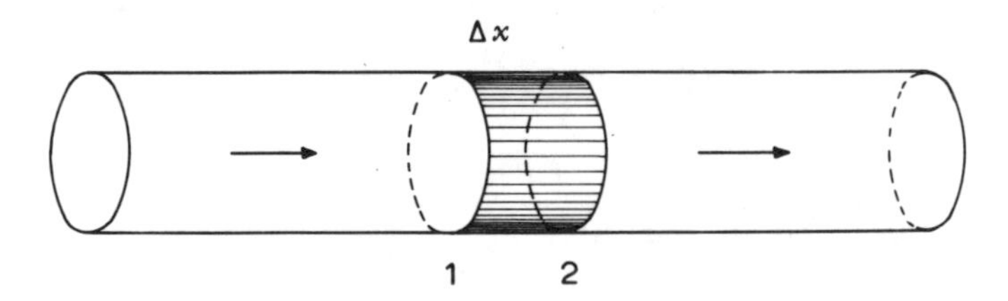

Fig. 52. Volume element of thickness Δx in capillary.

We consider a cylindrical capillary with radius r (Fig. 52), and write x for the distance along it, $y = y(x)$ for the concentration of solute at the point x; $z = z(x)$ is hydrostatic pressure, p is the osmotic pressure at this point, and q is the mean velocity of flow along the capillary, averaged over its cross section. The rate of volume flow from the capillary into the surrounding tissue, in a small element of the capillary of length Δx, is proportional to its surface area and to $z - p$. If the rate constant is taken to be S, then for this element the rate of flow out of this volume is

$$2\pi r \Delta x (z - p) S. \tag{1}$$

We now write down the condition that no fluid is lost. The rate of volume flow along the capillary into this volume element is $\pi r^2 q(x)$ and the corresponding rate of flow out of it is $\pi r^2 q(x+\Delta x)$.

No solute leaves the capillary, and under steady state conditions the rate of solute flow along the capillary is constant. If this constant is Q per second per unit area of cross section, then

$$Q = qy. \tag{2}$$

If T is the concentration of solute in the tissue, then the osmotic pressure p is given by

$$p = a(y - T) \tag{3}$$

where a is a constant. We can put $T = 0$ without loss of generality.

Then since no fluid is lost,

$$\pi r^2 q(x) = \pi r^2 q(x + \Delta x) + 2\pi r S(z - p)\Delta x \tag{4}$$

which is equivalent to

$$-2\pi r S(z - ay)\Delta x = \pi r^2 \{q(x + \Delta x) - q(x)\} .$$

Letting Δx tend to 0 we obtain the equation

$$-2\pi r S(z - ay) = \pi r^2\, \mathrm{d}q/\mathrm{d}x .$$

Since

$$q = \frac{Q}{y}, \quad \frac{\mathrm{d}q}{\mathrm{d}x} = -\frac{Q}{y^2}\frac{\mathrm{d}y}{\mathrm{d}x},$$

we have

$$\frac{\mathrm{d}y}{\mathrm{d}x} = \frac{2S}{rQ} y^2 (z - ay). \tag{5}$$

Since we assume Poiseuille flow, the mean rate of fluid flow along the capillary is proportional to the pressure gradient. Hence $q = -k\,\mathrm{d}z/\mathrm{d}x$, when k is a constant, equal to $r^2/(8\eta)$ where η is the coefficient of viscosity, or:

$$\frac{Q}{ky} = -\frac{dz}{dx}. \tag{6}$$

We regard (5) and (6) as the basic equations. They can be most easily handled if we regard z as the independent variable.* We have

$$\frac{dx}{dz} = -\frac{ky}{Q}, \tag{7}$$

and multiplying this by (5) gives:

$$\frac{dy}{dz} = -\frac{2Sk}{rQ^2} y^3 (z - ay). \tag{8}$$

* See page 62. Obtaining (8) from (5) and (7) depends on the *chain rule*, see Chapter 4, Theorems 15 and 16. Equation (8) is actually of the general type discussed in Chapter 9, § 2.

CHAPTER 4

THE DIFFERENTIAL CALCULUS

In the last chapter we defined the derivative of a function and found the derivatives of a few simple functions. We shall now set up rules for the differentiation of complicated functions in terms of the derivatives of these simpler functions. This set of ideas, taken with those of the last chapter, form what is known as the *differential calculus*. We shall conclude the chapter with a brief account of some applications of the differential calculus.

1. The Differentiation of Products and Quotients

We suppose $f(x)$ and $g(x)$ are functions of x whose derivatives can be calculated easily and that we wish to find the derivatives of the functions $f(x)g(x)$ and $f(x)/g(x)$ formed from them.

By definition, the derivative of $f(x)g(x)$ is given by

$$\frac{\mathrm{d}}{\mathrm{d}x}\{f(x)g(x)\} = \lim_{h\to 0}\left\{\frac{f(x+h)\,g(x+h) - f(x)\,g(x)}{h}\right\}. \tag{1}$$

Now we can write the numerator of the ratio on the right in the form

$$\begin{aligned} f(x+h)g(x+h) &- f(x+h)g(x) + f(x+h)g(x) - f(x)g(x) \\ &= f(x+h)\{g(x+h) - g(x)\} + g(x)\{f(x+h) - f(x)\}. \end{aligned}$$

If we put this expression in the equation (1) and make use of Theorems 1 and 4 (p. 125 of the last chapter) we find that

$$\begin{aligned} \frac{\mathrm{d}}{\mathrm{d}x}\{f(x)\,g(x)\} = \lim_{h\to 0} f(x+h) \lim_{h\to 0} &\frac{g(x+h) - g(x)}{h} \\ + g(x) \lim_{h\to 0} &\frac{f(x+h) - f(x)}{h}. \end{aligned}$$

Now $f(x+h) \to f(x)$ as $h \to 0$ and the other two limits are easily identifiable from the definition of the derivative of a function. It follows that we can deduce:

THEOREM 12: $\dfrac{\mathrm{d}}{\mathrm{d}x}\{f(x)\,g(x)\} = f(x)\,g'(x) + f'(x)\,g(x).$ (2)

To illustrate the use of this theorem we have:-

EXAMPLE 1: *Find the derivative of* $x^n \sin(ax+b)$.

In this case $y = f(x)g(x)$ with $f(x) = x^n$, $g(x) = \sin(ax+b)$. For these functions $f'(x) = nx^n$, $g'(x) = a\cos(ax+b)$ (cf. example 9 of Chapter 3) so that

$$\frac{\mathrm{d}y}{\mathrm{d}x} = x^n \cdot a\cos(ax+b) + nx^{n-1}\sin(ax+b)$$

$$= x^{n-1}\{ax\cos(ax+b) + n\sin(ax+b)\}.$$

If we put $n = 1$, $a = 1$, $b = 0$ in this equation we get the special result

$$\frac{\mathrm{d}}{\mathrm{d}x}(x\sin x) = x\cos x + \sin x$$

in agreement with that found directly previously (Problem 3 of Section 6 of the last chapter).

A geometrical interpretation of Theorem 12, due to Robert Simson, is given in Fig. 53. In this diagram the line OF represents the function

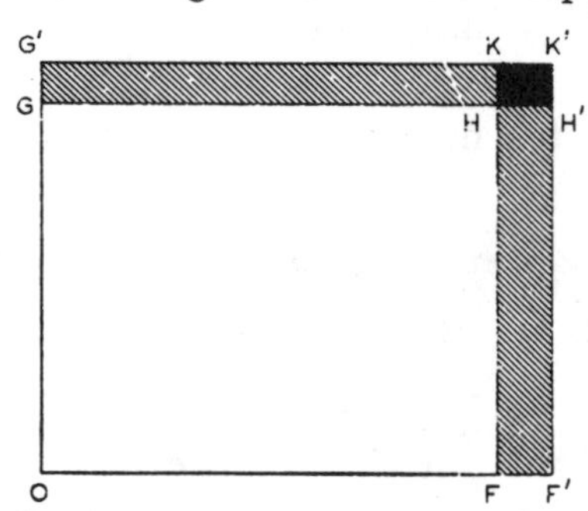

Fig. 53. Differentiation of a product.

f and FF' represents a small increment δf in the function f; similarly the lines OG and GG', which are perpendicular to OF, represent g and δg respectively. The rectangle $OFHG$ then represents fg and the rectangle $OF'K'G'$ represents $fg + \delta(fg)$. The increment $\delta(fg)$ is therefore made up of the two shaded rectangles which are a measure of $f\delta g$ and $g\delta f$ and a black rectangle of "area" $\delta f \cdot \delta g$. Now if δf and

δg are both small, the black area will be very much less than the areas of the shaded portions and may be neglected in comparison with them. We therefore have

$$\delta(fg) = f\delta g + g\delta f.$$

Dividing by the increment δx which was responsible for the changes δf, δg in f, g respectively, and then letting δx tend to zero we get equation (2).

If we wish to find the derivative of the product of three functions $f(x)$, $g(x)$ and $k(x)$ we write

$$y = f(x)g(x)k(x); \tag{3}$$

then grouping $g(x)$ and $k(x)$ together we find that

$$\begin{aligned}\frac{dy}{dx} &= f(x)\frac{d}{dx}\{g(x)\,k(x)\} + f'(x)\,g(x)\,k(x)\\ &= f(x)\{g(x)\,k'(x) + g'(x)\,k(x)\} + f'(x)\,g(x)\,k(x).\end{aligned}$$

If we now divide both sides of this equation by $y = f(x)g(x)k(x)$ we find that

$$\frac{1}{y}\frac{dy}{dx} = \frac{1}{f}\frac{df}{dx} + \frac{1}{g}\frac{dg}{dx} + \frac{1}{k}\frac{dk}{dx}. \tag{4}$$

The result expressed by equations (3) and (4) can easily be generalized to the case of y being the product of n functions $f_1(x)$, $f_2(x)$, ... $f_n(x)$:-

THEOREM 13: *If* $y = f_1(x)f_2(x)\cdots f_n(x)$, *then*

$$\frac{1}{y}\frac{dy}{dx} = \frac{1}{f_1}\frac{df_1}{dx} + \frac{1}{f_2}\frac{df_2}{dx} + \cdots + \frac{1}{f_n}\frac{df_n}{dx}. \tag{5}$$

The corresponding result for the quotient of two functions is given in:-

THEOREM 14: $$\frac{d}{dx}\left\{\frac{f(x)}{g(x)}\right\} = \frac{f'(x)\,g(x) - f(x)\,g'(x)}{[g(x)]^2}. \tag{6}$$

We can prove this theorem either from Theorem 12 or directly from first principles. To prove it by means of Theorem 12 we write

$$y(x) = \frac{f(x)}{g(x)};$$

then $f(x) = y(x)g(x)$ and it follows from Theorem 12 that

$$f'(x) = y'(x)\,g(x) + y(x)\,g'(x)$$

so that

$$y'(x) = \frac{f'(x)}{g(x)} - \frac{y(x)\,g'(x)}{g(x)}.$$

Substituting $f(x)/g(x)$ for $y(x)$ in the second term on the right we obtain equation (6).

The theorem can also be derived directly from the definition of a derivative. By this definition and Theorem 6 we have

$$\frac{d}{dx}\left\{\frac{f(x)}{g(x)}\right\} = \lim_{h\to 0} \frac{1}{h}\left\{\frac{f(x+h)}{g(x+h)} - \frac{f(x)}{g(x)}\right\}$$
$$= \frac{1}{[g(x)]^2} \lim_{h\to 0} \frac{f(x+h)\,g(x) - f(x)\,g(x+h)}{h}.$$

Now we can write

$$f(x+h)g(x) - f(x)g(x+h) = f(x+h)g(x) - f(x)g(x) + f(x)g(x) - f(x)g(x+h)$$
$$= [f(x+h) - f(x)]g(x) - f(x)[g(x+h) - g(x)]$$

so that

$$\frac{d}{dx}\left\{\frac{f(x)}{g(x)}\right\} = \frac{1}{[g(x)]^2}\left\{g(x) \lim_{h\to 0} \frac{f(x+h) - f(x)}{h} - f(x) \lim_{h\to 0} \frac{g(x+h) - g(x)}{h}\right\}$$

and equation (6) follows immediately.

EXAMPLE 2: *Use Theorem 9 to find the derivative of* $\tan x$.

We can write $\tan x = f(x)/g(x)$ where $f(x) = \sin x$, $g(x) = \cos x$. For these functions $f'(x) = \cos x$, $g'(x) = -\sin x$ so that

$$\frac{d}{dx}(\tan x) = \frac{\cos x \cdot \cos x - \sin x(-\sin x)}{\cos^2 x} = \frac{\sin^2 x + \cos^2 x}{\cos^2 x}.$$

If we recall that $\sin^2 x + \cos^2 x = 1$ and that $1/\cos x = \sec x$ we see that

$$\frac{d}{dx}(\tan x) = \sec^2 x.$$

in agreement with equation (3) of Section 6 of the last chapter.

A special case of Theorem 14 is worth noting. If we put $f(x) = 1$, $f'(x) = 0$ into equation (6) we find that

$$\frac{\mathrm{d}}{\mathrm{d}x}\left\{\frac{1}{g(x)}\right\} = -\frac{g'(x)}{[g(x)]^2}. \tag{7}$$

As an illustration we have:

EXAMPLE 3: *Use equation* (7) *to find the derivative of* $\sec x$.

We can write $\sec x = 1/g(x)$ where $g(x) = \cos x$, $g'(x) = -\sin x$. Hence

$$\frac{\mathrm{d}}{\mathrm{d}x}(\sec x) = -\frac{-\sin x}{\cos^2 x} = \sin x \cdot \sec^2 x$$

as in equation (4) of Section 6 of the last chapter.

PROBLEMS

1. Find the derivatives of the functions

$$\text{(i)}\ \frac{\cos x}{1 + x^2}, \quad \text{(ii)}\ \frac{\sin x}{x^3 + 2}, \quad \text{(iii)}\ x \tan x, \quad \text{(iv)}\ \frac{x \tan x}{1 + x + x^2}.$$

2. Find the first and second derivatives of

$$y = \frac{A}{x^2 + 1} + B(x^2 + 1)$$

and show that

$$x(x^2 + 1)^2 \frac{\mathrm{d}^2 y}{\mathrm{d}x^2} + (x^4 - 1)\frac{\mathrm{d}y}{\mathrm{d}x} - 4x^3 y = 0.$$

2. Function of a Function: The Chain Rule

The derivative of a function such as $y = (1 + x^3)^{\frac{1}{3}}$ cannot be found by any of the rules we have so far established. However, since the derivative of x^n with respect to x is known, we can find the derivative of $z = 1 + x^3$ and if we write $y = z^{\frac{1}{3}}$ we can, by the same rule, find the derivative of y with respect to z. We shall derive a formula (the "chain rule") which will enable us to calculate the derivative of y with respect to x in terms of $\mathrm{d}y/\mathrm{d}z$ and $\mathrm{d}z/\mathrm{d}x$.

In the general case we consider a function y of x which is expressed in the form $f(z)$ where $z = \phi(x)$ is a function of x. In the above example $f(z) = z^{\frac{1}{3}}$ and $\phi(x) = 1 + x^3$. Since y is expressed as a function of z which is itself a function of x we say that y is a "function of a function" of x. The chain rule may be stated as:-

THEOREM 15: *If y is a function of z and z is a function of x, then*

$$\frac{dy}{dx} = \frac{dy}{dz} \cdot \frac{dz}{dx}.$$

Suppose that $z = \phi(x)$. Then $y = f(z) = f\{\phi(x)\}$. If x takes an increment h then z will take an increment k given by the equation

$$z + k = \phi(x + h)$$

so that

$$k = \phi(x + h) - \phi(x). \tag{1}$$

and $k \to 0$ as $h \to 0$.
Now $y = f(z)$ so that, if z takes an increment k due to an increment h in x, the increment in y will be $f(z + k) - f(z)$. Hence, regarded as a function of x, the derivative of y is

$$\frac{dy}{dx} = \lim_{h \to 0} \frac{f(z + k) - f(z)}{h}. \tag{2}$$

Now we can write

$$\frac{f(z + k) - f(z)}{h} = \frac{f(z + k) - f(z)}{k} \cdot \frac{k}{h}$$

and using equation (1) we find that

$$\frac{f(z + k) - f(z)}{h} = \frac{f(z + k) - f(z)}{k} \cdot \frac{\phi(x + h) - \phi(x)}{h}. \tag{3}$$

Now since $k \to 0$ whenever $h \to 0$ we find on substituting from (3) into (2) and using Theorem 4 that

$$\frac{dy}{dx} = \lim_{k \to 0} \frac{f(z + k) - f(z)}{k} \cdot \lim_{h \to 0} \frac{\phi(x + h) - \phi(x)}{h}$$

so that

$$y'(x) = f'(z)\phi'(x)$$

i.e.

$$\frac{dy}{dx} = \frac{dy}{dz} \cdot \frac{dz}{dx}. \tag{4}$$

The full significance of the chain rule is best grasped by working out some examples.

The reader should study these carefully and then try to solve the problems at the end of this section.

EXAMPLE 4: *Find the derivative of* $(x^3 + 1)^{\frac{1}{3}}$.

Here we have $y = z^{\frac{1}{3}}$, with $z = x^3 + 1$ so that

$$\frac{dy}{dz} = \tfrac{1}{3}z^{-\frac{2}{3}}, \quad \frac{dz}{dx} = 3x^2$$

and

$$\frac{dy}{dx} = \tfrac{1}{3}z^{-\frac{2}{3}} \cdot 3x^2 = \frac{x^2}{(x^3 + 1)^{\frac{2}{3}}}.$$

EXAMPLE 5: *Find the derivative of* $\sqrt{(2ax - x^2)}$.

In this case $y = z^{\frac{1}{2}}$, with $z = 2ax - x^2$ for which

$$\frac{dy}{dz} = \tfrac{1}{2}z^{-\frac{1}{2}}, \quad \frac{dz}{dx} = 2a - 2x$$

and so

$$\frac{dy}{dx} = \tfrac{1}{2}z^{-\frac{1}{2}}\,(2a - 2x) = \frac{a - x}{\sqrt{(2ax - x^2)}}.$$

EXAMPLE 6: *Find the derivative of* $\sin^n x$.

In this example $y = z^n$ with $z = \sin x$ and for these functions

$$\frac{dy}{dz} = nz^{n-1}, \qquad \frac{dz}{dx} = \cos x$$

giving

$$\frac{dy}{dx} = n \sin^{n-1} x \cdot \cos x.$$

EXAMPLE 7: *Find the derivative of* $\sin(\sin x)$.

Here $y = \sin z$ with $z = \sin x$. We then have

$$\frac{dy}{dz} = \cos z, \qquad \frac{dz}{dx} = \cos x$$

giving the result

$$\frac{dy}{dx} = \cos(\sin x) \cdot \cos x.$$

EXAMPLE 8. *Model of the cardiovascular system.*

In a theoretical study of the cardiovascular system* the following empirical equation occurs:

* J.G. Defares, J.J. Osborn and H.H. Hara, Acta Physiol. Pharm. Neerl. 12 (1963) 189–265; see also: Bulletin of Mathematical Biophysics **27** (1965) 71–83, Special Issue.

$$s = \{a + (c - c_5 p_5)v\}\{3t^2/t_s^2 - 2t^3/t_s^3\}, \tag{1}$$

where s is stroke volume, p_5 is aortic pressure, v is end-diastolic left ventricular volume, written as v_{4d}^{e} in the paper, t_s is the duration of systole, t is time (during systole) and a, c and c_5 (d in the paper) are constants. Note that both s and p_5 are functions of time, t.

Differentiation of equation (1) to obtain the (required) flow, $\mathrm{d}s/\mathrm{d}t$, yields the formula

$$\frac{\mathrm{d}s}{\mathrm{d}t} = \frac{\{6a + (c - c_5 p_5)\}}{t_s^2}\left(t - \frac{t^2}{t_c}\right) - \left(\frac{3t^2}{t_s^2} - \frac{2t^3}{t_s^3}\right) c_5 v \frac{\mathrm{d}p_5}{\mathrm{d}t}\,. \tag{2}$$

If the student is willing to accept on faith the solution of the set of differential equations by means of the analog computer, he may be able to follow the (mathematical) reasoning in the extensive Acta paper referred to above.

PROBLEMS

Find the derivatives (with respect to x) of the following functions:-

1. $(ax + b)^n$ **2.** $(ax^2 + bx + c)^n$ **3.** $\sqrt{(2ax)}$

4. $\sqrt{(k - x)}$ **5.** $(a^2 - x^2)^4$ **6.** $\cos^n x$

7. $\sin(x^3 + 3)$ **8.** $\tan(ax + b)$ **9.** $\dfrac{a + x}{a - x}$

10. $\sqrt{\dfrac{a + x}{a - x}}$ **11.** $x\sqrt{(a + x)}$ **12.** $\sin^m(x^n)$

13. $\dfrac{1 + x^2}{1 + (1 + x^2)^n}$ **14.** $\dfrac{1}{x + \sqrt{(x^2 - 1)}}$ **15.** $\left(\sqrt{x} - \dfrac{1}{\sqrt{x}}\right)^5$

16. The volume of a sphere is increasing at the uniform rate of 10 cubic metres per second. Find the rate at which the surface area is increasing when the radius is 3 metres.

17. A circular disc of radius 10 cm is lowered (with its plane vertical) into water. When the lowest point is at a depth of x cm the chord separating the wet and dry areas is $2y$ cm. Show that

$$y^2 = 20x - x^2.$$

If the chord increases at the constant rate of 2 cm per second, find the speed of the descending disc when $y = 8$.

18. A right circular cone, altitude 20 cm and base-radius 6 cm is lowered with uniform speed 4 cm/sec into a tank full of water, its axis being vertical and its apex downwards. Prove that, when the cone is immersed to a depth of 15 cm the water is overflowing at the rate of 81π cm^3/sec.

19. Prove that the curves $3x^2 - y^2 + 4x + 1 = 0$ and $(36 - 7y)^3 = 8(1 - x^2)^2$ intersect orthogonally at the point $(-3, 4)$.

20. Show that the curves $y = x^n - 2a$ and $y = -a^2x^{-n}$ touch one another at a point on the curve $y = -x^n$ for all values of n and a.

21. Water is poured at a constant rate v into a hollow sphere of internal radius r through a small hole at the top. Show that, when the depth of water in the sphere is z, the surface is rising at the rate

$$\frac{v}{\pi z(2r - z)}.$$

Show also that, when $z = \frac{3}{2}r$, the upward acceleration of the surface of the water is $64v^2/27\pi^2r^5$.

3. The Differentiation of Inverse Functions

The concept of an inverse function has already been introduced (in the case of the trigonometrical functions) in Section 7 of Chapter 2. In general if y is a function of x symbolized by the equation

$$y = f(x) \tag{1}$$

then x will be a function of y and we may write

$$x = \phi(y). \tag{2}$$

We may now apply Theorem 15 to this situation to find that

$$\frac{dx}{dx} = \frac{dx}{dy} \cdot \frac{dy}{dx}.$$

Now $dx/dx = 1$ so that we have

THEOREM 16: *If y is a given function of x, then regarding x as a function of y we find that its derivative is given by the equation*

$$\frac{dx}{dy} = \frac{1}{\dfrac{dy}{dx}}.$$

We shall now use this theorem to find the derivatives of the inverse trigonometrical functions. If $y = \sin^{-1}x$ then it follows that $x = \sin y$ so that

$$\frac{\mathrm{d}}{\mathrm{d}x}(\sin^{-1} x) = \frac{1}{\dfrac{\mathrm{d}}{\mathrm{d}y}\sin y} = \frac{1}{\cos y} = \frac{1}{\pm\sqrt{(1 - \sin^2 y)}}$$

so that, taking the positive sign since $\cos y$ is positive for $-\frac{1}{2}\pi \leqq y \leqq \frac{1}{2}\pi$, we have

$$\frac{\mathrm{d}}{\mathrm{d}x}(\sin^{-1} x) = \frac{1}{\sqrt{(1 - x^2)}}. \tag{3}$$

Similarly, if $y = \cos^{-1} x$, we have $x = \cos y$ and $\mathrm{d}x/\mathrm{d}y = -\sin y$ so that

$$\frac{\mathrm{d}}{\mathrm{d}x}(\cos^{-1} x) = -\frac{1}{\sin y} = \frac{1}{\pm\sqrt{(1 - \cos^2 y)}}.$$

Here we take the negative sign, since $\sin y$ is positive for $0 \leqq y \leqq \pi$, and hence find that

$$\frac{\mathrm{d}}{\mathrm{d}x}(\cos^{-1} x) = -\frac{1}{\sqrt{(1 - x^2)}}. \tag{4}$$

It will be observed that

$$\frac{\mathrm{d}}{\mathrm{d}x}(\cos^{-1} x + \sin^{-1} x) = 0$$

in agreement with the fact that $\cos^{-1} x + \sin^{-1} x = \frac{1}{2}\pi$.

If $y = \tan^{-1} x$ then $x = \tan y$ and $\mathrm{d}x/\mathrm{d}y = \sec^2 y$ giving

$$\frac{\mathrm{d}}{\mathrm{d}x}(\tan^{-1} x) = \frac{1}{\sec^2 y} = \frac{1}{1 + \tan^2 y} = \frac{1}{1 + x^2}. \tag{5}$$

By a straightforward application of the chain rule we see that we can write equations (3), (4), (5) in the slightly more general forms

$$\frac{\mathrm{d}}{\mathrm{d}x}\cos^{-1}\left(\frac{x}{a}\right) = -\frac{1}{\sqrt{(a^2 - x^2)}}, \tag{6}$$

$$\frac{\mathrm{d}}{\mathrm{d}x}\sin^{-1}\left(\frac{x}{a}\right) = \frac{1}{\sqrt{(a^2 - x^2)}}, \tag{7}$$

$$\frac{\mathrm{d}}{\mathrm{d}x}\tan^{-1}\left(\frac{bx}{a}\right) = \frac{ab}{a^2 + b^2x^2}. \tag{8}$$

Another use of Theorem 11 is shown by:-

EXAMPLE 9: *If* $x(y + 1) = y^2$ *find* $\mathrm{d}y/\mathrm{d}x$.

Here we write

$$x = \frac{y^2}{y + 1}$$

and, by Theorem 9, we find that

$$\frac{\mathrm{d}x}{\mathrm{d}y} = \frac{2y(y + 1) - y^2}{(y + 1)^2} = \frac{y(y + 2)}{(y + 1)^2}$$

so that

$$\frac{\mathrm{d}y}{\mathrm{d}x} = \frac{(y + 1)^2}{y(y + 2)}.$$

Tabulating the results on differentiation we have derived so far we get Table 9.

TABLE 9
Derivatives of elementary functions

$f(x)$	$f'(x)$
$ax + b$	a
x^n	nx^{n-1}
$(ax + b)^n$	$na(ax + b)^{n-1}$
$\sin x$	$\cos x$
$\cos x$	$-\sin x$
$\tan x$	$\sec^2 x$
$\operatorname{cosec} x$	$-\operatorname{cosec} x \cdot \cot x$
$\sec x$	$\sec x \cdot \tan x$
$\cot x$	$-\operatorname{cosec}^2 x$
$\sin^{-1}(x/a)$	$(a^2 - x^2)^{-\frac{1}{2}}$
$\cos^{-1}(x/a)$	$-(a^2 - x^2)^{-\frac{1}{2}}$
$\tan^{-1}(x/a)$	$a(a^2 + x^2)^{-1}$
$f[u(x)]$	$f'(u)u'(x)$
$f(x)g(x)$	$f'(x)g(x) + f(x)g'(x)$
$\dfrac{f(x)}{g(x)}$	$\dfrac{f'(x)g(x) - f(x)g'(x)}{[g(x)]^2}$

PROBLEMS

1. Find the derivatives of the functions

$$\text{(i)} \quad \tan^{-1}\left(\frac{3x-1}{4}\right), \qquad \text{(ii)} \quad x\tan^{-1}x,$$

$$\text{(iii)} \quad \sin^{-1}x^2, \qquad \text{(iv)} \quad \sin^{-1}\sqrt{x}.$$

2. Show that the derivative of

$$\cos^{-1}\left(\frac{b+a\cos x}{a+b\cos x}\right)$$

is

$$\frac{\sqrt{(a^2-b^2)}}{a+b\cos x}.$$

3. Show that the derivative of

$$\tfrac{1}{2}x\sqrt{(a^2-x^2)}+\tfrac{1}{2}a^2\sin^{-1}\left(\frac{x}{a}\right)$$

is

$$\sqrt{(a^2-x^2)}.$$

4. If $y^3+3y^2=x+1$ show that

$$\frac{dy}{dx}=\frac{1}{3y(y+2)}$$

and hence find the equation of the tangent to this curve at the point (3, 1).

4. The Expansion of a Function in a Power Series

We can rewrite the result

$$1+x+x^2+\ldots+x^{n-1}=\frac{1-x^n}{1-x}$$

of Section 4 of Chapter 1 in the form

$$\frac{1}{1-x}=S_n(x)+R_n(x) \tag{1}$$

where $S_n(x)$ denotes the sum of a series of n terms

$$S_n(x)=1+x+x^2+\ldots+x^{n-1}=\sum_{r=0}^{n-1}x^r \tag{2}$$

and

$$R_n(x) = \frac{x^n}{1 - x}$$

is called the *remainder after n terms*. We saw that if $|x| < 1$, then $R_n(x) \to 0$ as $n \to \infty$ and it follows that the series $S_n(x)$ tends to the function $(1 - x)^{-1}$ as $n \to \infty$. We say that the function $(1 - x)^{-1}$ can be expanded in the form of the convergent infinite series

$$\sum_{r=0}^{\infty} x^r$$

provided that $|x| < 1$.

This idea can readily be generalized: if for all positive integral values of n, a function $f(x)$ can be expressed in the form

$$f(x) = S_n(x) + R_n(x)$$

where $S_n(x)$ denotes the sum of n functions, a_0, $a_1 x$, $a_2 x^2, \ldots, a_{n-1} x^{n-1}$,

$$S_n(x) = \sum_{r=0}^{n-1} a_r x^r$$

and $R_n(x)$ is a remainder term and if, when x lies in the interval $-c < x < c$, $R_n(x) \to 0$, as $n \to \infty$, we say that the function $f(x)$ can be expanded in the form of the convergent *power series*

$$\sum_{r=0}^{\infty} a_r x^r \tag{4}$$

provided $-c < x < c$. We say that (4) is the *power series expansion* of $f(x)$ and that c is the *radius of convergence* of the power series.

For example, the binomial theorem states in effect that the function $(1 + x)^n$ has the power series expansion

$$(1 + x)^n = \sum_{r=0}^{\infty} \frac{n(n - 1) \ldots (n - r + 1)}{r!} x^r$$

with radius of convergence unity.

It is natural to enquire if it is possible to find such an expansion for functions $f(x)$ other than the simple ones stated above. Suppose we *assume* that it is possible to expand an arbitrary function in this way so that, in some sense,

$$f(x) = a_0 + a_1 x + a_2 x^2 + a_3 x^3 + \ldots + a_n x^n + \ldots \tag{5}$$

and that we wish to determine the coefficients $a_0, a_1, a_2, \ldots, a_n, \ldots$ in the expansion. If we differentiate both sides of equation (5) repeatedly with respect to x we find that

$$\begin{aligned}
f'(x) &= a_1 + 2a_2x + 3a_3x^2 + \ldots + na_nx^{n-1} + \ldots \\
f''(x) &= 2a_2 + 3 \cdot 2a_3x + \ldots + n(n-1)a_nx^{n-2} + \ldots \\
f'''(x) &= 3 \cdot 2 \cdot a_3 + \ldots + n(n-1)(n-2)a_nx^{n-3} + \ldots \\
&\cdots\cdots\cdots \\
f^{(n)}(x) &= n!\,a_n + \ldots
\end{aligned}$$

and putting $x = 0$ in equation (5) and in these relations we find that

$$a_0 = f(0),\ a_1 = f'(0),\ a_2 = \frac{f''(0)}{2!},\ a_3 = \frac{f'''(0)}{3!}, \ldots, a_n = \frac{f^{(n)}(0)}{n!}. \tag{6}$$

If we substitute from equations (6) into equation (5) we find that, if the assumptions we have made are valid

$$f(x) = f(0) + f'(0)x + \frac{f''(0)}{2!}x^2 + \frac{f'''(0)}{3!}x^3 + \ldots + \frac{f^{(n)}(0)}{n!}x^n + \ldots \tag{7}$$

This result is known as MACLAURIN'S THEOREM and the series occurring on the right is called the *Maclaurin expansion of the function.*

It is emphasized that this is not a proper proof of Maclaurin's theorem because it makes two important assumptions — that a power series expansion of the type (5) must always exist, and that it may be differentiated term by term — which may not be justifiable. We shall delay giving a more rigorous proof of this theorem until we have developed the elements of the integral calculus. For the moment we shall illustrate how it is used, postponing discussion of its validity until later (Section 9 of Chapter 7).

EXAMPLE 10: *Find the Maclaurin expansion of* $\cos x$.

If we write $f(x) = \cos x$, then $f'(x) = -\sin x$, $f''(x) = -\cos x$, $f'''(x) = \sin x$, $f^{iv}(x) = \cos x, \ldots, f^{(2r)}(x) = (-1)^r \cos x$, $f^{(2r+1)}(x) = (-1)^{r+1} \sin x$ so that $f(0) = 1$, $f'(0) = 0$, $f''(0) = -1$, $f'''(0) = 0$, $f^{iv}(0) = 1, \ldots, f^{(2r)}(0) = (-1)^r$, $f^{(2r+1)}(0) = 0$.

Substituting these values into equation (7) we find that

$$\cos x = 1 - \frac{x^2}{2!} + \frac{x^4}{4!} + \ldots + (-1)^r \frac{x^{2r}}{(2r)!} + \ldots. \tag{8}$$

It can be shown that this expression is valid for all real values of x.

Maclaurin expansions have many uses in mathematics and we shall encounter some of them later. For the moment we shall merely illustrate their use in the evaluation of limits by

EXAMPLE 11: *Evaluate the limit* $\lim_{x \to 0} \dfrac{1 - \cos x}{x^2}$.

From equation (8) we see that if x is small we may replace $1 - \cos x$ by $1 - (1 - \frac{1}{2}x^2 + \frac{1}{24}x^4) = \frac{1}{2}x^2(1 - \frac{1}{12}x^2)$ so that

$$\frac{1 - \cos x}{x^2} = \tfrac{1}{2}(1 - \tfrac{1}{12}x^2) \to \tfrac{1}{2}$$

as $x \to 0$ (in agreement with example 7 of the last chapter).

If we assume an expansion of $f(x)$, not of the form (5) but of the form

$$f(x) = a_0 + a_1(x - a) + a_2(x - a)^2 + \ldots + a_n(x - a)^n + \ldots$$

then proceeding as before, we find that

$$\begin{aligned} f'(x) &= a_1 + 2a_2(x - a) + \ldots + na_n(x - a)^{n-1} + \ldots \\ f''(x) &= 2a_2 + \ldots + n(n - 1)a_n(x - a)^{n-2} + \ldots \\ &\cdots\cdots\cdots\cdots \\ f^{(n)}(x) &= n!\,a_n + \ldots \end{aligned}$$

and putting $x = a$ we have the equations

$$a_0 = f(a), \quad a_1 = f'(a), \quad a_2 = \frac{f''(a)}{2!}, \quad \ldots, \quad a_n = \frac{f^{(n)}(a)}{n!}, \quad \ldots$$

for the coefficients. Hence we obtain the expansion

$$f(x) = f(a) + f'(a)(x-a) + \frac{f''(a)}{2!}(x-a)^2 + \ldots + \frac{f^{(n)}(a)}{n!}(x-a)^n + \ldots. \quad (9)$$

This result is known as Taylor's theorem. Maclaurin's theorem is obviously a special case of Taylor's theorem obtained by putting $a = 0$ in equation (9).

We shall require later some approximations obtained from equation (9). If $|x - a|$ is small it follows from equation (9) that we may approximate to $f(x)$ by the expansion

$$f(x) = f(a) + f'(a)(x - a) + \tfrac{1}{2}f''(a)(x - a)^2 + \tfrac{1}{6}f'''(a)(x - a)^3.$$

If $f'(a) = 0$ but $f''(a) \neq 0$ then a first approximation to $f(x)$ would be given by the equation

$$f(x) = f(a) + \tfrac{1}{2}f''(a)(x - a)^2 \tag{10}$$

but if $f'(a) = f''(a) = 0$ then we must take

$$f(x) = f(a) + \tfrac{1}{6}f'''(a)(x - a)^3. \tag{11}$$

PROBLEMS

1. Show that the Maclaurin expansion of $\sin x$ is

$$x - \frac{x^3}{3!} + \frac{x^5}{5!} - \frac{x^7}{7!} + \dots + (-1)^n \frac{x^{2n+1}}{(2n+1)!} + \dots.$$

2. Show that the first two terms in the Maclaurin expansion of $\tan^{-1} x$ are $x - \frac{1}{3}x^3$ and that those of $\sin^{-1} x$ are $x - \frac{1}{6}x^3$. Hence evaluate

$$\lim_{x \to 0} \frac{\sin^{-1} x - \tan^{-1} x}{x^3}.$$

3. Show that the Maclaurin expansion of $\sin (x + \alpha)$ is

$$\sin \alpha + x \cos \alpha - \frac{x^2}{2} \sin \alpha - \frac{x^3}{3!} \cos \alpha + \dots.$$

4. Show that if x is nearly equal to 1,

$$\sqrt{x} = 1 + \tfrac{1}{2}(x - 1) - \tfrac{1}{8}(x - 1)^2,$$

approximately.

5. From the Maclaurin series for $\sin x$ and $\cos x$, show that

$$\tan x = x + \tfrac{1}{3}x^3 + \tfrac{2}{15}x^5 + \dots.$$

Prove that, when ml is small, the formula

$$\delta = \frac{Fl}{P}\left\{\frac{\tan ml}{ml} - 1\right\}$$

where $m^2 = P/EI$ gives approximately

$$\delta = \frac{1}{3}\frac{Fl^3}{EI}\left(1 + \frac{2}{5}\frac{Pl^2}{EI}\right).$$

5. Maxima and Minima

The problem of determining the maximum (or minimum) value of a given function $f(x)$ frequently arises in physics and physiology. We shall therefore briefly discuss this application of the differential calculus.

First of all we must define what we mean by the terms 'maximum' and 'minimum'. We say that a *maximum turning point* occurs on a curve when there is one point on it lying higher than all neighbouring points on the curve. The point B on the curve in Fig. 54 is such a

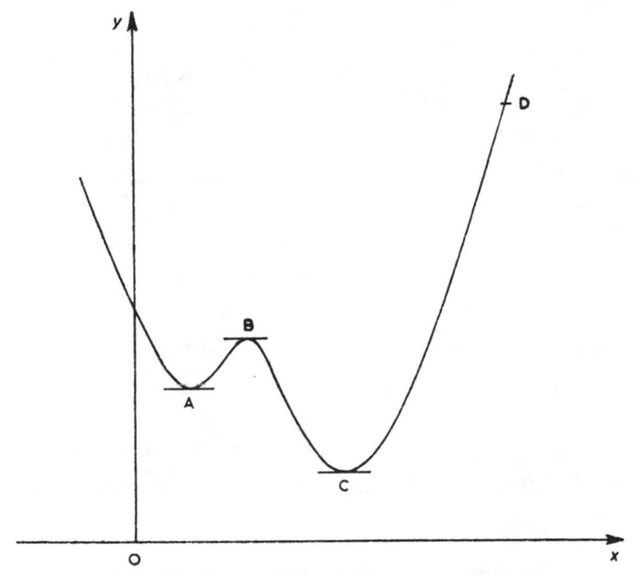

Fig. 54. Maximum and minimum turning values.

point. At the point B the gradient of the curve is zero. The function represented by the curve increases to a maximum value and then decreases from it; at the maximum point B it is neither increasing nor decreasing — it is *stationary*. From the diagram we see that a characteristic of a maximum turning point is that immediately to the left of the point the gradient of the curve is positive and immediately to the right the gradient is negative. Analytically, we can write that the condition that a point should be a maximum turning point on the graph $y = f(x)$ is that

$$f(a + h) < f(a) \tag{1}$$

for all h such that $|h| < \varepsilon$, where ε is some positive real number. The value $f(a)$ is called a maximum value of the function.

A ***minimum turning point*** on a curve lies below all the neighbouring points on the curve. The points A and C on the curve in Fig. 54 are minimum turning points. At both of these points the gradient to the curve is zero. The function represented by the curve must decrease to a minimum and then increase; immediately to the left of the minimum turning point the gradient is negative and immediately to the right it is positive. Analytically: $x = a$ is a minimum turning point of $y = f(x)$ if

$$f(a + h) > f(a) \tag{2}$$

for all h such that $|h| < \varepsilon$, some positive real number. The value $f(a)$ is called a minimum value of the function.

It should be observed that these are purely *local* properties. At a maximum turning point the curve lies higher than at neighbouring points, but not necessarily higher than at all other points. For example in the curve shown in Fig. 54, B is called a maximum turning point although the curve is higher at the point D, and A is called a minimum turning point although the curve is lower at C.

We observed that, in the case of maximum and minimum points, the gradient of the curve was zero. All the points of the curve $y = f(x)$ at which $f'(x) = 0$ are called the *stationary points* of the curve, and the corresponding values of y are called *stationary values* of the function. At first sight it might appear that a stationary point must be

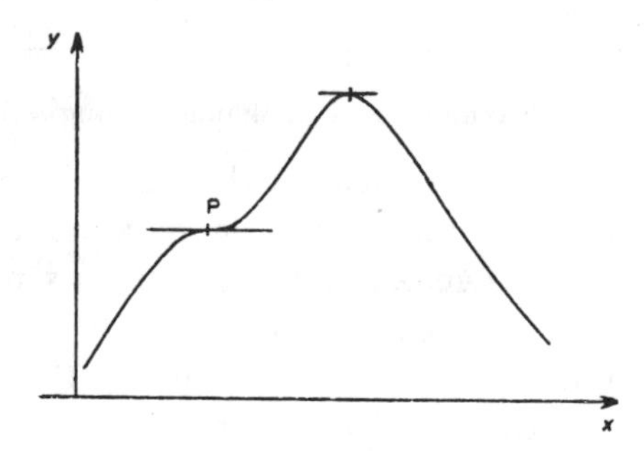

Fig. 55. Horizontal point of inflexion on a curve.

either a maximum or a minimum turning point, but it is obvious from looking at the shape of the curve shown in Fig. 55 that the gradient of the curve is zero at the point P, but that P is neither a maximum nor a minimum turning point. Such a point, where the curve crosses a horizontal tangent is called a *horizontal point of inflexion* on the curve.

We shall now use Taylor's theorem to derive conditions for stationary points on a curve. From equation (9) of the last section we see that if h is small then $f(a+h)$ is given by the equation

$$f(a+h) = f(a) + hf'(a) + \tfrac{1}{2}h^2f''(a) + \tfrac{1}{6}h^3f'''(a). \tag{3}$$

If $x = a$ is a stationary point of the curve then $f'(a) = 0$ and

$$f(a+h) - f(a) = \tfrac{1}{2}h^2f''(a) + \tfrac{1}{6}h^3f'''(a). \tag{4}$$

In cases in which $f''(a) \neq 0$ the first term on the right hand side of (4) will be the dominant one (since h is small) and we have

$$f(a+h) - f(a) = \tfrac{1}{2}h^2f''(a).$$

Since h^2 is always positive it follows that:

$$f(a+h) - f(a) > 0 \text{ if } f''(a) > 0;$$
$$f(a+h) - f(a) < 0 \text{ if } f''(a) < 0.$$

It follows from (1) that $x = a$ is a maximum turning point if $f''(a) < 0$, and from (2) that $x = a$ is a minimum turning point if $f''(a) > 0$. Hence we have:-

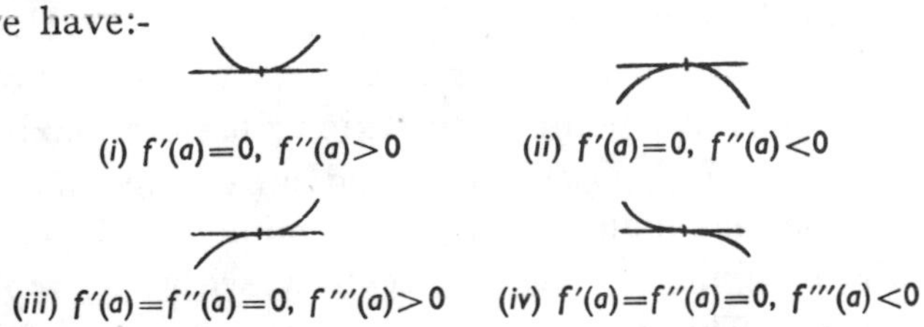

(i) $f'(a)=0,\ f''(a)>0$ (ii) $f'(a)=0,\ f''(a)<0$

(iii) $f'(a)=f''(a)=0,\ f'''(a)>0$ (iv) $f'(a)=f''(a)=0,\ f'''(a)<0$

Fig. 56. Shape of a curve in the neighbourhood of a turning value.

THEOREM 17: *A continuous function $y = f(x)$ with continuous first and second derivative near and at $x=a$ has a maximum at $x=a$ if $f'(a)=0$ and $f''(a) < 0$. It is a minimum if $f'(a) = 0$ and $f''(a) > 0$.*

The argument we have used in establishing the theorem is rather loose, since we talk of one term of the Maclaurin expansion being 'dominant' and reject the rest, but the proof can be made rigorous by replacing this kind of reasoning by a use of an extension of the mean value theorem. Theorem 17 is illustrated graphically in (i) and (ii) of Fig. 56.

Returning to equation (4) we see that if $f'(a) = f''(a) = 0$ then

$$f(a + h) - f(a) = \tfrac{1}{6}h^3 f'''(a). \tag{5}$$

If $f'''(a) > 0$ then

$$f(a + h) - f(a) < 0 \qquad \text{if } h < 0,$$

and

$$f(a + h) - f(a) > 0 \qquad \text{if } h > 0,$$

so that we get a point of inflexion of the kind shown in Fig. 56 (iii). On the other hand, if $f'''(a) < 0$ then

$$f(a + h) - f(a) > 0 \qquad \text{if } h < 0$$

and

$$f(a + h) - f(a) < 0 \qquad \text{if } h > 0$$

and we get a point of inflexion of the type shown in Fig. 56 (iv).

6. Points of Inflexion

Figs. 56 (iii) and (iv) show *horizontal* points of inflexion. If we simply rotate these curves either way about the points shown, through any angle less than a right angle, there is no longer a stationary point where $x = a$, but there is still a point of inflexion.

Geometrically, and rather loosely, as x increases, $f(x)$ curves one way when $x < a$, the other way when $x > a$, and is straight at $x = a$. Defined rigorously, at a point of inflexion $f'(a)$ is a maximum or a minimum. In Fig. 57 (i) $f'(a)$ – the gradient of the curve at $x = a$, (cf. Fig. 39) is a maximum, in (ii) it is a minimum. In fact we can apply all the reasoning of the last section, with $f(x)$ replaced by $f'(x)$ everywhere. This shows that if $x = a$ at a point of inflexion $f''(a) = 0$. Conversely, and by analogy with Theorem 17, if $f''(a) = 0$ there is a point of inflexion at $x = a$ whenever $f'''(a)$ is not equal to 0.

(i) (ii)

Fig. 57. Points of inflexion of a curve and the tangents at these points.

We shall not consider in detail the more complex cases where $f''(a)$ and $f'''(a)$ are both zero; for example at $x = 0$, x^4 has a minimum at the origin, whilst x^5 has a horizontal point of inflexion. But the above rules are adequate for most physiological applications.

7. Illustrations of Maxima, Minima, and Points of Inflexion

To illustrate the application of these ideas in curve tracing we consider:

EXAMPLE 12. *Find the stationary points of the curve*

$$y = x^4 - 4x^3 + 16x - 10$$

and determine whether they are maxima, minima, or points of inflexion. Find all the points of inflexion. Find the gradient of the curve at each point of inflexion and at integral values of x between -3 *and* $+3$. *Illustrate the results by drawing a graph for* $-3 \leqq x \leqq +3$.

If we write $f(x) = x^4 - 4x^3 + 16x - 10$ it follows that

$$f'(x) = 4x^3 - 12x^2 + 16 = 4(x+1)(x-2)^2$$
$$f''(x) = 4(3x^2 - 6x) = 12x(x-2).$$

We therefore get the table of values:-

x	-3	-2	-1	0	1	2	3
$f(x)$	131	6	−21	−10	3	6	11
$f'(x)$	−200	−64	0	16	8	0	16
$f''(x)$	180	96	36	0	−12	0	36
Nature			Min.			P.I.	

The tangent at the point of inflexion at $x = -1, y = 0$ is

$$y = -10 + 16x.$$

The distances of points on the curve from this straight line are given by

$$f(x) - y = x^4 - 4x^3$$

so that the curve is very close to this tangent for all small values of x.

Using all this information, we get the curve shown in Fig. 58.

We shall now consider one or two problems to illustrate the use of these principles in determining maximum (or minimum) values of

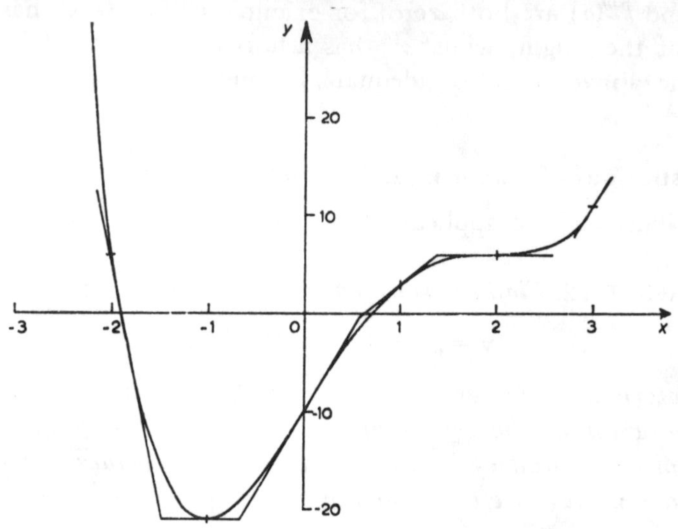

Fig. 58. Graph of the function $y = x^4 - 4x^3 + 16x - 10$.

functions occurring in some simple systems. We begin with a simple geometrical problem.

EXAMPLE 13: *An open tank, square in plan and of uniform depth, is to be made to have a prescribed capacity, the sides and the bottom being made of concrete of prescribed uniform thickness. Show that the quantity of concrete needed is a minimum if the inside dimensions are chosen so that the depth is one half of a side.*

We may suppose that the prescribed capacity of the tank is V and that the thickness of the concrete is t. We take x and y to be respectively the internal depth of the tank and the length of a side of the internal cavity. Since the capacity of the tank is V, we have $xy^2 = V$ from which it follows that $x = V/y^2$. Now the quantity of concrete required to make the tank is

$$Q = (x + t)(y + 2t)^2 - V.$$

Replacing x by V/y^2, we see that

$$Q = \left(\frac{V}{y^2} + t\right)(y + 2t)^2 - V$$

from which we can deduce that

$$Q = ty^2 + 4t^2y + \frac{4Vt}{y} + \frac{4Vt^2}{y^2} + 4t^3.$$

In this expression V and t are fixed so that Q is a function of y alone. This function has the derivative

$$Q'(y) = 2ty + 4t^2 - \frac{4Vt}{y^2} - \frac{8Vt^2}{y^3} = 2t(y + 2t)\left(1 - \frac{2V}{y^3}\right)$$

so that $dQ/dy = 0$ when $y = -2t$ and when $y = (2V)^{\frac{1}{3}}$. The former value has no physical significance so we consider only the second one. Since

$$Q''(y) = 2t + \frac{8Vt}{y^3} + \frac{24Vt^2}{y^4} = 2t\left[1 + \frac{4V}{y^3} + \frac{12Vt}{y^4}\right]$$

it follows that $Q''(2^{\frac{1}{3}}V^{\frac{1}{3}}) > 0$. Hence $y = (2V)^{\frac{1}{3}}$ gives a minimum value of Q. The corresponding value of x is

$$x = \frac{V}{y^2} = \tfrac{1}{2}(2V)^{\frac{1}{3}} = \tfrac{1}{2}y$$

showing that the minimum quantity of concrete is given when the depth is one half of a side.

As illustrations of the use of the method in biological problems we have:-

EXAMPLE 14: *Hoorweg's Law*

Hoorweg investigated the relationship between various electrical units in the stimulation of nerves by condenser discharges at threshold level. He found that the potential P required to elicit a minimal response (muscle contraction) was related to the capacity C of the condenser, by the following empirical formula:-

$$P = aR + \frac{b}{C} \tag{6}$$

where R is the resistance (assumed constant) and a, b and C are positive constants. Now if the capacity C is measured in microfarads and the potential P of the condensor is measured in volts, it is a well known result in elementary physics that the electrical energy E associated with the charge is (in ergs) $5CP^2$. Substituting the value of P given by (6) we find that

$$E = 5C\left(aR + \frac{b}{C}\right)^2 = 5a^2R^2C + 10abR + \frac{5b^2}{C}.$$

E is a function of C alone, and

$$E'(C) = 5a^2R^2 - \frac{5b^2}{C^2} = \frac{5a^2R^2}{C^2}\left(C^2 - \frac{b^2}{a^2R^2}\right)$$

$$E''(C) = \frac{10b^2}{C^3}.$$

Thus when $C = b(aR)$, $E'(C) = 0$ and $E''(C) > 0$ showing that when $C = b(aR)$ the energy E is a minimum.

EXAMPLE 15: *Maximum velocity in the airways during coughing* With the aid of röntgenograms it can be demonstrated that during coughing the diameter of the traches and main bronchi decreases. The following theoretical reasoning is due to B. F. Visser.

We assume that after a deep inspiration the glottis is closed. During the next phase a positive pressure develops which causes a decrease of the diameter of the airways. We shall make the simplest assumption, namely that the radius is a linear function of the pressure:-

$$r = r_0 - aP \tag{7}$$

where r is the radius at a pressure P above atmospheric pressure, r_0 is the radius at atmospheric pressure and a is a constant. From Poiseuille's law we know that the resistance R, offered by the air passages is inversely proportional to the fourth power of the radius r, i.e.

$$R = \frac{k}{r^4}. \tag{8}$$

After opening the glottis the rate of flow is given by

$$\dot{V} = \frac{P}{R} = \frac{Pr^4}{k} = \frac{r^4(r_0 - r)}{ak} \tag{9}$$

where $\dot{V}$ is the amount of gas passing a certain point per unit time. It should be noted that the rate of flow is not the same as the velocity v which is related to $\dot{V}$ and r by the equation

$$v = \frac{\dot{V}}{\pi r^2}. \tag{10}$$

Substituting for $\dot{V}$ from the last of equations (9) we find that

$$v = \frac{r^2(r_0 - r)}{\pi a k}. \tag{11}$$

Differentiating (11) with respect to r we find that

$$\frac{dv}{dr} = \frac{2r_0 r - 3r^2}{\pi a k} = -\frac{3r}{\pi a k}\left(r - \tfrac{2}{3}r_0\right)$$

and

$$\frac{d^2v}{dr^2} = \frac{2r_0 - 6r}{\pi a k}$$

from which it follows that when $r = \frac{2}{3}r_0$,

$$\frac{dv}{dr} = 0, \quad \frac{d^2 v}{dr^2} < 0$$

showing that v has a maximum value when $r = \frac{2}{3}r_0$.

This result may be interpreted as follows: with decreasing radius the velocity increases during coughing until a maximum velocity is reached at a radius having a value two-thirds that of the resting value. When there is a further decrease of the radius the velocity decreases (because of the increased resistance offered by the smaller radius).

PROBLEMS

1. Find the turning points and points of inflexion on the curve

$$16y = x^4 - 4x^3 - 18x^2 - 20x - 7.$$

Sketch the form of the curve in the neighbourhood of each point of inflexion and give a rough graph of the complete curve.

2. Investigate the maximum and minimum values of the function $\cos^4 x + 3\sin^4 x$ for values of x between 0 and π, and sketch the graph of the function for the same range.

3. According to Borel and Deltheil, the probability p that two bodies inside a spherical space of radius R would be separated by a distance smaller than a is given by the equation

$$p = x^3 - \tfrac{9}{16}x^4 + \tfrac{1}{32}x^6,$$

where $x = R/a$*. Draw the curve for x in the interval $0 \leqq x \leqq 2$.

4. A certain article costs \$ 1 to manufacture. The number sold per annum varies inversely as the square of the selling price, over a certain range of price. If 2 500 are sold when the selling price is \$ 1, how many must be sold to give the maximum profit, and what is this maximum profit?

5. A variable triangle ABC, of constant area Δ has sides AB and AC equal. If $BC = x$, and the sum of the squares on the three sides is y, show that

$$y = \frac{3x^4 + 16\Delta^2}{2x^2},$$

and that y is a minimum when the triangle is equilateral.

8. Differentials

We saw in Fig. 39, p. **120**, that it was possible to give a simple geometrical interpretation of the idea of the derivative of a function. We shall now use the same interpretation to introduce the idea of the differential of a function. Fig. 59 shows a portion of the graph of the function $y = f(x)$ in the neighbourhood of the point P whose abscissa OM is x. The point Q is a neighbouring point with abscissa OU equal to $x + \delta x$. Through P we draw the tangent PT to the curve cutting QU in T and the line PR parallel to the x-axis and cutting QU in R. Since UQ is the ordinate of a point on the curve with abscissa $x + \delta x$ it follows that $UQ = f(x + \delta x)$ and since $UR = MP = f(x)$ it follows that

$$RQ = f(x + \delta x) - f(x)$$

and since this is the increment in y due to an increment δx in x, we denote RQ by δy.

Since $f'(x)$ is the value of the gradient to the curve at the point P, it follows that

$$\frac{RT}{PR} = f'(x),$$

and since $PR = MU = \delta x$, we find that $RT = f'(x)\delta x$. We now imagine that a point is moving along the curve but that, when it comes to the point P, it moves not along the arc PQ but along the tangent PT.

* For the use of this formula in a biological problem, see G. Knaysi, "Elements of Bacterial Cytology" 2nd. ed. (Cornstock Pub. Co., Ithaca N.Y., 1951), p. 110.

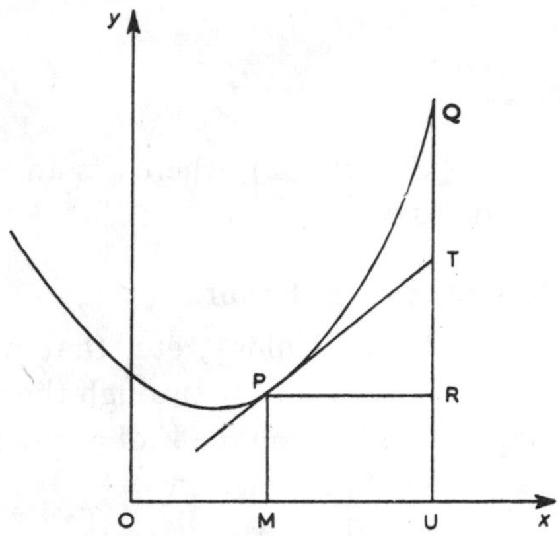

Fig. 59. The differential of a function.

As the abscissa of the point changes from x to $x + \delta x$ the ordinate will, on this assumption, take the increment $RT = f'(x)\delta x$, not the increment RQ. This *purely hypothetical* increment RT is called the *differential* of y and is denoted by $\mathrm{d}y$ or $\mathrm{d}f(x)$ so that

$$\mathrm{d}y = f'(x)\delta x. \tag{1}$$

In particular we can talk about the differential $\mathrm{d}x$ of the function $f(x) = x$. Since in this case $f'(x) = 1$ it follows that

$$\mathrm{d}x = \delta x. \tag{2}$$

It follows from equations (1) and (2) that, if $y = f(x)$, the differentials $\mathrm{d}x$ and $\mathrm{d}y$ satisfy the equation

$$\mathrm{d}y = f'(x)\mathrm{d}x \tag{3}$$

which may be regarded as a symbolic way of writing the equation

$$\frac{\mathrm{d}y}{\mathrm{d}x} = f'(x).$$

It should be noted that the equation $\mathrm{d}y = f'(x)\delta x$ is exact (it is merely a consequence of the definition of $\mathrm{d}y$) but that the equation $\delta y = f'(x)\delta x$ is only approximate. If δx is small the difference between δy (i.e. RQ) and $\mathrm{d}y$ (i.e. RT) is small but it is not zero.

PROBLEMS

1. Show that

(i) $\mathrm{d}(2x^3 - x^2 + 1) = (6x^2 - 2x)\mathrm{d}x$;

(ii) $\mathrm{d}fg = f\mathrm{d}g + g\mathrm{d}f$;

(iii) $\mathrm{d} \sin x = \cos x \, \mathrm{d}x$.

2. Show that

(i) $x^n \, dx = d\left(\frac{x^{n+1}}{n+1}\right)$ if $n \neq -1$;

(ii) $(x^3 - 1)dx = d(\frac{1}{4}x^4 - x + c)$, where c is an arbitrary constant;

(iii) $\sin x \, dx = - d \cos x$.

9. The Calculation of Small Errors

It often occurs in medicine, biology, etc. that we cannot measure a prescribed quantity directly, but only through the medium of another quantity, e.g. we may obtain the volume V of a sphere from a measurement of the radius r, using the relation

$$V = \tfrac{4}{3}\pi r^3. \tag{1}$$

Now suppose we have a large sphere having a radius which is assumed to be 100 cm but that, in fact, we have made an error of 1 % so that the radius is not 100 cm but 101 cm. Then the volume instead of having the "correct" value

$$V = \tfrac{4}{3}\pi(100)^3 = \tfrac{4}{3}\pi \times 1\,000\,000$$

will have the "false" value

$$V + \Delta V = \tfrac{4}{3}\pi(101)^3 = \tfrac{4}{3}\pi \times 1\,030\,301$$

so that the error is

$$\Delta V = \tfrac{4}{3}\pi \times 30\,301$$

and the percentage error is approximately

$$p(V) = \frac{\Delta V}{V} \times 100 = \frac{\tfrac{4}{3}\pi \times 30\,301}{\tfrac{4}{3}\pi \times 1\,000\,000} \times 100 = 3.0301.$$

In other words an error of 1 % in the measurement of the radius leads to an error of 3 % in the assessed value of the volume.

We can readily generalise this to the case in which we make a percentage error ε in the measurement of the radius r, i.e. the radius is actually $r(1 + \varepsilon/100)$ so that the volume has the false value

$$V + \Delta V = \tfrac{4}{3}\pi\left\{r\left(1 + \frac{\varepsilon}{100}\right)\right\}^3 = \tfrac{4}{3}\pi r^3\left(1 + \frac{\varepsilon}{100}\right)^3$$

and the percentage error in the assessment of the volume is

$$p(V) = \frac{\Delta V}{V} \times 100 = \left\{\left(1 + \frac{\varepsilon}{100}\right)^3 - 1\right\} 100.$$

Expanding $(1 + \varepsilon/100)^3$ by the binomial theorem we find that

$$p(V) = 3\varepsilon + \frac{3\varepsilon^2}{100} + \frac{\varepsilon^3}{10\,000}.$$

If ε is small we can neglect all but the first term on the right hand side of this equation and obtain

$$p(V) = 3\varepsilon. \tag{2}$$

When $\varepsilon = 1$, i.e. when the error in r is 1 %, we have an error of 3 % in V.

We have made these estimates by purely algebraic methods. We shall now show the same result by using the idea of differentials. The basic idea is that the differential $\mathrm{d}V$ represents a good approximation to the increment (here the absolute error) of V resulting from the increment (absolute error) $\mathrm{d}r$ in r. From equation (1) we have

$$\mathrm{d}V = 4\pi r^2 \mathrm{d}r \tag{3}$$

from which it follows that if the error in the measurement of r is $\mathrm{d}r$ then the error in the measurement of V is (to the first order) $\mathrm{d}V$ and the percentage error is

$$p(V) = \frac{\mathrm{d}V}{V} \times 100 = \frac{4\pi r^2 \mathrm{d}r}{\frac{4}{3}\pi r^3} \times 100 = 3\frac{\mathrm{d}r}{r} \times 100.$$

Now $(\mathrm{d}r/r) \times 100$ is the percentage error in the measurement of r. If we denote this by $p(r)$ we find that

$$p(V) = 3p(r) \tag{4}$$

in agreement with equation (2).

The corresponding formula in the general case in which two quantities x and y are related through an equation

$$y = f(x) \tag{5}$$

is readily derived. If the error in x is $\mathrm{d}x$ then the error in y, $\mathrm{d}y$ say, is given by the equation

$$\mathrm{d}y = f'(x)\mathrm{d}x$$

and it follows that

$$\frac{\mathrm{d}y}{y} = \frac{xf'(x)}{f(x)} \cdot \frac{\mathrm{d}x}{x}.$$

If we denote the percentage errors in x and y by $p(x)$ and $p(y)$ respectively then

$$p(y) = \frac{xf'(x)}{f(x)} \cdot p(x). \tag{6}$$

In particular if $y = kx^n$, $f'(x) = knx^{n-1}$ so that

$$p(y) = np(x). \tag{7}$$

When $n = 3$ we get equation (4) above.

It must be stressed that these results are not "correct" from the point of view of "the calculus of errors". We have confined our attention to the discussion of errors whose magnitudes are known precisely; we have not considered the nature of systematic errors or their method of variation. However, these observations serve as an introduction to the calculus of errors which is of great importance in any science, such as physics or physiology, in which deductions are made from the precise measurements of physical variables. For the extension of these very simple ideas to more complicated situations the reader is referred to p. 385 below; for a discussion of the calculus of errors interested readers should consult W. M. Smart's "The Calculus of Observations" (Cambridge University Press, 1958).

PROBLEMS

1. A right circular cylinder open at one end and with a given volume has to be made so that its total surface area is a minimum. Show that its radius must equal its length.

If there is a small increase of n % in the diameter, find the percentage change which must be made in the length in order to keep the volume constant.

2. If

$$y = \frac{x}{\sqrt{(3 - x^2)}},$$

and if x is a measurement which is subject to error, show that the percentage error in y, due to a small error h in x, is approximately

$$\frac{300h}{x(3 - x^2)}.$$

Hence find the value of x for which the percentage error in y due to a given small error in x is a minimum.

3. The area Δ of a triangle of sides a, b, c and angles A, B, C is calculated from the formula

$$\Delta = \tfrac{1}{2}bc \sin A.$$

Show that if the values of b and c are exact but there is an error of amount δA in A then the error in the value of Δ is $\Delta \cot A\,\delta A$.

If A is 45° show that to obtain a percentage error in area of less than 1 %, the error in measuring A must be less than 35′.

4. The volume V of the largest sphere which can be contained in a cone of volume v and semi-vertical angle θ is

$$V = \frac{4v(\sin\theta)(1-\sin\theta)}{(1+\sin\theta)^2}.$$

If δV is the change in V produced by a change $\delta\theta$ in θ, v remaining constant, show that

$$\frac{\delta V}{V} = \frac{1-3\sin\theta}{\sin\theta\cos\theta}\,\delta\theta.$$

10. Geometrical Applications

We have already interpreted the derivative of a function $f(x)$ as the slope of the tangent to the curve with equation $y = f(x)$. If $P(x_1, y_1)$ is the point of contact of the tangent, then $f'(x_1)$ is the gradient of the tangent, and the equation of the tangent at (x_1, y_1) is

$$y - y_1 = f'(x_1)(x - x_1). \tag{1}$$

The *normal* at the point P on the curve is the line through P perpendicular to the tangent. It is easily shown that the gradients m_1, m_2 of two perpendicular lines are related by the equation $m_1 m_2 = -1$ so that the gradient of the normal will be

$$m = -\frac{1}{f'(x_1)}.$$

Substituting this value into the equation $y - y_1 = m(x - x_1)$ we see that the equation of the normal is

$$f'(x_1)(y - y_1) + (x - x_1) = 0. \tag{2}$$

If the tangent at the point P cuts the x-axis in T and the normal cuts it in G, and if PN is perpendicular to the x-axis, (cf. Fig. 58), the line TN is called the *subtangent* of P and the line NG the *subnormal*. We shall now derive simple expressions for these lengths.

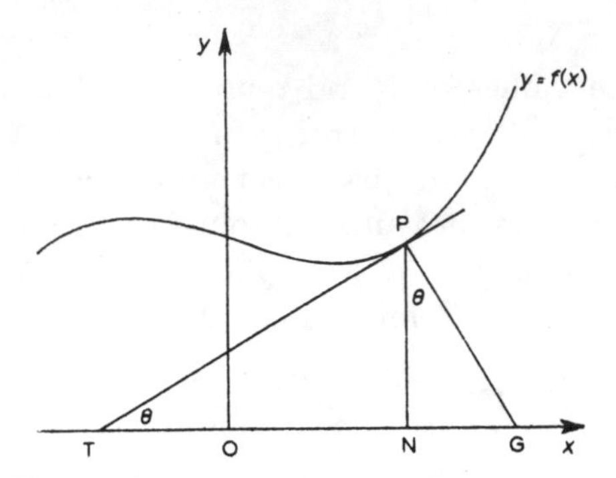

Fig. 60. The subtangent and the subnormal.

If θ is the angle which TP makes with the x-axis then, since the gradient of TP is $\tan\theta$, it follows that $\tan\theta = f'(x_1)$. In the triangle TNP, $NP = y_1$ and the angle NTP is θ so that

$$\tan\theta = \frac{NP}{TN} = \frac{y_1}{TN}$$

from which it follows that

$$TN = \frac{y_1}{\tan\theta} = \frac{y_1}{f'(x_1)}$$

or since $y_1 = f(x_1)$,

$$TN = \frac{f(x_1)}{f'(x_1)}. \tag{3}$$

Since the angles TPG and GNP are both right angles it follows that $\angle NTP + \angle TPN = \frac{1}{2}\pi$ and $\angle TPN + \angle NPG = \frac{1}{2}\pi$ from which we find that the angle NPG is θ. Now in the triangle PNG, $PN = y_1$ so that

$$\tan\theta = \frac{NG}{NP} = \frac{NG}{y_1}$$

and it follows at once that the subnormal is given by the equation

$$NG = y_1 \tan\theta = f'(x_1)\, f(x_1). \tag{4}$$

EXAMPLE 16: *Find the equations of the tangent and normal at the point $(a, 2a)$ on the parabola $y^2 = 4ax$. Show that the subtangent and subnormal of this point are equal.*

For this curve $f(x) = 2(ax)^{\frac{1}{2}}$ so that $f'(x) = a^{\frac{1}{2}}x^{-\frac{1}{2}}$ and $f'(a) = 1$. If we put $x_1 = a$, $y_1 = 2a$, $f'(x_1) = 1$ in equations (1) and (2) we find that the equation of the tangent is

$$y - 2a = x - a$$

i.e.

$$x - y + a = 0,$$

and the equation of the normal is

$$(y - 2a) + (x - a) = 0$$

i.e.

$$x + y - 3a = 0.$$

Since $f(a) = 2a$, $f'(a) = 1$ it follows from equations (3) and (4) that the subtangent and the subnormal of this point are both of length $2a$.

In certain applications of the calculus we are interested in the length s of the arc AP measured from a fixed point A to a point P on the curve with equation $y = f(x)$. We shall now derive an expression for the derivative of s with respect to x. We suppose that P and Q are points with abscissae x and $x + \delta x$ respectively, both points lying on the curve with equation $y = f(x)$; cf. Fig. 61. If we denote

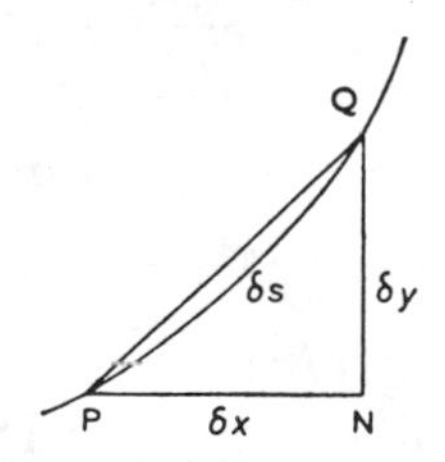

Fig. 61. The derivative of an arc.

the length of the small arc PQ by δs we then have the relation

$$(\delta s)^2 = \left(\frac{\text{arc } PQ}{\text{chord } PQ}\right)^2 \{(\delta x)^2 + (\delta y)^2\} \tag{5}$$

since the square of the length of the chord PQ is, by Pythagoras' theorem, equal to $PN^2 + NQ^2$ i.e. to $(\delta x)^2 + (\delta y)^2$. Dividing both sides of equation (5) by $(\delta x)^2$ we find that

$$\left(\frac{\delta s}{\delta x}\right)^2 = \left(\frac{\text{arc } PQ}{\text{chord } PQ}\right)^2 \left\{1 + \left(\frac{\delta y}{\delta x}\right)^2\right\}.$$

Now in the limit as Q tends to P, $\delta x \to 0$ and the ratio of the arc PQ to the chord PQ tends to unity so that

$$\left(\frac{ds}{dx}\right)^2 = 1 + \left(\frac{dy}{dx}\right)^2. \tag{6}$$

EXAMPLE 17: *Show that, for the parabola* $y^2 = 4ax$,

$$\frac{ds}{dx} = \sqrt{1 + \frac{a}{x}}.$$

This follows immediately from equation (6) and the fact that, since $y = 2a^{\frac{1}{2}}x^{\frac{1}{2}}$

$$\frac{dy}{dx} = \frac{a^{\frac{1}{2}}}{x^{\frac{1}{2}}}.$$

In many problems in geometry and physics it is possible to give a functional relation between two quantities x and y, by means of a third variable. For example for a body falling freely under gravity and starting from rest the relation between the distance x it has fallen and its velocity at that distance is of the form

$$y^2 = 2gx \tag{7}$$

where g is a constant. Alternatively we could specify this relationship by giving the values of x and y in terms of the time t at which they are measured; it is known from elementary mechanics that

$$x = \tfrac{1}{2}gt^2, \quad y = gt. \tag{8}$$

As we said at the beginning of this paragraph, we may express the functional relation between x and y not by the single equation (7) but by the pair of equations (8). In general it is possible to express a relation of the type

$$y = F(x) \tag{9}$$

by a pair of equations of the type

$$x = f(t), \quad y = g(t) \tag{10}$$

involving a new variable t. In some cases we are able to eliminate the variable t from the equations (10), but in other cases the resulting form of (9) is so complicated that it is preferable to retain the pair of equations (10). Equations of the type (10) are called *parametric equations* of the relation (9), the variable t being usually referred to as a *parameter.*

EXAMPLE 18: *Show that $x = at^2$, $y = 2at$ are parametric equations for the parabolic relation $y^2 = 4ax$.*

We can write the given equations in the form

$$t = \frac{y}{2a}, \quad t^2 = \frac{x}{a}.$$

Eliminating t from this pair of equations we have

$$\frac{x}{a} = \left(\frac{y}{2a}\right)^2$$

from which it follows that $y^2 = 4ax$.

Parametric equations are of the greatest value in specifying the coordinates of any point on a curve. Thus (by virtue of example 18) we speak of the parabola as the locus of points $(at^2, 2at)$.

To find dy/dx when x and y are given by the parametric equations (10) we let t take a small increment τ resulting in an increment h in x. Then h will be given by the relation

$$h = f(t + \tau) - f(t)$$

and the corresponding change in y will be

$$k = g(t + \tau) - g(t).$$

Now

$$\frac{dy}{dx} = \lim_{h\to 0} \frac{k}{h} = \lim_{\tau\to 0} \frac{g(t + \tau) - g(t)}{f(t + \tau) - f(t)} = \lim_{\tau\to 0} \frac{\{g(t + \tau) - g(t)\}/\tau}{\{f(t + \tau) - f(t)\}/\tau}$$

and by Theorem 6 it follows that

$$\frac{dy}{dx} = \lim_{\tau\to 0} \frac{g(t + \tau) - g(t)}{\tau} \div \lim_{\tau\to 0} \frac{f(t + \tau) - f(t)}{\tau}$$

so that we obtain the result

$$\frac{dy}{dx} = \frac{g'(t)}{f'(t)} \tag{11}$$

which is sometimes written in the form

$$\frac{\mathrm{d}y}{\mathrm{d}x} = \frac{\dfrac{\mathrm{d}y}{\mathrm{d}t}}{\dfrac{\mathrm{d}x}{\mathrm{d}t}}. \tag{12}$$

Similarly if a change δt in t produces changes δx, δy and δs in x, y and s respectively we can rewrite equation (5) in the form

$$\left(\frac{\delta s}{\delta t}\right)^2 = \left(\frac{\text{arc } PQ}{\text{chord } PQ}\right)^2 \left\{\left(\frac{\delta x}{\delta t}\right)^2 + \left(\frac{\delta y}{\delta t}\right)^2\right\}.$$

Letting δt tend to zero we then obtain the formula

$$\left(\frac{\mathrm{d}s}{\mathrm{d}t}\right)^2 = \left(\frac{\mathrm{d}x}{\mathrm{d}t}\right)^2 + \left(\frac{\mathrm{d}y}{\mathrm{d}t}\right)^2. \tag{13}$$

EXAMPLE 19: *Find* $\mathrm{d}y/\mathrm{d}x$ *and* $\mathrm{d}s/\mathrm{d}t$ *for the parametric equations of the parabola given in example* 18.

Here

$$\frac{\mathrm{d}x}{\mathrm{d}t} = 2at, \quad \frac{\mathrm{d}y}{\mathrm{d}t} = 2a.$$

Substituting these values in equations (12) and (13) we find that

$$\frac{\mathrm{d}y}{\mathrm{d}x} = \frac{1}{t}, \qquad \frac{\mathrm{d}s}{\mathrm{d}t} = 2a\sqrt{(1 + t^2)}.$$

PROBLEMS

1. Find the equation of the tangent at the point (1, 1) to the curve $y = 2x^4 - 3x^3 + 5x^2 - 2x - 1$.

2. Show that the subtangent and the subnormal of the point (a, a) of the curve* $ay^2 = x^3$ are $\frac{2}{3}a$ and $\frac{3}{2}a$ respectively.

Show also that for this curve

$$\frac{\mathrm{d}s}{\mathrm{d}x} = \sqrt{1 + \frac{9x}{4a}}.$$

Verify that the function

$$s = \frac{8a}{27}\left(1 + \frac{9x}{4a}\right)^{\frac{3}{2}} - \frac{8a}{27}$$

* This curve is called a *semi-cubical parabola*.

has this derivative, and is zero when $x = 0$.

3. Show that $x = a \sin t$, $y = a \cos t$ are parametric equations of the circle $x^2 + y^2 = a^2$, and prove that

$$\frac{dy}{dx} = -\tan t, \quad \frac{ds}{dt} = a.$$

4. Show that

$$x = \frac{1 - t^2}{1 + t^2} a, \quad y = \frac{2t}{1 + t^2} a$$

are parametric equations of the circle $x^2 + y^2 = a^2$, and prove that

$$\frac{dy}{dx} = \frac{t^2 - 1}{2t}, \quad \frac{ds}{dt} = \frac{2a}{1 + t^2}.$$

11. Newton's Method of Finding Approximate Values of Roots of Equations

We often have to find the value of a real root of an equation of the type

$$f(x) = 0 \tag{1}$$

where $f(x)$ is quite a complicated function of x. An approximate value can be found by drawing the graph of the function $y = f(x)$ and finding where it cuts the x-axis, since on the x-axis we have $y = 0$. Thus, if we measure OX in Fig. 62 we have an approximate value of a real root of equation (1). The difficulty is that we cannot draw graphs accurately enough nor measure distances such as OX accurately. What we are more likely to do is to measure a distance OQ which corresponds not to a zero value of y but to the non-zero value QP. Newton devised a method for getting a better approximation to the correct value once an approximate value is known. This depends on the fact that, in general, if we draw the tangent PT to the curve and that tangent cuts the x-axis in R, then $OR = x_1$ will be a better approximation to the exact value OX than is the first approximation $OQ = \xi$.

Now we saw in the last section that if ξ is the abscissa of P the subtangent RQ is of length $f(\xi)/f'(\xi)$ so that

$$x_1 = OR = OQ - RQ = \xi - \frac{f(\xi)}{f'(\xi)}.$$

This is only a very crude argument on geometrical considerations but

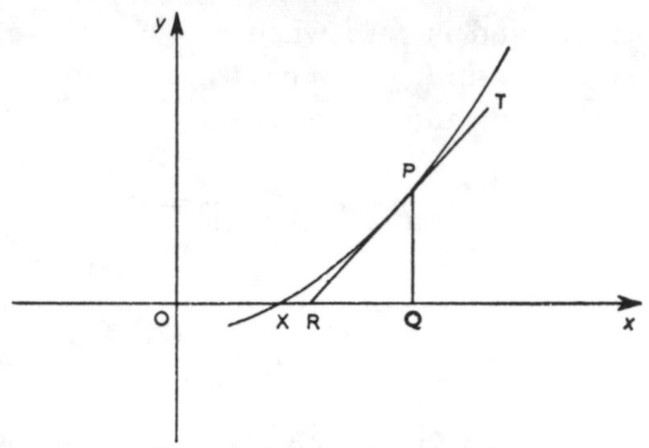

Fig. 62. Newton's method.

it can be shown rigorously that if $f''(x)$ has the same sign as $f(\xi)$ in the neighbourhood of $x = \xi$ then the approximation

$$x_1 = \xi - \frac{f(\xi)}{f'(\xi)} \tag{2}$$

is a better approximation to the correct value of the root of the equation (1) than ξ is itself. For ordinary functions x_1 is always a better approximation than ξ if the approximate value ξ is itself reasonably good.

Equation (2) is known as *Newton's formula*. We illustrate its use by

EXAMPLE 20: *Show that the equation*

$$x^3 + 5x^2 + 3x - 1 = 0$$

has only one positive real root and find that root correct to four places of decimals.

If we let $f(x) = x^3 + 5x^2 + 3x - 1$, then $f(0) = -1$, $f(1) = 8$ and $f'(x) = 3x^2 + 10x + 3$. Since $f'(x)$ is positive for all positive values of x it follows that $f'(x)$ increases steadily as x increases. Hence the curve $y = f(x)$ will cut the positive x-axis at only one point. Since $f(0) = -1$ and $f(0.5) = 1.125$ it follows that this root lies somewhere between 0 and 0.5. To find an approximate value ξ we use the process of linear interpolation shown in Fig. 63. At the point $x = 0$ we erect a perpendicular of length 1 pointing downwards and at $x = 0.5$ we erect one of length 1.125 pointing upwards. The straight line joining the extremities A, B of these lines cuts the x-axis at the point $x = 0.23$, and this would give us an approximate value ξ.

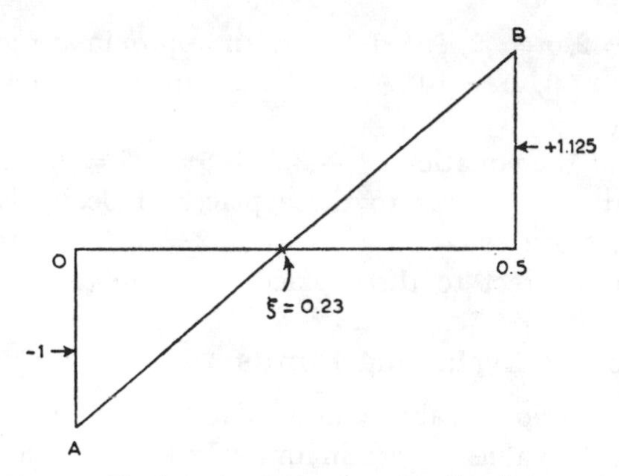

Fig. 63. Finding the first approximation to a root.

Substituting the present $f(x)$ in equation (2) we find that

$$x_1 = \xi - \frac{\xi^3 + 5\xi^2 + 3\xi - 1}{3\xi^2 + 10\xi + 3}. \tag{3}$$

If we put $\xi = 0.23$ we have

$$\begin{aligned} x_1 &= 0.2300 - \frac{0.0122 + 0.2645 + 0.6900 - 1.0000}{0.1587 + 2.3000 + 3.0000} \\ &= 0.2300 + \frac{0.0333}{5.4587} \\ &= 0.2361. \end{aligned}$$

If we now put $\xi = 0.2361$ and repeat the process we find that

$$\begin{aligned} x_1 &= 0.2361 - \frac{0.013\,16 + 0.278\,72 + 0.708\,30 - 1.000\,00}{0.167\,23 + 2.361\,00 + 3.000\,00} \\ &= 0.2361 - \frac{0.000\,18}{5.528\,23}, \end{aligned}$$

showing that, to the accuracy required, the positive root is 0.2361.

PROBLEMS

1. If ξ is an approximate value of the m-th root of a positive number n, show that a more accurate value is

$$\left(1 - \frac{1}{m}\right)\xi + \frac{n}{m\xi^{m-1}}.$$

Taking $m = 2$, $n = 2$, $\xi = 1.4$ find an approximate value for $\sqrt{2}$, and taking $m = 3$, $n = 10$, $\xi = 2.2$ find an approximate value for $\sqrt[3]{10}$.

2. Show that the equation $x^3 - 3x^2 + 5x - 7 = 0$ has a root near 2.2. Find that root correct to three places of decimals.

3. Show that the equation $x^3 - 5x - 2 = 0$ has a root near 2.4. Find that root correct to three places of decimals.

12. A Method of Evaluating Limits

We saw in Section 4 above how Maclaurin's theorem could be used to derive the values of certain limits. In this section we shall show how Taylor's theorem may be used to derive a quite general result.

Suppose that $f(a) = 0$ and $g(a) = 0$ and that we wish to evaluate the limit

$$\lim_{x \to a} \frac{f(x)}{g(x)}.$$

Then, by Taylor's theorem, since $f(a) = 0$, we may write

$$f(x) = (x - a)f'(a) + \tfrac{1}{2}(x - a)^2 f''(a) + \tfrac{1}{6}(x - a)^3 f'''(a)$$

and similarly

$$g(x) = (x - a)g'(a) + \tfrac{1}{2}(x - a)^2 g''(a) + \tfrac{1}{6}(x - a)^3 g'''(a)$$

so that

$$\frac{f(x)}{g(x)} = \frac{f'(a) + \frac{1}{2}(x - a)f''(a) + \frac{1}{6}(x - a)^2 f'''(a)}{g'(a) + \frac{1}{2}(x - a)g''(a) + \frac{1}{6}(x - a)^2 g'''(a)}. \tag{1}$$

If $f'(a)$ and $g'(a)$ are not both zero then it follows that

$$\lim_{x \to a} \frac{f(x)}{g(x)} = \frac{f'(a)}{g'(a)}. \tag{2}$$

If both $f'(a)$ and $g'(a)$ are zero then we can write equation (1) in the form

$$\frac{f(x)}{g(x)} = \frac{f''(a) + \frac{1}{3}(x - a)f'''(a)}{g''(a) + \frac{1}{3}(x - a)g'''(a)}$$

and it follows that, if $f''(a)$ and $g''(a)$ are not both zero,

$$\lim_{x \to a} \frac{f(x)}{g(x)} = \frac{f''(a)}{g''(a)}. \tag{3}$$

Equation (2) is known as *l'Hospital's rule.*

EXAMPLE 21: *Evaluate* $\lim_{x\to 1} \dfrac{x - \tan\dfrac{\pi x}{4}}{1 - x}$.

Since

$$\frac{d}{dx}\left(x - \tan\frac{\pi x}{4}\right) = 1 - \frac{\pi}{4}\sec^2\left(\frac{\pi x}{4}\right)$$

we see that

$$\lim_{x\to 1} \frac{x - \tan\dfrac{\pi x}{4}}{1 - x} = \frac{\pi}{4}\sec^2\frac{\pi}{4} - 1 = \frac{\pi}{2} - 1.$$

It is interesting to note that l'Hospital's rule can be derived directly from the definition of a derivative. We may write the limit

$$\lim_{x\to a} \frac{f(x)}{g(x)}$$

in the form

$$\lim_{h\to 0} \frac{f(a+h)}{g(a+h)}$$

and since $f(a) = g(a) = 0$ we see that this can be written in the form

$$\lim_{h\to 0} \frac{f(a+h) - f(a)}{g(a+h) - g(a)} = \lim_{h\to 0} \frac{[f(a+h) - f(a)]/h}{[g(a+h) - g(a)]/h}.$$

It follows at once from the definition of a derivative and the fact that the limit of a ratio is the ratio of the limits (Theorem 6) that this last limit is $f'(a)/g'(a)$. If it happens that not both $f'(a)$ and $g'(a)$ are zero then we obtain equation (2). If both $f'(a)$ and $g'(a)$ are zero but $f''(a)$ and $g''(a)$ are not we can repeat the above process to obtain the result (3), and so on.

PROBLEMS

1. Evaluate the limits

(i) $\lim_{x\to 1} \dfrac{x^4 + 3x^2 - 6x + 2}{x^3 - 2x^2 + 5x - 4}$; (iii) $\lim_{x\to 0} \dfrac{\tan^{-1} x - x}{x^3}$;

(ii) $\lim_{x\to 2} \dfrac{x^3 - 3x^2 + 4}{x^3 - 5x^2 + 8x - 4}$; (iv) $\lim_{x\to 1} \dfrac{1 - 4\sin^2 \frac{1}{6}\pi x}{1 - x^2}$.

2. By putting $x = 1/y$ and thus transforming to limits in which $x \to 0$, evaluate

(i) $\lim_{y\to\infty} \left(y \tan \dfrac{1}{y}\right)$; (ii) $\lim_{y\to\infty} \left(\cot \dfrac{1}{y} - y\right)$.

13. Higher Derivatives of a Product

The actual calculation of derivatives of high order is apt to be a tedious business and there are very few general theorems available to lighten the labour. One such, known as *Leibniz's theorem*, is useful for finding the n-th derivative of the product of two functions $f(x)$ and $g(x)$. If we denote $\mathrm{d}/\mathrm{d}x$ by D then we can write Theorem 12 in the form

$$\mathrm{D}(fg) = f\mathrm{D}g + g\mathrm{D}f \tag{1}$$

and if we differentiate both sides of this equation with respect to x we find that

$$\mathrm{D}^2(fg) = \mathrm{D}(f\mathrm{D}g) + \mathrm{D}(g\mathrm{D}f).$$

Now, by Theorem 12,

$$\mathrm{D}(f\mathrm{D}g) = f\mathrm{D}(\mathrm{D}g) + \mathrm{D}g\mathrm{D}f = f\mathrm{D}^2g + \mathrm{D}f\mathrm{D}g$$

and similarly

$$\mathrm{D}(g\mathrm{D}f) = g\mathrm{D}^2f + \mathrm{D}f\mathrm{D}g$$

so that

$$\mathrm{D}^2(fg) = f\mathrm{D}^2g + 2\mathrm{D}f\mathrm{D}g + g\mathrm{D}^2f. \tag{2}$$

To find the third derivative we differentiate both sides of equation (2) with respect to x and make use of the results

$$\begin{aligned} \mathrm{D}(f\mathrm{D}^2g) &= f\mathrm{D}^3g + \mathrm{D}f\mathrm{D}^2g, \\ \mathrm{D}(\mathrm{D}f\mathrm{D}g) &= \mathrm{D}f\mathrm{D}^2g + \mathrm{D}^2f\mathrm{D}g, \\ \mathrm{D}(g\mathrm{D}^2f) &= g\mathrm{D}^3f + \mathrm{D}g\mathrm{D}^2f. \end{aligned}$$

We then obtain the result

$$\mathrm{D}^3(fg) = f\mathrm{D}^3g + 3\mathrm{D}f\mathrm{D}^2g + 3\mathrm{D}^2f\mathrm{D}g + g\mathrm{D}^3f. \tag{3}$$

Now if we look at equations (1), (2) and (3) we see that the coefficients of the terms follow the same rule as the coefficients in the binomial theorem. In other words, these results suggest that, if n is a positive integer,

$$\mathrm{D}^n(fg) = \sum_{r=0}^{n} {}^nC_r \mathrm{D}^r f \cdot \mathrm{D}^{n-r} g \tag{4}$$

where ${}^nC_r$ denotes the binomial coefficient

$$\frac{n!}{r!(n-r)!}.$$

This is Leibnitz's theorem and we shall now prove it.

We have already shown the theorem to be valid when $n = 2$. We now establish the general result by induction, i.e. we assume the theorem to be true for a general value n and then prove that it holds for the next value $n + 1$. This general result means that having established the validity of the theorem for $n = 2$ we can assert that it is true for $n = 3, 4, \ldots$, i.e. for *all* integral values of n.

Differentiating both sides of (4) with respect to x we find that

$$\mathrm{D}^{n+1}(fg) = \sum_{r=0}^{n} {}^nC_r \mathrm{D}[\mathrm{D}^r f \cdot \mathrm{D}^{n-r} g].$$

Now

$$\mathrm{D}[\mathrm{D}^r f \cdot \mathrm{D}^{n-r} g] = \mathrm{D}^{r+1} f \cdot \mathrm{D}^{n-r} g + \mathrm{D}^r f \cdot \mathrm{D}^{n+1-r} g$$

so that

$$\mathrm{D}^{n+1}(fg) = \sum_{r=0}^{n} {}^nC_r \cdot \mathrm{D}^{r+1} f \cdot \mathrm{D}^{n-r} g + \sum_{r=0}^{n} {}^nC_r \cdot \mathrm{D}^r f \cdot \mathrm{D}^{n+1-r} g$$

which can be written in the form

$$\mathrm{D}^{n+1}(fg) = f \cdot \mathrm{D}^{n+1} g + \sum_{r=1}^{n} \{{}^nC_{r-1} + {}^nC_r\} \mathrm{D}^r f \cdot \mathrm{D}^{n+1-r} g + \mathrm{D}^{n+1} f \cdot g.$$

But we saw that

$${}^nC_{r-1} + {}^nC_r = {}^{n+1}C_r$$

so that we have

$$\mathrm{D}^{n+1}(fg) = \sum_{r=0}^{n+1} {}^{n+1}C_r \cdot \mathrm{D}^r f \cdot \mathrm{D}^{n+1-r} g,$$

which is identical with equation (4) with n replaced by $(n + 1)$. The general result (4) is thus established.

EXAMPLE 22: *Find the n-th derivative of* $x^2 \cos(ax)$.
Suppose n is even and is equal to $2r$, say. Then we have

$$\begin{aligned} f &= x^2, & D^{2r}\cos(ax) &= (-1)^r a^{2r}\cos(ax) \\ Df &= 2x, & D^{2r-1}\cos(ax) &= (-1)^r a^{2r-1}\sin(ax) \\ D^2 f &= 2, & D^{2r-2}\cos(ax) &= (-1)^{r-1} a^{2r-2}\cos(ax). \end{aligned}$$

Now by (4) we have

$$\begin{aligned} D^{2r}[x^2\cos(ax)] &= x^2 D^{2r}\cos ax + 2r\,D(x^2)D^{2r-1}\cos(ax) \\ &\qquad + \tfrac{1}{2}2r(2r-1)D^2(x^2)D^{2r-2}\cos(ax) \\ &= (-1)^r a^{2r-2}\{a^2x^2\cos(ax) + 4rax\sin(ax) \\ &\qquad - 2r(2r-1)\cos(ax)\}. \end{aligned}$$

On the other hand if n is odd (equal, say, to $2r - 1$) we have

$$\begin{aligned} f &= x^2 & D^{2r-1}\cos(ax) &= (-1)^r a^{2r-1}\sin(ax) \\ Df &= 2x & D^{2r-2}\cos(ax) &= (-1)^{r-1} a^{2r-2}\cos(ax) \\ D^2 f &= 2 & D^{2r-3}\cos(ax) &= (-1)^{r-1} a^{2r-3}\sin(ax) \end{aligned}$$

from which it follows that

$$\begin{aligned} D^{2r-1}[x^2\cos(ax)] &= (-1)^r a^{2r-3}\{a^2x^2\sin(ax) - 2(2r-1)ax\cos(ax) \\ &\qquad -2(2r-1)(r-1)\sin(ax)\}. \end{aligned}$$

FURTHER PROBLEMS

The problems in Chapter 10, § 14, pp. 596 and 599 illustrate many points in this chapter; see also Chapter 10, § 16, pp. 611–614. All, however, also involve logarithms and exponentials (Chapter 6).

CHAPTER 5

INTEGRATION

1. Definition of the Integral of a Continuous Function

Suppose that the function $f(u)$ is continuous for all values of u lying between the values a and b and that $a < b$. If we restrict our attention, in the first instance, to functions which are positive over the range considered, the graph of $f(u)$ will lie entirely above the u-axis, as shown in Fig. 64. If we take any point with abscissa $u = x$

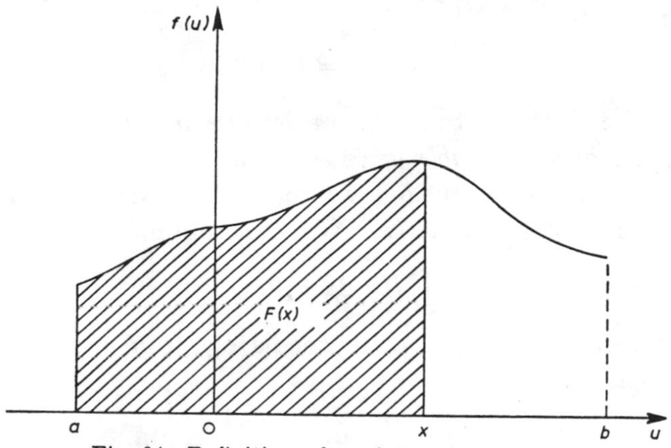

Fig. 64. Definition of an integral as an area.

lying on the u-axis between a and b and if we draw the ordinates through the points a and x, these lines together with the u-axis and the graph of the function will define an area which is shown shaded in Fig. 64. If we fix the number a then this area will vary as the position of the point x varies — in other words this area will be a function of x which we may denote by $F(x)$. Since the area vanishes when x coincides with a we have that

$$F(a) = 0 \tag{1}$$

and, since the area increases as x increases we see that $F(x)$ is a positive monotonic increasing function of x. To show the connection of this "area" function $F(x)$ with the original function $f(u)$ we write

$$F(x) = \int_a^x f(u)\mathrm{d}u \tag{2}$$

and we call $F(x)$ *the definite integral of* $f(u)$ *from* $u = a$ *to* $u = x$, or simply the *integral of* f *from* a *to* x. a is called the lower limit and x the upper limit of the integral.

Three simple results follow immediately from the definition. If we draw the graph of the function $f(v)$ as a function of v then the diagram we obtain is an exact copy of Fig. 64. All we have done is to re-label the axes by replacing u by v. If a and x have the same numerical values as before then

$$\int_a^x f(v)\mathrm{d}v$$

will represent precisely the same area as before. We therefore obtain the result

$$\int_a^x f(u)\mathrm{d}u = \int_a^x f(v)\mathrm{d}v \tag{3}$$

which may be stated in the form: *a definite integral is a function of its limits, not of the variable of integration.*

The second simple result follows merely by writing equation (1) in the notation of equation (2). We then obtain the equation

$$\int_a^a f(u)\mathrm{d}u = 0. \tag{4}$$

Another result which follows immediately from our definition is that if $c > x > a$ then

$$\int_x^c f(u)\mathrm{d}u = \int_a^c f(u)\mathrm{d}u - \int_a^x f(u)\mathrm{d}u. \tag{5}$$

EXAMPLE 1: *If* k *is a positive constant, evaluate the integrals*

$$\int_a^x k\mathrm{d}u, \int_a^x ku\mathrm{d}u,$$

when $x > a > 0$.

In the first integral $f(u) = k$ the graph of which is a straight line parallel to the u-axis at a height k above it (cf. Fig. 65). In this case the shaded area is that of a rectangle of length $x - a$ and height k so that it has the value $k(x - a)$. Hence we have that

$$\int_a^x k\mathrm{d}u = k(x - a). \tag{6}$$

In the second integral $f(u) = ku$ the graph of which is a straight line of gradient k passing through the origin O (cf. Fig. 66). Here the shaded area is the difference between the area of a triangle of base x and height kx and that of a triangle of base a and height ka:

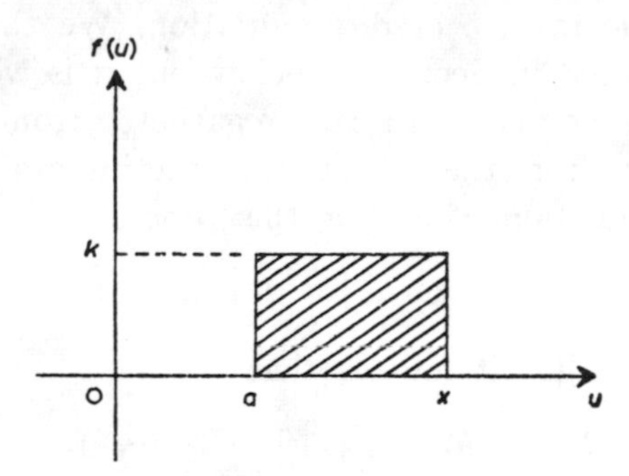

Fig. 65. The integral $\int_a^x k\mathrm{d}u$.

it therefore has the value $\frac{1}{2}x \times kx - \frac{1}{2}a \times ka$ and we obtain the result

$$\int_a^x ku\mathrm{d}u = \tfrac{1}{2}k(x^2 - a^2). \tag{7}$$

We have introduced the idea of the integral of a positive function as the area under the curve. In addition to the drawbacks inherent

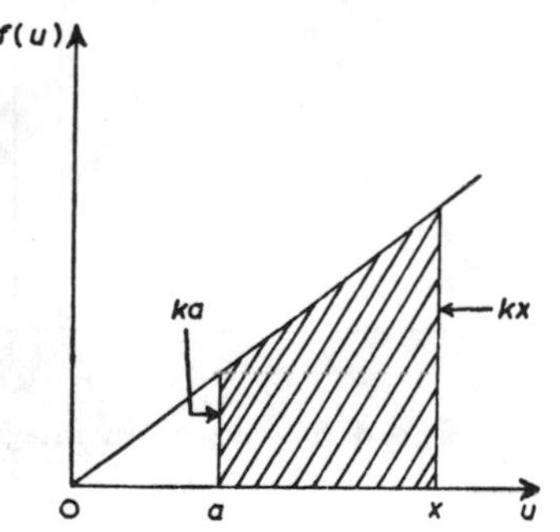

Fig. 66. The integral $\int_a^x ku\mathrm{d}u$.

in any geometrical definition of a numerical quantity, this definition is not particularly fruitful of results. In the next section we shall replace the geometrical definition by a purely numerical one (which is however, suggested by the geometrical concept) which is useful for the derivation of general theorems, of which equation (5) is a particularly simple example. We shall find that this definition, valuable as it is for these purposes, does not help us greatly when it comes to the actual calculation of the integrals of particular functions. These purely practical results are best achieved by regarding the process of integration as the inverse of differentiation. We shall consider the proof of this method in Section 5 below but it is worth mentioning briefly now since the main idea can be gathered from the area definition. It is obvious from the definition of the integral in equation (2) that $F(x)$ is a function of x. For this function

$$F(x+h) - F(x) = \int_a^{x+h} f(u)\mathrm{d}u - \int_a^x f(u)\mathrm{d}u\,.$$

and from equation (5) it follows that

$$F(x+h) - F(x) = \int_x^{x+h} f(u)\mathrm{d}u.$$

Now if h is small we see from a diagram such as Fig. 67 that the integral is an area which is nearly equal to that of a trapezium with

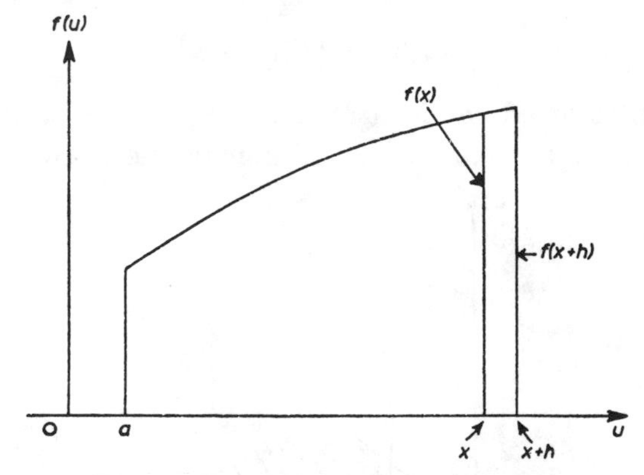

Fig. 67. The derivative of an integral.

parallel sides of length $f(x)$ and $f(x+h)$ and breadth h. Hence for small h we would have

$$F(x+h) - F(x) = \tfrac{1}{2}h[f(x) + f(x+h)]$$

that is

$$\frac{F(x+h) - F(x)}{h} = \tfrac{1}{2}[f(x) + f(x+h)].$$

If $f(x)$ is continuous then $f(x+h)$ will tend to $f(x)$ as $h \to 0$ and so we find that

$$\frac{d}{dx} F(x) = f(x). \tag{8}$$

This result, which we shall return to in Section 5, is known as *the fundamental theorem of the integral calculus.* It provides the link between the differential calculus and the integral calculus, and leads the way to the systematic evaluation of integrals.

A minor difficulty that arises here is due to the fact that if $F(x)$ is one function with the property (8), then, since the derivative of a constant k is zero, then the function $F(x) + k$ will also have the property (8). Any function $F(x)$ satisfying equation (8) is called an *indefinite integral* of $f(x)$. At first sight it seems as though we would have an infinite number of possible functions but then we remember that, if a is the lower limit of the integral, the integral must be zero when $x = a$ and this means that we must choose the constant k to have the value $-F(a)$. We therefore find that if $F(x)$ is any function satisfying equation (8)

$$\int_a^x f(u)du = F(x) - F(a). \tag{9}$$

The difference $F(x) - F(a)$ is often written as $[F(u)]_a^x$.

For example in the first part of example 1 p. 194 if we take $F(x)=kx$, then $F'(x) = k$ and $F(a) = ka$ so that

$$\int_a^x k du = kx - ka = k(x - a).$$

In the second part, if we take $F(x) = \frac{1}{2}kx^2$ then $F'(x) = kx$, $F(a) = \frac{1}{2}ka^2$ and

$$\int_a^x ku du = \tfrac{1}{2}kx^2 - \tfrac{1}{2}ka^2 = \tfrac{1}{2}k(x^2 - a^2).$$

Using equations and the results of the differential calculus (Table 9 p. 159) we get the list of elementary integrals shown in Table 10.

The reader who did some calculus at school (where, almost always integration is treated in this way as the inverse of differentiation) will find himself on more familiar ground and may well wonder why we bother to formulate the definition given above in terms of area — or, worse still!, the definition given in the next section. It should be emphasized again that this more 'unfamiliar' definition (in the school sense) makes the derivation of general theorems much easier. It is interesting to note in this context that the first integrations were carried out by Archimedes, using methods not unlike those of the next section, centuries before the differential calculus was developed. Logically the correct procedure would be to start with the purely numerical definition of an integral contained in the next section and

TABLE 10

Integrals of elementary functions

$$\int_a^x f(u)\,du = F(x) - F(a)$$

$f(u)$	$F(x)$
$u^n \quad (n \neq -1)$*	$\dfrac{x^{n+1}}{n+1}$
$\cos(ku)$	$\dfrac{1}{k}\sin(kx)$
$\sin(ku)$	$-\dfrac{1}{k}\cos(kx)$
$\sec^2(ku)$	$\dfrac{1}{k}\tan(kx)$
$\operatorname{cosec}^2(ku)$	$-\dfrac{1}{k}\cot(kx)$
$\sec(ku)\tan(ku)$	$\dfrac{1}{k}\sec(kx)$
$\operatorname{cosec}(ku)\cot(ku)$	$-\dfrac{1}{k}\operatorname{cosec}(kx)$
$(\alpha^2 - u^2)^{-\frac{1}{2}}$	$\sin^{-1}\left(\dfrac{x}{\alpha}\right)$
$(\alpha^2 + u^2)^{-1}$	$\dfrac{1}{\alpha}\tan^{-1}\left(\dfrac{x}{\alpha}\right)$

* The case $n = -1$ is covered by the analysis of the next chapter.

then prove that the area under a curve is an integral but it was felt that a reader who is interested in the applications of mathematics rather than in purely theoretical considerations might find it easier to follow the course we have adopted and to think of integrals as areas throughout and to use proofs based on the proper analytical definition as a safeguard that his geometrical intuition has not played him false. A reader who is prepared to take the results on trust can pass straight to p. 215.

PROBLEMS

1. Show that the graph of the function $f(u) = +\sqrt{c^2 - u^2}$ is the upper half of a circle with centre at the origin and radius c and, by interpreting the integrals as areas show that

(i) $\int_{-c}^{c} \sqrt{(c^2 - u^2)}\mathrm{d}u = \frac{1}{2}\pi c^2$;

(ii) $F(x) = \int_{0}^{x} \sqrt{(c^2-u^2)}\mathrm{d}u = \frac{1}{2}c^2 \sin^{-1}\frac{x}{c} + \frac{1}{2}x\sqrt{(c^2-x^2)}, \quad (0<x<c).$

Verify, that in case (ii),

$$F'(x) = \sqrt{(c^2 - x^2)}.$$

2. By differentiating the functions in the right hand column, verify that the entries in Table **10** are correct.

3. Using the entries in Table **10** evaluate the integrals

(i) $\int_{1}^{3} u^4 \mathrm{d}u$; (ii) $\int_{2}^{4} u^3 \mathrm{d}u$;

(iii) $\int_{0}^{\frac{1}{4}\pi} \frac{\mathrm{d}u}{\cos^2 u}$; (iv) $\int_{0}^{a} \frac{\mathrm{d}u}{\sqrt{(a^2 - u^2)}}$;

(v) $\int_{0}^{\frac{1}{2}\pi} \sin u\mathrm{d}u$; (vi) $\int_{0}^{1} \frac{\mathrm{d}u}{1 + u^2}$.

2. The Integral as the Limit of a Sum

We shall now turn our geometrical definition of an integral into a purely arithmetical one but even now we shall be guided by geometrical

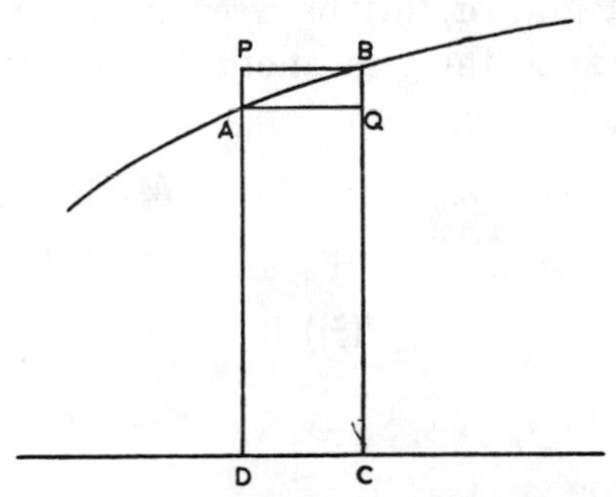

Fig. 68. The element of area.

intuition. In calculating the area under the curve in Fig. 68 we could divide the area up into strips a typical one of which is $ABCD$ shown in the diagram. Now, for the curve shown, the area $ABCD$ lies between the areas of the rectangles $PBCD$ and $AQCD$. If we denote the width x by Δx_i, the height BC by M_i and the height AD by m_i (regarding this as the i-th strip) then the area $ABCD$ will lie between $M_i \Delta x_i$ and $m_i \Delta x_i$, and the smaller Δx_i is the less will be the difference between these two estimates. It is this idea which we now formalize.

We suppose that $f(u)$ is a function of the independent variable u defined for all values of u between a and x. We shall assume that $x > a$ and divide the interval $a < u < x$ into n small parts, not necessarily equal, by the $n + 1$ points $u_0, u_1, u_2, \ldots, u_n$ chosen in such a way that

$$a = u_0 < u_1 < u_2 < \ldots < u_{n-1} < u_n = x.$$

We shall let Δu_i denote the length $u_i - u_{i-1}$ of the i-th division of the interval. Denoting the least and the greatest values respectively of $f(u)$ for the small interval $u_{i-1} < u < u_i$ by m_i and M_i we calculate the sums

$$s_n = \sum_{i=1}^{n} m_i \Delta u_i, \quad S_n = \sum_{i=1}^{n} M_i \Delta u_i \tag{1}$$

which geometrically give a lower and an upper bound to the value of the integral. If as $n \to \infty$ in such a way that each $\Delta u_i \to 0$, s_n and S_n tend respectively to the finite limits s and S, then if $s = S$ we say that $f(u)$ is *integrable* in the interval $a < u < x$ and we write

$$s = S = \int_a^x f(u) \mathrm{d}u \tag{2}$$

for the definite integral of $f(u)$ between a and x.

We can say the same thing in another way. If ξ_i is any value of u such that $u_{i-1} \leqq \xi_i \leqq u_i$ then

$$m_i \Delta u_i \leqq f(\xi_i) \Delta u_i \leqq M_i \Delta u_i$$

and so

$$s_n \leqq \sum_{i=1}^{n} f(\xi_i) \Delta u_i \leqq S_n,$$

when $f(x)$ is integrable $s_n \to s$, $S_n \to S$ and $s = S$ so that we may take

$$\int_a^x f(u) \mathrm{d}u = \lim_{\substack{n \to \infty \\ \text{all } \Delta u_i \to 0}} \sum_{i=1}^{n} f(\xi_i) \Delta u_i \tag{3}$$

where ξ_i is any point in the small interval $u_{i-1} \leqq u \leqq u_i$.

It can be shown that, if $s = S$ for any division of the interval, the value of the integral is quite independent of the choice of the points $u_1, u_2, \ldots, u_{n-1}$ provided that when the number of intervals becomes infinite the length of each of them tends to zero.

EXAMPLE 2: *Evaluate the definite integral*

$$\int_0^1 u^2 \, du.$$

Since the method of dividing the interval does not affect the final result we can divide the interval $0 < u < 1$ into n equal parts of length $1/n$ with endpoints $u_i = i/n \quad (i = 0, 1, \ldots, n)$.

In the interval $u_{i-1} < u < u_i$, $u_i = 1/n$ and the function u^2 increases steadily from $u_{i-1}^2 = (i-1)^2/n^2$ to $u_i^2 = i^2/n^2$ so that

$$s_n = \sum_{i=1}^{n} \frac{(i-1)^2}{n^2} \cdot \frac{1}{n}, \qquad S_n = \sum_{i=1}^{n} \frac{i^2}{n^2} \cdot \frac{1}{n}.$$

Now by (12) of Section 4 of Chapter 1

$$S_n = \frac{1}{n^3} \sum_{i=1}^{n} i^2 = \frac{1}{n^3} \{\tfrac{1}{6} n(n+1)(2n+1)\} = \frac{1}{3}\left(1 + \frac{1}{n}\right)\left(1 + \frac{1}{2n}\right).$$

Similarly

$$s_n = \frac{1}{n^3} \sum_{i=1}^{n-1} i^2 = \frac{1}{n^3} \{\tfrac{1}{6}(n-1)n(2n-1)\} = \frac{1}{3}\left(1 - \frac{1}{n}\right)\left(1 - \frac{1}{2n}\right).$$

It follows immediately that as $n \to \infty$, both s_n and S_n tend to the value $\frac{1}{3}$. Hence

$$\int_0^1 u^2 \, du = \tfrac{1}{3}.$$

If the upper limit of the integral is not 1 but x then $u_i = x/n$ and we have

$$s_n = \sum_{i=1}^{n} \frac{(i-1)^2 x^2}{n^2} \cdot \frac{x}{n}, \qquad S_n = \sum_{i=1}^{n} \frac{x^2 i^2}{n^2} \cdot \frac{x}{n}.$$

Proceeding as before we then find that

$$\int_0^x u^2 \, du = \tfrac{1}{3} x^3. \tag{4}$$

To illustrate how integral expressions arise in biological problems we shall consider a simple problem in the theory of populations*. The annual female births at time t are the daughters of the females who at that time lie within the limits α to ω of the female reproductive period. If $N(t)$ is the total female population and $N(t)c(a)\Delta a$ is the number of those aged a to $a + \Delta a$ and if $m(a)$ is the annual rate at which females of age a give birth to daughters then the annual female births will be made of the sum of terms like $N(t)c(a)m(a)\Delta a$ taken over the values of a lying between α and ω. If we take the intervals Δa sufficiently small we see that this sum is given by the integral

$$B(t) = \int_\alpha^\omega N(t)c(a)m(a)\mathrm{d}a. \tag{5}$$

Now, in a closed population, i.e. one free from immigration and emigration, the females of age a to $a + \Delta a$ at the time t will be the survivors of those born in the time interval $t - a$ to $t - a + \Delta a$. Hence if $p(a)$ is the probability at birth of a female surviving to attain the age a,

$$N(t)c(a)\Delta a = B(t - a)p(a)\Delta a.$$

Hence $B(t)$ will be the sum of terms of the type $N(t)c(a)m(a)\Delta a = B(t - a)p(a)m(a)\Delta a$, so that, as before it may be represented by the integral

$$B(t) = \int_\alpha^\omega B(t - a)p(a)m(a)\mathrm{d}a. \tag{6}$$

An equation of this kind which involves the unknown function $B(t)$ under the sign of integration as well as on the left-hand side is called an *integral equation*.

PROBLEMS

1. By finding the limit of the appropriate sum, prove that

(i) $\int_0^x u\mathrm{d}u = \frac{1}{2}x^2$;

(ii) $\int_0^x u^3\mathrm{d}u = \frac{1}{4}x^4$.

2. Show that

$$\sin\frac{\pi}{2n}\sin\frac{i\pi}{2n} = \frac{1}{2}\left\{\cos\frac{(i-1)\pi}{2n} - \cos\frac{(i+1)\pi}{2n}\right\}.$$

* A. J. Lotka, "Population Analysis as a Chapter in the Mathematical Theory of Evolution" in "Essays on Growth and Form" (Oxford, 1945) p. 358.

Deduce that

$$\sum_{i=0}^{n-1} \sin \frac{i\pi}{2n} = \frac{1}{2} \frac{\cos(\pi/2n)}{\sin(\pi/2n)}.$$

Hence, by dividing the interval into n equal parts and using equation (3) with $\xi_i = u_i = i\pi/2n$, prove that

$$\int_0^{\frac{1}{2}\pi} \sin u du = 1.$$

3. The Sign of an Area Interpreted as an Integral

If a function $f(u)$ can take negative values in the interval $a \leqq u \leqq b$ then certain terms in the summation

$$\sum_{i=1}^{n} f(\xi_i)\Delta u_i$$

will be negative. For simplicity we shall consider a function $f(u)$

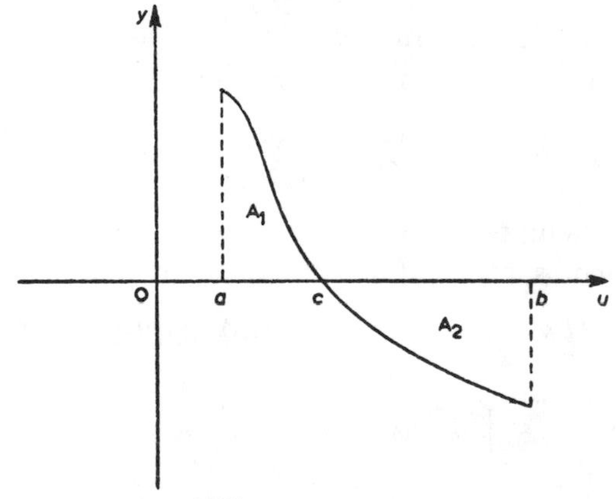

Fig. 69. The sign of an area.

which changes sign at the point $u = c$. Then the "area" under the curve between the ordinates $u = a$ and $u = b$ is

$$\int_a^b f(u)du = \int_a^c f(u)du + \int_c^b f(u)du.$$

Of the two terms on the right hand side of this equation the first integral will be positive, but the second will be negative. If A_1 and A_2 are the numerical values of the areas as shown in Fig. 69, these are both positive but

$$\int_a^b f(u)\mathrm{d}u = A_1 - A_2.$$

The definite integral of a function whose sign changes therefore gives the algebraic value of the area bounded by the curve, the u-axis, and the ordinates at $u = a$ and $u = b$. In any problem in which it is required to find the numerical value of an area, it is advisable to divide the range of integration into stretches for which the function is always of the same sign, e.g. the sub-intervals $a \leqq u \leqq c$ and $c \leqq u \leqq b$ in the above illustration.

4. General Properties of Integrals

We can derive a number of results directly from the definition of an integral.

The first result:-

THEOREM 18: $$\int_a^b f(u)\mathrm{d}u = \int_a^b f(v)\mathrm{d}v,$$

which we discussed geometrically in Section 1 above can be proved in purely arithmetical terms by observing that

$$\sum_{r=1}^{n} a_r = \sum_{s=1}^{n} a_s$$

(since each of them represents $a_1 + a_2 + \ldots + a_n$).

A result which is of great practical use is

THEOREM 19: *If k is a constant and $f(u)$ is integrable over the interval $a \leqq u \leqq b$ then*

$$\int_a^b kf(u)\mathrm{d}u = k\int_a^b f(u)\mathrm{d}u.$$

This follows from the fact that if

$$\lim_{n\to\infty} \sum_{i=1}^{n} f(\xi_i)\Delta u_i = S$$

then

$$\lim_{n\to\infty} \sum_{i=1}^{n} \{kf(\xi_i)\}\Delta u_i = kS.$$

The result

THEOREM 20: $$\int_a^a f(u)\mathrm{d}u = 0,$$

follows directly from the definition.

To prove

THEOREM 21: *If $f(u)$ is integrable over $a \leqq u \leqq b$ and c lies between a and b then*

$$\int_a^b f(u)du = \int_a^c f(u)du + \int_c^b f(u)du \tag{1}$$

we observe that if $f(u)$ is integrable over $a \leqq u \leqq b$ then it is integrable over both $a \leqq u \leqq c$ and $c \leqq u \leqq b$ if $a < c < b$, so that the integrals on the right do in fact exist. If we now form sums like $\sum_{r=0}^{n-1}$ for each of these intervals then their sum can be written in the form $\sum_{r=0}^{2n-1}$ over $a \leqq u \leqq b$ and when $n \to \infty$ the result follows.

In our definition of an integral we made the assumption that the upper limit was greater than the lower limit. We now remove this restriction by adopting the convention

$$\int_b^a f(u)du = -\int_a^b f(u)du, \tag{2}$$

which is consistent with Theorems 20 and 21.

A theorem which is useful for combining integrals is:-

THEOREM 22: *If $f(u)$ and $g(u)$ are integrable over the interval $a \leqq u \leqq b$, then $f(u) + g(u)$ is integrable over the same interval and*

$$\int_a^b \{f(u) + g(u)\}du = \int_a^b f(u)du + \int_a^b g(u)du. \tag{3}$$

To prove this theorem we note that if s_1, S_1, and s_2, S_2 denote the sums of $f(u)$ and $g(u)$ respectively for the same division of the interval it is clear that

$$s_1 + s_2 \leqq s \leqq S \leqq S_1 + S_2$$

where s, S are the sums of $f(u) + g(u)$ for the same division of the interval. If s_1 and S_1 both tend to L_1 as $n \to \infty$ and s_2 and S_2 both tend to L_2 it follows that s and S will both tend to $L_1 + L_2$, so that $f(u) + g(u)$ is an integrable function. Again, using equation (3) of Section 2, we have that

$$\lim_{n\to\infty} \sum_{i=0}^{n-1} \{f(\xi_i) + g(\xi_i)\}\Delta u_i = \lim_{n\to\infty} \sum_{i=0}^{n-1} f(\xi_i)\Delta u_i + \lim_{n\to\infty} \sum_{i=0}^{n-1} g(\xi_i)\Delta u_i$$

leading to the theorem.

By means of Theorems 19 and 22 and the first entry of Table 10 we can find the integrals of polynomials. Thus

$$\begin{aligned}
&\int_a^x (c_n u^n + c_{n-1}u^{n-1} + \ldots + c_1 u + c_0)\mathrm{d}u \\
&= \int_a^x c_n u^n \,\mathrm{d}u + \int_a^x c_{n-1}u^{n-1}\mathrm{d}u + \ldots + \int_a^x c_1 u\mathrm{d}u + \int_a^x c_0\,\mathrm{d}u \\
&= c_n \int_a^x u^n\,\mathrm{d}u + c_{n-1}\int_a^x u^{n-1}\mathrm{d}u + \ldots + c_1\int_a^x u\mathrm{d}u + c_0\int_a^x \mathrm{d}u \\
&= \frac{c_n}{n+1}(x^{n+1} - a^{n+1}) + \frac{c_{n-1}}{n}(x^n - a^n) + \ldots + \frac{c_1}{2}(x^2 - a^2) + c_0(x-a).
\end{aligned}$$

EXAMPLE 3: *Evaluate*

$$\int_1^2 (x^3 + 2x^2 + 3x + 1)\mathrm{d}x.$$

This integral has the value

$$\begin{aligned}
&\tfrac{1}{4}(2^4 - 1^4) + \tfrac{2}{3}(2^3 - 1^3) + \tfrac{3}{2}(2^2 - 1^2) + 1(2 - 1) \\
&= \tfrac{15}{4} + \tfrac{14}{3} + \tfrac{9}{2} + 1 \\
&= \tfrac{167}{12}.
\end{aligned}$$

THEOREM 23: *If* $f(u) > 0$ *for* u *in the interval* $a \leqq u \leqq b$ *then*

$$\int_a^b f(u)\mathrm{d}u > 0.$$

To prove this we take $\Delta u_i = (b-a)/n > 0$ and if ξ_i is in $u_{i-1} < \xi_i < u_i$ which is itself contained in $a \leqq u \leqq b$ we have $f(\xi_i) > 0$. Every term of the series

$$\sum_{i=1}^{n} f(\xi_i)\Delta u_i$$

is therefore positive, so that the sum is positive and the result follows immediately. The theorem obviously remains true if $f(u)$ is zero for some (but not all!) of the values of u in $a \leqq u \leqq b$.

From this result we deduce

THEOREM 24: *If* $\phi(u) < f(u) < \psi(u)$ *for all* u *in* $a \leqq u \leqq b$, *then*

$$\int_a^b \phi(u)\mathrm{d}u < \int_a^b f(u)\mathrm{d}u < \int_a^b \psi(u)\mathrm{d}u$$

merely by applying it to the positive functions $f(u) - \phi(u)$, $\psi(u) - f(u)$.

As a special (but very useful) case of this theorem we have:-

THEOREM 25: *If* $m < f(u) < M$ *in* $a \leqq u \leqq b$ *then*

$$m(b - a) < \int_a^b f(u)\mathrm{d}u < M(b - a)$$

obtained by putting $\psi(u) = M$, $\phi(u) = m$ in the statement of Theorem 24.

Geometrically this means that the area under the curve lies between the rectangle formed by a segment of the u-axis of length $b - a$ and the highest ordinate M and the rectangle formed by the same part of the u-axis and the lowest ordinate m.

From this theorem we in turn deduce

THEOREM 26: *If $f(u)$ is continuous in $a \leqq u \leqq b$ there exists a number ξ lying between a and b such that*

$$\int_a^b f(u)du = (b - a)f(\xi).$$

It follows from Theorem 25 that there exists a number μ lying between m and M such that

$$\int_a^b f(u)du = (b - a)\mu.$$

Since $f(u)$ is continuous in $a \leqq u \leqq b$ and there exist bounds m and M such that $m < f(u) < M$, it follows from the fundamental theorem on continuous functions (Theorem 8 on p. 140 above) that there exists at least one value ξ in (a, b) such that $f(\xi) = \mu$. The theorem follows immediately.

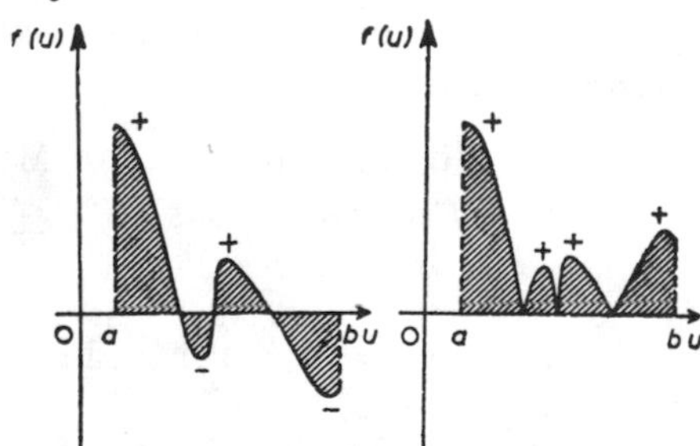

Fig. 70. Geometrical interpretation of Theorem 27.

Another result, which follows immediately from Theorem 24, is

THEOREM 27: *If $f(u)$ is integrable over the range $a \leqq u \leqq b$ then*

$$\int_a^b |f(u)|\, du \geqq \left|\int_a^b f(u)du\right|.$$

That the result is true is obvious from the geometrical interpretation of the integral (cf. Fig. 70). The formal proof follows from the fact (which is easily proved) that if $f(u)$ is integrable for $a \leqq u \leqq b$

then $|f(u)|$ is integrable over the same range, and the fact that $|f(u)| \geqq f(u)$ for all values of u, the theorem then being a simple consequence of Theorem 24.

In the estimation of the value of an integral we often make use of

THEOREM 28: *If* $a < b$, *if* $m < f(u) < M$ *and* $g(u) > 0$ *in* $a \leqq u \leqq b$ *then*

$$m \int_a^b g(u)du < \int_a^b f(u)g(u)du < M \int_a^b g(u)du.$$

For, if $m < f(u)$ and $g(u) > 0$ in $a \leqq u \leqq b$ then

$$\{f(u) - m\}g(u) > 0,$$

so that if $a < b$,

$$\int_a^b \{f(u) - m\}g(u)du > 0$$

and the first inequality follows. The second inequality is established in the same way by observing that $M - f(u) > 0$ for $a \leqq u \leqq b$.

It follows from this theorem that

$$\int_a^b f(u)g(u)du = \mu \int_a^b g(u)du \tag{4}$$

where $m < \mu < M$. Now if $f(u)$ is continuous in $a \leqq u \leqq b$ and if $m < f(u) < M$ in that range, then it follows from Theorem 8 that there exists at least one real number ξ between a and b such that $f(\xi) = \mu$. We therefore have

THEOREM 29: (FIRST INTEGRAL THEOREM OF MEAN VALUE).
If $a < b$ *and if* $g(u) > 0$, $f(u)$ *continuous in* $a \leqq u \leqq b$ *then there is a value* ξ *lying between* a *and* b *such that*

$$\int_a^b f(u)g(u)du = f(\xi) \int_a^b g(u)du.$$

A result which we shall use but not prove is:-

THEOREM 30: *If the function* $f(x)$ *is integrable and bounded for* $a \leqq x \leqq b$, *then the function*

$$F(x) = \int_a^x f(u)du$$

is continuous for $a \leqq x \leqq b$.

For example, this theorem will be used in the proof of

THEOREM 31: *If* $g(u) > 0$ *and if* $f(u)$ *is a positive decreasing function for* $a < u < b$ *then there is a value* ξ *between* a *and* b *such that*

$$\int_a^b f(u)g(u)du = f(a)\int_a^\xi g(u)du.$$

On the other hand if $f(u)$ is a positive increasing function there is a value η between a and b such that

$$\int_a^b f(u)g(u)du = f(b)\int_\eta^b g(u)du.$$

We shall prove the first result and leave the second as an excercise to the reader.

Since $f(x)$ is a positive decreasing function,

$$0 \leqq f(u) \leqq f(a)$$

for $u > a$, and since $g(u) > 0$ for $a < u < b$, it follows that

$$0 \leqq f(u)g(u) \leqq f(a)g(u)$$

so that by Theorem 24

$$0 \leqq \int_a^b f(u)g(u)du \leqq f(a)\int_a^b g(u)du.$$

Now since $g(u) > 0$ the function

$$\int_a^x g(u)du$$

will be an increasing, continuous function of x. Hence there exists a number ξ between a and b with the property that

$$\int_a^b f(u)g(u)du = f(a)\int_a^\xi g(u)du.$$

The second part of the theorem can be proved in a similar fashion.

An important corollary of this theorem is:-

THEOREM 32: (SECOND INTEGRAL THEOREM OF MEAN VALUE).

If $f(u)$ is either an increasing or a decreasing function of u for $a \leqq u \leqq b$, then there is a value ξ between a and b such that

$$\int_a^b f(u)g(u)du = f(a)\int_a^\xi g(u)du + f(b)\int_\xi^b g(u)du.$$

For, if $f(u)$ is an increasing function of u in $a \leqq u \leqq b$, then $f(b) - f(u)$ will be a positive decreasing function in the same range so that, by Theorem 31, there exists a value ξ with the property $a < \xi < b$ and such that

$$\int_a^b \{f(b) - f(u)\}g(u)du = \{f(b) - f(a)\}\int_a^\xi g(u)du.$$

We therefore have

$$\int_a^b f(u)g(u)\mathrm{d}u = f(a)\int_a^\xi g(u)\mathrm{d}u + f(b)\left\{\int_a^b g(u)\mathrm{d}u - \int_a^\xi g(u)\mathrm{d}u\right\}$$

from which it follows by Theorem 21 that

$$\int_a^b f(u)g(u)\mathrm{d}u = f(a)\int_a^\xi g(u)\mathrm{d}u + f(b)\int_\xi^b g(u)\mathrm{d}u,$$

proving the theorem in the first case. The second case ($f(u)$ a decreasing function of u) may be proved similarly by considering $f(a) - f(u)$ and using the second part of Theorem 31.

Theorems 18 to 32 are mainly of theoretical interest — they are the tools we use to prove general results about integrals*. We shall conclude this section with a theorem which can be of great practical use.

THEOREM 33: $\int_0^a f(u)\mathrm{d}u = \int_0^a f(a-u)\mathrm{d}u.$

By the definition (contained in equation (3) of § 2 above) of an integral as the limit of a sum we have

$$\int_0^a f(a-u)\mathrm{d}u = \lim_{n\to\infty} \sum_{i=1}^n f(a-\xi_i)\Delta u_i,$$

and we may take

$$\Delta u_i = \frac{a}{n}$$

so that

$$(i-1)\frac{a}{n} \leqq \xi_i \leqq i\frac{a}{n}.$$

Putting $j = n - i + 1$, $a - \xi_i = \eta_j$ we have the relation

$$\sum_{i=1}^n f(a-\xi_i) = \sum_{j=1}^n f(\eta_j), \qquad \frac{(j-1)a}{n} \leqq \eta_j \leqq \frac{ja}{n}.$$

Hence, taking $\Delta u_i = \Delta u_j = a/n$, we have

$$\sum_{i=1}^n f(a-\xi_i)\Delta u_i = \sum_{j=1}^n f(\eta_j)\Delta u_j.$$

Letting $n \to \infty$ we have

* For this reason they may be skipped over at a first reading and returned to as required.

$$\int_0^a f(a-u)\mathrm{d}u = \int_0^a f(u)\mathrm{d}u.$$

That the theorem is true for areas is immediately obvious from an inspection of the two relevant areas in Fig. 71. The curve $f(a-u)$ is a reflexion of $f(u)$ and the area is the same in both cases.

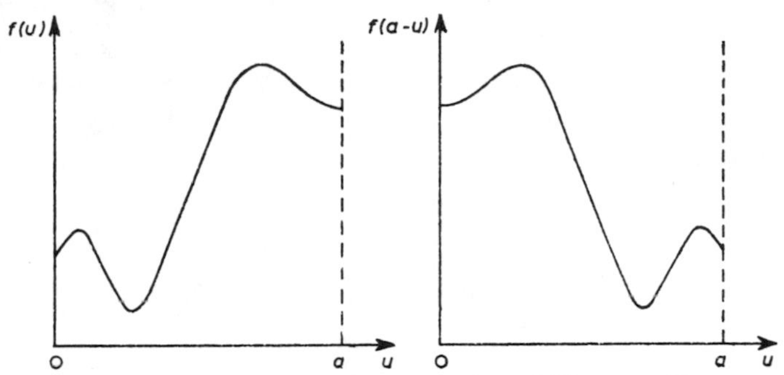

Fig. 71. Geometrical interpretation of Theorem **33**.

To illustrate the use of this theorem we have

EXAMPLE 4: *Evaluate*

$$\int_0^1 x^2(1-x)^9\,\mathrm{d}x.$$

By Theorem 33

$$\begin{aligned}\int_0^1 x^2(1-x)^9\,\mathrm{d}x &= \int_0^1 (1-x)^2x^9\,\mathrm{d}x\\ &= \int_0^1 (x^9 - 2x^{10} + x^{11})\mathrm{d}x\\ &= \tfrac{1}{10} - \tfrac{2}{11} + \tfrac{1}{12}\\ &= \tfrac{1}{660}.\end{aligned}$$

5. The Fundamental Theorem of the Integral Calculus

So far we have not established the relationship between the operations of differentiation and integration indicated in equation (8) of Section 1. In this section we shall prove that if

$$F(x) = \int_a^x f(u)\mathrm{d}u,$$

where a is a constant, then

$$\frac{\mathrm{d}F}{\mathrm{d}x} = f(x),$$

or, what is the same thing, if

$$\frac{d\phi}{dx} = f(x),$$

then

$$\int_a^b f(u)du = \phi(b) - \phi(a).$$

When these results are established we may (as we have already shown) evaluate definite integrals by using well-known results in differentiation, instead of by computing sums and then carrying out a limiting process.

The main result is

THEOREM 34: *If $f(u)$ is continuous for $a \leqq u \leqq b$ then the function*

$$F(x) = \int_a^x f(u)du$$

possesses a derivative which is the function $f(x)$ for $a \leqq x \leqq b$.

From Theorem 21 we have

$$F(x+h) - F(x) = \int_x^{x+h} f(u)du,$$

and it follows from Theorem 26 that, since $f(u)$ is continuous,

$$\int_x^{x+h} f(u)du = hf(\xi)$$

where $x \leqq \xi \leqq x + h$. Hence we have the relation

$$\frac{F(x+h) - F(x)}{h} = f(\xi), \qquad x < \xi < x + h$$

from which we find that

$$\lim_{h \to 0} \frac{F(x+h) - F(x)}{h} = \lim_{h \to 0} f(\xi).$$

The limit on the left-hand side is merely the definition of the derivative of the function $F(x)$, and as $h \to 0$, $\xi \to x$ so that the right-hand side tends to $f(x)$ and the theorem is proved.

As a corollary to this theorem we have:-

THEOREM 35: *If $\phi(x)$ is a continuous function of x whose derivative is a continuous function $f(x)$, then*

$$\int_a^x f(u)du = \phi(x) - \phi(a).$$

To prove this theorem we consider the function

$$\psi(x) = \int_a^x f(u)du - \phi(x).$$

By Theorem 34 and the hypothesis that $\phi'(x) = f(x)$ we find that

$$\psi'(x) = 0$$

and it follows that $\psi(x)$ is a constant, c, say. Hence for all values of x

$$\int_a^x f(u)du - \phi(x) = c.$$

To determine c, we let $x = a$ and use Theorem 20; we then find that

$$c = -\phi(a)$$

and the theorem is proved.

To simplify our expressions we often denote $\phi(x) - \phi(a)$ by the symbol $[\phi(u)]_a^x$.

It will be observed from Theorem 35 that a change in the lower limit a will affect the integral

$$\int_a^x f(u)du$$

only to the extent of an additive constant. For by Theorem 21

$$\int_c^x f(u)du = \int_a^x f(u)du + \int_c^a f(u)du = \int_a^x f(u)du + C_1$$

where $C_1 = \phi(a) - \phi(c)$. Another way of putting this is to say that the integral is determined to within the addition of an arbitrary constant. For that reason we sometimes write an integral of this type as

$$\int^x f(u)du$$

or as

$$\int f(x)dx. \tag{1}$$

An expression of the kind (1) is called an *indefinite integral,* and is a function of x, the variable of integration. We could give a formal definition of it as follows:-

If $\phi'(x) = f(x)$ then we write $\phi(x) = \int f(x)dx$. (This is in fact the definition of integration given in most school textbooks.)

Once we have found such an indefinite integral $\phi(x)$ then we can evaluate a definite integral by the formula

$$\int_a^b f(x)\mathrm{d}x = [\phi(x)]_a^b. \tag{2}$$

(Theorem 35 above.)

EXAMPLE 5: *Show that*

$$\int \frac{\mathrm{d}x}{\sqrt{(1-x^2)}} = \sin^{-1} x$$

and evaluate the integral

$$\int_0^1 \frac{\mathrm{d}x}{\sqrt{(1-x^2)}}.$$

Since

$$\frac{\mathrm{d}}{\mathrm{d}x}(\sin^{-1} x) = \frac{1}{\sqrt{(1-x^2)}}$$

it follows that

$$\int \frac{\mathrm{d}x}{\sqrt{(1-x^2)}} = \sin^{-1} x$$

so that

$$\int_0^1 \frac{\mathrm{d}x}{\sqrt{(1-x^2)}} = [\sin^{-1} x]_0^1.$$

Now $\sin^{-1} 1 = \frac{1}{2}\pi$, $\sin^{-1} 0 = 0$, so that

$$\int_0^1 \frac{\mathrm{d}x}{\sqrt{(1-x^2)}} = \tfrac{1}{2}\pi.$$

PROBLEMS

1. Find the indefinite integral of the function $1 + 2x + 6x^2$ and hence evaluate the integrals

$$\int_0^1 (1 + 2x + 6x^2)\mathrm{d}x, \qquad \int_2^3 (1 + 2x + 6x^2)\mathrm{d}x.$$

2. Show that

$$\int \frac{x\mathrm{d}x}{1+x^4} = \tfrac{1}{2}\tan^{-1}(x^2)$$

and hence evaluate

$$\int_0^1 \frac{x\mathrm{d}x}{1+x^4}, \qquad \int_0^\infty \frac{x\mathrm{d}x}{1+x^4}.$$

3. Show that

$$\int x \sin x \, dx = \sin x - x \cos x$$

and hence show that

$$\int_0^{\pi} x \sin x \, dx = \pi.$$

Show that this last result can be established by means of Theorem 33.

6. Change of Variable in an Integral

If $f(x)$ is a function of x, and x is itself a function of a variable t, say $x = \phi(t)$, then if $F(x)$ is a function of x such that $F'(x) = f(x)$ it follows from Theorem 15 (p. 154 above) that

$$\frac{dF}{dt} = \phi'(t) f(x) = \phi'(t) f[\phi(t)],$$

so that, by the fundamental theorem of the integral calculus we should expect

$$\int_a^b f(x) dx = \int_\alpha^\beta \phi'(t) f[\phi(t)] dt \tag{1}$$

where $a = \phi(\alpha)$, $b = \phi(\beta)$.

This is indeed true, the formal statement of the result being contained in:-

THEOREM 36: *If the integrals*

$$\int_a^b f(x) dx, \qquad \int_\alpha^\beta \phi'(t) dt$$

exist and if $a = \phi(\alpha)$, $b = \phi(\beta)$ *and* $\phi'(t) > 0$ *(or* $\phi'(t) < 0$*) for all* t *satisfying* $\alpha \leqq t \leqq \beta$ *then*

$$\int_a^b f(x) dx = \int_\alpha^\beta \phi'(t) f\{\phi(t)\} dt.$$

The proof of this theorem is rather long and we shall omit it here. We shall however illustrate the procedure by an example here and return to it later in Chapter 7 (pp. 274—295 below).

The method consists essentially in "spotting" that an integrand $F(t)$ can be written in the form $f\{\phi(t)\}\phi'(t)$ when the substitution $x = \phi(t)$ will reduce the integral

$$\int F(t) dt = \int f\{\phi(t)\}\phi'(t) dt$$

to the form

$$\int f(x)\mathrm{d}x$$

which may be much easier to evaluate.

EXAMPLE 6: *Evaluate* $\int_0^a t^2(a^3 - t^3)^{\frac{1}{3}}\,\mathrm{d}t$.

We note that $t^2 = \frac{1}{3}\mathrm{d}/\mathrm{d}t(t^3)$ so that the integrand is

$$(a^3 - t^3)^{\frac{1}{3}} \times \frac{-1}{3}\frac{\mathrm{d}}{\mathrm{d}t}(a^3 - t^3).$$

We therefore make the substitution

$$x = a^3 - t^3.$$

When $t = 0$, $x = a^3$ and when $t = a$, $x = 0$ so

$$\int_0^a t^2(a^3 - t^3)^{\frac{1}{3}}\,\mathrm{d}t = -\tfrac{1}{3}\int_{a^3}^0 x^{\frac{1}{3}}\,\mathrm{d}x = \tfrac{1}{3}\int_0^{a^3} x^{\frac{1}{3}}\,\mathrm{d}x = \tfrac{1}{3}[\tfrac{3}{4}x^{\frac{4}{3}}]_0^{a^3} = \tfrac{1}{4}a^4.$$

PROBLEMS

1. Evaluate the integral

$$\int_0^1 (x + 1)(x^2 + 2x + 5)^7\,\mathrm{d}x.$$

2. Show that

$$\int_0^a f(ax + b)\mathrm{d}x = \frac{1}{a}\int_b^{a+b} f(u)\mathrm{d}u.$$

3. Show that

$$\int_0^1 \frac{x^2\,\mathrm{d}x}{1 + x^6} = \frac{\pi}{12}.$$

7. Infinite Integrals

The definition we have given of a definite integral as a limit of a sum and its interpretation as an area requires that the limits of integration be finite. We define integrals with infinite limits in the following ways:-

We define

$$\int_a^\infty f(x)\mathrm{d}x$$

to be

$$\lim_{b\to\infty}\int_a^b f(x)\mathrm{d}x$$

if the limit exists and is finite. Otherwise the integral does not exist.

For example,

$$\int_1^\infty \frac{\mathrm{d}x}{x^2} = \lim_{b\to\infty}\int_1^b \frac{\mathrm{d}x}{x^2} = \lim_{b\to\infty}\left(1-\frac{1}{b}\right) = 1.$$

Similarly we define

$$\int_{-\infty}^a f(x)\mathrm{d}x = \lim_{b\to-\infty}\int_b^a f(x)\mathrm{d}x;$$

$$\int_{-\infty}^\infty f(x)\mathrm{d}x = \lim_{\alpha,\beta\to\infty}\int_{-\alpha}^\beta f(x)\mathrm{d}x,$$

the α and β tending to infinity independently. As examples of integrals of the last two types we have

$$\int_{-\infty}^1 \frac{\mathrm{d}x}{(2-x)^2} = \lim_{b\to-\infty}\int_b^1 \frac{\mathrm{d}x}{(2-x)^2} = \lim_{b\to-\infty}\left(1-\frac{1}{2-b}\right) = 1;$$

$$\int_{-\infty}^\infty \frac{\mathrm{d}x}{1+x^2} = \lim_{\alpha,\beta\to 0}\int_{-\alpha}^\beta \frac{\mathrm{d}x}{1+x^2} = \lim_{\alpha,\beta\to 0}\{\tan^{-1}\beta - \tan^{-1}(-\alpha)\}$$
$$= \lim_{\alpha,\beta\to\infty}(\tan^{-1}\beta + \tan^{-1}\alpha) = \pi.$$

Such integrals are known as *improper integrals of the first kind. Improper integrals of the second kind* have an integrand which becomes infinite at some point within the range of integration. If, for instance, the function $f(x)$ becomes infinite at $x = c$ where $a < c < b$, but is finite continuous everywhere else in the range, we define

$$\int_a^b f(x)\mathrm{d}x$$

to be

$$\lim_{\epsilon_1\to 0}\int_a^{c-\epsilon_1} f(x)\mathrm{d}x + \lim_{\epsilon_2\to 0}\int_{c+\epsilon_2}^b f(x)\mathrm{d}x$$

provided both these limits exist and are finite. If $c = a$ or b, one of these integrals will not be present; for instance if $f(x) \to \infty$ as $x \to b$ we define

$$\int_a^b f(x)\mathrm{d}x = \lim_{\epsilon\to 0}\int_a^{b-\epsilon} f(x)\mathrm{d}x.$$

As an example of an improper integral of the second kind we have

$$\int_0^1 \frac{dx}{\sqrt{(1-x)}}.$$

In this case the integrand tends to infinity as $x \to 1$, but

$$\int_0^{1-\varepsilon} \frac{dx}{\sqrt{(1-x)}} = [-2\sqrt{(1-x)}]_0^{1-\varepsilon} = 2(1-\sqrt{\varepsilon}),$$

which tends to 2 as $\varepsilon \to 0$ so that

$$\int_0^1 \frac{dx}{\sqrt{(1-x)}} = 2.$$

Let us now consider the integral

$$\int_0^1 x^{m-1}(1-x)^{n-1}\,dx. \tag{1}$$

If $m < 1$ the integrand is infinite at $x = 0$ and if $n < 1$ it is infinite at $x = 1$, so that

$$\int_0^1 x^{m-1}(1-x)^{n-1}\,dx = \lim_{\substack{\varepsilon_1\to 0\\ \varepsilon_2\to 0}} \int_{\varepsilon_1}^{1-\varepsilon_2} x^{m-1}(1-x)^{n-1}\,dx.$$

By Theorem 21 we can write

$$\int_0^1 x^{m-1}(1-x)^{n-1}\,dx = \lim_{\substack{\varepsilon_1\to 0\\ \varepsilon_2\to 0}} \left\{\int_{\varepsilon_1}^{c_1} + \int_{c_1}^{c_2} + \int_{c_2}^{1-\varepsilon_2}\right\} x^{m-1}(1-x)^{n-1}\,dx$$

where c_1 and c_2 are fixed, c_1 being small and c_2 being nearly 1. Then the integral

$$\int_{c_1}^{c_2} x^{m-1}(1-x)^{n-1}\,dx$$

will not be improper, i.e. it will have a finite value. Also if c_1 is small and $\varepsilon_1 < c_1$, $x^{m-1}(1-x)^{n-1}$ will behave like x^{m-1} when $\varepsilon_1 < x < c_1$, i.e.,

$$\int_{\varepsilon_1}^{c_1} x^{m-1}(1-x)^{n-1}\,dx \simeq \int_{\varepsilon_1}^{c_1} x^{m-1}\,dx = \frac{c_1^m - \varepsilon_1^m}{m} \qquad (m \neq 0)$$

and this tends to a definite limit c_1^m/m as $\varepsilon_1 \to 0$ if and only if $m > 0$. Similarly

$$\int_{c_2}^{1-\varepsilon_2} x^{m-1}(1-x)^{n-1}\,dx \simeq \int_{c_2}^{1-\varepsilon_2} (1-x)^{n-1}\,dx = \frac{(1-c_2)^n - \varepsilon_2^n}{n} \quad (n \neq 0)$$

which tends to a definite limit as $\varepsilon_2 \to 0$ if and only if $n > 0$. In other words the integral (1) exists if and only if both m and n are positive.

8. Approximate Integration

The area under a given curve between two fixed ordinates can be found by integration when the equation of the curve is known, provided of course that we can do the necessary integrations. In many practical cases it happens either that we do not know the equation of the curve (but only the coordinates of some points on it) or that, knowing the equation, we cannot carry out the integrations. In such cases we carry out the integrations approximately by means of a numerical procedure.

We can consider the integral

$$\int_a^b f(x)\mathrm{d}x$$

to be the area bounded by the curve $y = f(x)$, the x-axis, and the ordinates $x = a$, and $x = b$. If we can estimate this area, we will obtain an estimate of the value of the integral. We arrive at an

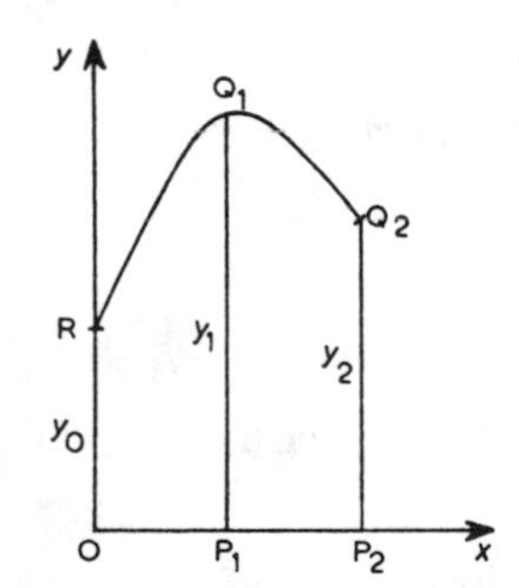

Fig. 72. Simpson's 3-ordinate rule.

estimate of the area by cutting it into strips of equal width, estimating the area of each strip and adding. Suppose that in the course of such a sub-division we have three ordinates OR, P_1Q_1, P_2Q_2 of heights y_0, y_1, y_2 respectively and at a distance h apart. We can construct a parabola to pass through the points R, Q_1, Q_2. Taking O as the origin, OP_1P_2 as the x-axis and OR as the y-axis we can assume that this parabola has equation

$$y = px^2 + qx + r$$

then if $OP_1 = P_1P_2 = h$, the area $ORQ_1Q_2P_2O$ is

$$A = \int_0^{2h} (px^2 + qx + r)\mathrm{d}x = \tfrac{1}{3}p(2h)^3 + \tfrac{1}{2}q(2h)^2 + r(2h)$$
$$= \tfrac{8}{3}ph^3 + 2qh^2 + 2rh.$$

We must now determine p, q, r in terms of the date. Since R, Q_1, Q_2 have respectively the coordinates (O, y_0), (h, y_1), $(2h, y_2)$ we have

$$y_0 = r$$
$$y_1 = ph^2 + qh + r$$
$$y_2 = 4ph^2 + 2qh + r.$$

From the first equation we have $r = y_0$. If we subtract twice the second equation from the third we find that

$$y_2 - 2y_1 = 2ph^2 - y_0$$

from which it follows that

$$2ph^2 = y_2 - 2y_1 + y_0.$$

Substituting this value in the last equation we have

$$2qh = 4y_1 - y_2 - 3y_0.$$

Now

$$A = \tfrac{1}{3}h(8ph^2 + 6qh + 6r)$$
$$= \tfrac{1}{3}h(4y_2 - 8y_1 + 4y_0 + 12y_1 - 3y_2 - 9y_0 + 6y_0),$$

that is,

$$A = \tfrac{1}{3}h(y_0 + 4y_1 + y_2). \tag{1}$$

The approximation given by equation (1) is known as *Simpson's three-ordinate rule.*

EXAMPLE 7: *Estimate the value of π by applying Simpson's three-ordinate rule to the formula*

$$\pi = 4\int_0^1 \frac{\mathrm{d}x}{1 + x^2}.$$

Dividing the range 0 to 1 into two strips of equal width we have $h = \frac{1}{2}$,

$$y_0 = \frac{1}{1 + 0} = 1, \quad y_1 = \frac{1}{1 + \frac{1}{4}} = \tfrac{4}{5}, \quad y_2 = \frac{1}{1 + 1} = \tfrac{1}{2}$$

so that by Simpson's three-ordinate rule

$$\pi \simeq 4 \times \tfrac{1}{3} \times \tfrac{1}{2}(1 + \tfrac{16}{5} + \tfrac{1}{2}) = \tfrac{94}{30}.$$

This gives a value 3.1333 for π. This is in error by less than $\frac{1}{2}$%.

We can improve the accuracy, of course, by taking a finer division of the range of integration. Suppose, for instance, that we have five equidistant ordinates as shown in Fig. 73. We can think of the total

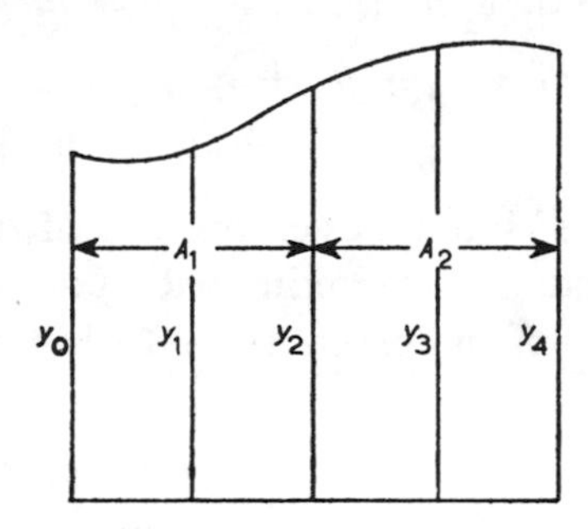

Fig. 73. Simpson's 5-ordinate rule.

area A being made up of two parts A_1 and A_2, each of which can be calculated by means of equation (1). We find that

$$A_1 = \tfrac{1}{3}h(y_0 + 4y_1 + y_2), \quad A_2 = \tfrac{1}{3}h(y_2 + 4y_3 + y_4)$$

so that

$$A = \tfrac{1}{3}h\{y_0 + y_4 + 4(y_1 + y_3) + 2y_2\}. \tag{2}$$

This is known as *Simpson's five-ordinate rule.*

EXAMPLE 8: *Estimate the value of π by applying Simpson's five-ordinate rule to the formula of the last example.*

Here $h = \frac{1}{4}$ so that $\frac{4}{3}h = \frac{1}{3}$.

$$\begin{aligned}
\text{Also } y_0 &= 1 & &= 1.00000\\
y_1 &= \tfrac{16}{17},\ 4y_1 = \tfrac{64}{17} & &= 3.76471\\
y_2 &= \tfrac{4}{5},\ 2y_2 = \tfrac{8}{5} & &= 1.60000\\
y_3 &= \tfrac{16}{25},\ 4y_3 = \tfrac{64}{25} & &= 2.56000\\
y_4 &= \tfrac{1}{2} & &= 0.50000\\
& & \textit{sum} &= 9.42471\\
& & \tfrac{1}{3} \times \textit{sum} &= 3.14157.
\end{aligned}$$

We therefore obtain a value 3.14157 for π. The value given by Barlow's tables* is 3.14159265...

We can easily extend Simpson's rule if the number of strips is even. Taking the strips in pairs we have

$$A = \tfrac{1}{3}h\{(y_0 + 4y_1 + y_2) + (y_2 + 4y_3 + y_4) + (y_4 + 4y_5 + y_6) + \dots + (y_{n-2} + 4y_{n-1} + y_n)$$

$$= \tfrac{1}{3}h\{y_0 + y_n + 4(y_1 + y_3 + y_5 + \dots + y_{n-1}) + 2(y_2 + y_4 + \dots + y_{n-2})\} \quad (3)$$

i.e. if the area is divided into an even number of strips by equidistant ordinates, the area can be approximated by:-
$\frac{1}{3}$ (width of strip) {sum of end ordinates +4 (sum of even ordinates) + 2(sum of odd ordinates)}.

* L. J. Comrie, editor, "Barlow's Tables" (4th ed., Spon. London, 1965).

CHAPTER 6

THE LOGARITHMIC AND EXPONENTIAL FUNCTIONS

1. The Logarithmic Function

The fundamental concepts of the integral calculus provide a much more satisfying theory of the logarithmic and exponential functions than does the 'elementary' procedure adopted in most schools. The latter approach omits some delicate points when the arguments of the functions are irrational numbers. In this chapter the elementary order is reversed; we define the logarithm as a definite integral, develop its properties and then deduce those of the exponential function. If $x > 0$, the function $\log x$ is defined by the equation

$$\log x = \int_1^x \frac{d\xi}{\xi} \tag{1}$$

so that it is the area bounded by the lines $\xi = 1$, $\xi = x$, $\eta = 0$ and the curve $\xi\eta = 1$ i.e. $\eta = 1/\xi$ (cf. Fig. 74). It is meaningful to study

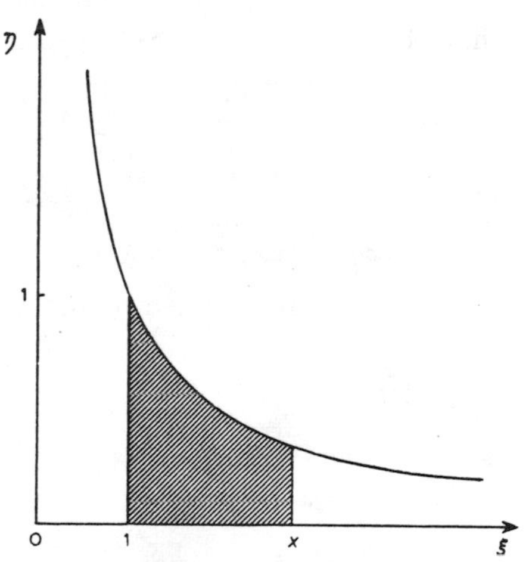

Fig. 74. The definition of $\log x$; the area of the shaded portion gives the value of $\log x$.

the function defined by equation (1), since we know of no function whose derivative is x^{-1}. We might therefore expect that the integration on the right of equation (1) would lead to an entirely new kind of function. The connection between this definition of the logarithm and the more elementary definition will emerge later. It is interesting to note, however, that the definition we have adopted is very close to that originally framed by Napier of Merchiston.

We shall now show that the function defined by equation (1) has all the properties usually associated with logarithm. Firstly we observe that

$$\begin{aligned} \log x &< 0 \quad \text{if} \quad 0 < x < 1; \\ \log x &= 0 \quad \text{if} \quad x = 1; \\ \log x &> 0 \quad \text{if} \quad x > 1. \end{aligned}$$

Also, directly from the definition, we have

$$\log (xy) = \int_1^{xy} \frac{\mathrm{d}\xi}{\xi}$$

which by Theorem 21 of Chapter 5 becomes

$$\log (xy) = \int_1^{x} \frac{\mathrm{d}\xi}{\xi} + \int_x^{xy} \frac{\mathrm{d}\xi}{\xi}.$$

Changing the variable in the second integral on the right from ξ to ζ where $\xi = x\zeta$ we find that

$$\log (xy) = \int_1^{x} \frac{\mathrm{d}\xi}{\xi} + \int_1^{y} \frac{\mathrm{d}\zeta}{\zeta}$$

giving the well-known rule

$$\log (xy) = \log x + \log y. \tag{2}$$

In particular, putting $y = 1/x$ and remembering that $\log 1 = 0$ we see that

$$\log (1/x) = - \log x. \tag{3}$$

The result (2) can be generalized to

$$\log (x^n) = n \log x \tag{4}$$

where $x > 0$ and n is rational; for

$$\log (x^n) = \int_1^{x^n} \frac{\mathrm{d}\xi}{\xi} = n \int_1^{x} \frac{\mathrm{d}\xi}{\xi}$$

where we have made the substitution $\xi = \zeta^n$. Also, combining (2) and (3) we see that

$$\log (x/y) = \log x - \log y. \tag{5}$$

We must now investigate the behaviour of $\log x$ for large values of x. For large x we can find an integer n such that $x > 2^n$ and

$$\log x > n \log 2.$$

Now as $n \to \infty$, $x \to \infty$ and we get the result

$$\lim_{x\to\infty} \log x = \infty \tag{6}$$

since $\log 2 > 0$. If we write $y = 1/x$, so that $\log y = -\log x$, we have the corollary

$$\lim_{y\to 0} \log y = -\lim_{x\to\infty} \log x = -\infty, \tag{7}$$

by equation (6).

Furthermore if $x > y$ we have

$$\int_y^x \frac{\mathrm{d}\xi}{\xi} > 0$$

which is equivalent to the relation

$$\log x > \log y$$

showing that $\log x$ is a monotonic increasing function of x. This fact is also shown by the relation

$$\frac{\mathrm{d}}{\mathrm{d}x} \log x = \frac{1}{x} > 0 \tag{8}$$

which follows from the definition by applying the fundamental theorem of the integral calculus.

With the help of all these facts we can draw the graph of the function $\log x$; such a curve is shown in Fig. 75.

If we use the chain rule we see that

$$\frac{\mathrm{d}}{\mathrm{d}x} \log f(x) = \frac{\mathrm{d}}{\mathrm{d}u} (\log u) \frac{\mathrm{d}u}{\mathrm{d}x}$$

where $u = f(x)$. Now

$$\frac{\mathrm{d}}{\mathrm{d}u} \log u = \frac{1}{u} = \frac{1}{f(x)}$$

so that

$$\frac{\mathrm{d}}{\mathrm{d}x}\log f(x) = \frac{f'(x)}{f(x)}. \tag{9}$$

From this it follows that

$$\int_a^x \frac{f'(u)}{f(u)}\,\mathrm{d}u = \log\frac{f(x)}{f(a)}. \tag{10}$$

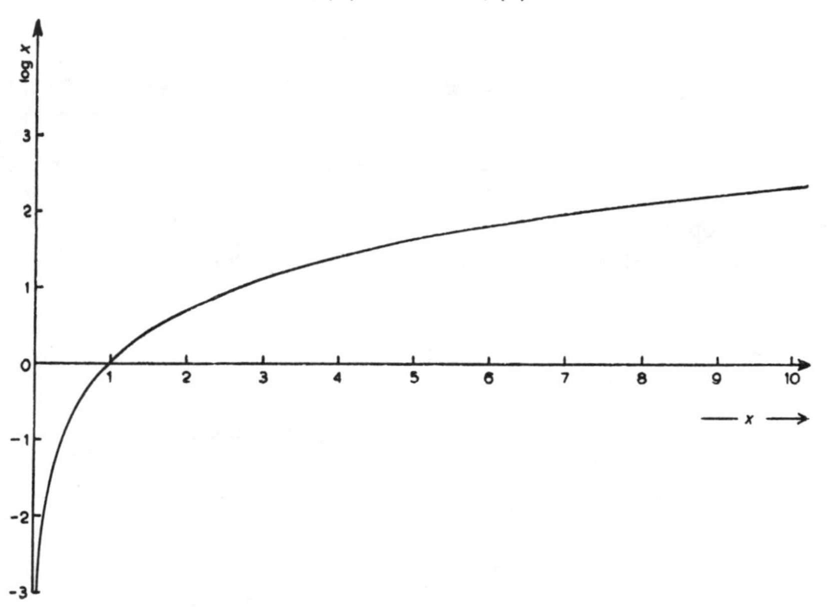

Fig. 75. Graph of the function $\log x$.

As a special case of equation (9) we notice that the specific growth rate

$$\frac{1}{W}\frac{\mathrm{d}W}{\mathrm{d}t}$$

which occurs in certain biological applications can be written as

$$\frac{\mathrm{d}}{\mathrm{d}t}(\log W).$$

The rule for differentiating a product of functions can be proved easily by using logarithms.

If

$$f(x) = u(x)v(x)w(x) \ldots z(x) \tag{11}$$

then

$$\log f(x) = \log u(x) + \log v(x) + \log w(x) + \ldots + \log z(x).$$

Differentiating both sides with respect to x and using (9) we find that

$$\frac{f'(x)}{f(x)} = \frac{u'(x)}{u(x)} + \frac{v'(x)}{v(x)} + \frac{w'(x)}{w(x)} + \ldots + \frac{z'(x)}{z(x)}. \tag{12}$$

PROBLEMS

1. Differentiate with respect to x the following functions

(i) $\log(1 + x^2)$;

(ii) $\log \dfrac{1+x}{1-x}$;

(iii) $\log \sqrt{\dfrac{x^2-1}{x^2+1}}$;

(iv) $\left(1 + \dfrac{1}{\log x}\right)^n$;

(v) $\log \tan(\frac{1}{4}\pi + \frac{1}{2}x)$;

(vi) $\log\left(\dfrac{1+e^{-ax}}{1-e^{-ax}}\right)$;

(vii) $\log\left(\dfrac{2+\cos x}{2-\cos x}\right)$.

2. Find the turning value of $x \log x$.

3. If $\log y = \frac{1}{2}\log\dfrac{1+x}{1-x} - x$, show that $(1-x^2)\dfrac{dy}{dx} = x^2 y$.

4. Prove that

$$\frac{d}{dx}\left[\sin^{-1}\frac{x}{a} - \log\left\{\frac{a + \sqrt{(a^2-x^2)}}{x}\right\}\right] = \frac{1}{x}\sqrt{\frac{a+x}{a-x}}.$$

5. If $y = A\cos(2\log x) + B\sin(2\log x)$, where A and B are constants, show that

$$x^2\frac{d^2y}{dx^2} + x\frac{dy}{dx} + 4y = 0.$$

6. If $y = \log (A + B/x)$ where A and B are constants, show that

$$x \frac{d^2 y}{dx^2} + x \left(\frac{dy}{dx}\right)^2 + 2 \frac{dy}{dx} = 0.$$

2. The Exponential Function

The function $\log x$ is a continuous function for all positive values of x, and we showed in the last section that $y = \log x$ is a strictly

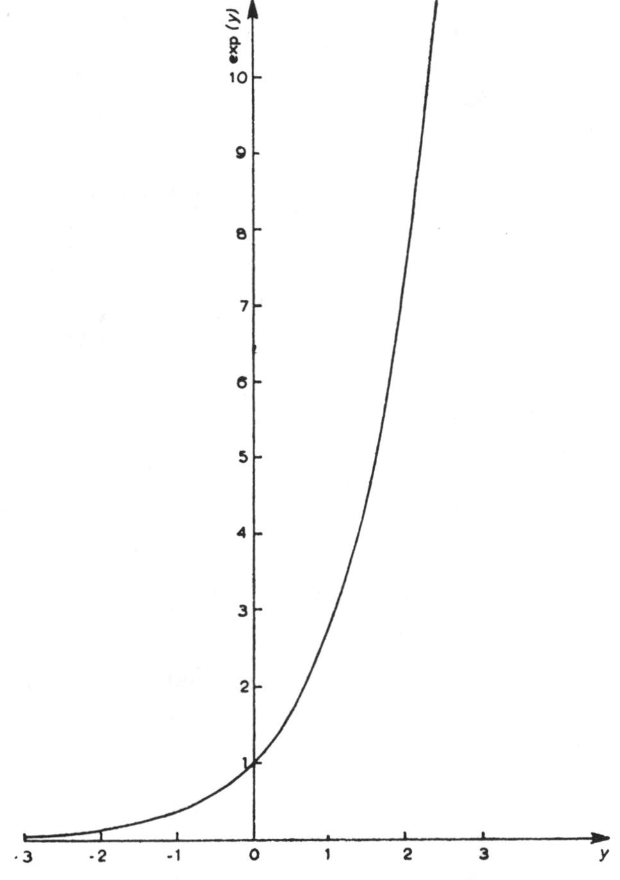

Fig. 76. Graph of the exponential function.

monotonic increasing function in the region $x > 0$. From this it follows that there exists a unique, single-valued, inverse function

$$x = \exp (y) \tag{1}$$

defined for all values of y. Its graph, obtained merely by re-labelling the axes in Fig. 75 is shown in Fig. 76. In addition there is only one number, e say, which has the property that

$$\log e = 1 \tag{2}$$

(cf. Fig. 75 above). Hence if

$$x = e^y$$

where y is a rational number it follows from equation (4) of the last section that

$$\log x = y \log e = y$$

showing that if y is a rational number

$$\exp(y) = e^y.$$

We have thus demonstrated fully that the logarithmic function defined in this chapter is identical with the logarithm to the base e defined in textbooks on ordinary algebra. In the notation of the latter we should write $\log_e x$ for our logarithmic function. However, in higher mathematics, unless the contrary is stated explicitly, all logarithms are taken to the base e. From numerical values of the integral (see Chapter 5, § 8)

$$\int_1^x \frac{d\xi}{\xi}$$

it is found that e is approximately 2.7. Actually e = 2.718281828...; the reader will soon be able to obtain this easily, by putting $x = 1$ in § 7, equation (2).

From the same integral we find that $\log_e 10$ is about 2.3 and it is actually 2.302585..., see example 5, p. 256. Then since

$$\log_{10} e \log_e n = \log_{10} n$$

we have

$$\log_e n = 2.302585 \log_{10} n.$$

If $y = \exp(x)$ it follows that

$$x = \log y$$

and hence by equation (8) of the last section we have

$$\frac{dx}{dy} = \frac{1}{y}$$

showing that

$$\frac{dy}{dx} = y$$

i.e.

$$\frac{d}{dx} \exp(x) = \exp(x). \tag{3}$$

An immediate generalization of this result is

$$\frac{d}{dx} \exp(\alpha x) = \alpha \exp(\alpha x). \tag{4}$$

Similarly if

$$y_1 = \exp(x_1), \quad y_2 = \exp(x_2)$$

we have

$$\log(y_1 y_2) = \log y_1 + \log y_2 = x_1 + x_2$$

showing that

$$\exp(x_1) \exp(x_2) = \exp(x_1 + x_2). \tag{5}$$

If we use the chain rule we see that

$$\frac{d}{dx} e^{f(x)} = \frac{d}{du} (e^u) \frac{du}{dx}$$

where $u = f(x)$. It follows that

$$\frac{d}{dx} e^{f(x)} = f'(x) e^{f(x)} \tag{6}$$

and hence that

$$\int_a^x f'(u) e^{f(u)} du = e^{f(x)} - e^{f(a)}. \tag{7}$$

PROBLEMS

1. Differentiate with respect to x the functions

(i) $(1 + x)^2 e^{-x}$;

(ii) $e^{-2x} \cos(2x + 3)$;

(iii) $e^{-\sin x}$;

(iv) xe^{-x^2}.

2. Find the coordinates of the turning points and of the points of inflexion on the curve $y = (1 - x)^2 e^x$, and sketch the curve.

3. Find the turning point and points of inflexion of the curve $y = x^3 e^{-\frac{1}{2}x}$ for $x \geqq 0$, and sketch it.

4. If $y = ae^{-x} \sin(x + \alpha)$, where a, α are constants, show that dy/dx can be written in the form $be^{-x} \sin(x + \beta)$, where b, β are constants. Find b and β in terms of a and α, and hence find d^2y/dx^2.

5. If y is a function of x and $x = e^z/(1 + e^z)$, show that

$$\frac{dy}{dz} = x(1 - x)\frac{dy}{dx}.$$

6. If y is a function of x and if $x = e^t$, show that

$$x\frac{dy}{dx} = \frac{dy}{dt}, \quad x^2\frac{d^2y}{dx^2} = \frac{d^2y}{dt^2} - \frac{dy}{dt}.$$

7. If $y = Ax^n e^{kx}$ where A is an arbitrary constant, prove that

$$\frac{dy}{dx} - ky = \frac{ny}{x}.$$

3. Some Important Limits

We shall now show that the definition of e given in the last section is in agreement with the more usual definition given in elementary textbooks. First of all we show that if $h \to 0$ and $n \to \infty$ in such a way that $hn \to \mu$, a fixed number, then

$$\lim_{h\to 0} n \log (1 + h) = \mu \tag{1}$$

for

$$\lim_{h\to 0} n \log (1 + h) = \mu \lim_{h\to 0} \frac{\log (1 + h)}{h} = \mu \lim_{h\to 0} \frac{\frac{1}{1+h}}{1} = \mu$$

by the formula (2) of Section 12 of Chapter 4.

Since e^x is a continuous function of x it follows that when $x \to \mu$, $e^x \to e^\mu$ so that if $h \to 0$, $n \to \infty$ so that $hn \to \mu$ we have

$$\lim_{h\to 0} (1 + h)^n = e^\mu \tag{2}$$

from which follow the well-known limits:-

$$\lim_{n\to\infty} \left(1 + \frac{1}{n}\right)^n = e \tag{3}$$

$$\lim_{n\to\infty} \left(1 + \frac{x}{n}\right)^n = e^x. \tag{4}$$

As an example of the formula (2) of Section 12 of Chapter 4 we see that if $n > 0$

$$\lim_{x\to 0} (x^n \log x) = \lim_{x\to 0} \frac{\log x}{x^{-n}} = \lim_{x\to 0} \frac{\frac{1}{x}}{-nx^{-n-1}} = -\frac{1}{n} \lim_{x\to 0} x^n$$

so that for $n > 0$

$$\lim_{x\to 0} (x^n \log x) = 0. \tag{5}$$

Putting $x = \mathrm{e}^{-y}$ we see that this result is equivalent to

$$\lim_{y\to\infty} y\mathrm{e}^{-y} = 0. \tag{6}$$

4. Applications of the Exponential Function in Biology

The exponential function occurs frequently in applications of mathematics to biology because of the property expressed by equation (4) of Section 2 above. This means that the function

$$f(x) = A\mathrm{e}^{kx} \tag{1}$$

has the property that

$$f'(x) = kf \tag{2}$$

so that any quantity whose rate of change is proportional to f is of the form (1). Situations of this kind arise frequently; we shall illustrate it with a few examples. Other examples are given in the next section.

(*a*) *Growth of an insect population*

Suppose that a population of insects is placed in an experimental environment and is allowed to grow for several days. Its net time-rate of increase is known to be proportional to its size, and its increases during the third and the fifth day are estimated (from counts of newly-hatched and dead insects) as 2455 and 4314 insects respectively. The problem is to estimate the population at a later time.

Since the net time-rate of increase is proportional to the size n it follows that n is of the form (1), i.e. that

$$n = A\mathrm{e}^{kt} \tag{3}$$

where t is the time (measured in days, say) and A and k are constants. At the beginning of the third day $t = 2$ and the size of the population is Ae^{2k}; at the end of that day $t = 3$ and the size is then Ae^{3k}. Hence the increase is

$$Ae^{3k} - Ae^{2k} = 2455$$

which can be written in the form

$$Ae^{2k}(e^k - 1) = 2455. \tag{4}$$

Similarly the increase in the fifth day will be

$$Ae^{4k}(e^k - 1) = 4314. \tag{5}$$

Dividing equation (5) by equation (4) we see that

$$e^{2k} = \tfrac{4314}{2455}$$

and taking the logarithm with respect to e of both sides we have

$$2k = \log_e 4314 - \log_e 2455$$

so that

$$k = \tfrac{1}{2} (\log_{10} 4314 - \log_{10} 2455) \times 2.3026.$$

Since $\log_{10} 4314 = 3.6349$, $\log_{10} 2455 = 3.3899$ it follows that $k = 0.282$. For this value of k, $e^{3k} = 2.3303$ and $e^{2k} = 1.7577$ so that from equation (4)

$$A = \frac{2455}{0.5726} = 4285.$$

Thus, if t is the time measured in days the size of the population at time t is given by the equation

$$n = 4285\, e^{0.282t}$$

and the increase in the size during the r-th day is

$$\begin{aligned} \Delta_r &= 4285\, \{e^{0.282r} - e^{0.282(r-1)}\} \\ &= 4285\, e^{0.282r}(1 - e^{-0.282}) \\ &= 1053\, e^{0.282r} \end{aligned}$$

which may be put in the convenient form

$$\log_{10} \Delta_r = 3.0224 + 0.12247r.$$

(b) The half-life of a radioactive substance

In a simple radioactive process the total number of atoms that survive after any given time t is proportional to $e^{-\lambda t}$ where λ is a constant, called the *decay constant*. Thus, if there are N atoms initially, the number after time t will be

$$n = N e^{-\lambda t}. \tag{6}$$

The time τ during which half the atoms disintegrate is called the half-life of the radioactive substance. If we put $n = \frac{1}{2}N$, $t = \tau$ in (6) we see that

$$\tfrac{1}{2}N = N e^{-\lambda \tau}$$

so that

$$e^{\lambda \tau} = 2$$

and hence

$$\lambda \tau = \log_e 2.$$

The half-life τ of a substance is therefore given in terms of the decay constant λ by the equation

$$\tau = \frac{1}{\lambda} \log_e 2 = \frac{0.6931}{\lambda}. \tag{7}$$

(*c*) *The concentration-time curve in the blood of drugs injected intramuscularly, subcutaneously or intra-peritoneally*

This important clinical application has been studied theoretically by E. Heinz in 1949.* He showed that the concentration y of the drug

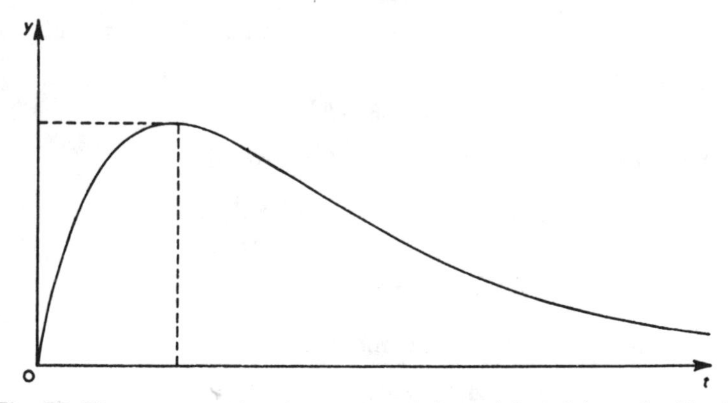

Fig. 77. The concentration-time curve of drugs injected into the blood.

in the blood at time t can be given by the formula

* E. Heinz, Biochem. Zeitschrift **319** (1949) 482.

$$y = \frac{A}{\sigma_2 - \sigma_1}(e^{-\sigma_1 t} - e^{-\sigma_2 t}) \tag{8}$$

where A, σ_1 and σ_2 are positive constants and $\sigma_2 > \sigma_1$.

The graph of the function y is shown (for typical values of the constants) in Fig. 77. This functional form has been verified in many particular cases.* From the curve it will be seen that the concentration has a maximum. We shall now determine the time at which the maximum concentration occurs and its magnitude.

Differentiating both sides of equation (8) with respect to t we obtain the equation

$$y' = \frac{A}{\sigma_2 - \sigma_1}(\sigma_2 e^{-\sigma_2 t} - \sigma_1 e^{-\sigma_1 t}) \tag{9}$$

so that $y' = 0$ when

$$\sigma_2 e^{-\sigma_2 t} = \sigma_1 e^{-\sigma_1 t},$$

that is, when

$$e^{(\sigma_2 - \sigma_1)t} = \frac{\sigma_2}{\sigma_1}.$$

Taking the natural logarithms of both sides we see that $y' = 0$ when

$$t = \frac{1}{\sigma_2 - \sigma_1}\log\left(\frac{\sigma_2}{\sigma_1}\right). \tag{10}$$

Hence at the value of t given by equation (10) the concentration has a turning value, which is obviously a maximum.

Now for the value (10) of t

$$e^{(\sigma_2 - \sigma_1)t} = \frac{\sigma_2}{\sigma_1}$$

and

$$e^{-\sigma_2 t} = \exp\left\{-\frac{\sigma_2}{\sigma_2 - \sigma_1}\log\frac{\sigma_2}{\sigma_1}\right\} = \exp\left\{\log\left(\frac{\sigma_1}{\sigma_2}\right)^{\sigma_2/(\sigma_2 - \sigma_1)}\right\} = \left(\frac{\sigma_1}{\sigma_2}\right)^{\sigma_2/(\sigma_2 - \sigma_1)}$$

so that, writing

$$y = \frac{A}{\sigma_2 - \sigma_1} e^{-\sigma_2 t}(e^{(\sigma_2 - \sigma_1)t} - 1)$$

we see that the maximum concentration is given by the formula

* For other possibilities see Chapter 10, §15, 16.

$$y_{\max} = \frac{A}{\sigma_2 - \sigma_1}\left(\frac{\sigma_1}{\sigma_2}\right)^{\sigma_2/(\sigma_2-\sigma_1)}\left(\frac{\sigma_2}{\sigma_1} - 1\right)$$

which reduces to

$$y_{\max} = A\sigma_1^{\sigma_1/(\sigma_2-\sigma_1)}\,\sigma_2^{-\sigma_2/(\sigma_2-\sigma_1)}. \tag{11}$$

It will be observed that in solving this problem we did not prove (by considering the sign of y'') that the turning value was a maximum. This latter test is usually omitted when it is obvious on physical grounds what the nature of the turning value is.

(*d*) *The normal error curve*

The function

$$y(x) = e^{-\frac{1}{2}x^2} \tag{12}$$

arises in the theory of errors and its graph is called a normal error curve. (It is treated more fully in Chapter 7, §13.)

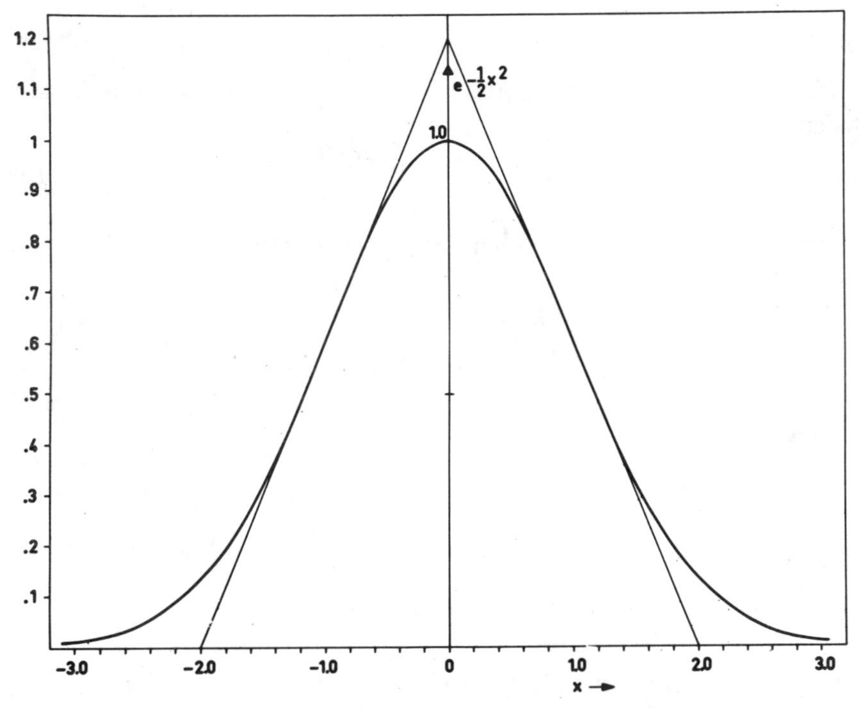

Fig. 78. The normal error curve.

It follows from (12) that, for all values of x,

$$y(-x) = y(x).$$

This means that the curve is symmetrical about the y-axis. Also, if we differentiate (12) with respect to x, we find that

$$y'(x) = -xe^{-\frac{1}{2}x^2}$$

so that $y'(x) = 0$ if and only if $x = 0$, when y has the value unity. Furthermore, as $x \to \pm\infty$, $y \to 0$. The shape of the curve is shown in Fig. 78.

PROBLEMS

1. Show that the points of inflexion are at $x = \pm 1$, $y = e^{-\frac{1}{2}}$.

2. Find the slopes of the tangents to the curve at these points of inflexion, and show that these tangents intersect each other at height $2e^{-\frac{1}{2}}$ on the y-axis, and intersect the x-axis at $x = \pm 2$.

5. The Common Growth Functions

As interesting example of the drawing of curves involving exponential functions is provided by the simpler forms of growth function in common use*. We usually describe a growth function by an equation of the type

$$W = \phi(t) \tag{1}$$

or if $W = W_0$ when $t = 0$ by one of the type

$$W = W_0 f(t) \tag{2}$$

with $f(0) = 1$. In equations (1) and (2) W usually symbolizes "weight" but it may mean "size" and in population studies may mean the number of individuals in the population. t usually refers to time, either the age of an individual or the time during which the population has been growing.

We shall consider the various types separately.

(a) *The exponential curve*

For this type of growth

* See, for instance, the article entitled "Size, Shape and Age" by P. B. Medewar in "Essays on Growth and Form" (Clarendon Press, Oxford, 1945) pp. 157–187.

$$W = W_0 e^{kt} \tag{3}$$

where k is a constant. The growth rate is given by

$$\frac{dW}{dt} = kW_0 e^{kt} \tag{4}$$

so that the specific growth rate

$$\frac{1}{W}\frac{dW}{dt} = k, \tag{5}$$

showing that it is a constant. The shape of this growth curve is easily found from Fig. 76 by a simple change of the scales.

If we wish to determine whether a certain growth process follows a law of this kind we plot $\omega = \log W$ against t. If the resulting graph is a straight line then W and t are connected by a formula of the kind (3) since that formula can be written in the form

$$\log W = \log W_0 + kt.$$

(b) The 'monomolecular' curve

This curve acquired its name because it first arose in the theory of monomolecular reactions in chemistry. It has the equation

$$W = \frac{W_0}{1-b}(1 - be^{-kt}) \tag{6}$$

where W_0, b and k are positive constants and $b < 1$. Since $e^{-kt} \to 1$ as $t \to 0$ and $e^{-kt} \to 0$ as $t \to \infty$ (since $k > 0$) it follows that $W = W_0$ when $t = 0$ and that $W \to W_0/(1 - b)$ when $t \to \infty$.

For this curve the growth rate is given by

$$\frac{dW}{dt} = \frac{bke^{-kt}}{1-b} W_0, \tag{7}$$

and the specific growth rate by

$$\frac{1}{W}\frac{dW}{dt} = \frac{bke^{-kt}}{1 - be^{-kt}}. \tag{8}$$

It follows from these equations that both the growth rate and the specific growth rate tend to zero as $t \to \infty$. From equation (7) we see that when $t = 0$, $dW/dt = bkW_0/(1 - b)$ and that as t increases dW/dt decreases steadily from this value to zero.

Differentiating equations (7) and (8) again with respect to t we find that

$$\frac{\mathrm{d}^2 W}{\mathrm{d}t^2} = -\frac{bk^2 \mathrm{e}^{-kt}}{1-b} W_0, \quad \frac{\mathrm{d}^2}{\mathrm{d}t^2}(\log W) = -\frac{bk^2 \mathrm{e}^{-kt}}{(1-b\mathrm{e}^{-kt})^2}.$$

The variation of W and $\log W$ and their derivatives is shown in Fig. 79.

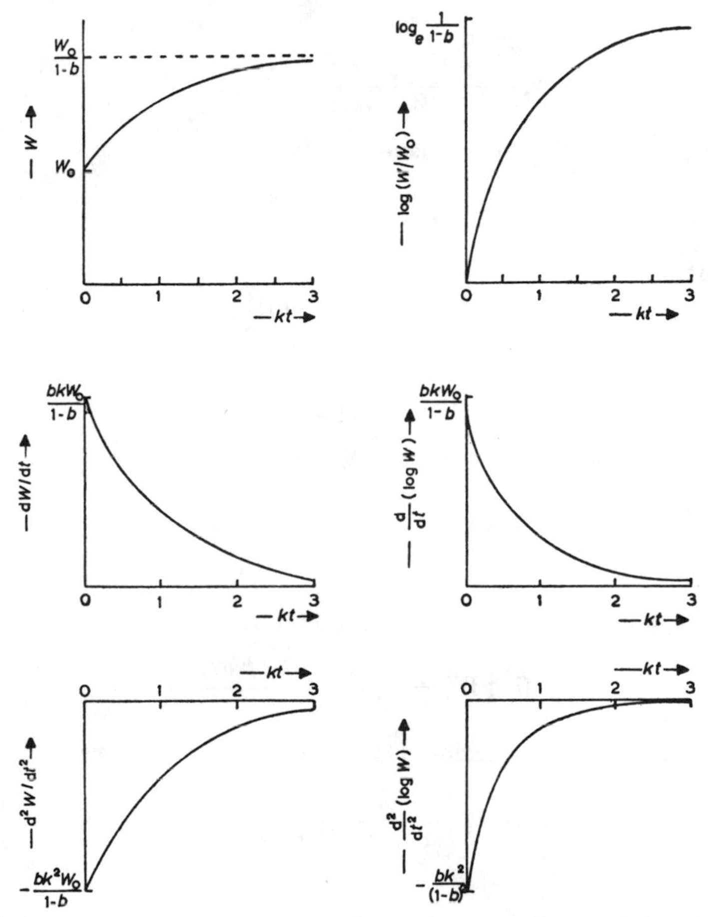

Fig. 79. The variation of the growth function and the specific growth function (with their first and second derivatives) in the case of 'monomolecular' growth.

(c) The logistic curve

The equation governing this type of growth is

$$W = W_0 \frac{1+b}{1+be^{-kt}} \qquad (b > 0). \tag{9}$$

If we solve this equation for be^{-kt} we find that

$$be^{-kt} = \frac{(1+b)W_0 - W}{W}$$

from which we find, on taking logarithms, that the logarithmic form of (9) is

$$\log \frac{(1+b)W_0 - W}{W} = \log b - kt. \tag{10}$$

Writing $1 + be^{-kt} = u$ we have

$$W = \frac{(1+b)W_0}{u}$$

so that

$$\frac{dW}{du} = -\frac{(1+b)W_0}{u^2}$$

and

$$\frac{dW}{dt} = \frac{dW}{du}\cdot\frac{du}{dt} = -\frac{(1+b)W_0}{u^2} \times -bke^{-kt}$$

showing that

$$\frac{dW}{dt} = \frac{(1+b)bkW_0 e^{-kt}}{(1+be^{-kt})^2}. \tag{11}$$

From equations (9) and (11) we therefore have

$$\frac{d}{dt}(\log W) = \frac{1}{W}\frac{dW}{dt} = \frac{bke^{-kt}}{1+be^{-kt}}. \tag{12}$$

If we differentiate equation (11) with respect to t we have

$$\frac{d^2W}{dt^2} = (1+b)bkW_0\left\{\frac{-ke^{-kt}}{(1+be^{-kt})^2} + e^{-kt}\frac{d}{dt}(1+be^{-kt})^{-2}\right\}.$$

Now

$$\frac{d}{dt}(1+be^{-kt})^{-2} = -2(1+be^{-kt})^{-3}(-bke^{-kt})$$
$$= 2bke^{-kt}(1+be^{-kt})^{-3}$$

so that

$$\begin{aligned}\frac{d^2W}{dt^2} &= (1+b)bkW_0 \frac{-ke^{-kt}(1+be^{-kt})+2bke^{-2kt}}{(1+be^{-kt})^3} \\ &= \frac{(1+b)bk^2W_0e^{-kt}(be^{-kt}-1)}{(1+be^{-kt})^3}.\end{aligned} \tag{13}$$

It is obvious from equation (11) that dW/dt is never zero for any finite value of t, and from equation (13) that d^2W/dt^2 is zero when

$$be^{-kt} = 1,$$

that is, when

$$t = \frac{\log b}{k}$$

the corresponding value of W being $\frac{1}{2}(1+b)W_0$. Further, from equation (9) we note that $W = W_0$ when $t = 0$ and that $W \to (1+b)W_0$

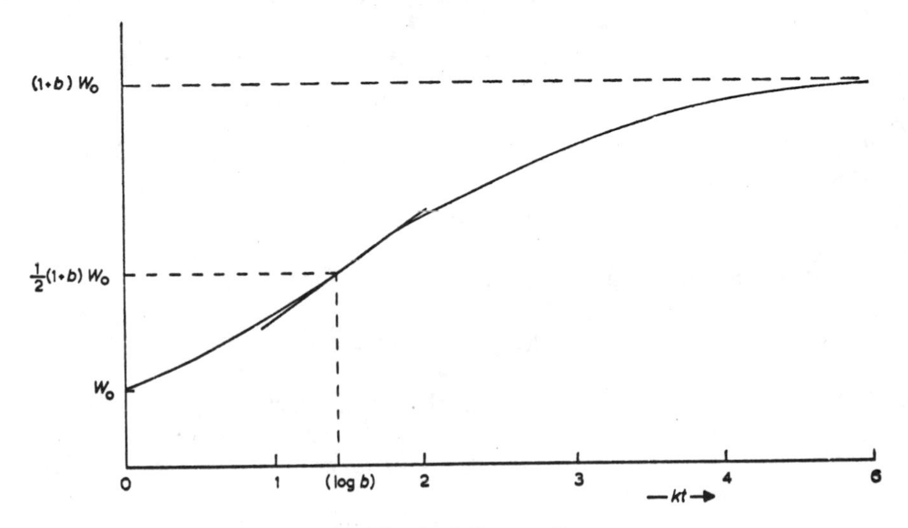

Fig. 80. The logistic growth curve.

as $t \to \infty$. These facts imply that the curve of logistic growth rises steadily from the value W_0 to the value $(1+b)W_0$ having no maxima or minima but a point of inflexion at the point $t = (\log b)/k$, $W = \frac{1}{2}(1+b)W_0$. (Cf. Fig. 80.)

The expression for the derivative of W can be put in a neater form. From equations (11) and (9) we see that

$$\frac{dW}{dt} = kbe^{-kt}\left(\frac{1+b}{1+be^{-kt}}\right)^2 \frac{W_0}{1+b}$$

$$= k\left\{\frac{(1+b)W_0 - W}{W}\right\}\left(\frac{W}{W_0}\right)^2 \frac{W_0}{1+b}$$

from which we find the result

$$\frac{dW}{dt} = kW\left\{1 - \frac{W}{(1+b)W_0}\right\} \tag{14}$$

and its corollary

$$\frac{d}{dt}\log W = k\left\{1 - \frac{W}{(1+b)W_0}\right\}. \tag{15}$$

For applications of the logistic curve the reader is referred to the literature in biological journals.*

(d) The Gompertz growth curve

The equation of the Gompertz growth curve** is

$$W = ae^{-be^{-kt}}. \tag{16}$$

Taking logarithms of both sides of this equation we find that

$$\log\frac{W}{a} = -be^{-kt} \tag{17}$$

an equation which can be written in the form

$$\log\frac{a}{W} = be^{-kt}.$$

Taking logarithms again we find that

$$\log\log\frac{a}{W} = \log b - kt. \tag{18}$$

In this case it is easier to calculate the specific growth rates. Writing

$$\log\frac{W}{a} = \log W - \log a$$

* T. B. Robertson, Arch. Entw. Mech. Org. **25** (1908) 4, 581; **26** (1908) 108;
W. J. Crozier, J. Gen. Physiol. **10** (1926) 53;
H. W. Titus, J. Washington Acad. Sci. **20** (1930) 357.

** For the occurrence of the Gompertz growth curve in biology see:-
F. W. Weymouth, H. C. McMillin and W. H. Rich, J. Exper. Biol. **8** (1931) 228;
C. P. Winsor, Proc. Nat., Acad. Sci. **18** (1932) 1.

in equation (17) and differentiating with respect to t we have

$$\frac{\mathrm{d}}{\mathrm{d}t}(\log W) = bk\mathrm{e}^{-kt} \tag{19}$$

and differentiating again with respect to t we find that

$$\frac{\mathrm{d}^2}{\mathrm{d}t^2}(\log W) = -\,bk^2\mathrm{e}^{-kt}. \tag{20}$$

If we write the relation

$$\frac{\mathrm{d}}{\mathrm{d}t}(\log W) = \frac{1}{W}\frac{\mathrm{d}W}{\mathrm{d}t}$$

in the form

$$\frac{\mathrm{d}W}{\mathrm{d}t} = W\frac{\mathrm{d}}{\mathrm{d}t}(\log W)$$

we see, from equations (16) and (19) that

$$\frac{\mathrm{d}W}{\mathrm{d}t} = bk\mathrm{e}^{-kt}a\mathrm{e}^{-b\mathrm{e}^{-kt}} = abk\mathrm{e}^{-kt-b\mathrm{e}^{-kt}}. \tag{21}$$

If we put

$$u = kt + b\mathrm{e}^{-kt}$$

this can be written as

$$\frac{\mathrm{d}W}{\mathrm{d}t} = abk\mathrm{e}^{-u}.$$

Writing

$$\frac{\mathrm{d}^2W}{\mathrm{d}t^2} = abk\frac{\mathrm{d}}{\mathrm{d}u}(\mathrm{e}^{-u})\frac{\mathrm{d}u}{\mathrm{d}t} = -abk\mathrm{e}^{-u}\frac{\mathrm{d}u}{\mathrm{d}t}$$

and making use of the fact that

$$\frac{\mathrm{d}u}{\mathrm{d}t} = k - bk\mathrm{e}^{-kt}$$

we find that

$$\frac{\mathrm{d}^2W}{\mathrm{d}t^2} = -abk(k - bk\mathrm{e}^{-kt})\mathrm{e}^{-kt-b\mathrm{e}^{-kt}}. \tag{22}$$

As a consequence of equations (21) and (22) we see that $\mathrm{d}W/\mathrm{d}t$ is not zero for any finite value of t but that $\mathrm{d}^2W/\mathrm{d}t^2 = 0$ when

$$k = bk\mathrm{e}^{-kt}$$

i.e. when

$$e^{kt} = b$$

or

$$t = \frac{1}{k} \log b$$

and this is positive if $b > 1$. The Gompertz growth curve therefore has no maxima or minima but has (when $b > 1$) a point of inflexion at $t = (\log b)/k$ as is shown in Fig. 81.

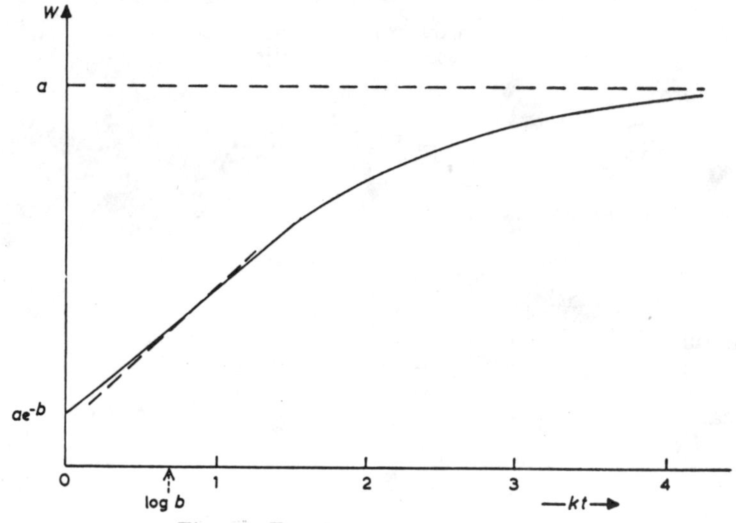

Fig. 81. The Gompertz growth curve.

PROBLEM

1. Show that, in general

$$\frac{d^2}{dt^2}(\log W) = \frac{1}{W}\frac{d^2 W}{dt^2} - \frac{1}{W^2}\left(\frac{dW}{dt}\right)^2$$

and verify the relation for each of the growth curves considered above.

6. The Hyperbolic Functions

Closely related to the exponential function and of great value in mathematical analysis are the hyperbolic functions, which we shall now define and whose simple properties we shall derive.

We define the *hyperbolic sine* by the equation

$$\sinh x = \tfrac{1}{2}(e^x - e^{-x}) \tag{1}$$

and the *hyperbolic cosine* by the equation

$$\cosh x = \tfrac{1}{2}(e^x + e^{-x}). \tag{2}$$

By analogy with the trigonometrical functions we define the *hyperbolic tangent* by

$$\tanh x = \frac{\sinh x}{\cosh x} = \frac{e^x - e^{-x}}{e^x + e^{-x}}. \tag{3}$$

The values of the hyperbolic functions can be derived easily from tables of values of the exponential function, and tables of values of sinh x, cosh x, tanh x are to be found in many books of tables

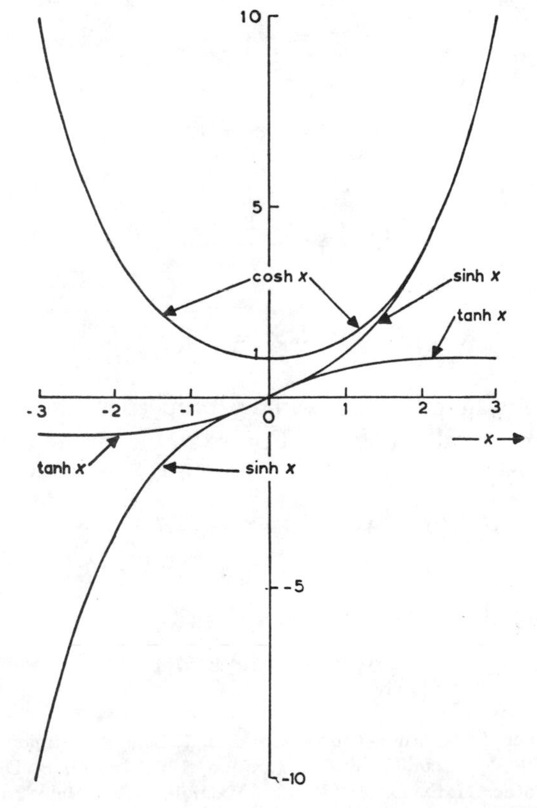

Fig. 82. Graphs of the functions sinh x, cosh x, tanh x.

of mathematical functions.* Using such tables we can easily draw the graphs of these functions. A set of such graphs is shown in Fig. 82. It should be observed (either from these graphs or the definitions) that

$$\sinh 0 = 0, \quad \cosh 0 = 1, \quad \tanh 0 = 0,$$

that $\cosh x$ is an even function whereas $\sinh x$ and $\tanh x$ are odd functions.** We know that for large positive values of x, e^x is large but e^{-x} is small, so that for large positive values of x

$$\sinh x \simeq \cosh x \simeq \tfrac{1}{2}e^x.$$

Similarly for large negative values of x we have

$$\sinh x \simeq -\tfrac{1}{2}e^{-x}, \quad \cosh x \simeq \tfrac{1}{2}e^{-x}.$$

If we write the second of equations (3) in the form

$$\tanh x = \frac{1 - e^{-2x}}{1 + e^{-2x}},$$

we see that

$$\lim_{x \to \infty} \tanh x = 1, \tag{4}$$

while if we write

$$\tanh x = -\frac{1 - e^{2x}}{1 + e^{2x}},$$

we see that

$$\lim_{x \to -\infty} \tanh x = -1. \tag{5}$$

The hyperbolic functions have many properties analogous to those of the trigonometrical functions. For example, from the definitions (1), (2) we have

$$\begin{aligned}\cosh^2 x - \sinh^2 x &= \tfrac{1}{4}\{(e^x + e^{-x})^2 - (e^x - e^{-x})^2\} \\ &= \tfrac{1}{4}\{(e^{2x} + 2 + e^{-2x}) - (e^{2x} - 2 + e^{-2x})\}\end{aligned}$$

from which it follows that for all values of x

$$\cosh^2 x - \sinh^2 x = 1 \tag{6}$$

* See, for instance, L. M. Milne-Thomson and L. J. Comrie, "Standard Four Figure Mathematical Tables" (Macmillan & Co., London, 1957), or H. B. Dwight, "Tables of Integrals and other Mathematical Data" (Macmillan Co., New York, 1947).

** A function $f(x)$ is *even* if $f(-x) = f(x)$ and is *odd* if $f(-x) = -f(x)$.

which is the analogue of the relation $\cos^2 x + \sin^2 x = 1$ for trigonometrical functions.

It follows from equation (6) that the point with coordinates $(a \cosh t, a \sinh t)$ always lies on the hyperbola with equation

$$x^2 - y^2 = a^2,$$

whatever the value of t. This is why the adjective "hyperbolic" is applied to these functions. The connection between cosh and cos and between sinh and sin will be discussed at the end of the next section.

It is a matter of simple algebra to show that

$$(e^x + e^{-x})(e^y + e^{-y}) + (e^x - e^{-x})(e^y - e^{-y}) = 2(e^{x+y} + e^{-x-y})$$

from which we can deduce that

$$\cosh(x + y) = \cosh x \cosh y + \sinh x \sinh y. \tag{7}$$

Similarly we can show that

$$\sinh(x + y) = \sinh x \cosh y + \cosh x \sinh y. \tag{8}$$

The derivatives of the hyperbolic functions are readily derived from the equations

$$\frac{d}{dx}(e^x) = e^x, \quad \frac{d}{dx}(e^{-x}) = -e^{-x}.$$

Applied to equation (1) these results show that

$$\frac{d}{dx}(\sinh x) = \frac{1}{2}\frac{d}{dx}(e^x) - \frac{1}{2}\frac{d}{dx}(e^{-x}) = \tfrac{1}{2}(e^x + e^{-x})$$

showing that

$$\frac{d}{dx}(\sinh x) = \cosh x \tag{9}$$

while applied to equation (2) they yield the result

$$\frac{d}{dx}(\cosh x) = \sinh x. \tag{10}$$

We can define secondary functions precisely in the same way as in elementary trigonometry; we define a *hyperbolic secant, cosecant* and *cotangent* by the equations

$$\operatorname{sech} x = \frac{1}{\cosh x}, \quad \operatorname{cosech} x = \frac{1}{\sinh x}, \quad \coth x = \frac{\cosh x}{\sinh x} \tag{11}$$

respectively. If we divide both sides of equation (6) by $\cosh^2 x$ and make use of these definitions we find that

$$\operatorname{sech}^2 x = 1 - \tanh^2 x \tag{12}$$

whereas if we divide both sides of the same equation by $\sinh^2 x$ we obtain the result

$$\operatorname{cosech}^2 x = \coth^2 x - 1. \tag{13}$$

The values of $\operatorname{sech} x$, $\operatorname{cosech} x$, $\coth x$ can easily be derived from those of $\cosh x$ and $\sinh x$ and the graphs of these functions drawn.

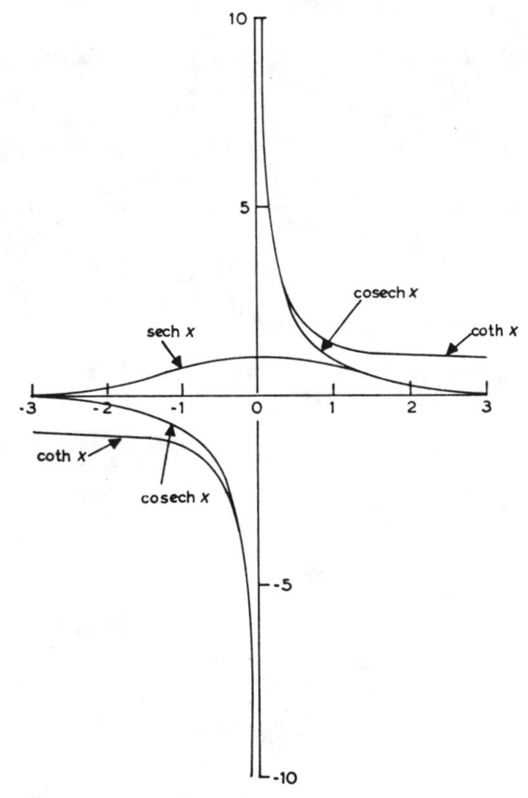

Fig. 83. Graphs of the functions $\operatorname{sech} x$, $\operatorname{cosech} x$, $\coth x$.

Such graphs are shown in Fig. 83. It will be observed that

$$\lim_{x\to 0} \operatorname{cosech} x = \infty$$
$$\lim_{x\to 0} \coth x = \infty$$
$$\lim_{x\to \infty} \coth x = 1$$

and that for large positive values of x

$$\operatorname{cosech} x \simeq \operatorname{sech} x \simeq 2e^{-x}$$

(from the behaviour of $\cosh x$, $\sinh x$ for large positive values of x).

EXAMPLE 1. *Show that*

$$\frac{d}{dx}(\tanh x) = \operatorname{sech}^2 x.$$

From the rule for differentiating a quotient we have

$$\frac{d}{dx}(\tanh x) = \frac{d}{dx}\frac{\sinh x}{\cosh x} = \frac{\cosh x \dfrac{d}{dx}(\sinh x) - \sinh x \dfrac{d}{dx}(\cosh x)}{\cosh^2 x}$$

so that by equations (9) and (10)

$$\frac{d}{dx}(\tanh x) = \frac{\cosh^2 x - \sinh^2 x}{\cosh^2 x} = \frac{1}{\cosh^2 x}$$

if we make use of equation (6). By the first of the definitions (11) we therefore have

$$\frac{d}{dx}(\tanh x) = \operatorname{sech}^2 x. \tag{14}$$

We can define *inverse hyperbolic functions* in exactly the same way as we defined inverse trigonometrical functions. If for $x > 1$ we can find a number y such that

$$x = \cosh y \tag{15a}$$

we say that y is the *inverse hyperbolic cosine* of x and

$$y = \cosh^{-1} x. \tag{15b}$$

Similarly if, for any x, we can find a number y such that

$$x = \sinh y \tag{16a}$$

we say that

$$y = \sinh^{-1} x. \tag{16b}$$

The functions $\tanh^{-1} x$, $\operatorname{sech}^{-1} x$, $\operatorname{cosech}^{-1} x$, $\coth^{-1} x$ can be defined similarly.

The inverse hyperbolic functions are related to the logarithmic function as is shown by

EXAMPLE 2. *If* $x > 1$ *show that*

$$\cosh^{-1} x = \log \{x + \sqrt{(x^2 - 1)}\}$$

and that for all values of x

$$\sinh^{-1} x = \log \{x + \sqrt{(x^2 + 1)}\}.$$

If $y = \cosh^{-1} x$ then $x = \cosh y$, so that

$$x = \tfrac{1}{2}(e^y + e^{-y}).$$

Multiplying both sides of this equation by $2e^y$ we find that

$$e^{2y} - 2xe^y + 1 = 0.$$

Solving this quadratic equation for e^y we find that

$$e^y = x + \sqrt{(x^2 - 1)}$$

and hence that

$$y = \log \{x + \sqrt{(x^2 - 1)}\}$$

proving the first result.

To prove the second result we note that if $y = \sinh^{-1} x$ then $x = \sinh y$ and

$$x = \tfrac{1}{2}(e^y - e^{-y})$$

from which we have

$$e^{2y} - 2xe^y - 1 = 0.$$

The appropriate solution of this equation is

$$e^y = x + \sqrt{(x^2 + 1)},$$

since the solution $x - \sqrt{(x^2 + 1)}$ would yield a negative value of e^y, which is impossible. Hence, taking logarithms, we find that

$$y = \log \{x + \sqrt{(x^2 + 1)}\}$$

and we have the required result.

PROBLEMS

1. Prove that, for all real values of x,

$$4 \cosh^3 x - 3 \cosh x - \cosh 3x = 0.$$

2. Show that if $y = C \cosh m(x - h)$, where C is arbitrary, then

$$\frac{d^2 y}{dx^2} - m^2 y = 0$$

and $dy/dx = 0$ when $x = h$.

3. Prove that

$$\frac{\cosh 3x}{\cosh x} - \frac{\sinh 3x}{\sinh x} + 2 = 0.$$

4. If $\tanh x = \frac{4}{5}$, prove that $x = \log 3$.

5. Show that

$$\tanh^{-1} x = \tfrac{1}{2} \log \left(\frac{1 + x}{1 - x}\right).$$

6. Show that

$$\frac{d}{dx} \sinh^{-1} x = \frac{1}{\sqrt{(x^2 + 1)}},$$

$$\frac{d}{dx} \cosh^{-1} x = \frac{1}{\sqrt{(x^2 - 1)}},$$

$$\frac{d}{dx} \tanh^{-1} x = \frac{1}{1 - x^2}.$$

(See § 3 of Chapter 4, where the derivatives of inverse trigonometrical functions are discussed.)

7. Series Expansions of Exponential and Logarithmic Functions

In this section we shall derive series expansions for the logarithmic and exponential functions, using the form of Maclaurin's theorem stated in Section 4 of Chapter 4. This leaves undecided questions of convergence of the series we derive. At this point we shall merely *state* the conditions under which the series are convergent; the proofs of these statements will be given in Section 9 of the next chapter.

(*a*) *The exponential series*

Since

$$\frac{\mathrm{d}}{\mathrm{d}x}(\mathrm{e}^x) = \mathrm{e}^x$$

and $\mathrm{e}^x = 1$ when $x = 0$ it follows that, if $f(x) = \mathrm{e}^x$ then

$$f(x) = f'(x) = f''(x) = \ldots = f^{(n)}(x) = \ldots = \mathrm{e}^x$$

and

$$f(0) = f'(0) = f''(0) = \ldots = f^{(n)}(0) = \ldots = 1.$$

Substituting these values in Maclaurin's theorem

$$f(x) = f(0) + f'(0)x + f''(0)\frac{x^2}{2!} + \ldots + f^{(n)}(0)\frac{x^n}{n!} + \ldots \tag{1}$$

(see Section 4 of Chapter 4 above), we see that

$$\mathrm{e}^x = 1 + \frac{x}{1!} + \frac{x^2}{2!} + \ldots + \frac{x^n}{n!} + \ldots. \tag{2}$$

It can be shown that the series on the right hand side of equation (2) converges for *all* real values of x to the function e^x. Replacing x by $-x$ in equation (2) we find that

$$\mathrm{e}^{-x} = 1 - \frac{x}{1!} + \frac{x^2}{2!} + \ldots + (-1)^n\frac{x^n}{n!} + \ldots. \tag{3}$$

We can deduce an important inequality from the series (2). If $x > 0$, it follows from equation (2) that for any positive integer n,

$$\mathrm{e}^x > \frac{x^n}{n!}. \tag{4}$$

From this inequality we deduce that

$$\mathrm{e}^{-x} < \frac{n!}{x^n}, \quad (n > 0). \tag{5}$$

This inequality in turn enables us to establish that

$$\lim_{x\to\infty} x^m \mathrm{e}^{-x} = 0.$$

For by (5) we can choose a positive integer $n \geqq m$ such that

$$x^m \mathrm{e}^{-x} < \frac{n!}{x^{n-m}}$$

and if $x > 0$, $x^m \mathrm{e}^{-x} > 0$ so that

$$0 < x^m \mathrm{e}^{-x} < \frac{n!}{x^{n-m}}$$

from which it follows that

$$\lim_{x\to\infty} x^m \mathrm{e}^{-x} = 0. \tag{6}$$

To illustrate the manipulation of the exponential series we have:-

EXAMPLE 3. *Show from the series expansion of the exponential function that*

(i) $$\frac{1+x}{x}(\mathrm{e}^x - 1) = 1 + \frac{3}{2}\frac{x}{1} + \frac{4}{3}\frac{x^2}{2!} + \frac{5}{4}\frac{x^3}{3!} + \ldots$$

(ii) $$(ax+b)\mathrm{e}^x = \sum_{r=0}^{\infty}(ar+b)\frac{x^r}{r!}\cdot$$

(i) For the first part we note that

$$\mathrm{e}^x - 1 = 1 + \frac{x}{1!} + \frac{x^2}{2!} + \frac{x^3}{3!} + \ldots + \frac{x^n}{n!} + \ldots - 1$$

so that

$$\frac{\mathrm{e}^x - 1}{x} = \frac{1}{1!} + \frac{x}{2!} + \frac{x^2}{3!} + \ldots + \frac{x^{n-1}}{n!} + \ldots$$

and

$$\begin{aligned}\frac{1+x}{x}(\mathrm{e}^x - 1) &= (1+x)\left\{\frac{1}{1!} + \frac{x}{2!} + \frac{x^2}{3!} + \ldots + \frac{x^{n-1}}{n!} + \ldots\right\} \\ &= 1 + \frac{x}{2!} + \frac{x^2}{3!} + \frac{x^3}{4!} + \ldots + \frac{x^n}{(n+1)} + \ldots \\ &\quad + \frac{x}{1!} + \frac{x^2}{2!} + \frac{x^3}{3!} + \ldots + \frac{x^n}{n!} + \ldots.\end{aligned}$$

Now

$$\frac{x^n}{(n+1)!} + \frac{x^n}{n!} = \frac{x^n}{n!}\left(1 + \frac{1}{n+1}\right) = \frac{n+2}{n+1}\frac{x^n}{n!}$$

so that

$$\frac{1+x}{x}(e^x-1) = 1 + \frac{3}{2}\frac{x}{1!} + \frac{4}{3}\frac{x^2}{2!} + \ldots + \frac{n+2}{n+1}\frac{x^n}{n!} + \ldots.$$

(ii) For the second part, we have

$$(ax+b)e^x = (ax+b)\left(1 + \frac{x}{1!} + \frac{x^2}{2!} + \ldots + \frac{x^r}{r!} + \ldots\right)$$

$$= ax + a\frac{x^2}{1!} + a\frac{x^3}{2!} + \ldots + a\frac{x^r}{(r-1)!} + \ldots$$

$$+ b + b\frac{x}{1!} + b\frac{x^2}{2!} + b\frac{x^3}{3!} + \ldots + b\frac{x^r}{r!} + \ldots$$

so that the coefficient of x^r is

$$\frac{a}{(r-1)!} + \frac{b}{r!} = \frac{ar+b}{r!}.$$

Therefore

$$(ax+b)e^x = \sum_{r=0}^{\infty}(ar+b)\frac{x^r}{r!}.$$

(b) Series for hyperbolic functions

If we substitute the series (1) and (2) in the definition

$$\cosh x = \tfrac{1}{2}(e^x + e^{-x})$$

we find that

$$\cosh x = 1 + \frac{x^2}{2!} + \frac{x^4}{4!} + \ldots + \frac{x^{2m}}{(2m)!} + \ldots. \tag{7}$$

Similarly, from the definition

$$\sinh x = \tfrac{1}{2}(e^x - e^{-x})$$

we derive the series

$$\sinh x = x + \frac{x^3}{3!} + \frac{x^5}{5!} + \ldots + \frac{x^{2m+1}}{(2m+1)!} + \ldots. \tag{8}$$

Since the series for e^x and e^{-x} are valid for all real values of x it follows that the series for $\sinh x$ and $\cosh x$ are also valid for all real values of x.

EXAMPLE 4. *Prove that*

$$\lim_{x\to 0}\frac{x(\cosh x - \cos x)}{\sinh x - \sin x} = 3.$$

From the Maclaurin series for $\cosh x$ and $\cos x$ we see that, for small values of x,

$$x(\cosh x - \cos x) \simeq x\left\{1 + \frac{x^2}{2} - \left(1 - \frac{x^2}{2}\right)\right\} = x^3$$

while

$$\sinh x - \sin x \simeq x + \frac{x^3}{6} - \left(x - \frac{x^3}{6}\right) = \frac{x^3}{3}$$

from which the limit follows.

(c) *The logarithmic series*

If we put

$$f(x) = \log(1 + x)$$

then

$$f'(x) = \frac{1}{1 + x}$$

and

$$f''(x) = -\frac{1}{(1 + x)^2}.$$

Repeating these differentiations we see that

$$f'''(x) = \frac{2!}{(1 + x)^3}$$

$$f^{iv}(x) = -\frac{3!}{(1 + x)^4}$$

and so on, it being obvious that the general result is

$$f^{(n)}(x) = (-1)^{n-1}\frac{(n - 1)!}{(1 + x)^n}.$$

Putting $x = 0$ in these results we see that

$$f(0) = 0, \quad f'(0) = 1, \quad f''(0) = -1, \quad f'''(0) = 2$$

and, in general,

$$f^{(n)}(0) = (-1)^{n-1}(n - 1)!$$

from which it follows that

$$\frac{f^{(n)}(0)}{n!} = \frac{(-1)^{n-1}}{n}.$$

From Maclaurin's theorem — equation (i) above — we obtain the formula

$$\log (1 + x) = x - \frac{x^2}{2} + \frac{x^3}{3} + \ldots + (-1)^{n-1}\frac{x^n}{n} + \ldots. \tag{9}$$

It can be shown that the series on the right converges if $-1 < x < 1$. We shall return to this point in Section 9 of the next chapter.

If we replace x by $-x$ in equation (9) we find that

$$\log (1 - x) = -x - \frac{x^2}{2} - \frac{x^3}{3} - \ldots - \frac{x^n}{n} - \ldots. \tag{10}$$

Subtracting equations (9) and (10) and dividing by 2 we obtain the formula

$$\tfrac{1}{2} \log \frac{1+x}{1-x} = x + \frac{x^3}{3} + \frac{x^5}{5} + \ldots + \frac{x^{2n+1}}{2n+1} + \ldots. \tag{11}$$

We make use of these formulae in:-

EXAMPLE 5. *Show that*

$$\log 2 = 1 - \tfrac{1}{2} + \tfrac{1}{3} - \ldots + \frac{(-1)^n}{n+1} + \ldots$$

$$\tfrac{1}{2} \log 2 = \tfrac{1}{3} + \tfrac{1}{81} + \tfrac{1}{1215} + \ldots + \frac{1}{(2n+1)\,3^{2n+1}} + \ldots$$

$$\tfrac{1}{2} \log 1.5 = \tfrac{1}{5} + \tfrac{1}{375} + \tfrac{1}{15625} + \ldots + \frac{1}{(2n+1)\,5^{2n+1}} + \ldots$$

$$\log 10 = \log 9 + \tfrac{1}{10} + \tfrac{1}{200} + \tfrac{1}{3000} + \ldots \frac{1}{10^n n} + \ldots$$

$$\log 7 = \log 20 - \log 3 + \tfrac{1}{20} - \tfrac{1}{800} + \ldots \frac{(-1)^{n+1}}{20^n n} + \ldots$$

and hence show how the logarithms of the positive integers from 1 *to* 10 *may be calculated.*

Since the expansion on the right hand side of (9) converges for $x = 1$ we get a series for log 2 by putting $x = 1$ in the series on the right of (9). But we would have to sum hundreds of terms of this series to calculate log 2 to a few places of decimals.

If we put $x = \frac{1}{3}$ then $(1 + x)/(1-x) = 2$, so that if we put $x = \frac{1}{3}$ in equation (11) we get the series stated for $\frac{1}{2}$ log 2; from the first few terms of this series we easily obtain for log 2 the value 0.693147, as for the half-life formula in §4, equation (7).

We can now calculate the logarithms of the other integers up to 10 in terms of one another and of log 2 (more or less as in Chapter 1, example 2). If

$$x = \tfrac{1}{5}, \qquad \frac{1+x}{1-x} = 1.5,$$

so that we obtain a series for $\log 1.5 = \log 3 - \log 2$ by putting $x = \frac{1}{5}$ in equation (11); this gives $\log 1.5 = 0.405465$. Then putting $x = 0.1$ in equation (10) we get the series for $\log 0.9 = \log 9 - \log 10$; $\log 9 = \log 3^2 = 2 \log 3$, so this gives log 10. Then, putting $x = \frac{1}{20}$ in equation (9) we get $\log 7 = \log 21 - \log 3$. Similarly we calculate $\log 4 = 2 \log 2$, $\log 5 = \log 10 - \log 2$, $\log 6 = \log 2 + \log 3$ and $\log 8 = 3 \log 2$.

As a further example of the use of the logarithmic series we have:-

EXAMPLE 6. *Show that, for small values of x,*

$$\log (1 + e^{-x}) \simeq \log 2 - \tfrac{1}{2}x + \tfrac{1}{8}x^2.$$

From the exponential series (2) we have the series

$$1 + e^{-x} = 1 + (1 - x + \tfrac{1}{2}x^2) = 2(1 - \xi)$$

where

$$\xi = \tfrac{1}{2}x - \tfrac{1}{4}x^2.$$

Hence, using the logarithmic series (9), we have

$$\log (1 + e^{-x}) = \log 2 + \log (1 - \xi) \simeq \log 2 - \xi - \tfrac{1}{2}\xi^2$$

and substituting the value of ξ we have (since powers higher than the second are neglected)

$$\log (1 + e^{-x}) \simeq \log 2 - (\tfrac{1}{2}x - \tfrac{1}{4}x^2) - \tfrac{1}{2} \cdot \tfrac{1}{4}x^2 = \log 2 - \tfrac{1}{2}x + \tfrac{1}{8}x^2.$$

(d) The relation between hyperbolic and trigonometrical functions

There is a marked similarity between the series (4) for $\cosh x$ and the series

$$\cos x = 1 - \frac{x^2}{2!} + \frac{x^4}{4!} + \dots (-1)^m \frac{x^{2m}}{(2m)!} + \dots \tag{12}$$

for $\cos x$ derived in Section 4 of Chapter 4. If we replace x in equation (7) by ix, where i denotes the imaginary quantity $\sqrt{(-1)}$, then we see that

$$\cosh(ix) = 1 + \frac{(ix)^2}{2!} + \frac{(ix)^4}{4!} + \ldots + \frac{(ix)^{2m}}{(2m)!} + \ldots. \tag{13}$$

If we use the facts that $i^2 = -1$, $i^4 = 1$, $i^6 = -1$, $i^8 = 1, \ldots$, we see that $(ix)^2 = -x^2$, $(ix)^4 = x^4, \ldots, (ix)^{2m} = i^{2m}x^{2m} = (i^2)^m x^{2m} = (-1)^m x^{2m}$, and hence that

$$\cosh(ix) = 1 - \frac{x^2}{2!} + \frac{x^4}{4!} + \ldots + (-1)^m \frac{x^{2m}}{(2m)!} + \ldots. \tag{14}$$

Comparing this series with the right-hand side of equation (12) we see that, formally at least, we may write

$$\cosh(ix) = \cos x. \tag{15}$$

If we replace x by ix in equation (15) we find that

$$\cos(ix) = \cosh(i^2 x) = \cosh(-x).$$

Using the fact that $\cosh(-x) = \cosh x$, we find that

$$\cos(ix) = \cosh x, \tag{16}$$

so that we can think of $\cosh x$ as being the cosine of the imaginary quantity (ix).

Similarly from equation (8) we have

$$\sinh(ix) = (ix) + \frac{(ix)^3}{3!} + \frac{(ix)^5}{5!} + \frac{(ix)^7}{7!} + \ldots.$$

Since $(ix)^3 = -ix^3$, $(ix)^5 = ix^5$, $(ix^7) = -ix^7$, we see that

$$\sinh(ix) = i\left\{x - \frac{x^3}{3!} + \frac{x^5}{5!} - \frac{x^7}{7!} + \ldots\right\}.$$

The series within the brace is the Maclaurin expansion for $\sin x$, so that we have found that

$$\sinh(ix) = i\sin x. \tag{17}$$

Replacing x by ix in this equation we find that

$$i\sin(ix) = \sinh(-x).$$

Now $\sinh(-x) = -\sinh x$ and we can write this as $i^2 \sinh x$ so that dividing both sides of the equation by an i we obtain the result

$$\sin(ix) = i\sinh x. \tag{18}$$

Similarly if we replace x by ix in equation (2) we have

$$e^{ix} = 1 + \frac{ix}{1!} + \frac{(ix)^2}{2!} + \frac{(ix)^3}{3!} + \frac{(ix)^4}{4!} + \ldots$$

$$= \left(1 - \frac{x^2}{2!} + \frac{x^4}{4!} \ldots\right) + i\left(x - \frac{x^3}{3!} + \frac{x^5}{5!} - \ldots\right).$$

Identifying the two series we see that

$$e^{ix} = \cos x + i \sin x. \tag{19}$$

Using the fact that $\cos \pi = -1$, $\sin \pi = 0$ we obtain the startling result

$$e^{i\pi} + 1 = 0, \tag{20}$$

due to Euler.*

Replacing x by $-x$ in (19) and using the fact that $\cos(-x) = \cos x$, $\sin(-x) = -\sin x$, we have

$$e^{-ix} = \cos x - i \sin x. \tag{21}$$

From (19) and (21) we can readily deduce that

$$\cos x = \tfrac{1}{2}(e^{ix} + e^{-ix}), \quad \sin x = \frac{1}{2i}(e^{ix} - e^{-ix}) \tag{22}$$

a pair of equations which are consistent with equations (15) and (17).

As an example of the use of these formulae we have:-

EXAMPLE 7. *Prove Demoivre's theorem that, if n is rational*

$$(\cos x + i \sin x)^n = \cos nx + i \sin nx. \tag{23}$$

Deduce that

(i) $\cos 6x = \cos^6 x - 15 \cos^4 x \sin^2 x + 15 \cos^2 x \sin^4 x - \sin^6 x$;

(ii) $\sin 6x = 6 \sin x \cos^5 x - 20 \cos^3 x \sin^3 x + 6 \cos x \sin^5 x$.

From equation (19) we have

$$(\cos x + i \sin x)^n = (e^{ix})^n = e^{inx} = \cos nx + i \sin nx$$

and the theorem is proved.

Putting $n = 6$ we have

$$\cos 6x + i \sin 6x = (\cos x + i \sin x)^6. \tag{24}$$

* Readers familiar with E. T. Bell's "Men of Mathematics" will remember his remark that "anyone who can perceive this mystery intuitively will not need to square the circle"!

Expanding the right-hand side by the binomial theorem we see that it takes form

$$\begin{aligned}\cos^6 x &+ 6\cos^5 x\,(i\sin x) + 15\cos^4 x\,(i\sin x)^2 + 20\cos^3 x\,(i\sin x)^3 \\ &\qquad + 15\cos^2 x\,(i\sin x)^4 + 6\cos x\,(i\sin x)^5 + (i\sin x)^6 \\ &= \cos^6 x + 6i\sin x\cos^5 x - 15\cos^4 x\sin^2 x - 20i\cos^3 x\sin^3 x \\ &\qquad + 15\cos^2 x\sin^4 x + 6i\cos x\sin^5 x - \sin^6 x \\ &= \cos^6 x - 15\cos^4 x\sin^2 x + 15\cos^2 x\sin^4 x - \sin^6 x \\ &\qquad + i\,\{6\sin x\cos^5 x - 20\cos^3 x\sin^3 x + 6\cos x\sin^5 x\}.\end{aligned}$$

Identifying the real and imaginary parts of both sides it follows from equation (24) that $\cos 6x$ and $\sin 6x$ have the values stated.

As an illustration of a similar procedure, sometimes useful in the evaluation of integrals involving trigonometrical functions, we have

EXAMPLE 8. *Prove that*

$$\cos^6 x = \tfrac{1}{32}(\cos 6x + 6\cos 4x + 15\cos 2x + 10)$$

and deduce the value of the integral

$$\int_0^{\frac{1}{2}\pi} \cos^6 x\mathrm{d}x.$$

From the first of equations (22) we have

$$\cos^6 x = (\tfrac{1}{2})^6 (\mathrm{e}^{ix} + \mathrm{e}^{-ix})^6$$

and expanding by the binomial theorem we have

$$\begin{aligned}\cos^6 x &= \tfrac{1}{64}\{\mathrm{e}^{6ix} + 6\mathrm{e}^{5ix}\cdot\mathrm{e}^{-ix} + 15\mathrm{e}^{4ix}\cdot\mathrm{e}^{-2ix} + 20\mathrm{e}^{3ix}\cdot\mathrm{e}^{-3ix} \\ &\qquad + 15\mathrm{e}^{2ix}\cdot\mathrm{e}^{-4ix} + 6\mathrm{e}^{ix}\cdot\mathrm{e}^{-5ix} + \mathrm{e}^{-6ix} \\ &= \tfrac{1}{64}\{(\mathrm{e}^{6ix} + \mathrm{e}^{-6ix}) + 6(\mathrm{e}^{4ix} + \mathrm{e}^{-4ix}) + 15(\mathrm{e}^{2ix} + \mathrm{e}^{-2ix}) + 20\} \\ &= \tfrac{1}{32}(\cos 6x + 6\cos 4x + 15\cos 2x + 10).\end{aligned}$$

Hence

$$\int_0^{\frac{1}{2}\pi} \cos^6 x\mathrm{d}x = \tfrac{1}{32}\int_0^{\frac{1}{2}\pi} (\cos 6x + 6\cos 4x + 15\cos 2x + 10)\mathrm{d}x.$$

Using the fact that, if n is an integer,

$$\int_0^{\frac{1}{2}\pi} \cos 2nx\mathrm{d}x = \frac{1}{2n}[\sin 2nx]_0^{\frac{1}{2}\pi} = 0,$$

(since both $\sin n\pi$ and $\sin 0$ are zero) we have

$$\int_0^{\frac{1}{2}\pi} \cos^6 x\mathrm{d}x = \tfrac{10}{32}\int_0^{\frac{1}{2}\pi} \mathrm{d}x = \tfrac{5}{32}\pi.$$

PROBLEMS

1. If x is so small that the fourth powers of αx and βx can be neglected, show that

$$\log \frac{(1 + \alpha x)(1 + \beta x)}{1 + (\alpha + \beta)x} = \alpha\beta x^2\{1 - (\alpha + \beta)x\}.$$

2. Using the power series expansions for $\sinh x$ and $\cosh x$, show that, for small values of x,

$$\coth x - \frac{1}{x} \simeq \tfrac{1}{3}x.$$

3. Expand $\log(1 + \sin x)$ in ascending powers of x as far as the term in x^4.

4. Show that

$$\log \cos x = -\frac{x^2}{2} - \frac{x^4}{12} - \frac{x^6}{45} - \dots$$

and deduce that

$$\log(1 + \cos 2x) = \log 2 - x^2 - \frac{x^4}{6} - \frac{2x^6}{45} - \dots.$$

5. Show that if the eighth and higher powers of x can be neglected

$$\cos x \cosh x = 1 - \tfrac{1}{6}x^4.$$

6. Write down the series for e^x and show that, if terms of degree higher than $n + 1$ in x are neglected

$$1 + x + \frac{x^2}{2!} + \dots + \frac{x^n}{n!} = e^x\left\{1 - \frac{x^{n+1}}{(n+1)!}\right\}.$$

Deduce that, to the same degree of approximation,

$$\log\left(1 + x + \frac{x^2}{2!} + \dots + \frac{x^n}{n!}\right) = x - \frac{x^{n+1}}{(n+1)!}.$$

8. The General Power

The function a^n is defined in elementary algebra only when the index n is rational. We now consider what is meant by a symbol of this kind when n is *any* real number. If n is a rational number

$$\log(a^n) = n \log a$$

so that we may write

$$a^n = e^{n \log a}. \tag{1}$$

Now exp ($n \log a$) has been defined as a single-valued continuous function for *all* real values of n if $a > 0$. Equation (1) is therefore employed to define the real value of a^n when n is *irrational* ($a > 0$); we write

$$a^n = \exp(n \log a) \qquad (a > 0). \tag{2}$$

We shall now show that this definition leads to the well-known index laws. From equation (2) we have

$$\begin{aligned} a^m a^n &= \exp(m \log a) \cdot \exp(n \log a) \\ &= \exp\{(m+n) \log a\} \\ &= a^{m+n} \end{aligned} \tag{3}$$

and

$$\begin{aligned} (a^m)^n &= \{\exp(m \log a)\}^n \\ &= \exp(mn \log a) \\ &= a^{mn}. \end{aligned} \tag{4}$$

Furthermore

$$\frac{d}{dx} a^x = \frac{d}{dx}\{\exp(x \log a)\} = \log a \cdot \exp(x \log a)$$

showing that

$$\frac{d}{dx} a^x = (\log a) a^x. \tag{5}$$

The formula

$$\frac{d}{dx}(x^n) = nx^{n-1}$$

holds for irrational as well as rational values of n; for

$$\frac{d}{dx}(x^n) = \frac{d}{dx} \exp(n \log x) = \frac{n}{x} \exp(n \log x) = nx^{n-1}.$$

9. Logarithms to any Base

If

$$x = a^y$$

we say that y is the logarithm of x to the base a and denote it by $\log_a x$.

From equation (2) of the last section we have that

$$x = e^{y \log a}$$

so that

$$\log x = y \log a$$

from which we deduce that if $a > 0$, $x > 0$

$$\log_a x = \frac{\log x}{\log a}. \tag{1}$$

It is an immediate consequence of this result that

$$\log_x a = \frac{1}{\log_a x}. \tag{2}$$

10. Simple Differential Equations Associated with the Exponential Function

We saw in equation (4) of § 2 above that, if α is a constant,

$$\frac{d}{dx} e^{\alpha x} = \alpha e^{\alpha x}.$$

It follows from this that, if

$$y = Ae^{\alpha x}, \tag{1}$$

with A and α constants, then

$$\frac{dy}{dx} = \alpha y. \tag{2}$$

A relation of the type (2) connecting the derivative of a function with the function itself is called a *differential equation.* We have already had two biological examples of differential equations, in Chapter 3, §9 and Chapter 4, §2, without going into details. We shall discuss them more fully in Chapter 9, and their applications to biological and medical problems in Chapter 10. At this point we shall merely make some remarks about two simple differential equations related to the exponential function. What we have shown above is that the function (1) satisfies the differential equation (2); since (1) involves an arbitrary constant A we say that (1) is the *general solution* of the differential equation.

EXAMPLE 9. *A function $c(t)$ satisfies the differential equation*

$$\frac{dc}{dt} = -\frac{c}{\tau}, \tag{3}$$

where τ is a constant, and has the value c_0 when $t = 0$. Find the form of $c(t)$.

Here y is replaced by c, x by t and α has the value $-1/\tau$ so that the general solution of the differential equation is

$$c = A\mathrm{e}^{-t/\tau},$$

where A denotes an arbitrary constant. To find the value of A appropriate to our problem we use the fact that when $t = 0$, $\mathrm{e}^{-t/\tau} = 1$ and $c(t) = c_0$. We therefore have

$$c_0 = A.$$

Substituting this value for A into the general solution we find that the solution we want is

$$c(t) = c_0 \mathrm{e}^{-t/\tau}. \tag{4}$$

Another differential equation which arises frequently in practical applications is

$$\frac{\mathrm{d}y}{\mathrm{d}x} + \lambda y = \mu \tag{5}$$

where λ and μ are constants. If we write

$$y = \frac{\mu}{\lambda} + z \tag{6}$$

then, since μ/λ is a constant,

$$\frac{\mathrm{d}y}{\mathrm{d}x} = \frac{\mathrm{d}z}{\mathrm{d}x},$$

so that substituting in equation (5)

$$\frac{\mathrm{d}z}{\mathrm{d}x} + \lambda\left(\frac{\mu}{\lambda} + z\right) = \mu$$

which can be written in the form

$$\frac{\mathrm{d}z}{\mathrm{d}x} = -\lambda z.$$

It follows from the first part of this section that

$$z = A\mathrm{e}^{-\lambda x}$$

where A is an arbitrary constant. Putting this expression for z into the right-hand side of (6) we find that the general solution of (5) is

$$y = \frac{\mu}{\lambda} + Ae^{-\lambda x}. \tag{7}$$

EXAMPLE 10. *A function* $P_A(t)$ *satisfies the differential equation*

$$\frac{dP_A}{dt} + KP_A = KP_v, \tag{8}$$

where K *and* P_v *are constants, and has the value* $P_A^{(0)}$ *when* $t = 0$. *Find the form of* $P_A(t)$.

In the notation of equation (5) $\lambda = K$, $\mu = KP_v$ and P_A and t take the roles of y and x respectively. The general solution of the equation (8) is therefore

$$P_A = P_v + Ae^{-Kt} \tag{9}$$

where A is an arbitrary constant. To determine the value of A appropriate to our problem, we let $t = 0$ in equation (9). In this way we find that

$$P_A^{(0)} = P_v + A$$

so that

$$A = P_A^{(0)} - P_v$$

and hence

$$P_A(t) = P_v + (P_A^{(0)} - P_v)e^{-Kt}. \tag{10}$$

To illustrate the use of these formulae we shall consider an example from respiration and one from circulation.

11. The CO_2 Time Course in the Lung During Breath-holding*

It is obvious that, when the breath is held, CO_2 diffuses from the blood into the lungs at a steadily decreasing rate. We denote by P_A, P_a, P_v the pressure of CO_2 in alveolar gas, arterial blood (leaving the lung) and in mixed venous blood (entering the lung) respectively and by C_a, C_v, C_A the concentration of CO_2 in arterial blood, mixed venous blood and alveolar gas respectively. According to the well-known Fick principle the rate of change of the volume V of CO_2 given off by the lungs is equal to the difference of the flow rate of CO_2 in the blood entering the lungs and the flow rate of CO_2 in the blood leaving the lungs so that

* Numbered references [1,2,3] are given at the end of the chapter.

$$\frac{dV}{dt} = (C_v - C_a)F \tag{19}$$

where F is the blood flow.

We can obtain a similar expression in terms of partial pressures instead of concentrations. To change from concentration to pressure we need a relation between the CO_2 concentration and CO_2 pressure in the blood. This is provided by the well-known CO_2 dissociation curve of the blood

$$C = aP + b \tag{20}$$

where a, b are constants. Hence on the venous side we have $C_v = aP_v + b$ and on the arterial side $C_a = aP_a + b$, so that $C_v - C_a = a(P_v - P_a)$ and so

$$\frac{dV}{dt} = aF(P_v - P_a).$$

Now the change in CO_2 concentration in the lung resulting from the addition of an amount dV in time dt is obviously $C_A = dV/L$, where L is the effective lung volume (which includes the blood in the lungs). Hence

$$\frac{dC_A}{dt} = \frac{1}{L}\frac{dV}{dt}$$

or

$$\frac{dV}{dt} = L\frac{dC_A}{dt}. \tag{21}$$

The relation between the concentration C_A, the pressure P_A, and the barometric pressure B is given by

$$P_A = \frac{BC_A}{100}. \tag{22}$$

From equations (19), (21) and (22) we find that

$$\frac{dP_A}{dt} = \frac{B}{100}\frac{dC_A}{dt} = \frac{B}{100L}\frac{dV}{dt} = \frac{aBF}{100L}(P_v - P_a)$$

where L is assumed to be constant. If we write

$$K = \frac{abF}{100L} \tag{23}$$

we can write this equation in the form

$$\frac{dP_A}{dt} = K(P_v - P_a). \tag{24}$$

In order to integrate this equation we make use of the well-established result that the CO_2 pressure in the arterial blood is equal to the CO_2 tension in the alveolar space, so that $P_a = P_A$. Hence we have the equation

$$\frac{dP_A}{dt} + KP_A = KP_v. \tag{25}$$

This is a linear equation of exactly the same type as equation (8) above. If P_v and K are constants we find, from example 10 above, that

$$P_A = P_v - (P_v - P_A^{(0)})e^{-Kt}. \tag{26}$$

As $t \to \infty$, $e^{-Kt} \to 0$ so that $P_A \to P_v$. This means that during breath-holding the partial pressure of CO_2 in the lungs rises exponentially

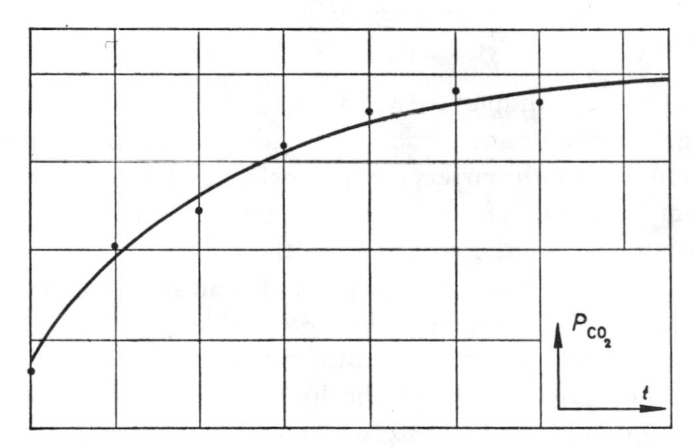

Fig. 84.

towards the limiting value P_v, the pressure of CO_2 in the mixed venous blood entering the lung.

Formula (26) has clinical significance since it offers a simple means to calculate the cardiac output on the so-called "indirect Fick principle". The experimental graph taken from a paper of Defares[3] is shown in Fig. 84.

12. The Measurement of Cardiac Output by the Dye-Dilution Method

The dye-dilution method first proposed by Hamilton and Stewart[1,2] has in recent years become, together with the direct Fick method, the standard method of measuring cardiac output.

Two forms of the method are in use, viz. the instantaneous injection method and the continuous injection method. We only discuss the instantaneous injection method.

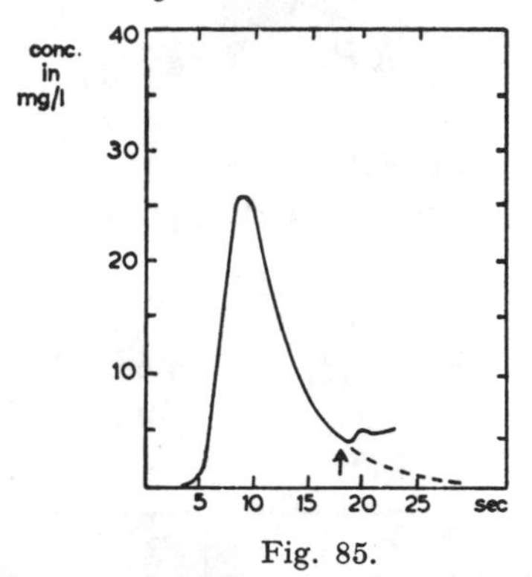

Fig. 85.

The procedure is simple: within a negligibly short time, 1 sec, say, a dye, e.g. Evans blue, or some other indicator, is injected into the cubital vein. From the artery of the other arm (the radial artery) blood samples are taken at frequent intervals (every second, say). The concentration of the indicator in these samples is determined.

When these concentrations are plotted against the corresponding times, a curve like that shown in Fig. 85 is obtained.

When the substance is injected into a vein it will pass through the right heart, the vascular bed of the lungs, the left heart and thence into the aorta and systemic arteries as shown very schematically in Fig. 86. Since the injected particles will traverse different pathways

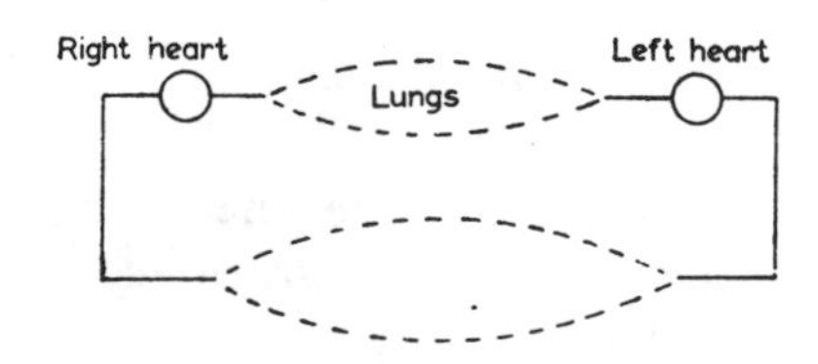

Fig. 86. Scheme of the circulatory system as a closed system of tubes.

in the pulmonary vessels, etc. and will travel with different linear velocities, the particles of dye entering at the vein will require varying amounts of time to reach the sampling site.

This circumstance contributes to the particular shape of the concentration time curve at the sampling site.

The principle is easiest to understand if we sample from the aorta instead of from the brachial artery (in practice sampling from the aorta is by no means necessary). Let q units of indicator be injected into the vein at $t = 0$, and let the concentration at the sampling site be denoted by $c(t)$, since it varies with time.

Under "steady state" conditions, a constant proportion θ of the blood that leaves the left heart will pass this site; if the site is in the aorta, θ will be 1. Then if F is the cardiac output, $F\theta$ will be the rate of volume flow past this site.

So if we write $Q = Q(t)$ for the total amount of indicator that has passed the site up to time t from injection, then

$$\frac{\mathrm{d}Q}{\mathrm{d}t} = F\theta c(t) \tag{1}$$

since all the indicator enters and leaves the heart.

Now the total amount of indicator that passes the site is

$$\int_0^\infty \frac{\mathrm{d}Q}{\mathrm{d}t}\,\mathrm{d}t = [Q]_0^\infty = q\theta. \tag{2}$$

Hence from (1)

$$F\int_0^\infty c(t)\mathrm{d}t = q \tag{3}$$

and the cardiac output F can be computed by dividing the amount of dye injected by the total area under the concentration time curve when this is expressed in appropriate units.

A complicating factor is that there is a secondary maximum in nearly all such curves, such as the one in Fig. 85. This is produced by dye that has *recirculated*; it has passed through the heart twice before reaching the point of measurement, having returned to the heart by a shorter route after its first passage. This means that in equation (1) θ increases to a new value. For calculating F, θ must be taken to have its first (constant) value, or, what amounts to the same thing, $c(t)$ must only include dye that has passed through the heart once; and for calculating F in (3) $c(t)$ is estimated by ignoring the secondary maximum and extrapolating the primary curve.

This is often done by assuming that the decrease in concentration would have been exponential without recirculation[2,3]. This extrapolation (see broken line) is best done on log-linear paper. However very little of the observed curve can be used – it can only begin to be exponential well past its maximum, so that the estimated area under this part of the curve is not often reproducible; it is very sensitive to small changes in the times of arrival of the first recirculating particles.

A number of investigators have produced methods that depend on fitting the whole of the observed curve except its upper (right-hand) tail to a known mathematical form, and calculating the area of the completed curve. For this mathematical form, Sheppard[4,5] chose the well-known curve defined in Chapter 7, Section 15, equation (6); all three forms shown in Fig. 98 (in the same section) have been fitted by different authors[6,8,9,10] and many others have been tried.[7] In nearly all cases the extrapolated area comes out *less* than that for the corresponding exponential extrapolation, so that the cardiac output F in (3) will be larger than the one from an exponential. On Plate IV, opposite p. 433, a set-up is shown.

References for Chapter 6, § 11 and 12

Section 11

1. A.B. Chilton and R.W. Stacy, Bull. Math. Biophys. **14** (1952) 1.
2. A.B. Du Bois, A.G. Britt and W.O. Fenn, J. Appl. Physiol. **4** (1952) 535.
3. J.G. Defares, J. Appl. Physiol. **13** (1958) 159.

Section 12

1. G.N. Stewart, J. Physiol. **22** (1897) 159.
2. W.F. Hamilton, J.W. Moore, J.M. Kinsman and R.G. Spurling, Amer. J. Physiol. **84** (1928) 338.
3. K.L. Zierler, in: Handbook of Physiology, section 2 (Circulation) **1** (1962) 585.
4. C.W. Sheppard and L.J. Savage, Phys. Rev. **83** (1951) 489.
5. C.W. Sheppard and M.B. Uffer, J. Theor. Biol. **22** (1969) 188.

6. R.W. Stow and P.S. Hetzel, J. Appl. Physiol. **7** (1954) 161.
7. P. Dow, J. Appl. Physiol. **7** (1955) 399.
8. H.K. Thompson, C.F. Starmer, A.F. Whalen and H.D. McIntosh, Circulation Research **14** (1964) 502.
9. K.H. Norwich and S. Zelin, Bull. Math. Biophys. **32** (1970) 25.
10. M.E. Wise, Statistica Neerlandica **20** (1966) 119, see also Fig. 98, page 343.

CHAPTER 7

TECHNIQUES OF INTEGRATION

1. Introduction

In Chapter 5 we introduced the basic ideas of integration, and then, in Chapter 6, we introduced the logarithmic and exponential functions. We shall now consider the methods available for the evaluation of integrals.

From what we have already done it is obvious that if we can find the indefinite integral $F(x)$ of a function $f(x)$, defined by

$$F'(x) = f(x), \quad \int f(x)\mathrm{d}x = F(x) \tag{1}$$

then the evaluation of a definite integral is performed by the formula

$$\int_a^x f(u)\mathrm{d}u = F(x) - F(a). \tag{2}$$

The real problem of integration is therefore the discovery of the appropriate indefinite integral. If the reader wishes to become proficient in the handling of this process then:

(*a*) he must know the elementary integrals by heart;

(*b*) he must be familiar with the elementary rules of integration.

With respect to (*a*) the reader must memorize Table 10 (p. 198 above) and the results of Chapter 6 relating to the integration of logarithmic and exponential functions. To facilitate the use of these results they are collected together in Table 11. In this table, entries 1 and 2 are given by equations (8) and (10) of § 1, entries 3 and 4 by equations (4) and (6) of § 2, entries 5 and 6 by equations (9) and (10) of § 6, and entries 7, 8, 9 by Problems 5 and 6 of § 6*. The contents of Table 11 should also be memorized.

As regards (*b*) the reader should be thoroughly familiar with the basic theorems of Chapter 6, especially Theorems 19, 21, 22, 33, 35 and 36. The ideas of § 7 on infinite integrals should also be regarded as essential.

We might say that the evaluation of integrals consists in bringing

* All the references are to Chapter 6.

any given integral to one of the elementary forms listed in Tables 10 and 11. Apart from very special devices (each applicable to only one or two integrals) we have at our disposal two powerful methods to effect this reduction:

(*a*) the method of change of variable;

(*b*) the method of integration by parts.

In this chapter we shall discuss both of these methods. §§ 2, 3, 4 and 5 are concerned with the method of change of variable, while

TABLE 11

$$\int_a^x f(u)du = F(x) - F(a)$$

	$f(u)$	$F(x)$
1	$\dfrac{1}{u}$	$\log x$
2	$\dfrac{\psi'(u)}{\psi(u)}$	$\log [\psi(x)]$
3	e^{ku}	$\dfrac{1}{k} e^{kx}$
4	$\chi'(u)e^{\chi(u)}$	$e^{\chi(x)}$
5	$\sinh (ku)$	$\dfrac{1}{k} \cosh (kx)$
6	$\cosh (ku)$	$\dfrac{1}{k} \sinh (kx)$
7	$\dfrac{1}{\sqrt{(u^2 - 1)}}$	$\cosh^{-1} x = \log [x + \sqrt{(x^2 - 1)}]$
8	$\dfrac{1}{\sqrt{(u^2 + 1)}}$	$\sinh^{-1} x = \log [x + \sqrt{(x^2 + 1)}]$
9	$\dfrac{1}{1 - u^2}$	$\tanh^{-1} x = \frac{1}{2} \log \dfrac{1 + x}{1 - x}$

§§ 6, 7, 8 deal with the method of integration by parts. With the exception of § 15, the remaining sections of the chapter deal with the applications of these two methods to the study of certain special problems. In § 9 the theorem for integrating by parts is applied to provide a proof of Maclaurin's theorem; any reader who merely wants to use Maclaurin's theorem and is prepared to accept the proof of § 4 of Chapter 4 may omit this section. The next three sections

(§§ 10, 11, 12) are concerned with the definitions of gamma and beta functions and may be omitted by the reader until he needs to refer to them. In § 13 some of the integrals encountered in the theory of statistics are considered. These lead directly to § §14 and 15. It is hoped that these sections will be of value to the reader if he (she) needs to study statistics; if not §13 and §14 may be neglected. Parts of §15 are of more general interest. Similarly the reading of §16, on the Laplace transform, may be postponed until it is needed, e.g. in the study of differential equations, §12 of Chapter 9 below. All readers should, however, read §17 which deals with a matter of real practical importance – the evaluation of integrals with the aid of published tables of integrals.

2. Integration by Change of Variable

We shall now describe how Theorem 36 is used in the evaluation of definite and indefinite integrals. It will be recalled from § 6 of Chapter 5 that the method consists essentially in spotting that an integrand $f(x)$ can be written in the form

$$\psi[\theta(x)]\theta'(x)$$

and then making the substitution

$$t = \theta(x). \tag{1}$$

The expression $\psi[\theta(x)]$ then becomes $\psi(t)$ and $\theta'(x)\mathrm{d}x$ is simply $\mathrm{d}t$ and we have

$$\int f(x)\mathrm{d}x = \int \psi[\theta(x)]\theta'(x)\mathrm{d}x = \int \psi(t)\mathrm{d}t. \tag{2}$$

In many cases the t-integration is easier to carry out than the x-integration.

In the case of definite integrals the limits of integration have also to be adjusted. If

$$\alpha = \theta(a),\ \beta = \theta(b) \tag{3}$$

then

$$\int_a^b f(x)\mathrm{d}x \equiv \int_a^b \psi[\theta(x)]\theta'(x)\mathrm{d}x = \int_\alpha^\beta \psi(t)\mathrm{d}t. \tag{4}$$

These results can be written the other way round. If we make the substitution

$$x = \phi(t) \tag{5}$$

then

$$\int f(x)\mathrm{d}x = \int f[\phi(t)]\phi'(t)\mathrm{d}t \tag{6}$$

and

$$\int_a^b f(x)\mathrm{d}x = \int_A^B f[\phi(t)]\phi'(t)\mathrm{d}t \tag{7}$$

where A and B are defined by the relations

$$a = \phi(A), \; b = \phi(B). \tag{8}$$

The only difficulty in this method is that it is not always obvious what simple substitution will facilitate an integration. There are, however, many cases where such a substitution can be easily found. As an example of a very simple case we have:-

EXAMPLE 1. *Evaluate*

$$\int_0^1 \frac{x^2\,\mathrm{d}x}{\sqrt{(1+x^3)}}.$$

It is obvious that if we write $t = 1 + x^3$ the denominator of this expression will assume the simple form $t^{\frac{1}{2}}$. Now

$$\frac{\mathrm{d}t}{\mathrm{d}x} = 3x^2$$

so that we may write

$$x^2\,\mathrm{d}x = \tfrac{1}{3}\mathrm{d}t.$$

Further, when $x = 1$, $t = 2$, and when $x = 0$, $t = 1$ so that

$$\int_0^1 \frac{x^2\,\mathrm{d}x}{\sqrt{(1+x^3)}} = \frac{1}{3}\int_1^2 \frac{\mathrm{d}t}{t^{\frac{1}{2}}} = \tfrac{2}{3}[t^{\frac{1}{2}}]_1^2 = \tfrac{2}{3}(\sqrt{2}-1).$$

It frequently happens that we have to evaluate integrals which can be written in the form

$$\int f(ax+b)\mathrm{d}x.$$

In these cases the obvious substitution to try is

$$t = ax + b,$$

for which $\mathrm{d}x = \mathrm{d}t/a$, and hence

$$\int f(ax+b)\mathrm{d}x = a^{-1}\int f(t)\mathrm{d}t.$$

EXAMPLE 2. *Find the indefinite integral of* $(2x + 3)^5$.
Here we let $t = 2x + 3$, $dx = \frac{1}{2}dt$ so that

$$\int (2x + 3)^5 \, dx = \tfrac{1}{2}\int t^5 \, dt = \tfrac{1}{2}\cdot\tfrac{1}{6}t^6 = \tfrac{1}{12}(2x + 3)^6.$$

It often happens that we have to evaluate integrals in which the integrand can be written as the ratio of the derivative of a function to the function itself, i.e. can be put into the form

$$\frac{f'(x)}{f(x)}.$$

In this case we make the substitution

$$t = f(x)$$

for which

$$dt/dx = f'(x).$$

We therefore find that

$$\int \frac{f'(x)dx}{f(x)} = \int \frac{dt}{t} = \log |t| + C' = \log |f(x)| + C. \tag{9}$$

In particular

$$\int \frac{dx}{ax + b} = \frac{1}{a}\log |ax + b| + C, \tag{10}$$

$$\int \frac{2ax + b}{ax^2 + bx + c}\, dx = \log |ax^2 + bx + c| + C. \tag{11}$$

To illustrate the use of this method we have two examples:

EXAMPLE 3. *Find the indefinite integral of*

$$\frac{2x^3 + 2x^2 + 1}{3x^4 + 4x^3 + 6x + 5}.$$

If we write $f(x) = 3x^4 + 4x^3 + 6x + 5$, then $f'(x) = 12x^3 + 12x^2 + 6$, so that we can put this function into the form

$$\frac{1}{6}\,\frac{f'(x)}{f(x)}$$

and it follows that the indefinite integral is

$$\tfrac{1}{6}\log f(x) = \tfrac{1}{6}\log (3x^4 + 4x^3 + 6x + 5).$$

EXAMPLE 4. *Find the indefinite integral of*

$$\frac{1}{x \log x}.$$

We can write this function in the form

$$\frac{1/x}{\log x} = \frac{f'(x)}{f(x)}$$

where $f(x) = \log x$. The indefinite integral of the function is therefore

$$\log f(x) = \log \log x.$$

Another simple substitution suggests itself in the evaluation of certain trigonometrical integrals. In cases in which we have to evaluate integrals of the type

$$\int f(\sin x) \cos x \mathrm{d}x$$

it is obvious that, if we make the substitution $t = \sin x$, so that

$$\mathrm{d}t = \cos x \mathrm{d}x,$$

then the integral reduces to the form

$$\int f(t) \mathrm{d}t.$$

Similarly if we have integrals of the form

$$\int f(\cos x) \sin x \mathrm{d}x$$

we make the substitution $\cos x = t$ to reduce them to integrals of the type

$$-\int f(t) \mathrm{d}t$$

which are often more easily manageable.

As a simple example of this procedure we have

EXAMPLE 5. *Find the indefinite integral of* $\sin^3 x \cos x$.

If we write $t = \sin x$, then $\cos x \mathrm{d}x = \mathrm{d}t$ so that

$$\int \sin^3 x \cos x \mathrm{d}x = \int t^3 \mathrm{d}t = \tfrac{1}{4}t^4 = \tfrac{1}{4} \sin^4 x.$$

A less obvious example is provided by

EXAMPLE 6. *Evaluate the integral* $\int_0^{\frac{1}{2}\pi} \cos^5 x \sin^5 x \mathrm{d}x$.

The integrand can be written as

$$\begin{aligned}\sin^5 x \cdot \cos^4 x \cdot \cos x &= \sin^5 x \cdot (1 - \sin^2 x)^2 \cdot \cos x \\ &= (\sin^5 x - 2 \sin^7 x + \sin^9 x) \cdot \cos x\end{aligned}$$

which is of the form $f(x) \cos x$. If, therefore, we make the substitution $t = \sin x$, and note that $t = 1$ when $x = \frac{1}{2}\pi$ and $t = 0$ when $x = 0$ we find that

$$\int_0^{\frac{1}{2}\pi} \cos^5 x \sin^5 x \mathrm{d}x = \int_0^1 (t^5 - 2t^7 + t^9)\mathrm{d}t = \tfrac{1}{6} - \tfrac{2}{8} + \tfrac{1}{10} = \tfrac{1}{60}.$$

When the integrand $f(x)$ contains the factor $a^2 - x^2$ it is often helpful to make the substitution $x = a \sin t$, and when it contains the factor $a^2 + x^2$ to put $x = a \tan t$.

To illustrate the former procedure we have

EXAMPLE 7. *Find the indefinite integrals of* $(a^2 - x^2)^{-\frac{1}{2}}$ *and* $(a^2-x^2)^{\frac{1}{2}}$.

If we put $x = a \sin t$, $\mathrm{d}x = a \cos t\mathrm{d}t$, $(a^2 - x^2)^{\frac{1}{2}} = a \cos t$ we find that

$$\int \frac{\mathrm{d}x}{\sqrt{(a^2 - x^2)}} = \int \frac{a \cos t\mathrm{d}t}{a \cos t} = \int \mathrm{d}t = t + C = \sin^{-1}\frac{x}{a} + C, \quad (12)$$

$$\int \sqrt{(a^2 - x^2)}\mathrm{d}x = \int a^2 \cos^2 t\mathrm{d}t = \tfrac{1}{2}a^2 \int (1 + \cos 2t)\mathrm{d}t = \tfrac{1}{2}a^2(t + \tfrac{1}{2}\sin 2t).$$

Using the fact that

$$\tfrac{1}{2}\sin 2t = \sin t \cos t = \sin t \sqrt{(1 - \sin^2 t)} = \frac{1}{a^2}(a \sin t)\sqrt{(a^2 - a^2 \sin^2 t)}$$

we find that

$$\int \sqrt{(a^2 - x^2)}\mathrm{d}x = \tfrac{1}{2}a^2 \sin^{-1}\frac{x}{a} + \tfrac{1}{2}x\sqrt{(a^2 - x^2)} + C. \quad (13)$$

The use of the substitution $x = a \tan t$ is illustrated by

EXAMPLE 8. *Find the indefinite integral of the function*

$$\frac{x^2}{(a^2 + x^2)^{\frac{5}{2}}}.$$

If we let $x = a \tan t$, then $\mathrm{d}x = a \sec^2 t\mathrm{d}t$, $a^2 + x^2 = a^2 \sec^2 t$ so that

$$\int \frac{x^2 \mathrm{d}x}{(a^2+x^2)^{\frac{5}{2}}} = \int \frac{a^2 \tan^2 t \cdot a \sec^2 t \mathrm{d}t}{a^5 \sec^5 t}.$$

Now

$$\frac{\tan^2 t \cdot \sec^2 t}{\sec^5 t} = \frac{\sin^2 t}{\cos^2 t} \cdot \cos^5 t \frac{1}{\cos^2 t} = \sin^2 t \cdot \cos t$$

so that

$$\int \frac{x^2 \mathrm{d}x}{(a^2+x^2)^{\frac{5}{2}}} = \frac{1}{a^2} \int \sin^2 t \cdot \cos t \mathrm{d}t.$$

Making the substitution $u = \sin t$, $\mathrm{d}u = \cos t \mathrm{d}t$ we see that

$$\int \sin^2 t \cos t \mathrm{d}t = \int u^2 \mathrm{d}u = \tfrac{1}{3}u^3 + C = \tfrac{1}{3}\sin^3 t + C$$

where C is a constant of integration. Now since $\tan t = x/a$, $\sin t = x/\sqrt{(a^2+x^2)}$ so that

$$\int \sin^2 t \cos t \mathrm{d}t = \frac{1}{3} \frac{x^3}{(a^2+x^2)^{\frac{3}{2}}} + C$$

and

$$\int \frac{x^2 \mathrm{d}x}{(a^2+x^2)^{\frac{5}{2}}} = \frac{x^3}{3a^2(a^2+x^2)^{\frac{3}{2}}} + C.$$

PROBLEMS

1. Find the indefinite integrals of

(i) xe^{x^2}; (ii) $\dfrac{2x+3}{2x+1}$;

(iii) $\dfrac{ax+b}{x^2+c^2}$; (iv) $\dfrac{x}{1-x^4}$.

2. Find the indefinite integral of $\cos x/\sqrt{(\sin x)}$ and show that

$$\int_0^{\frac{1}{2}\pi} \frac{\cos x \mathrm{d}x}{\sqrt{(\sin x)}} = 2.$$

3. Using the substitution $x = a \sin^2 t$, show that

$$\int_0^a \sqrt{\left(\frac{x}{a-x}\right)} \mathrm{d}x = \tfrac{1}{2}\pi a.$$

4. Show that the indefinite integral of

$$\frac{1}{x\sqrt{(x^2+1)}}$$

is

$$\log\left\{\frac{1+\sqrt{(x^2+1)}}{x}\right\}.$$

(Let $t = 1/x$.) Hence, prove that

$$\int_1^\infty \frac{dx}{x\sqrt{(x^2+1)}} = \log(1+\sqrt{2}).$$

5. Use the substitution $x = u/\sqrt{(1-u^2)}$ to show that the indefinite integral of

$$\frac{1}{(5x^2+4)\sqrt{(1+x^2)}}$$

is

$$\tfrac{1}{2}\tan^{-1}\left\{\frac{x}{2\sqrt{(1+x^2)}}\right\}.$$

3. Integrals Involving Quadratic Functions and their Square Roots

We have already considered the basic results for very simple integrals involving quadratic functions and their square roots. For convenience the results are listed in Table 12. Entries 1 and 5 of this table are entries 9 and 7, 8 of Table 11; entries 2 and 3 are respectively the ultimate and penultimate entries of Table 10; entry 4 is given by example 7 of § 2. The only integral we have not yet encountered is entry 6.

If we write

$$x = \sqrt{k}\cdot\sinh t$$

then

$$\sqrt{(x^2+k)} = \sqrt{k}\cdot\cosh t, \quad dx = \sqrt{k}\cdot\cosh t\,dt$$

and

$$\begin{aligned}\int\sqrt{(x^2+k)}dx &= k\int\cosh^2 t\,dt = \tfrac{1}{2}k\int(1+\cosh 2t)dt\\ &= \tfrac{1}{2}kt + \tfrac{1}{4}k\sinh 2t\\ &= \tfrac{1}{2}k\sinh t\cdot\cosh t + \tfrac{1}{2}kt.\end{aligned}$$

Now

$$t \sinh^{-1}\left(\frac{x}{\sqrt{k}}\right) = \log\left(\frac{x}{\sqrt{k}} + \sqrt{1 + \frac{x^2}{k}}\right) = \log\left[x + \sqrt{(x^2 + k)}\right] - \tfrac{1}{2}\log k$$

(by example 2 of § 6 of Chapter 6) and

$$\sqrt{k} \cdot \sinh t = x, \quad \sqrt{k} \cdot \cosh t = \sqrt{(k + k \sinh^2 t)} = \sqrt{(k + x^2)}.$$

Dropping the term $-\frac{1}{4}k \log k$, which may be absorbed in the constant of integration, we find that

$$\int \sqrt{(x^2 + k)}\mathrm{d}x = \tfrac{1}{2}x\sqrt{(x^2 + k)} + \tfrac{1}{2}k \log\left[x + \sqrt{(x^2 + k)}\right] \quad (1)$$

in agreement with entry 6 of the table.

TABLE 12

Standard integrals involving quadratic functions and their square roots

1	$\int \frac{\mathrm{d}x}{x^2 - a^2} = \frac{1}{2a}\log\frac{x - a}{x + a}$, or $\frac{1}{2a}\log\frac{a - x}{a + x}$, according as $x^2 > a^2$ or $x^2 < a^2$.
2	$\int \frac{\mathrm{d}x}{x^2 + a^2} = \frac{1}{a}\tan^{-1}\frac{x}{a}$, or $-\frac{1}{a}\cot^{-1}\frac{x}{a}$.
3	$\int \frac{\mathrm{d}x}{\sqrt{(a^2 - x^2)}} = \sin^{-1}\frac{x}{a}$, or $-\cos^{-1}\frac{x}{a}$.
4	$\int \sqrt{(a^2 - x^2)}\mathrm{d}x = \frac{1}{2}x\sqrt{(a^2 - x^2)} + \frac{1}{2}a^2 \sin^{-1} x/a$.
5	$\int \frac{\mathrm{d}x}{\sqrt{(x^2 + k)}} = \log\{x + \sqrt{(x^2 + k)}\}$.
6	$\int \sqrt{(x^2 + k)}\mathrm{d}x = \frac{1}{2}x\sqrt{(x^2 + k)} + \frac{1}{2}k \log\{x + \sqrt{(x^2 + k)}\}$.

The constant k in Nos. 5, 6 may be either positive or negative, and may therefore be replaced by either $+a^2$ or $-a^2$.

Any quadratic function, $u \equiv ax^2 + bx + c$, can always be expressed in the form $a(x - p)^2 \pm q^2$, where a may be positive or negative; and hence the integrals of $1/u$, $1/\sqrt{u}$, $\sqrt{u}$ can always be reduced to one or other of the above standard forms.

We shall now consider how these results may be applied to slightly more complicated integrals.

We frequently require to evaluate indefinite integrals of the form

$$\frac{(ux + v)\mathrm{d}x}{px^2 + qx + r}$$

where p, q, r, u, v are constants and $p \neq 0$. (If $p = 0$ the evaluation is much simpler.) Now we saw in § 6 of Chapter 2 (p. 89 above) that we can write

$$px^2 + qx + r = p[(x - a)^2 + c]$$

where $a = -q/2p$, $c = (4pr - q^2)/4p^2$. It is easy to show also that we can write

$$ux + v = u(x - a) + (v + ua)$$

so that we may decompose the integral into two terms

$$\int \frac{(ux + v)\mathrm{d}x}{px^2 + qx + r} = \frac{u}{p} \int \frac{(x - a)\mathrm{d}x}{(x - a)^2 + c} + \frac{v + ua}{p} \int \frac{\mathrm{d}x}{(x - a)^2 + c}$$

and the problem is reduced to evaluating these two integrals.

If we make the substitution $\xi = x - a$ the first of these integrals becomes

$$\int \frac{\xi \mathrm{d}\xi}{\xi^2 + c} = \frac{1}{2} \int \frac{2\xi \mathrm{d}\xi}{\xi^2 + c} = \tfrac{1}{2} \log (\xi^2 + c) + k = \tfrac{1}{2} \log [(x - a)^2 + c] + k$$

where k is a constant of integration. Writing $(x - a)^2 + c = (px^2 + qx + r)/p$ and $K = k - \frac{1}{2} \log p$ we see that

$$\int \frac{(x - a)\mathrm{d}x}{(x - a)^2 + c} = \tfrac{1}{2} \log (px^2 + qx + r) + K$$

where K is a constant.

Making the same substitution in the second integral we see that it takes the form

$$\int \frac{\mathrm{d}\xi}{\xi^2 + c}.$$

The form of this integral will depend on whether c is zero, positive or negative.

Case i: $c = 0$. In this case the integral becomes

$$\int \frac{\mathrm{d}\xi}{\xi^2} = -\frac{1}{\xi}.$$

Case ii: $c > 0$. Here we may set $c = \alpha^2$ and the integral becomes

$$\int \frac{\mathrm{d}\xi}{\xi^2 + \alpha^2} = \frac{1}{\alpha} \tan^{-1} \left(\frac{\xi}{\alpha} \right).$$

Case iii: $c < 0$. In this case we have $c = -\beta^2$ and the integral takes the form

$$\int \frac{d\xi}{\xi^2 - \beta^2} = \frac{1}{2\beta}\left\{\int \frac{d\xi}{\xi - \beta} - \int \frac{d\xi}{\xi + \beta}\right\} = \frac{1}{2\beta} \log\left(\frac{\xi - \beta}{\xi + \beta}\right).$$

In evaluating integrals of this type it is best to work each case out from first principles. The above remarks should be taken to provide a guide to the procedure to be followed — not a set of ready-made formulae.

To illustrate the procedure we have

EXAMPLE 9. *Find the indefinite integral of*

$$\frac{x - 1}{3x^2 - 2x + 5}.$$

It is easily shown that

$$3(x^2 - \tfrac{2}{3}x + \tfrac{5}{3}) = 3\{(x - \tfrac{1}{3})^2 + \tfrac{14}{9}\}, \qquad x - 1 = (x - \tfrac{1}{3}) - \tfrac{2}{3}$$

so that

$$\frac{x - 1}{3x^2 - 2x + 5} = \frac{1}{3}\,\frac{u}{u^2 + \frac{14}{9}} - \frac{2}{9}\,\frac{1}{u^2 + \frac{14}{9}}$$

where $u = x - \frac{1}{3}$. Hence

$$\int \frac{(x - 1)dx}{3x^2 - 2x + 5} = \frac{1}{6}\int \frac{2u du}{u^2 + \frac{14}{9}} - \frac{2}{9}\int \frac{du}{u^2 + \frac{14}{9}}$$

$$= \tfrac{1}{6} \log\,(u^2 + \tfrac{14}{9}) - \frac{2}{9}\cdot\frac{3}{\sqrt{14}} \tan^{-1}\left(\frac{3u}{\sqrt{14}}\right) + k$$

$$= \tfrac{1}{6} \log\,(x^2 - \tfrac{2}{3}x + \tfrac{5}{3}) - \frac{\sqrt{14}}{21} \tan^{-1}\left(\frac{3x - 1}{\sqrt{14}}\right) + K,$$

where k and K are constants of integration.

A similar procedure is followed when a square root of a quadratic function is involved. As an example of the method, we have

EXAMPLE 10. *Find the indefinite integral of*

$$\frac{1}{\sqrt{(5 + 2x - 3x^2)}}.$$

We can write the expression under the square root in the form

$$5 - (3x^2 - 2x) = 5 - 3[(x - \tfrac{1}{3})^2 - \tfrac{1}{9}] = \tfrac{16}{3} - 3(x - \tfrac{1}{3})^2.$$

If we make the substitution $u = 3x - 1$ we find that

$$5 + 2x - 3x^2 = \tfrac{1}{3}(16 - u^2), \quad dx = \tfrac{1}{3}du$$

so that

$$\int \frac{dx}{\sqrt{(5 + 2x - 3x^2)}} = \frac{1}{\sqrt{3}} \int \frac{du}{\sqrt{(16 - u^2)}}$$

$$= \frac{1}{\sqrt{3}} \sin^{-1} \tfrac{1}{4}u = \frac{1}{\sqrt{3}} \sin^{-1} \left(\frac{3x - 1}{4}\right).$$

PROBLEMS

Show that

1. $\displaystyle\int \frac{dx}{x^2 - 6x + 5} = \tfrac{1}{4} \log \frac{x - 5}{x - 1}.$

2. $\displaystyle\int \frac{dx}{3x^2 - 2x + 4} = \frac{1}{\sqrt{11}} \tan^{-1} \frac{3x - 1}{\sqrt{11}}.$

3. $\displaystyle\int \frac{x\,dx}{x^2 - x + 1} = \tfrac{1}{2} \log (x^2 - x + 1) + \frac{1}{\sqrt{3}} \tan^{-1} \left(\frac{2x - 1}{\sqrt{3}}\right).$

4. $\displaystyle\int \frac{dx}{\sqrt{(3x^2 - 2x + 5)}} = \frac{1}{\sqrt{3}} \log \{3x - 1 + \sqrt{3(3x^2 - 2x + 5)}\}.$

5. $\displaystyle\int \sqrt{(x - x^2)}\,dx = \tfrac{1}{8}\{2(2x - 1)\sqrt{x - x^2} + \sin^{-1}(2x - 1)\}.$

6. $\displaystyle\int \sqrt{(5x^2 - 2x + 1)}\,dx = \tfrac{1}{10}(5x - 1)\sqrt{(5x^2 - 2x + 1)}$

$$+ \tfrac{2}{25}\sqrt{5} \log \left\{\frac{5x - 1}{\sqrt{5}} + \sqrt{(5x^2 - 2x + 1)}\right\}.$$

4. Use of Partial Fractions

In certain problems we can carry out an integration by "splitting" the integrand into two simpler parts. For example, if we wish to integrate

$$\frac{1}{x^2 - a^2}$$

we can do this by writing

$$\frac{1}{x^2 - a^2} = \frac{1}{2a}\left\{\frac{1}{x-a} - \frac{1}{x+a}\right\}$$

so that

$$\int \frac{dx}{x^2 - a^2} = \frac{1}{2a}\left\{\int \frac{dx}{x-a} - \int \frac{dx}{x+a}\right\} = \frac{1}{2a}\{\log(x-a) - \log(x+a)\}.$$

Combining the logarithms we have that

$$\int \frac{dx}{x^2 - a^2} = \frac{1}{2a}\log\left(\frac{x-a}{x+a}\right) + C,$$

where C is a constant.

An example of a decomposition of a different kind is illustrated by:-

EXAMPLE 11. *Find the indefinite integral of* $x^2/(1 + x)$.

Since we can write $x^2 = 1 - (1 - x^2)$ we see that

$$\frac{x^2}{1+x} = \frac{1}{1+x} - (1 - x)$$

so that

$$\int \frac{x^2 dx}{1+x} = \int \frac{dx}{1+x} - \int (1-x)dx = \log(1+x) - x + \tfrac{1}{2}x^2 + C$$

where C is a constant.

EXAMPLE 12. *Show that*

$$\int_2^3 \frac{(2x^2 + 4x)dx}{2x^3 + x^2 - 2x - 1} = \log 21 - \log 10.$$

We have already seen in example 14 of Chapter 1 (p. 41 above) that

$$\frac{2x^2 + 4x}{2x^3 + x^2 - 2x - 1} = \frac{1}{x + \frac{1}{2}} - \frac{1}{x+1} + \frac{1}{x-1}$$

so that

$$\int_2^3 \frac{(2x^2 + 4x)dx}{2x^3 + x^2 - 2x - 1} = \int_2^3 \frac{dx}{x + \frac{1}{2}} - \int_2^3 \frac{dx}{x+1} + \int_2^3 \frac{dx}{x-1}.$$

Now

$$\int_\alpha^\beta \frac{dx}{x+a} = \log\frac{\beta + a}{\alpha + a}$$

and it follows that

$$\int_2^3 \frac{(2x^2 + 4x)\mathrm{d}x}{2x^3 + x^2 - 2x - 1} = \log \frac{\frac{7}{2}}{\frac{5}{2}} - \log \tfrac{4}{3} + \log \tfrac{2}{1} = \log 21 - \log 10.$$

EXAMPLE 13. *Show that*

$$\int \frac{(3x^2 + 4x - 1)\mathrm{d}x}{(x + 2)^2(2x + 1)} = 2 \log (x + 2) - \tfrac{1}{2} \log (2x + 1) + \frac{1}{x + 2}.$$

From example 15 of Chapter 1 (p. 42 above) we see that

$$\int \frac{3x^2 + 4x - 1}{(x + 2)^2(2x + 1)} \mathrm{d}x = 2\int \frac{\mathrm{d}x}{x + 2} - \int \frac{\mathrm{d}x}{2x + 1} - \int \frac{\mathrm{d}x}{(x + 2)^2},$$

and the result follows immediately.

PROBLEMS

1. Using the results of Problems 1, 4 of § 8 of Chapter 1 find the indefinite integrals of

(i) $\dfrac{x^3 + 3x^2 + 4x + 3}{x^2 + 3x + 2}$;

(ii) $\dfrac{2x}{(x^2 + 1)(x + 1)^2}$;

(iii) $\dfrac{x^3 + 2x - 1}{(x^2 - 1)^2(x + 1)}$.

2. Show that the indefinite integral of

$$\frac{x^2 - 1}{x(ax - b)(bx - a)}$$

is

$$\frac{1}{ab} \log \left\{\frac{(ax - b)(bx - a)}{x}\right\} + C,$$

where C is a constant of integration.

3. Integrate the following functions with respect to x

(i) $\dfrac{x + 4}{3x^2 - x - 2}$; (ii) $\dfrac{2x^4 - 1}{x(x^2 - 1)}$;

(iii) $\dfrac{x^3 + 1}{x^2(x + 1)}$; (iv) $\dfrac{x + 1}{x(2x + 1)^2}$;

(v) $\dfrac{3x^2 + 2x + 1}{x(x^2 - 1)^2}$; (vi) $\dfrac{x^2 + 3x + 3}{(x + 1)^3}$.

5. Standard Substitutions

In this section we shall consider more complicated examples of the evaluation of integrals by the method of change of variable. The standard substitutions are listed in Table 13. When the integrand is of the form shown in the first column of the table, we change the variable from x to t by the formula shown in the second column.

TABLE 13
Integration by change of variable

	Integrand a rational function of	Substitution	$\dfrac{dx}{dt}$
1	$\sqrt{[(x \pm a)^2 + b^2]}$	$x = \mp a + b \tan t$	$b \sec^2 t$
2	$\sqrt{[(x \pm a)^2 - b^2]}$	$x = \mp a + b \sec t$	$b \sec t \tan t$
3	$\sqrt{[b^2 - (x \pm a)^2]}$	$x = \mp a + b \sin t$	$b \cos t$
4	$\sin x,\ \cos x$	$t = \tan \frac{1}{2}x$	$\dfrac{2}{1 + t^2}$
5	$\sinh x,\ \cosh x$	$t = \tanh \frac{1}{2}x$	$\dfrac{2}{1 - t^2}$
6	e^x	$t = e^x$	$\dfrac{1}{t}$
7	$x,\ \sqrt{[(x - a)(x - b)]}$	$t = \sqrt{\dfrac{x - a}{x - b}}$	$\dfrac{2(a - b)t}{(t^2 - 1)^2}$
8	$x,\ \sqrt{[(b-x)(x-a)]}$	$t = \sqrt{\dfrac{x - a}{b - x}}$	$\dfrac{2(b - a)t}{(1 + t^2)^2}$

In using entries 4 and 5 of the table it is useful to remember the following formulae:

If $\tan \frac{1}{2}x = t$ it is easily shown (by considering a right-angled triangle with sides $1,\ t,\ \sqrt{[1 + t^2]}$) that

$$\sin \tfrac{1}{2}x = \frac{t}{1 + t^2}, \quad \cos \tfrac{1}{2}x = \frac{1}{1 + t^2}.$$

Writing $\sin x = 2 \sin \frac{1}{2}x \cos \frac{1}{2}x$, $\cos x = \cos^2 \frac{1}{2}x - \sin^2 \frac{1}{2}x$, we have

$$\sin x = \frac{2t}{1+t^2}, \quad \cos x = \frac{1-t^2}{1+t^2}. \tag{1}$$

We have similar relations for the substitution 5 of Table 13. Since $1 - \operatorname{sech}^2 \frac{1}{2}x = \tanh^2 \frac{1}{2}x$ it follows that $\operatorname{sech} \frac{1}{2}x = \sqrt{(1-t^2)}$. Hence

$$\cosh \tfrac{1}{2}x = \frac{1}{\sqrt{(1-t^2)}}$$

and hence

$$\sinh \tfrac{1}{2}x = \sqrt{(\cosh^2 \tfrac{1}{2}x - 1)} = \frac{t}{\sqrt{(1-t^2)}}.$$

Using the fact that $\sinh x = 2 \sinh \frac{1}{2}x \cosh \frac{1}{2}x$ we find that

$$\sinh x = \frac{2t}{1-t^2} \tag{2}$$

and calculating $\cosh x$ from the fact that it is $\sqrt{(\sinh^2 x + 1)}$ we find

$$\cosh x = \frac{1+t^2}{1-t^2}. \tag{3}$$

We shall illustrate the use of Table 13 by some examples:-

EXAMPLE 14. *Find the indefinite integral of* $\operatorname{cosec} x$.

By the definition of $\operatorname{cosec} x$ and the use of the relation

$$\sin^2 x = 1 - \cos^2 x$$

we have

$$\operatorname{cosec} x = \frac{1}{\sin x} = \frac{\sin x}{\sin^2 x} = \frac{\sin x}{1-\cos^2 x},$$

so that

$$\int \operatorname{cosec} x dx = \int \frac{\sin x dx}{1-\cos^2 x} = -\int \frac{du}{1-u^2},$$

where we have made the substitution $u = \cos x$. Using the result

$$\int \frac{du}{1-u^2} = \tfrac{1}{2} \log \frac{1+u}{1-u}$$

we see that

$$\int \operatorname{cosec} x dx = -\tfrac{1}{2} \log \left(\frac{1+\cos x}{1-\cos x}\right) = \tfrac{1}{2} \log \left(\frac{1-\cos x}{1+\cos x}\right). \tag{4}$$

According to entry 4 of Table 13 we could also make the substitution

$$t = \tan \tfrac{1}{2}x$$

in which case

$$\mathrm{d}x = \frac{2\mathrm{d}t}{1+t^2}, \quad \sin x = \frac{2t}{1+t^2}$$

so that

$$\int \frac{\mathrm{d}x}{\sin x} = \int \frac{\mathrm{d}t}{t} = \log t = \log (\tan \tfrac{1}{2}x). \tag{5}$$

To show that the results (4) and (5) are equivalent we note that

$$\frac{1-\cos x}{1+\cos x} = \frac{2\sin^2 \frac{1}{2}x}{2\cos^2 \frac{1}{2}x} = \left(\frac{\sin \frac{1}{2}x}{\cos \frac{1}{2}x}\right)^2 = (\tan \tfrac{1}{2}x)^2.$$

EXAMPLE 15. *Find the indefinite integral of* sech x *and deduce that*

$$\int_0^\infty \operatorname{sech} x\mathrm{d}x = \tfrac{1}{2}\pi.$$

From entry 5 of Table 13 we know to write

$$t = \tanh \tfrac{1}{2}x$$

so that

$$\int \operatorname{sech} x\mathrm{d}x = \int \frac{\mathrm{d}x}{\cosh x} = \int \frac{1-t^2}{1+t^2}\,\frac{2\mathrm{d}t}{1-t^2} = 2\int \frac{\mathrm{d}t}{1+t^2} = 2\tan^{-1} t$$

from which it follows that the indefinite integral of sech x is

$$2\tan^{-1} (\tanh \tfrac{1}{2}x).$$

Now as $x \to \infty$, $\tanh \frac{1}{2}x \to 1$ and hence $\tan^{-1} (\tanh \frac{1}{2}x) \to \frac{1}{4}\pi$. Hence

$$\int_0^\infty \operatorname{sech} x\mathrm{d}x = \tfrac{1}{2}\pi.$$

EXAMPLE 16. *Find the value of the integral*

$$\int_0^\infty \frac{\mathrm{d}x}{1+\mathrm{e}^x}.$$

This is obviously an example of entry 6 of the table.
If we let $t = \mathrm{e}^x$ then $\mathrm{d}x = \mathrm{d}t/t$ and

$$\int \frac{dx}{1+e^x} = \int \frac{1}{1+t} \cdot \frac{dt}{t} = \int \frac{dt}{t} - \int \frac{dt}{1+t} = \log t - \log(1+t)$$

$$= \log \frac{t}{1+t} = \log \frac{1}{1+t^{-1}} = \log \frac{1}{1+e^{-x}}$$

therefore

$$\int_0^\infty \frac{dx}{1+e^x} = \left[\log \frac{1}{1+e^{-x}}\right]_0^\infty = \log_e 2.$$

EXAMPLE 17. *Find the indefinite integral of the function*

$$\frac{1}{x\sqrt{(x^2-5x+4)}}.$$

The expression $x^2 - 5x + 4$ factorizes to the form $(x-4)(x-1)$ so that we have the situation envisaged in entry 7 of the table, with $a = 4$, $b = 1$. The appropriate substitution is

$$t = \sqrt{\frac{x-4}{x-1}}$$

for which

$$x = \frac{t^2-4}{t^2-1}$$

and

$$(x-1)(x-4) = \frac{9t^2}{(t^2-1)^2}, \qquad dx = \frac{6t dt}{(t^2-1)^2}$$

so that

$$\int \frac{dx}{x\sqrt{(x^2-5x+4)}} = \int \frac{(t^2-1)(t^2-1)6t dt}{(t^2-4)3t(t^2-1)^2}$$

$$= 2\int \frac{dt}{t^2-4} = \frac{1}{2}\int \frac{dt}{2+t} + \frac{1}{2}\int \frac{dt}{2-t}$$

$$= \tfrac{1}{2}\log\left(\frac{2+t}{2-t}\right) = \tfrac{1}{2}\log\left\{\frac{2\sqrt{(x-1)}+\sqrt{(x-4)}}{2\sqrt{(x-1)}-\sqrt{(x-4)}}\right\}$$

EXAMPLE 18. *Show that*

$$\int_a^b \frac{dx}{\sqrt{[(b-x)(x-a)]}} = \pi.$$

(i) By entry 8 of Table 13 we can make the substitution

$$t = \sqrt{\frac{x-a}{b-x}}.$$

When $x = a$, $t = 0$ and when $x = b$, $t = \infty$. Squaring both sides of this equation and solving for x we find that

$$x = \frac{a + bt^2}{1 + t^2}.$$

Writing this equation in the form

$$x = b - \frac{b-a}{1+t^2},$$

and differentiating with respect to t, we find that

$$\frac{dx}{dt} = \frac{2(b-a)t}{(1+t^2)^2}$$

(in agreement with the entry in the last column). Also

$$b - x = \frac{b-a}{1+t^2}, \quad x - a = \frac{(b-a)t^2}{1+t^2}, \quad \sqrt{(b-x)(x-a)} = \frac{(b-a)t}{1+t^2}.$$

Hence

$$\int_a^b \frac{dx}{\sqrt{[(b-x)(x-a)]}} = \int_0^\infty \frac{2dt}{1+t^2} = 2[\tan^{-1} t]_0^\infty = \pi.$$

(ii) Making the second substitution

$$x = b \sin^2 t + a \cos^2 t,$$

we note that when $x = b$, $t = \frac{1}{2}\pi$ and that $x = a$, $t = 0$. Also

$$b - x = (b-a)\cos^2 t,$$
$$x - a = (b-a)\sin^2 t,$$
$$dx/dt = 2(b-a)\sin t \cos t$$

so that

$$\int_a^b \frac{dx}{\sqrt{[(b-x)(x-a)]}} = 2\int_0^{\frac{1}{2}\pi} dt = \pi.$$

Integration of a class of irrational functions

The integration of functions of the type

$$\frac{lx + m}{(ax^2 + bx + c)\sqrt{(px^2 + qx + r)}} \tag{6}$$

is a fairly complicated business. Integrals of this kind do however arise in practice, so we shall consider some simple cases here.* A reader who is not interested in technical details can postpone reading this section until he encounters a problem in which he has to evaluate the integral of a function of the type (6).

The appropriate substitutions are listed in Table 14.

TABLE 14

Substitutions for evaluation of integrals of some irrational functions

	Integrand	Substitution
1	$\dfrac{1}{(ax + b)\sqrt{(p + qx)}}$	$t^2 = p + qx$
2	$\dfrac{x}{(ax + b)\sqrt{(px^2 + q)}}$	$t^2 = px^2 + q$
3	$\dfrac{1}{(ax^2 + b)\sqrt{(px^2 + q)}}$	$u = \dfrac{1}{x^2}$ reduces integral to case 1.
4	$\dfrac{1}{(x - a)\sqrt{(px^2 + 2qx + r)}}$	$t = \dfrac{1}{x - a}$

EXAMPLE 19. *Find the indefinite integral of*

$$\frac{1}{(1 + x)\sqrt{(2 + x)}}.$$

Entry 1 of Table 14 gives the substitution

$$t^2 = 2 + x$$

* The complete theory is given in G. H. Hardy, "The Integration of Functions of a Single Variable", 2nd edition (Cambridge University Press, 1926).

so that

$$\int \frac{dx}{(1+x)\sqrt{(2+x)}} = 2\int \frac{tdt}{(t^2-1)t} = 2\int \frac{dt}{t^2-1} = \log\left(\frac{t-1}{t+1}\right) + C,$$

where C is an arbitrary constant. Returning to the original variable we find that

$$\int \frac{dx}{(1+x)\sqrt{(2+x)}} = \log\left\{\frac{\sqrt{(2+x)}-1}{\sqrt{(2+x)}+1}\right\}.$$

EXAMPLE 20. *Integrate*

(i) $\dfrac{x}{(x^2+1)\sqrt{(x^2+2)}}$; (ii) $\dfrac{1}{(x^2+1)\sqrt{(x^2+2)}}$.

(i) From entry 2 of Table 14 we know to put

$$t^2 = x^2 + 2.$$

Then

$$\int \frac{xdx}{(x^2+1)\sqrt{(x^2+2)}} = \int \frac{tdt}{(t^2-1)t} = \int \frac{dt}{t^2-1} = \tfrac{1}{2}\log\left(\frac{t-1}{t+1}\right) + C$$

so that the required indefinite integral is

$$\tfrac{1}{2}\log\left\{\frac{\sqrt{(x^2+2)}-1}{\sqrt{(x^2+2)}+1}\right\}.$$

(ii) We can write

$$\int \frac{dx}{(x^2+1)\sqrt{(x^2+2)}} = \int \frac{dx/x^3}{(1+1/x^2)\sqrt{(1+2/x^2)}} = -\frac{1}{2}\int \frac{du}{(1+u)\sqrt{(1+2u)}}$$

where, in accordance with entry 3 of Table 14, we have put

$$u = 1/x^2, \qquad du = -2dx/x^3.$$

The integral is now of type 1 and we put

$$v^2 = 1 + 2u, \qquad vdv = du$$

to give

$$-\frac{1}{2}\int \frac{dv}{\frac{1}{2}(v^2+1)} = C - \tfrac{1}{2}\tan^{-1} v = C - \tfrac{1}{2}\tan^{-1}\sqrt{1+2u}$$

showing that

$$\int \frac{dx}{(x^2+1)\sqrt{(x^2+2)}} = C - \tfrac{1}{2}\tan^{-1}\left\{\frac{\sqrt{(x^2+2)}}{x}\right\}$$

EXAMPLE 21. *Find the indefinite integral of*

$$\frac{1}{(x-1)\sqrt{(3x^2-10x+3)}}.$$

According to entry 4 of the table we write

$$t = \frac{1}{x-1}, \quad x = 1 + \frac{1}{t}$$

so that $dx = dt/t^2$ and

$$3x^2 - 10x + 3 = 3\left(1 + \frac{2}{t} + \frac{1}{t^2}\right) - 10\left(1 + \frac{1}{t}\right) + 3 = \frac{3 - 4t - 4t^2}{t^2}$$

$$= \frac{4}{t^2}\{1 - (t + \tfrac{1}{2})^2\}.$$

Hence

$$\int \frac{dx}{(x-1)\sqrt{(3x^2-10x+3)}} = -\frac{1}{2}\int \frac{dt}{\sqrt{\{1-(t+\frac{1}{2})^2\}}} = C - \tfrac{1}{2}\sin^{-1}(t+\tfrac{1}{2}).$$

Replacing t by $(x-1)^{-1}$ we see that the required indefinite integral is

$$C - \tfrac{1}{2}\sin^{-1}\left\{\frac{x+1}{2(x-1)}\right\}$$

where C is an arbitrary constant.

PROBLEMS

1. Find the indefinite integrals of

(i) $\sqrt{(x^2 + 4x + 8)}$;
(ii) $\sqrt{(x^2 + 4x + 3)}$;
(iii) $\sqrt{(5 - 4x - x^2)}$.

2. Integrate with respect to x

(i) $\dfrac{1}{5 + 4\cos x}$; (ii) $\dfrac{\cos x}{(1 + \sin x)^2}$; (iii) $\dfrac{\tan x}{1 + \tan x}$;

(iv) $\operatorname{cosech} x$; (v) $\dfrac{1}{e^{-x} + 1}$; (vi) $\dfrac{1}{2 + 3\cosh x}$;

(vii) $\dfrac{1}{3 + 2\cosh x}$.

3. Show that

$$\int_a^b x\sqrt{[(b-x)(x-a)]}\,dx = \tfrac{1}{8}\pi(b+a)(b-a)^2.$$

4. Show that

(i) $$\int \frac{dx}{(x-1)\sqrt{(x-4)}} = \tfrac{2}{3}\tan^{-1}\left\{\frac{\sqrt{(x-4)}}{3}\right\} + C;$$

(ii) $$\int \frac{dx}{(5x^2+4)\sqrt{(1+x^2)}} = \tfrac{1}{2}\tan^{-1}\left\{\frac{x}{2\sqrt{(1+x^2)}}\right\} + C.$$

5. Show that, if $b > 1$

$$\int_0^\infty \frac{dx}{(1+bx^2)\sqrt{(1+x^2)}} = \frac{\pi}{2\sqrt{(b-1)}}.$$

6. Evaluate

(i) $$\int_0^3 \frac{dx}{(16+x^2)\sqrt{(9-x^2)}};$$ (ii) $$\int_1^2 \frac{dx}{x^2\sqrt{(5x^2-4)}};$$

(iii) $$\int_0^3 \frac{x^2\,dx}{\sqrt{(9-x^2)}}.$$

6. Integration by Parts

There is no simple formula for the integral of the product of two functions. By means of a procedure known as *integration by parts*, we can, however, reduce a complicated integral with an integrand consisting of the product of two functions to a simpler one of the same type. The method depends essentially on the formula

$$\frac{d}{du}F(u)g(u) = F'(u)g(u) + F(u)g'(u). \tag{1}$$

Suppose we have found the indefinite integral $F(u)$ of the function $f(u)$ so that $F'(u) = f(u)$. Then we can write equation (1) in the form

$$\frac{d}{du}F(u)g(u) = f(u)g(u) + F(u)g'(u)$$

which can be rewritten as

$$f(u)g(u) = \frac{d}{du}\{F(u)g(u)\} - F(u)g'(u).$$

If we integrate both sides of this equation with respect to u from a to x we find that

$$\int_a^x f(u)\, g(u) \mathrm{d}u = \int_a^x \frac{\mathrm{d}}{\mathrm{d}u} \{F(u)\, g(u)\} \mathrm{d}u - \int_a^x F(u)\, g'(u) \mathrm{d}u.$$

Using the fact that

$$\int_a^x \frac{\mathrm{d}}{\mathrm{d}u} \{F(u)\, g(u)\} \mathrm{d}u = [F(u)\, g(u)]_a^x$$

we see that we have proved:-

THEOREM 37. (*Integration by Parts*)

If $F'(x) = f(x)$ *and if* $f(x)$ *and* $g'(x)$ *are continuous for* $a < x < b$ *then*

$$\int_a^x f(u) g(u) \mathrm{d}u = [F(u) g(u)]_a^x - \int_a^x F(u) g'(u) \mathrm{d}u \tag{2}$$

where $a \leqq x \leqq b$.

In the case of indefinite integrals we can write the result in the form

$$\int f(x)\, g(x) \mathrm{d}x = F(x)\, g(x) - \int F(x)\, g'(x) \mathrm{d}x \tag{3}$$

where $F(x)$ is the indefinite integral of $f(x)$.

To illustrate the use of the theorem we have

EXAMPLE 22. *Integrate* $x \sin x$.

Since the integral of $\sin x$ is $-\cos x$ and the derivative of x is 1 we have

$$\int x \sin x \mathrm{d}x = x\, (-\cos x) - \int 1 \cdot (-\cos x) \mathrm{d}x$$

and since the integral of $\cos x$ is $\sin x$ we find that

$$\int x \sin x \mathrm{d}x = \sin x - x \cos x.$$

EXAMPLE 23. *Integrate* $\log x$.

At first sight this appears to have strayed into this section by accident since $\log x$ is not the product of two functions, but is just one function. We can think of $\log x$ as $1 \cdot \log x$ and use the fact that the integral of 1 is x and the derivative of $\log x$ is $1/x$. We therefore have from equation (3)

$$\int \log x \mathrm{d}x = x \log x - \int x \frac{1}{x} \mathrm{d}x = x \log x - x. \tag{4}$$

EXAMPLE 24. *Integrate* $\sqrt{(a^2 - x^2)}$.

Here again we write the integrand as $1 \cdot \sqrt{(a^2 - x^2)}$ and use the fact that

$$\frac{d}{dx}\sqrt{(a^2 - x^2)} = -\frac{x}{\sqrt{(a^2 - x^2)}}$$

to obtain the result

$$\int \sqrt{(a^2 - x^2)}dx = x\sqrt{(a^2 - x^2)} - \int x \frac{-x}{\sqrt{(a^2 - x^2)}} dx$$

$$= x\sqrt{(a^2 - x^2)} + \int \frac{x^2 \, dx}{\sqrt{(a^2 - x^2)}}. \tag{5}$$

Now the integral on the right-hand side looks more difficult to evaluate than the one we began with. We get over this difficulty by a trick. If we write

$$x^2 = a^2 - (a^2 - x^2)$$

we see that

$$\int \frac{x^2 \, dx}{\sqrt{(a^2 - x^2)}} = a^2 \int \frac{dx}{\sqrt{(a^2 - x^2)}} - \int \sqrt{(a^2 - x^2)}dx \tag{6}$$

and if we use the fact that

$$\int \frac{dx}{\sqrt{(a^2 - x^2)}} = \sin^{-1}(x/a)$$

and substitute from equation (6) into equation (5) we obtain the relation

$$\int \sqrt{(a^2 - x^2)}dx = x\sqrt{(a^2 - x^2)} + a^2 \sin^{-1}(x/a) - \int \sqrt{(a^2 - x^2)}dx$$

from which it follows that

$$\int \sqrt{(a^2 - x^2)}dx = \tfrac{1}{2}x\sqrt{(a^2 - x^2)} + \tfrac{1}{2}a^2 \sin^{-1}(x/a). \tag{7}$$

EXAMPLE 25. *Find the indefinite integral of* $\sin^{-1} x$.

If we write $\sin^{-1} x = 1 \cdot \sin^{-1} x$ and use the fact that the indefinite integral of 1 is x and the derivative of $\sin^{-1} x$ is $(1 - x^2)^{-\frac{1}{2}}$ we find that

$$\int \sin^{-1} x dx = x \sin^{-1} x - \int \frac{x}{\sqrt{(1 - x^2)}} dx = x \sin^{-1} x + \sqrt{(1 - x^2)}. \tag{8}$$

Important results are contained in:-

EXAMPLE 26. *Find the indefinite integrals of* $e^{ax}\cos(bx+c)$, $e^{ax}\sin(bx+c)$ *and deduce that, if* $\alpha < 0$,

$$\text{(i)} \quad \int_0^\infty e^{-\alpha x}\cos\beta x\,dx = \frac{\alpha}{\alpha^2+\beta^2},$$

$$\text{(ii)} \quad \int_0^\infty e^{-\alpha x}\sin\beta x\,dx = \frac{\beta}{\alpha^2+\beta^2}.$$

If we now let $f(x) = e^{ax}$, $g(x) = \cos(bx+c)$, $F(x) = a^{-1}e^{ax}$, $g'(x) = -b\sin(bx+c)$ in Theorem 37 we see that

$$\int e^{ax}\cos(bx+c)dx = \frac{1}{a}e^{ax}\cos(bx+c) + \frac{b}{a}\int e^{ax}\sin(bx+c)dx$$

which can be written in the form

$$C = \frac{1}{a}e^{ax}\cos(bx+c) + \frac{b}{a}S, \tag{9}$$

where

$$C = \int e^{ax}\cos(bx+c)dx, \quad S = \int e^{ax}\sin(bx+c)dx. \tag{10}$$

Similarly if we let $f(x) = e^{ax}$, $g(x) = \sin(bx+c)$, $F(x) = a^{-1}e^{ax}$, $g'(x) = b\cos(bx+c)$ in Theorem 37 we find that

$$S = \frac{1}{a}e^{ax}\sin(bx+c) - \frac{b}{a}C. \tag{11}$$

Solving equations (9) and (11) for C and S we find that

$$C = \frac{e^{ax}}{a^2+b^2}[a\cos(bx+c) + b\sin(bx+c)],$$

$$S = \frac{e^{ax}}{a^2+b^2}[a\sin(bx+c) - b\cos(bx+c)]. \tag{12}$$

It follows that

$$\int_0^N e^{-\alpha x}\cos\beta x\,dx = \frac{e^{-\alpha N}}{\alpha^2+\beta^2}\{\beta\sin\beta N - \alpha\cos\beta N\} + \frac{\alpha}{\alpha^2+\beta^2},$$

$$\int_0^N e^{-\alpha x}\sin\beta x\,dx = \frac{e^{-\alpha N}}{\alpha^2+\beta^2}\{-\alpha\sin\beta N - \beta\cos\beta N\} + \frac{\beta}{\alpha^2+\beta^2}.$$

Letting $N \to \infty$ and using the fact that $e^{-\alpha N} \to 0$ we get the values stated.

PROBLEMS

1. Prove that:-

(i) $\int x \cos x dx = x \sin x + \cos x$;

(ii) $\int x \log x dx = \frac{1}{2}x^2 \log x - \frac{1}{4}x^2$;

(iii) $\int \tan^{-1} x dx = x \tan^{-1} x - \frac{1}{2} \log (1 + x^2)$;

(iv) $\int xe^x dx = xe^x - e^x$;

(v) $\int x^2 e^x dx = (x^2 - 2x + 2)e^x$;

(vi) $\int (\log x)^2 dx = x (\log x)^2 - 2x \log x + 2x$.

2. If $p > 0$ and $f(x)$ remains finite as $x \to \infty$, show that

$$\int_0^\infty f'(x)e^{-px} dx = -f(0) + p\int_0^\infty f(x)e^{-px} dx.$$

Deduce that

$$\int_0^\infty f''(x)e^{-px} dx = -f'(0) - pf(0) + p^2\int_0^\infty f(x)e^{-px} dx.$$

3. Show, by integrating by parts, that

$$\int \sqrt{(x^2 + a^2)}dx = \tfrac{1}{2}[x\sqrt{(x^2 + a^2)} + a^2 \log \{x + \sqrt{(x^2 + a^2)}\}].$$

7. Integration by Successive Reduction

In certain problems, a single use of the theorem for integrating by parts often fails to provide the value of the integral, but does reduce the integral to a slightly simpler integral of the same type. When this happens, this second integral can be treated in exactly the same way to reduce it to an even simpler integral. The process of reduction is carried on until we arrive at an integral which can be evaluated simply. This process, which is known as *integration by successive reduction*, is best illustrated by a specific example.

Suppose, for instance, that we consider the integral

$$I(n) = \int \tan^n x dx. \tag{1}$$

If we write

$$\tan^n x = \tan^{n-2} x \tan^2 x = \tan^{n-2} x (\sec^2 x - 1),$$

then we see that

$$I(n) = \int \tan^{n-2} x \cdot \sec^2 x \, dx - \int \tan^{n-2} x \, dx. \qquad (2)$$

Now the second integral on the right-hand side is the same as (1) but with n replaced by $n - 2$, i.e., it is the integral $I(n - 2)$. To evaluate the first integral on the right hand side of (2) we make the substitution $t = \tan x$ for which $dt = \sec^2 x dx$ so that

$$\int \tan^{n-2} x \sec^2 x dx = \int t^{n-2} dt = \frac{1}{n-1} t^{n-1} = \frac{1}{n-1} \tan^{n-1} x.$$

Substituting this expression into equation (2) we find that

$$I(n) = \frac{\tan^{n-1} x}{n-1} - I(n-2). \qquad (3)$$

This equation does not yield the value of the integral (1) but it does reduce the problem of finding it by making it depend upon an integral in which the index of $\tan x$ is reduced by 2. An equation of the type (3) is for this reason called a *reduction formula*.

We shall now show how the reduction formula can be used to evaluate an indefinite integral. Suppose we wish to find the indefinite integral of $\tan^6 x$, i.e., in the notation of equation (1) we wish to find $I(6)$. Putting $n = 6$ in equation (3) we have

$$I(6) = \tfrac{1}{5} \tan^5 x - I(4).$$

Putting $n = 4$ in equation (3) we have

$$I(4) = \tfrac{1}{3} \tan^3 x - I(2)$$

so that

$$I(6) = \tfrac{1}{5} \tan^5 x - \tfrac{1}{3} \tan^3 x + I(2).$$

Putting $n = 2$ in equation (3) and using the fact that

$$I(0) = \int 1 \, dx = x,$$

we find that

$$I(2) = \tan x - x.$$

Hence, finally we find that

$$I(6) = \tfrac{1}{5} \tan^5 x - \tfrac{1}{3} \tan^3 x + \tan x - x.$$

This procedure is particularly useful in the evaluation of definite integrals. We shall use this method a great deal in later sections

(§§ 8, 10, and 11 below). We shall illustrate it here by

EXAMPLE 27. *If n is a positive integer, and*

$$I(n) = \int_0^1 x^n e^x \, dx,$$

prove that

$$I(n) = e - nI(n-1) \tag{4}$$

and hence evaluate

$$\int_0^1 x^3 e^x \, dx.$$

If we write $f(x) = e^x$, $g(x) = x^n$, $F(x) = e^x$, $g'(x) = nx^{n-1}$ in Theorem 37 we find that

$$\int_0^1 x^n e^x \, dx = [x^n e^x]_0^1 - n \int_0^1 x^{n-1} e^x \, dx.$$

If $n > 0$, $x^n e^x \to 0$ as $x \to 0$ and the result is established. Putting $n = 3, 2, 1$ successively in equation (4) we find that

$$\begin{aligned} I(3) &= e - 3I(2) \\ I(2) &= e - 2I(1) \\ I(1) &= e - I(0). \end{aligned}$$

Also

$$I(0) = \int_0^1 e^x \, dx = [e^x]_0^1 = e - 1.$$

Hence

$$\begin{aligned} I(1) &= e - (e - 1) = 1 \\ I(2) &= e - 2 \\ I(3) &= e - 3(e - 2) \end{aligned}$$

showing that

$$\int_0^1 x^3 e^x \, dx = 2(3 - e).$$

EXAMPLE 28. *If*

$$I(n) = \int_0^a \frac{x^n \, dx}{\sqrt{(a^2 - x^2)}}$$

show that

$$I(n) = \frac{n-1}{n} a^2 I(n-2)$$

and hence find the value of the integral

$$\int_0^a \frac{x^7 \, dx}{\sqrt{(a^2 - x^2)}}.$$

We can write the integrand in this case as $f(x) \cdot g(x)$ where

$$f(x) = \frac{x}{\sqrt{(a^2 - x^2)}}, \quad g(x) = x^{n-1}$$

and, in the notation of Theorem 37,

$$F(x) = -\sqrt{(a^2 - x^2)}, \quad g'(x) = (n-1)x^{n-2},$$

$$I(n) = [-x^{n-1}\sqrt{(a^2 - x^2)}]_0^a + (n-1)\int_0^a x^{n-2}\sqrt{(a^2 - x^2)}dx.$$

The expression in the square bracket vanishes at both limits; in the integral on the right we may write

$$x^{n-2}\sqrt{(a^2 - x^2)} = \frac{x^{n-2}(a^2 - x^2)}{\sqrt{(a^2 - x^2)}} = a^2\frac{x^{n-2}}{\sqrt{(a^2 - x^2)}} - \frac{x^n}{\sqrt{(a^2 - x^2)}}$$

so that the integral has the value $I(n-2) - I(n)$. Hence we find that

$$I(n) = (n-1)[a^2 I(n-2) - I(n)]$$

from which it follows that

$$I(n) = \frac{n-1}{n}a^2 I(n-2).$$

Putting $n = 7$ we have

$$I(7) = \tfrac{6}{7}a^2 I(5) = \tfrac{6}{7}a^2 \cdot \tfrac{4}{5}a^2 I(3) = \tfrac{24}{35}a^4 \cdot \tfrac{2}{3}a^2 I(1).$$

Now

$$I(1) = \int_0^a \frac{x dx}{\sqrt{(a^2 - x^2)}} = [-\sqrt{(a^2 - x^2)}]_0^a = a$$

so that

$$I(7) = \tfrac{16}{35}a^7.$$

PROBLEMS

1. Show that

$$\int x^n \cos x dx = x^n \sin x - n\int x^{n-1} \sin x dx,$$

$$\int x^n \sin x dx = x^n \cos x + n\int x^{n-1} \cos x dx.$$

Deduce that, if

$$C(n) = \int x^n \cos x dx, \quad S(n) = \int x^n \sin x dx,$$

then

$$C(n) = \quad x^n \sin x + nx^{n-1} \cos x - n(n-1)C(n-2),$$
$$S(n) = -x^n \cos x + nx^{n-1} \sin x - n(n-1)S(n-2),$$

and show that

$$\int_0^{\frac{1}{2}\pi} x^4 \cos x\mathrm{d}x = \tfrac{1}{16}\pi^4 - 3\pi^2 + 24, \quad \int_0^{\frac{1}{2}\pi} x^4 \sin x\mathrm{d}x = \tfrac{1}{2}\pi^3 - 12\pi + 24.$$

2. If

$$I(n) = \int (\log x)^n \,\mathrm{d}x,$$

show that

$$I(n) = x\,(\log x)^n - nI(n-1).$$

Deduce the indefinite integral of $(\log x)^3$.

3. If

$$I(n) = \int_0^N x^{n-1}\,\mathrm{e}^{-x}\,\mathrm{d}x, \quad (n > 0),$$

show that

$$I(n) = -N^{n-1}\mathrm{e}^{-N} + (n-1)\,I(n-1).$$

Hence show that if

$$\Gamma(n) = \int_0^\infty x^{n-1}\,\mathrm{e}^{-x}\,\mathrm{d}x;$$

then

$$\Gamma(n) = (n-1)\Gamma(n-1), \quad (n > 1)$$

(cf. § 10 below).

8. Some Standard Trigonometrical Integrals

In this section we shall derive expressions for indefinite and definite integrals, involving trigonometrical functions, which occur frequently. The first set of integrals depends on the formulae

$$\sin\,(m+n)x + \sin\,(m-n)x = 2\sin mx \cos nx, \tag{1}$$

$$\cos\,(m+n)x + \cos\,(m-n)x = 2\cos mx \cos nx, \tag{2}$$

$$\cos\,(m-n)x - \cos\,(m+n)x = 2\sin mx \sin nx. \tag{3}$$

(Cf. Problem 1, § 7 of Chapter 2.)
If $m = n$ these formulae are replaced by

$$\sin 2nx = 2\sin nx \cos nx, \tag{4}$$

$$1 + \cos 2nx = 2\cos^2 nx, \tag{5}$$

$$1 - \cos 2nx = 2\sin^2 nx. \tag{6}$$

For example, it follows from equation (1), that if $m \neq n$

$$\int \sin mx \cos nx dx = \tfrac{1}{2}\int [\sin (m+n)x + \sin (m-n)x]dx$$
$$= -\frac{1}{2}\left\{\frac{\cos (m+n)x}{m+n} + \frac{\cos (m-n)x}{m-n}\right\} + C \qquad (7)$$

while if $m = n$, the integral is evaluated by means of equation (4)

$$\int \sin nx \cos nx dx = \frac{1}{2}\int \sin 2nx dx = -\frac{1}{4n}\cos 2nx + C. \qquad (8)$$

The insertion of limits in these integrals gives

$$\int^{\alpha} \sin mx \cos nx dx = \frac{1}{2}\left\{\frac{1-\cos (m+n)\alpha}{m+n} + \frac{1-\cos (m-n)\alpha}{m-n}\right\} \ (m \neq n), \qquad (9)$$

$$\int_0^{\alpha} \sin nx \cos nx dx = \frac{1}{4n}(1 - \cos 2n\alpha). \qquad (10)$$

If m and n are positive integers, then using the fact that $\cos 2p\pi = 1$ when p is an integer (positive or negative), we see that

$$\int_0^{2\pi} \sin mx \cos nx dx = 0 \qquad (11)$$

for all integers m and n.

Similarly from equation (2) we have that, if $m \neq n$,

$$\int \cos mx \cos nx dx = \tfrac{1}{2}\int [\cos (m+n)x + \cos (m-n)x]dx$$
$$= \frac{1}{2}\left\{\frac{\sin (m+n)x}{m+n} + \frac{\sin (m-n)x}{m-n}\right\} + C. \qquad (12)$$

On the other hand, when $m = n$,

$$\int \cos^2 nx dx = \frac{1}{2}\int (1 + \cos 2nx)dx = \tfrac{1}{2}x + \frac{\sin 2nx}{4n} + C. \qquad (13)$$

The corresponding definite integrals are

$$\int_0^{\alpha} \cos mx \cos nx dx = \frac{\sin (m+n)\alpha}{2(m+n)} + \frac{\sin (m-n)\alpha}{2(m-n)}, \qquad (14)$$

$$\int_0^{\alpha} \cos^2 nx dx = \tfrac{1}{2}\alpha + \frac{\sin 2n\alpha}{4n}. \qquad (15)$$

Inserting the limits 0, 2π and using the fact that $\sin 2p\pi = 0$ if p is an integer, we find that

$$\int_0^{2\pi} \cos mx \cos nx dx = \begin{cases} 0 \text{ if } m \neq n; \\ \pi \text{ if } m = n. \end{cases} \tag{16}$$

The equations (3) and (6) lead to the integrals

$$\int \sin mx \sin nx dx = \frac{1}{2}\int [\cos (m-n)x - \cos (m+n)x] dx \quad (m \neq n)$$

$$= \frac{1}{2}\left\{\frac{\sin (m-n)x}{m-n} - \frac{\sin (m+n)x}{m+n}\right\} + C, \tag{17}$$

$$\int \sin^2 nx dx = \frac{1}{2}\int [1 - \cos 2nx] dx = \tfrac{1}{2}x - \frac{\sin 2nx}{4n} + C \tag{18}$$

from which we can derive definite integrals

$$\int_0^{\alpha} \sin mx \sin nx dx = \frac{\sin (m-n)\alpha}{2(m-n)} - \frac{\sin (m+n)\alpha}{2(m+n)}, \tag{19}$$

$$\int_0^{\alpha} \sin^2 nx dx = \tfrac{1}{2}\alpha - \frac{\sin 2n\alpha}{4n} \tag{20}$$

of which

$$\int_0^{\pi} \sin mx \sin nx dx = \begin{cases} 0 \text{ if } m \neq n \\ \pi \text{ if } m = n \end{cases} \tag{21}$$

is an important special case.

To evaluate integrals of the type

$$I(n) = \int \cos^n x dx \tag{22}$$

we make use of the technique of reduction formulae developed in the last section. We write

$$I(n) = \int \cos^{n-1} x \cos x dx$$

and use the theorem for integration by parts with $f(x) = \cos x$, $g(x) = \cos^{n-1} x$, so that $F(x) = \sin x$, $g'(x) = -(n-1)\cos^{n-2} x \sin x$, we find that

$$I(n) = \sin x \cos^{n-1} x + (n-1)\int \cos^{n-2} x \sin x \sin x dx. \tag{23}$$

Now

$$\cos^{n-2} x \sin^2 x = \cos^{n-2} x(1 - \cos^2 x) = \cos^{n-2} x - \cos^n x$$

so that

$$\int \cos^{n-2} x \sin^2 x dx = \int \cos^{n-2} x dx - \int \cos^n x dx = I(n-2) - I(n).$$

Inserting this expression in equation (23) we have the equation

$$I(n) = \cos^{n-1} x \sin x + (n-1)[I(n-2) - I(n)]$$

which may be solved for $I(n)$ to give

$$I(n) = \frac{1}{n} \cos^{n-1} x \sin x + \frac{n-1}{n} I(n-2). \tag{24}$$

By repeated use of this formulae we can reduce integrals of this type, with integral values of n, ultimately to the evaluation of

$$I(1) = \int \cos x dx = \sin x$$

or

$$I(0) = \int dx = x.$$

This method is particularly valuable in the evaluation of definite integrals of the type

$$I_n = \int_0^{\frac{1}{2}\pi} \cos^n x dx.$$

For this integral, equation (24) reduces to

$$I_n = \frac{n-1}{n} I_{n-2} \tag{25}$$

since $\cos^{n-1} x \sin x = 0$ when $x = 0$ and when $x = \frac{1}{2}\pi$. Repeated application of (25) gives

$$I_n = \frac{(n-1)(n-3)}{n(n-2)} I_{n-4} = \frac{(n-1)(n-3)(n-5)}{n(n-2)(n-4)} I_{n-6} = \ldots.$$

If n is an *even* positive integer we can reduce the integral to

$$I_0 = \int_0^{\frac{1}{2}\pi} dx = \tfrac{1}{2}\pi$$

and we obtain the result

$$\int_0^{\frac{1}{2}\pi} \cos^n x dx = \frac{(n-1)(n-3)(n-5)\cdots 3\cdot 1}{n(n-2)(n-4)\cdots 4\cdot 2} \frac{\pi}{2} \quad (n \text{ even}) \tag{26}$$

on the other hand, if n is *odd*, the final integral we have to evaluate is

$$I_1 = \int_0^{\frac{1}{2}\pi} \cos x dx = [\sin x]_0^{\frac{1}{2}\pi} = 1$$

and we have

$$\int_0^{\frac{1}{2}\pi} \cos^n x dx = \frac{(n-1)(n-3)(n-5)\cdots 4\cdot 2}{n(n-2)(n-4)\cdots 5\cdot 3} \quad (n \text{ odd}). \qquad (27)$$

Readers will recall that in § 7 (*d*) of the last chapter (cf. example 8 on p. 260) we demonstrated an alternative method for evaluating integrals of the type (22).

We shall now illustrate the use of these formulae by two examples.

EXAMPLE 29. *Find the indefinite integral of* $\cos^5 x$.

From equation (24) putting $n = 5$ we have,

$$I_5 = \tfrac{1}{5}\cos^4 x \sin x = \tfrac{4}{5} I_3$$

and putting $n = 3$,

$$I_3 = \tfrac{1}{3}\cos^2 x \sin x + \tfrac{2}{3} I_1 = \tfrac{1}{3}\cos^2 x \sin x + \tfrac{2}{3}\sin x$$

since $I_1 = \sin x$. Substituting this value of I_3 in the previous expression we find that

$$\int \cos^5 x dx = \tfrac{1}{15}\sin x\,(3\cos^4 x + 4\cos^2 x + 8).$$

EXAMPLE 30. *Evaluate the definite integrals*

$$\int_0^{\frac{1}{2}\pi} \cos^7 x dx, \quad \int_0^{\frac{1}{2}\pi} \cos^6 x dx.$$

From equation (27)

$$\int_0^{\frac{1}{2}\pi} \cos^7 x dx = \frac{6\cdot 4\cdot 2}{7\cdot 5\cdot 3} = \frac{16}{35}$$

and from equation (26)

$$\int_0^{\frac{1}{2}\pi} \cos^6 x dx = \frac{5\cdot 3\cdot 1}{6\cdot 4\cdot 2}\,\frac{\pi}{2} = \frac{5\pi}{32}$$

(in agreement with example 8 of Chapter 6).

We could derive formulae for the evaluation of the integral

$$J_n = \int_0^{\frac{1}{2}\pi} \sin^n x dx$$

in precisely the same way (cf. Problem 3 below) but it is easier to make use of Theorem 33 (p. 210 above) which states that

$$J_n = \int_0^{\frac{1}{2}\pi} [\sin(\tfrac{1}{2}\pi - x)]^n dx = \int_0^{\frac{1}{2}\pi} \cos^n x dx = I_n$$

so that the integrals can be evaluated by means of equations (26) and (27).

These formulae may be regarded as special cases of the reduction formula for the more general integral

$$I(m, n) = \int \cos^m x \sin^n x dx. \tag{28}$$

If we write the integrand of this integral as $f(x)g(x)$ where

$$f(x) = \cos^m x \sin x, \quad g(x) = \sin^{n-1} x,$$

then, in the notation of Theorem **37**,

$$F(x) = -\frac{1}{m+1}\cos^{m+1} x, \quad g'(x) = (n-1)\sin^{n-2} x \cos x$$

so that, integrating by parts,

$$I(m, n) = -\frac{1}{m+1}\cos^{m+1} x \sin^{n-1} x + \frac{n-1}{m+1}\int \cos^{m+2} x \sin^{n-2} x dx.$$

Now

$$\cos^{m+2} x \sin^{n-2} x = \cos^m x \cos^2 x \sin^{n-2} x = \cos^m x(1 - \sin^2 x)\sin^{n-2} x$$

so that

$$\begin{aligned}\int \cos^{m+2} x \sin^{n-2} x dx &= \int \cos^m x \sin^{n-2} x dx - \int \cos^m x \sin^n x dx \\ &= I(m, n-2) - I(m, n).\end{aligned}$$

We therefore have the equation

$$I(m, n) = -\frac{1}{m+1}\cos^{m+1} x \sin^{n-1} x + \frac{n-1}{m+1}[I(m, n-2) - I(m, n)].$$

Multiplying both sides of this equation by $m + 1$ and solving for $I(m, n)$ we obtain the reduction formula

$$I(m, n) = \frac{n-1}{m+n} I(m, n-2) - \frac{1}{m+n}\cos^{m+1} x \sin^{n-1} x. \tag{29}$$

Alternatively we could write the integrand in the integral (28) in the form $f(x)g(x)$ where

$$f(x) = \sin^n x \cos x, \quad g(x) = \cos^{m-1} x$$

for which

$$F(x) = \frac{1}{n+1} \sin^{n+1} x, \quad g'(x) = -(m-1) \cos^{m-2} x \sin x.$$

Integrating by parts, we have

$$I(m, n) = \frac{1}{n+1} \sin^{n+1} x \cos^{m-1} x + \frac{m-1}{n+1} \int \sin^{n+2} x \cos^{m-2} x dx$$

and the integral on the right-hand side of this equation can be written as

$$I(m-2, n) - I(m, n).$$

Substituting this value for the integral and solving for $I(m, n)$ we find that

$$I(m, n) = \frac{m-1}{m+n} I(m-2, n) + \frac{1}{m+n} \cos^{m-1} x \sin^{n+1} x. \quad (30)$$

If we replace n by $n-2$ in (30) we find that

$$I(m, n-2) = \frac{m-1}{m+n-2} I(m-2, n-2) + \frac{1}{m+n-2} \cos^{m-1} x \sin^{n-1} x.$$

Substituting this expression for $I(m, n-2)$ into equation (29) we obtain the reduction formula

$$I(m, n) = \frac{(m-1)(n-1)}{(m+n)(m+n-2)} I(m-2, n-2) - \frac{1}{m+n} \cos^{m+1} x \sin^{n-1} x$$

$$+ \frac{n-1}{(m+n)(m+n-2)} \cos^{m-1} x \sin^{n-1} x. \quad (31)$$

By repeated application of the reduction formula (31) we can reduce any integral of the form $I(m, n)$, m and n positive integers, to the evaluation of $I(1, 1)$, $I(0, 1)$, $I(1, 0)$ or $I(0, 0)$ and it is easily shown that

$$I(1, 1) = \int \sin x \cos x dx = \tfrac{1}{2} \sin^2 x,$$

$$I(0, 1) = \int \sin x dx = -\cos x, \quad I(1, 0) = \int \cos x dx = \sin x,$$

$$I(0, 0) = x$$

so that the reduction is complete.

An important special case is the integral

$$I_{m,n} = \int_0^{\frac{1}{2}\pi} \cos^m x \sin^n x dx. \quad (32)$$

If we introduce the limits 0, $\frac{1}{2}\pi$ into equation (31) we find that

$$I_{m,n} = \frac{(m-1)(n-1)}{(m+n)(m+n-2)} I_{m-2,n-2}. \tag{33}$$

By repeated application of this formula, together with the special results

$$I_{0,0} = \tfrac{1}{2}\pi, \quad I_{0,1} = I_{1,0} = 1, \quad I_{1,1} = \tfrac{1}{2}$$

we see that if m and n are both even

$$I_{m,n} = \frac{(m-1)(m-3)\cdots 3\cdot 1\ (n-1)(n-3)\cdots 3\cdot 1}{(m+n)(m+n-2)(m+n-4)\cdots 4\cdot 2}\ \frac{\pi}{2} \tag{34}$$

while if m and n are both odd

$$I_{m,n} = \frac{(m-1)(m-3)\cdots 4\cdot 2\ (n-1)(n-3)\cdots 4\cdot 2}{(m+n)(m+n-2)\cdots 6\cdot 4}\ \frac{1}{2}. \tag{35}$$

If one of the indices is odd and the other even we have

$$I_{m,n} = \frac{(m-1)(m-3)\cdots (n-1)(n-3)\cdots}{(m+n)(m+n-2)\cdots 5\cdot 3\cdot 1}. \tag{36}$$

It will be observed that formula (35) is of the same form as (36) so that the case of m, n both even is the only one which is different.

EXAMPLE 31. *Evaluate the integrals*

(i) $\int_0^{\frac{1}{2}\pi} \cos^6 x \sin^4 x dx$; (ii) $\int_0^{\frac{1}{2}\pi} \cos^5 x \sin^4 x dx$; (iii) $\int_0^{\frac{1}{2}\pi} \cos^5 x \sin^3 x dx$.

(i) $\dfrac{5\cdot 3\cdot 1\cdot 3\cdot 1}{10\cdot 8\cdot 6\cdot 4\cdot 2}\ \dfrac{\pi}{2} = \dfrac{3\pi}{512}$;

(ii) $\dfrac{4\cdot 2\cdot 3\cdot 1}{9\cdot 7\cdot 5\cdot 3} = \dfrac{8}{315}$;

(iii) $\dfrac{4\cdot 2\cdot 1}{8\cdot 6\cdot 4\cdot 2} = \dfrac{1}{48}$.

PROBLEMS

1. Show that the indefinite integral of $\sin 4x \cos 2x$ is

$$C - \tfrac{1}{12}(\cos 6x + 3\cos 2x)$$

where C is a constant. Hence evaluate the integral

$$\int_0^{\frac{1}{6}\pi} \sin 4x \cos 2x dx.$$

2. By making the change of variable $x = au/2\pi$ in the integrals involved, deduce from equations (16), (21) and (11) that, if m and n are positive integers,

(i) $$\int_0^a \cos\frac{2\pi mu}{a}\cos\frac{2\pi nu}{a}\,du = \begin{cases} 0 & \text{if } m \neq n, \\ \frac{1}{2}a & \text{if } m = n; \end{cases}$$

(ii) $$\int_0^a \sin\frac{2\pi mu}{a}\sin\frac{2\pi nu}{a}\,du = \begin{cases} 0 & \text{if } m \neq n, \\ \frac{1}{2}a & \text{if } m = n; \end{cases}$$

(iii) $$\int_0^a \sin\frac{2\pi mu}{a}\cos\frac{2\pi nu}{a}\,du = 0.$$

3. If $J(n)$ denotes the integral $\int \sin^n x dx$, prove that

$$J(n) = -\frac{1}{n}\sin^{n-1} x\cos x + \frac{n-1}{n}J(n-2)$$

and deduce the value of the integral when the limits of integration are 0 and $\frac{1}{2}\pi$.

Show that the indefinite integral of $\sin^5 x$ is

$$-\tfrac{1}{15}\cos x\,(3\sin^4 x + 4\sin^2 x + 8)$$

and verify your answer by differentiating this expression.

4. Evaluate the integrals

(i) $\int_0^{\frac{1}{2}\pi} \sin^{10} x dx$; (ii) $\int_0^{\frac{1}{2}\pi} \cos^8 x dx$;

(iii) $\int_0^{\frac{1}{2}\pi} \cos^9 x dx$; (iv) $\int_0^{\frac{1}{2}\pi} \cos^7 x \sin^3 x dx$;

(v) $\int_0^{\frac{1}{2}\pi} \cos^6 x \sin^6 x dx$; (vi) $\int_0^{\frac{1}{2}\pi} \cos^4 x \sin^5 x dx$.

5. Show that, if n is an *even* positive integer,

$$\int_0^\pi \cos^n x dx = 2\int_0^{\frac{1}{2}\pi} \cos^n x dx,$$

but that, if n is an *odd* integer,

$$\int_0^\pi \cos^n x dx = 0.$$

6. Using the fact that

$$\sin nx - \sin (n-2)x = 2\cos (n-1)x \sin x$$

show that if

$$I(n) = \int \frac{\sin nx}{\sin x}\,dx$$

then

$$I(n) - I(n-2) = \frac{2}{n-1}\sin (n-1)x.$$

Hence show that the indefinite integral of

$$\frac{\sin 6x}{\sin x}$$

is

$$\tfrac{2}{15}(3 \sin 5x + 5 \sin 3x + 15 \sin x).$$

9. Proof of Maclaurin's Theorem *

We shall now make use of the theorem for integrating by parts to prove Maclaurin's theorem stated in equation (7) of § 4 of Chapter 4.

From the fundamental theorem of the integral calculus we know that

$$\int_0^1 f'(xu)du = \frac{1}{x}\int_0^x f'(t)dt = \frac{f(x) - f(0)}{x}$$

which leads to the relation

$$f(x) = f(0) + x\int_0^1 f'(xu)du. \tag{1}$$

Now we can write

$$\int_0^1 f'(xu)du = \int_0^1 1 \cdot f'(xu)du$$

where $\int 1du = -(1-u)$ and $d/du\, f'(xu) = xf''(xu)$ so that from the formula (2) of Section 6 for integrating by parts we find that

$$\begin{aligned}\int_0^1 f'(xu)du &= [-(1-u)f'(xu)]_0^1 + x\int_0^1 (1-u)f''(xu)du \\ &= f'(0) + \tfrac{1}{2}x\int_0^1 2(1-u)f''(xu)du.\end{aligned} \tag{2}$$

If we substitute from equation (2) into equation (1) we find that

$$f(x) = f(0) + f'(0)x + \frac{x^2}{2!}\int_0^1 2(1-u)f''(xu)du. \tag{3}$$

* This section may be omitted by any reader who is prepared to take the proof of Maclaurin's theorem on trust.

Similarly we can write

$$\int_0^1 2(1-u)f''(xu)\mathrm{d}u = \int_0^1 \frac{\mathrm{d}}{\mathrm{d}u}\{-(1-u)^2\}f''(xu)\mathrm{d}u$$

so that if we use the fact that $\mathrm{d}/\mathrm{d}u\, f''(xu) = xf'''(xu)$ the formula for integrating by parts gives

$$\int_0^1 2(1-u)f''(xu)\mathrm{d}u = [-(1-u)^2 f''(xu)]_0^1 + \int_0^1 (1-u)^2 f'''(xu)\mathrm{d}u.$$

Substituting this expression for the integral into equation (3) we find that

$$f(x) = f(0) + \frac{x}{1!}f'(0) + \frac{x^2}{2!}f''(0) + \frac{x^3}{3!}\int_0^1 3(1-u)^2 f'''(xu)\mathrm{d}u.$$

If we repeat this procedure of integrating by parts a further $n-3$ times we obtain the formula

$$f(x) = f(0) + \frac{x}{1!}f'(0) + \frac{x^2}{2!}f''(0) + \ldots + \frac{x^{n-1}}{(n-1)!}f^{(n-1)}(0) + R_n(x) \tag{4}$$

where the remainder after n terms is given by the equation

$$R_n(x) = \frac{x^n}{n!}\int_0^1 n(1-u)^{n-1} f^{(n)}(xu)\mathrm{d}u. \tag{5}$$

If therefore the function $f(x)$ is such that it possesses derivatives of all orders at $x = 0$ and if, for certain values of x, $R_n(x) \to 0$ as $n \to \infty$, then, for these values of x,

$$f(x) = \sum_{r=1}^{\infty} \frac{x^r}{r!} f^{(r)}(0). \tag{6}$$

The crucial step in the application of this result is the proof that $R_n(x) \to 0$ as $n \to \infty$. To facilitate this procedure we derive a simpler expression for $R_n(x)$. By the first integral theorem of mean value (Theorem 29 above) we can write

$$\int_0^1 n(1-u)^{n-1} f^{(n)}(xu)\mathrm{d}u = f^{(n)}(\theta x)\int_0^1 n(1-u)^{n-1}\mathrm{d}u = f^{(n)}(\theta x)$$

where $0 < \theta < 1$, so that from equation (5) we obtain the expression

$$R_n(x) = \frac{x^n}{n!}f^{(n)}(\theta x), \qquad 0 < \theta < 1, \tag{7}$$

for the remainder after n terms. This is known as *Lagrange's form* of the remainder.

(*a*) THE COSINE SERIES

From example 9 of Chapter 4 we know that

$$\cos x = 1 - \frac{x^2}{2!} + \frac{x^4}{4!} + \ldots + (-1)^{n-1}\frac{x^{2n-2}}{(2n-2)!} + R_{2n}(x) \tag{8}$$

where by equation (7) above

$$R_{2n}(x) = \frac{x^{2n}}{(2n)!}\cos(\theta x), \qquad 0 < \theta < 1. \tag{9}$$

Now $\cos(\theta x)$ is always numerically less than or equal to 1 so that numerically $R_{2n}(x)$ is always less than or equal to $x^{2n}/(2n)!$ and this tends to zero as $n \to \infty$ for all real values of x. Hence $R_n(x) \to 0$ as $n \to \infty$ for all real values of x. The Maclaurin expansion for $\cos x$ is therefore valid for all real values of x.

(*b*) THE LOGARITHMIC SERIES

We see immediately from the definition of the logarithmic function that

$$\log(1 + x) = \int_1^{1+x} \frac{\mathrm{d}t}{t} \tag{10}$$

if $x > -1$. If we make the substitution $t = 1 + u$ in this integral we find that

$$\log(1 + x) = \int_0^x \frac{\mathrm{d}u}{1 + u}. \tag{11}$$

Now

$$1 - u + u^2 + \ldots + (-1)^{n-1}u^{n-1} = \frac{1 - (-u)^n}{1 + u}$$

so that we may write

$$\frac{1}{1 + u} = \sum_{r=0}^{n-1}(-1)^{r-1}u^{r-1} + \frac{(-1)^n u^n}{1 + u}. \tag{12}$$

Substituting from equation (12) into equation (11) we find that

$$\log(1 + x) = \int_0^x \{\sum_{r=0}^{n-1}(-1)^r u^r\}\,\mathrm{d}u + (-1)^n\int_0^x \frac{u^n}{1 + u}\,\mathrm{d}u$$

which can be written in the form

$$\log(1+x) = \sum_{r=0}^{n-1} (-1)^r \frac{x^{r+1}}{r+1} + R_n(x) \tag{13}$$

where

$$R_n(x) = (-1)^n \int_0^x \frac{u^n \, du}{1+u}.$$

If $x > 0$,

$$|R_n(x)| = \int_0^x \frac{u^n \, du}{1+u} < \int_0^x u^n \, du = \frac{x^{n+1}}{n+1}$$

since in the interval $(0, x)$, $1 + u > 1$ and so $(1 + u)^{-1} < 1$. Therefore if $x \leqq 1$, $R_n(x) \to 0$ as $n \to \infty$.

On the other hand if $x < 0$, i.e. if $x = -\xi$ where $\xi > 0$ we have

$$R_n(x) = (-1)^n \int_0^{-\xi} \frac{u^n \, du}{1+u}.$$

Putting $u = -v$ in this integral we find that

$$R_n(x) = -\int_0^{\xi} \frac{v^n \, dv}{1-v}.$$

In the interval $0 \leqq v \leqq \xi$, $1 - v > 1 - \xi$ so that $(1 - v)^{-1} < (1-\xi)^{-1}$ and hence if $\xi < 1$

$$R_n(x) = -\frac{1}{1-\xi}\int_0^{\xi} v^n \, dv = -\frac{\xi^{n+1}}{(1-\xi)(n+1)}.$$

Therefore, if $0 < \xi < 1$, $R_n(x) \to 0$ as $n \to \infty$, i.e. if $-1 < x < 0$, $R_n(x) \to 0$ as $n \to \infty$.

Taking the two results together we see that $R_n(x) \to 0$ as $n \to \infty$ provided that $-1 < x \leqq 1$, in which case it follows from equation (13) that

$$\begin{aligned}\log(1+x) &= \sum_{r=0}^{\infty} \frac{(-1)^r x^{r+1}}{r+1} \\ &= x - \frac{x^2}{2} + \frac{x^3}{3} - \ldots + \frac{(-1)^r x^{r+1}}{r+1} + \ldots \end{aligned} \tag{14}$$

if $-1 < x \leqq 1$.

(*c*) THE EXPONENTIAL SERIES

If $f(x) = e^x$, then $f'(x) = e^x$ and so on, giving

$$f^{(r)}(x) = e^x, \qquad r \geqq 1$$

and hence

$$f^{(r)}(0) = 1, \qquad r \geqq 1.$$

Substituting these results together with $f(0) = 1$ in equation (4) we see that

$$e^x = 1 + \frac{x}{1!} + \frac{x^2}{2!} + \ldots + \frac{x^{n-1}}{(n-1)!} + R_n(x)$$

where, according to equation (7),

$$R_n(x) = \frac{x^n}{n!} e^{\theta x}, \qquad 0 < \theta < 1.$$

Now $e^{\theta x}$ is finite and $x^n/n! \to 0$ as $n \to \infty$ for all real values of x so that $R_n(x) \to 0$ as $n \to \infty$ for all real values of x. Hence

$$e^x = 1 + \frac{x}{1!} + \frac{x^2}{2!} + \ldots + \frac{x^{n-1}}{(n-1)!} + \ldots = \sum_{r=0}^{\infty} \frac{x^r}{r!} \qquad (15)$$

for all real values of x.

10. Gamma Functions

We saw in Section 7 of Chapter 6 that $e^x > x^p/p!$ for any positive integer p. It follows that

$$e^{-x} < p!\, x^{-p}$$

and hence that

$$x^{n-1} e^{-x} < p!\, x^{n-p-1}.$$

If we now choose p to be a positive integer greater than the integral part of n then if $a > 0$,

$$\int_a^b x^{n-1} e^{-x} dx < p! \int_a^b \frac{dx}{x^{p-n+1}} = p! \left[\frac{-1}{(p-n)x^{p-n}}\right]_a^b.$$

Since $p > n$ it follows that, for $a > 0$,

$$\int_a^{\infty} x^{n-1} e^{-x} dx$$

is finite for all values of n.

On the other hand if a is small $x^{n-1}e^{-x}$ will be very nearly equal to x^{n-1} when $\varepsilon < x < a$ and

$$\int_{\varepsilon}^{a} x^{n-1}\,dx = \frac{1}{n}(a^n - \varepsilon^n)$$

and this tends to a definite limit as $\varepsilon \to 0$ only if n is positive.

Hence the integral

$$\int_0^{\infty} x^{n-1}e^{-x}\,dx \tag{1}$$

exists if $n > 0$. The integral (1) is a function of n and is called the *gamma function*, $\Gamma(n)$. In other words the gamma function $\Gamma(n)$ is defined by the equation

$$\Gamma(n) = \int_0^{\infty} x^{n-1}e^{-x}\,dx, \qquad n > 0. \tag{2}$$

In particular we have

$$\Gamma(1) = \int_0^{\infty} e^{-x}\,dx = [-e^{-x}]_0^{\infty} = 1$$

so that

$$\Gamma(1) = 1. \tag{3}$$

If we use the formula for integrating by parts (noting that the indefinite integral of e^{-x} is $-e^{-x}$) we have the equation

$$\int_0^{\infty} x^{n-1}e^{-x}\,dx = [-x^{n-1}e^{-x}]_0^{\infty} + \int_0^{\infty}(n-1)x^{n-2}e^{-x}\,dx.$$

Because of the presence of e^{-x} the expression in the square bracket vanishes at the upper limit; it vanishes at the lower limit if $n - 1 > 0$. If this inequality is satisfied the integral on the right exists and has the value $(n-1)\Gamma(n-1)$. We have therefore proved that

$$\Gamma(n) = (n-1)\Gamma(n-1), \qquad n > 1. \tag{4}$$

The result (4) holds whether n is an integer or not, but if we now assume that n is a positive integer and use (4) repeatedly we find that

$$\begin{aligned}\Gamma(n) &= (n-1)\Gamma(n-1)\\ &= (n-1)(n-2)\Gamma(n-2)\\ &= (n-1)(n-2)\cdots 3\cdot 2\cdot 1\cdot \Gamma(1).\end{aligned}$$

Using equation (3) we see that we have proved that, when n is a positive integer,

$$\Gamma(n) = (n-1)!\,.$$

Some authors use (2) and (5) to *define* the factorial of any real number x, as

$$x! = \Gamma(x+1)$$

and avoid using the gamma notation.

EXAMPLE 32. *Prove that, if n is a positive integer*

$$\int_0^\infty x^{2n+1} e^{-x^2} dx = \tfrac{1}{2}(n!).$$

If we make the substitution $u = x^2$, so that $x dx = \frac{1}{2} du$ we find that the integral is equal to

$$\tfrac{1}{2}\int_0^\infty u^n e^{-u} du = \tfrac{1}{2}\Gamma(n+1) = \tfrac{1}{2}(n!)$$

if n is a positive integer.

When n is not an integer, the value of $\Gamma(n)$ can be determined by numerical integration. A table of values of $\Gamma(n)$ is given by L. M. Milne Thomson and L. J. Comrie in their "Standard Four-Figure Mathematical Tables" (Macmillan, London, 1957) p. 210. The variation

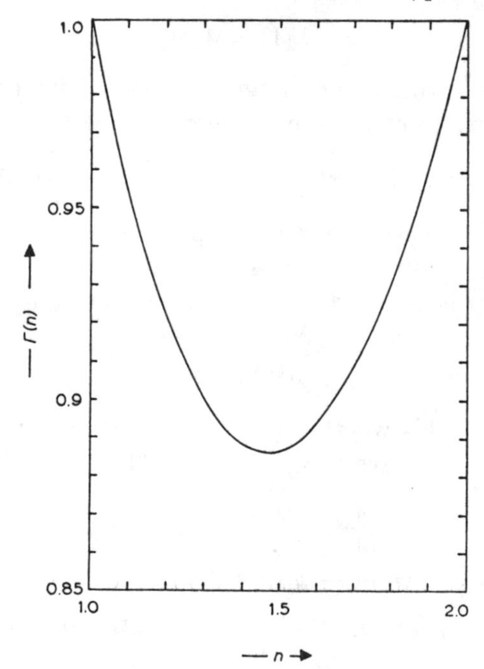

Fig. 87. The variation of $\Gamma(n)$ with n.

of $\Gamma(n)$ with n for $1 < n < 2$ is shown graphically in Fig. 87. By repeated application of equation (4) the value of $\Gamma(n)$ can be found for any value of n. Thus

$$\Gamma(3.4) = 2.4\Gamma(2.4) = 2.4 \times 1.4 \times \Gamma(1.4). \qquad (6)$$

Taking the value 0.8873 given by the table for $\Gamma(1.4)$ we find that

$$\Gamma(3.4) = 2.9814.$$

In practice, before making the numerical calculations, the integral in (2) would be manipulated to yield a more convenient form. Or instead of this, Stirling's approximation can be used (this is treated fully in textbooks on probability)

$$\log \Gamma(x+1) = \tfrac{1}{2}\log(2\pi) + (x+\tfrac{1}{2})\log x - x + 1/(12x) \qquad (7)$$

e.g. by applying (4) the other way round, with $x = 3.4$: even for this moderate value of x, (7) is extremely accurate. We shall not try to prove it.

Finally, an approximation for the ratio of two gamma functions can be very useful. A good one is*

$$\frac{\Gamma(n)}{\Gamma(n-x)} \simeq m^x \left\{1 - \frac{(x-1)x(x+1)}{24m^2}\right\} \qquad (8)$$

or

$$\log \Gamma(n) - \log \Gamma(n-x) \simeq x\left(\log m - \frac{x^2-1}{24m^2}\right) \qquad (9)$$

where $m = n - \frac{1}{2}x - \frac{1}{2}$. Here n can be any positive number, x may be either positive or negative and x^2 should be less than m^2.

11. Beta Functions

We saw in Section 7 of Chapter 5 that the improper integral of the second kind

$$\int_0^1 x^{m-1}(1-x)^{n-1}\,dx$$

has a finite value if and only if both m and n are positive. This integral

* The shortest derivations are a little more advanced than this book, and use complex variable theory or operators. See M.E. Wise, Proc. Koninklijke Nederlandse Akad. Wetenschappen, Amsterdam. Series A 57 (1954) 513.

therefore defines a function of m and n for positive values of these variables. This function is called the *beta function*. We define the beta function by the equation

$$B(m, n) = \int_0^1 x^{m-1}(1-x)^{n-1}\,dx, \quad (m > 0,\ n > 0). \tag{1}$$

Using Theorem 33 (p. 210 above) we see that

$$B(m, n) = \int_0^1 (1-x)^{m-1}x^{n-1}\,dx$$

from which it is immediately obvious that

$$B(m, n) = B(n, m). \tag{2}$$

If we use the fact that

$$-\frac{1}{n}(1-x)^n$$

is the indefinite integral of $(1-x)^{n-1}$ in the formula for integrating by parts we find that

$$\int_0^1 x^{m-1}(1-x)^{n-1}\,dx = \left[-\frac{1}{n}(1-x)^n x^{m-1}\right]_0^1 + \frac{m-1}{n}\int_0^1 x^{m-2}(1-x)^n\,dx.$$

If $n > 0$, $m > 1$ the quantity in the square bracket vanishes at both the upper and the lower limits, and writing

$$x^{m-2}(1-x)^n = x^{m-2}(1-x)^{n-1}(1-x) = x^{m-2}(1-x)^{n-1} - x^{m-1}(1-x)^{n-1}$$

we find that

$$n\int_0^1 x^{m-1}(1-x)^{n-1}dx = (m-1)\left\{\int_0^1 x^{m-2}(1-x)^{n-1}dx - \int_0^1 x^{m-1}(1-x)^{n-1}dx\right\}$$

which is equivalent to

$$nB(m, n) = (m-1)B(m-1, n) - (m-1)B(m, n)$$

and therefore to the relation

$$B(m, n) = \frac{m-1}{m+n-1}B(m-1, n). \tag{3}$$

Now using equations (2) and (3) we find that

$$B(m, n) = B(n, m) = \frac{n-1}{m+n-1}B(n-1, m),$$

and using (2) again we find that

$$B(m,\ n) = \frac{n-1}{m+n-1} B(m,\ n-1). \tag{4}$$

The relations (3) and (4) are true in general for real values of m. If we assume now that m and n are positive integers, then by (3)

$$B(m,\ n) = \frac{(m-1)(m-2)\cdots 2\cdot 1}{(m+n-1)(m+n-2)\cdots(n+2)(n+1)} B(1,\ n)$$

and by (4)

$$B(1,\ n) = \frac{(n-1)(n-2)\cdots 2\cdot 1}{n(n-1)(n-2)\cdots 2\cdot 1} B(1,\ 1).$$

Multiplying these two results together we find that

$$B(m,\ n) = \frac{(m-1)!(n-1)!}{(m+n-1)!} B(1,\ 1).$$

Now

$$B(1,\ 1) = \int_0^1 (1-x)^{1-1} x^{1-1}\,dx = \int_0^1 dx = 1$$

so that, if m and n are positive integers

$$B(m,\ n) = \frac{(m-1)!(n-1)!}{(m+n-1)!}. \tag{5}$$

EXAMPLE 33. *Evaluate* $\int_0^1 x^6(1-x)^9\,dx$.

This is derived from formula (5) by putting $m = 7$, $n = 10$. In this way we find that

$$\int_0^1 x^6(1-x)^9\,dx = \frac{6!\,9!}{16!} = \frac{1}{11\cdot 13\cdot 14\cdot 5\cdot 8} = \frac{1}{80080}.$$

A useful form of integral expression for the beta function is obtained by changing the variable in (1) from x to θ where $x = \sin^2\theta$. When $x = 0$, $\theta = 0$ and when $x = 1$, $\theta = \frac{1}{2}\pi$. Also $1 - x = 1-\sin^2\theta$ $= \cos^2\theta$ and $dx = 2\sin\theta\cos\theta d\theta$ so that

$$B(m,\ n) = 2\int_0^{\frac{1}{2}\pi} \sin^{2m-1}\theta\cos^{2n-1}\theta d\theta. \tag{6}$$

In particular

$$B(\tfrac{1}{2},\ \tfrac{1}{2}) = \pi. \tag{7}$$

EXAMPLE 34. *Prove that*

$$B(n,\ n) = 2^{1-2n} B(n,\ \tfrac{1}{2}). \tag{8}$$

From equation (6)

$$B(n, n) = 2\int_0^{\frac{1}{2}\pi} \sin^{2n-1}\theta \cos^{2n-1}\theta d\theta.$$

Now

$$\begin{aligned}\sin^{2n-1}\theta \cos^{2n-1}\theta &= (2\cos\theta\sin\theta)^{2n-1} \times 2^{1-2n} \\ &= 2^{1-2n}(\sin 2\theta)^{2n-1}\end{aligned}$$

so that

$$B(n, n) = 2^{2-2n}\int_0^{\frac{1}{2}\pi} \sin^{2n-1} 2\theta d\theta.$$

Putting $\phi = 2\theta$ we see that

$$\begin{aligned}B(n, n) &= 2^{1-2n}\int_0^{\pi} \sin^{2n-1}\phi d\phi \\ &= 2^{1-2n}\left\{2\int_0^{\frac{1}{2}\pi} \sin^{2n-1}\phi \cos^{2(\frac{1}{2})-1}\phi d\phi\right\} \\ &= 2^{1-2n}B(n, \tfrac{1}{2}).\end{aligned}$$

PROBLEMS

1. Show that if m and n are positive integers

$$\int_0^{\frac{1}{2}\pi} \sin^{2m-1}\theta \cos^{2n-1}\theta d\theta = \frac{1}{2}\frac{(m-1)!(n-1)!}{(m+n-1)!}.$$

Evaluate

$$\int_0^{\frac{1}{2}\pi} \sin^7\theta \cos^5\theta d\theta; \quad \int_0^{\frac{1}{2}\pi} \sin^7\theta \cos^7\theta d\theta.$$

2. Evaluate the integrals

$$\int_0^1 x^2(1-x)^7 dx; \quad \int_0^1 x^5(1-x)^5 dx.$$

3. Making the substitution $x = 1/(1+u)$ in (1) show that

$$B(m, n) = \int_0^{\infty} \frac{u^{n-1} du}{(1+u)^{m+n}}.$$

Writing this in the form

$$B(m, n) = \int_0^1 \frac{u^{n-1} du}{(1+u)^{m+n}} + \int_1^{\infty} \frac{u^{n-1} du}{(1+u)^{m+n}}$$

and making the substitution $u = 1/v$ in the second of these two integrals show that

$$B(m, n) = \int_0^1 \frac{v^{m-1} + v^{n-1}}{(1+v)^{m+n-1}} dv.$$

12. Relation between Beta and Gamma Functions

We showed in the last section that when m and n are positive integers

$$B(m, n) = \frac{(m-1)!(n-1)!}{(m+n-1)!}$$

and in Section 10 that when n is a positive integer

$$\Gamma(n) = (n-1)!$$

so that we can say that, *when m and n are positive integers,*

$$B(m, n) = \frac{\Gamma(m)\Gamma(n)}{\Gamma(m+n)}. \tag{1}$$

This result is in fact true for all positive values of m and n whether they are integers or not. The proof of this assertion involves the use of double integration, which we have not defined, so it is omitted here.*

If we put $m = n = \frac{1}{2}$ in equation (1) and use the results $B(\frac{1}{2}, \frac{1}{2}) = \pi$, $\Gamma(1) = 1$ we see that

$$\Gamma(\tfrac{1}{2}) = \sqrt{\pi}. \tag{2}$$

Now, using equation (4) of Section 10 we see that if p is a positive integer

$$\begin{aligned}\Gamma(p + \tfrac{1}{2}) &= (p - \tfrac{1}{2})(p - \tfrac{3}{2}) \cdots \tfrac{3}{2} \cdot \tfrac{1}{2}\Gamma(\tfrac{1}{2}) \\ &= \frac{(2p-1)(2p-3)\cdots 3 \cdot 1}{2^p}\sqrt{\pi} \\ &= \frac{(2p)(2p-1)(2p-2)(2p-3)\cdots 4 \cdot 3 \cdot 2 \cdot 1}{(2p)(2p-2)\cdots 4 \cdot 2 \cdot 2^p}\sqrt{\pi}\end{aligned}$$

i.e. that

$$\Gamma(p + \tfrac{1}{2}) = \frac{(2p)!}{2^{2p}p!}\sqrt{\pi}. \tag{3}$$

This is, in fact, a special case of a more general result. If we put $m = n$ in equation (1) and then use equation (8) of the last section

* For such a proof the reader is referred to R. P. Gillespie, "Integration" (Oliver & Boyd, Edinburgh, 1939) p. 87. See Problem 5 of § 16 below.

we find that

$$\frac{\Gamma(n)\,\Gamma(n)}{\Gamma(2n)} = B(n,\ n) = 2^{1-2n}B(n,\ \tfrac{1}{2}) = 2^{1-2n}\frac{\Gamma(n)\,\Gamma(\frac{1}{2})}{\Gamma(n+\frac{1}{2})}$$

from which we obtain the relation

$$\Gamma(2n)\Gamma(\tfrac{1}{2}) = 2^{2n-1}\,\Gamma(n)\Gamma(n+\tfrac{1}{2}) \tag{4}$$

of which (3) is the special case obtained by putting $n = p + \frac{1}{2}$. Equation (4) is called the *duplication formula* for the gamma function.

EXAMPLE 35. *Show that*

$$\int_0^{\frac{1}{2}\pi} \sin^p \theta \mathrm{d}\theta = \frac{\Gamma(\frac{3}{2})\Gamma(\frac{1}{2}+\frac{1}{2}p)}{\Gamma(1+\frac{1}{2}p)}, \qquad (p > -\tfrac{1}{2}).$$

Putting $m = \frac{1}{2} + \frac{1}{2}p$, $n = \frac{1}{2}$ in equation (6) of the last section we find that

$$\int_0^{\frac{1}{2}\pi} \sin^p \theta \mathrm{d}\theta = \tfrac{1}{2}B(\tfrac{1}{2}+\tfrac{1}{2}p,\ \tfrac{1}{2}) = \frac{\Gamma(\frac{1}{2})\Gamma(\frac{1}{2}+\frac{1}{2}p)}{2\Gamma(1+\frac{1}{2}p)}.$$

Using the fact that $\frac{1}{2}\Gamma(\frac{1}{2}) = \Gamma(\frac{3}{2})$ we get the stated result.

PROBLEMS

1. Show that if m and n are positive integers

$$\int_0^{\frac{1}{2}\pi} \sin^{2m}\theta \cos^{2m}\theta \mathrm{d}\theta = \frac{1}{2}\,\frac{\Gamma(m+\frac{1}{2})\Gamma(n+\frac{1}{2})}{\Gamma(m+n+1)}$$

$$= \frac{(2m)!\,(2n)!\,\pi}{2^{2m+2n+1}\,m!\,n!\,(m+n)!}.$$

2. Prove that, if n is a positive integer,

$$\int_0^1 \frac{v^n\,\mathrm{d}v}{(1+v)^{2n+1}} = \frac{(n!)^2}{(2n+1)!}.$$

3. Prove that

(i) $\int_{-\infty}^{\infty} \mathrm{e}^{-x^2}\,\mathrm{d}x = \sqrt{\pi}$;

(ii) $\int_{-\infty}^{\infty} x^{2n}\,\mathrm{e}^{-x^2}\,\mathrm{d}x = \Gamma(n+\frac{1}{2})$.

13. Integrals Associated with Continuous Distributions in Statistics *

In mathematical statistics it is convenient to introduce the idea of a frequency $f(x)$ of a continuous variable (synonym: density function). This may be interpreted in the sense of $f(x)dx$ being the probability that the variable takes a value between x and $x + dx$, dx being small. With this interpretation $f(x)$ obviously has the properties

$$f(x) \geqq 0, \tag{1}$$

$$\int_{-\infty}^{\infty} f(x)dx = 1, \tag{2}$$

the first being necessary since negative probability has no meaning, and the second corresponding to the requirement that the probability of an event that is certain to occur should be equal to unity. If the

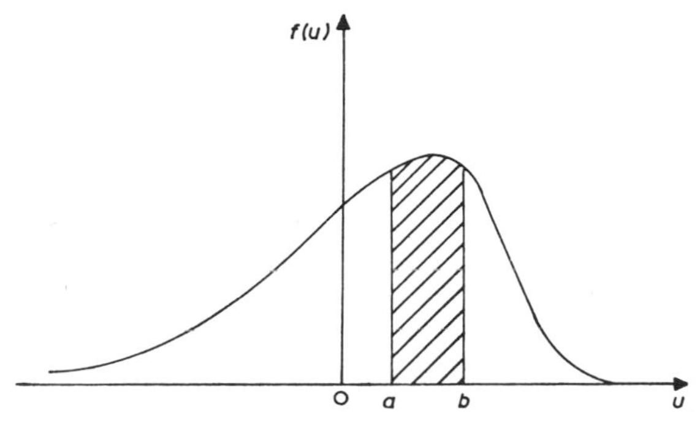

Fig. 88. Interpretation of $P(a, b)$ as an area under the frequency curve.

range of x is not $-\infty < x < \infty$ but some finite range, we take $f(x)$ to be zero for values of x outside the range of the variable. Further, if we denote by $P(a, b)$ the probability that x lies in the interval $a \leqq x \leqq b$, then

$$P(a, b) = \int_a^b f(u)du \tag{3}$$

and this has an obvious geometrical interpretation. (Cf. Fig. 88.)

It is also useful to introduce a distribution function $F(x)$ which is defined by the equation

* This section is an exercise in the use of the ideas of the last three sections. The reader may postpone reading it and §§ 14 and 15 until he needs them in the study of statistics. No knowledge of statistics is assumed in them.

$$F(x) = \int_{-\infty}^{x} f(u)\mathrm{d}u. \tag{4}$$

If we make use of Theorem 21, p. 205 we see that

$$\int_{a}^{b} f(u)\mathrm{d}u = \int_{-\infty}^{b} f(u)\mathrm{d}u - \int_{-\infty}^{a} f(u)\mathrm{d}u$$

so that

$$P(a, b) = F(b) - F(a). \tag{5}$$

The function $F(x)$ is the area under the frequency curve $y = f(u)$ up to the ordinate at the point $u = x$. (Cf. Fig. 89.)

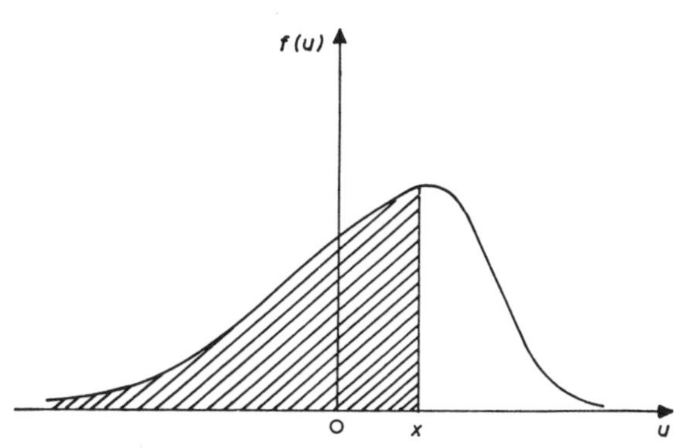

Fig. 89. Definition of the distribution function $F(x)$ in terms of the frequency function $f(x)$. The shaded area represents $F(x)$.

It is often called a *probability integral,* reasonably enough. Similarly, the curve in Fig. 89 is sometimes called a probability curve or a probability distribution, and its ordinates, cf. (3) are then probability densities.

Finally, the term "theoretical frequency distribution" is often used for a curve such as Fig. 88 with area N instead of unity. We consider N quantities (in this introduction N is supposed to be a large number). Then the shaded area in Fig. 88 can be taken to mean the predicted number, i.e. the frequency, of those quantities lying between a and b, and the ordinates of the curve are frequency densities. Similarly the area $F(x)$ in Fig. 89 then corresponds to the total (accumulated) number (frequency) of quantities less than x.*

* From all this, the reader will conclude, rightly, that these terms are not always used consistently! Fortunately it is usually clear from the context what is meant.

The most generally used statistical measure is perhaps the arithmetical mean. We define the arithmetic mean μ_1' about an arbitrary point $x = a$ by the equation

$$\mu_1' = \int_{-\infty}^{\infty} (x - a) f(x) \mathrm{d}x. \tag{6}$$

It will be observed that the value a is arbitrary; it is generally most convenient to choose it to be near a value giving a maximum to $f(x)$. If we calculate the arithmetic mean (call it $\tilde{\mu}_1$) about another point b, then

$$\int_{-\infty}^{\infty} (x - b) f(x) \mathrm{d}x = \int_{-\infty}^{\infty} (x - a) f(x) \mathrm{d}x - (b - a) \int_{-\infty}^{\infty} f(x) \mathrm{d}x,$$

and using equations (2) and (6) we see that

$$\tilde{\mu}_1 = \mu_1' - (b - a) \tag{7}$$

showing that we can easily calculate the mean about any point when we know the mean about any other.

We define higher moments in the same way. The quantities μ_2', μ_3', . . . are defined by the equation

$$\mu_r' = \int_{-\infty}^{\infty} (x - a)^r f(x) \mathrm{d}x. \tag{8}$$

Of special importance are the moments μ_2, μ_3, . . . about the arithmetic mean μ_1' which are defined by the equation

$$\mu_r = \int_{-\infty}^{\infty} (x - \mu_1')^r f(x) \mathrm{d}x. \tag{9}$$

The quantities μ_r, μ_r' $(r = 1, 2, \ldots)$ are not independent. For instance, we can write

$$\mu_2 = \int_{-\infty}^{\infty} [(x-a)+(a-\mu_1')]^2 f(x) \mathrm{d}x =$$

$$\int_{-\infty}^{\infty} (x-a)^2 f(x) \mathrm{d}x + 2(a-\mu_1') \int_{-\infty}^{\infty} (x-a) f(x) \mathrm{d}x + (a-\mu_1')^2 \int_{-\infty}^{\infty} f(x) \mathrm{d}x.$$

Using equations (8), (6) and (1) we see that

$$\mu_2 = \mu_2' + 2(a - \mu_1')\mu_1' + (a - \mu_1')^2$$

showing that

$$\mu_2 = \mu_2' + a^2 - \mu_1'^2. \tag{10}$$

The square root σ of the second moment is called the standard deviation, i.e.

$$\sigma^2 = \mu_2.$$

These are the two most important statistical measures after the mean.

The function

$$m(t) = \int_{-\infty}^{\infty} e^{xt} f(x) dx \tag{11}$$

has a very interesting property. If we differentiate it r times with respect to t and then put $t = 0$ we find that

$$m^{(r)}(0) = \int_{-\infty}^{\infty} x^r f(x) dx = \mu_r',$$

where μ_r' is the r-th moment about the origin. Also

$$m(0) = \int_{-\infty}^{\infty} f(x) dx = 1$$

by equation (2). Hence, by Maclaurin's theorem

$$m(t) = 1 + \mu_1' t + \frac{1}{2!}\mu_2' t^2 + \ldots + \frac{1}{r!}\mu_r' t^r + \ldots \tag{12}$$

showing that when the function $m(t)$, defined by equation (11), is expanded as a Maclaurin series, the coefficients are simply related to the moments about the origin. For this reason $m(t)$ is called the *moment generating function.*

We shall now illustrate these remarks by considering the simpler properties of some commonly occurring distributions.

(*a*) UNIFORM DISTRIBUTION

Here the frequency function has the form

$$f(x) = \begin{cases} 0, & x < a; \\ \dfrac{1}{b-a}, & a \leqq x \leqq b; \\ 0, & x > b \end{cases}$$

and the distribution function is

$$F(x) = \begin{cases} 0, & x < a; \\ \dfrac{x-a}{b-x}, & a \leqq x \leqq b; \\ 1, & x > b. \end{cases}$$

The r-th moment about the origin is

$$\mu'_r = \frac{1}{b-a}\int_a^b x^r\,dx = \frac{1}{r+1}\,\frac{b^{r+1}-a^{r+1}}{b-a}. \tag{13}$$

In particular

$$\mu'_1 = \frac{1}{2}\,\frac{b^2-a^2}{b-a} = \tfrac{1}{2}(b+a),$$

$$\mu'_2 = \frac{1}{3}\,\frac{b^3-a^3}{b-a} = \tfrac{1}{3}(b^2+ab+a^2)$$

and so, from equation (10)

$$\mu_2 = \mu'_2 - \mu'^2_1 = \tfrac{1}{12}(b-a)^2.$$

The moment generating function is easily seen to be

$$m(t) = \frac{1}{b-a}\int_a^b e^{xt}\,dx = \frac{e^{bt}-e^{at}}{(b-a)t}. \tag{14}$$

Both the frequency function and the distribution function are shown in Fig. 90.

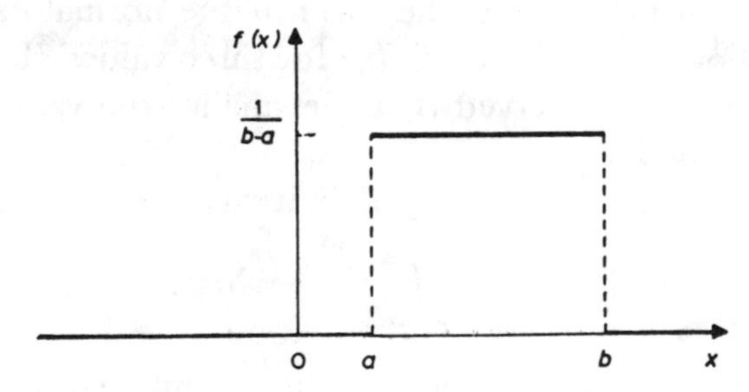

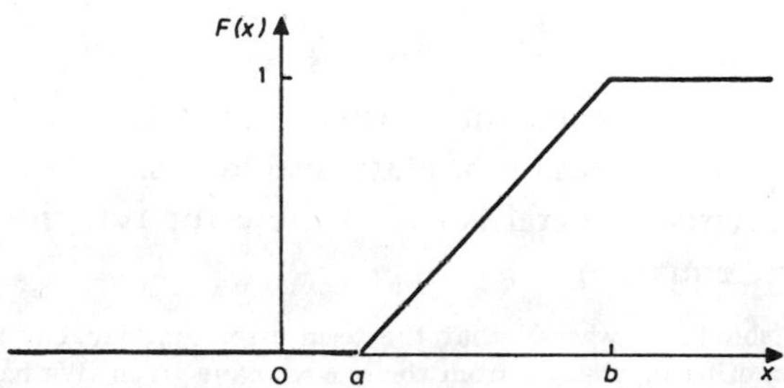

Fig. 90. Frequency function, $f(x)$, and distribution function, $F(x)$, for a uniform distribution.

(*b*) NORMAL DISTRIBUTION

The most commonly occurring distribution in statistics is the normal distribution whose frequency function is defined by the equation

$$f(x) = \frac{1}{\sigma\sqrt{2\pi}} e^{-(x-m)^2/2\sigma^2} \qquad (\sigma > 0). \qquad (15)$$

Since

$$f'(x) = -\frac{(x-m)}{\sigma^3\sqrt{2\pi}} e^{-(x-m)^2/2\sigma^2}$$

it follows that the only turning value of $f(x)$ is at the point $x = m$, and since $f \to 0$ as $x \to \pm\infty$ and

$$f(m) = \frac{1}{\sigma\sqrt{2\pi}} > 0$$

it follows that this is a maximum value. If we write

$$\xi = (x-m)/\sigma \qquad (16)$$

then

$$f = e^{-\frac{1}{2}\xi^2}$$

so that the graph of $f(x)$ has the form of the normal error curve. (Cf. Fig. 78 above.) The form of $f(x)$ for three values of σ is shown in Fig. 91. It will be observed that the smaller the value of σ, the narrower is the "bell".

The distribution function is, by definition,

$$F(x) = \frac{1}{\sigma\sqrt{2\pi}} \int_{-\infty}^{x} e^{-(x-m)^2/2\sigma^2} dx$$

and if we make the substitution (16) in the integral we have

$$F(x) = \frac{1}{\sqrt{2\pi}} \int_{-\infty}^{(x-m)/\sigma} e^{-\frac{1}{2}\xi^2} d\xi \; . \qquad (17)$$

In this form $F(x)$ is called the normal probability integral, and it is tabulated in every collection of statistical tables.

A closely related integral is called the error function, and this is defined by the equation*

* The reader should be warned that the term error function has been applied to various integrals, differing slightly from the one we have given. We have followed the notation adopted by the Mathematical Tables Project of the W. P. A. of the City of New York (Tables of Probability Functions, vols. I, II, 1941—42).

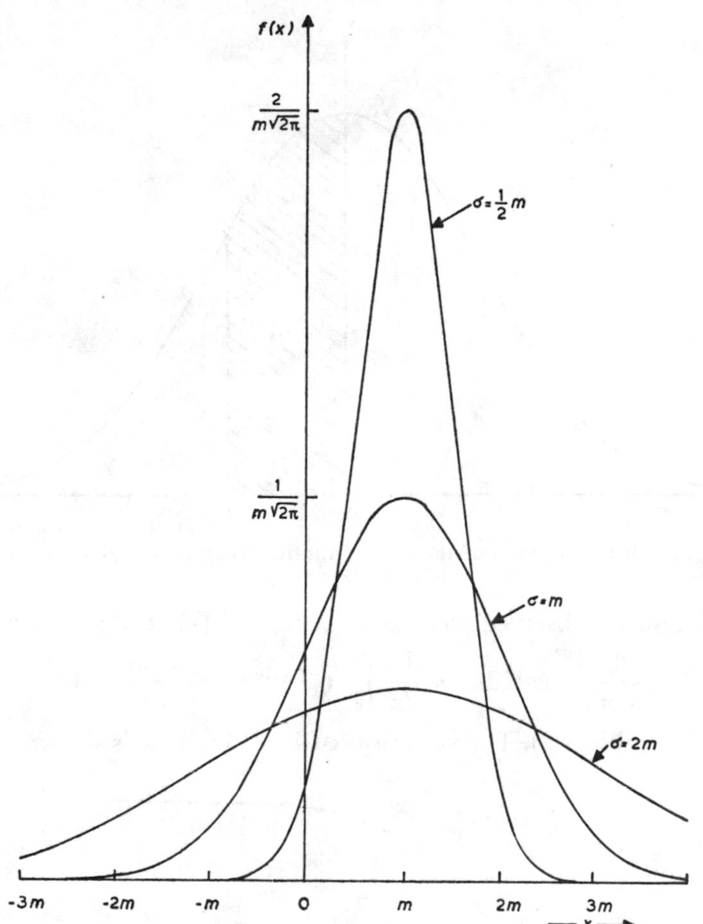

Fig. 91. Frequency curves for three normal distributions.

$$\mathrm{Erf}(x) = \frac{2}{\sqrt{\pi}} \int_0^x e^{-u^2}\, du \tag{18}$$

so that Erf(x) can be interpreted as an area under the curve

$$y = \frac{2}{\sqrt{\pi}}\, e^{-u^2}. \tag{19}$$

(Cf. Fig. 92.) Because of the symmetry of the curve about the y-axis, it is easily seen that if $x < 0$,

$$\frac{2}{\sqrt{\pi}} \int_{-x}^{0} e^{-u^2} du = \frac{2}{\sqrt{\pi}} \int_0^x e^{-u^2} du = \mathrm{Erf}\ (x). \tag{20}$$

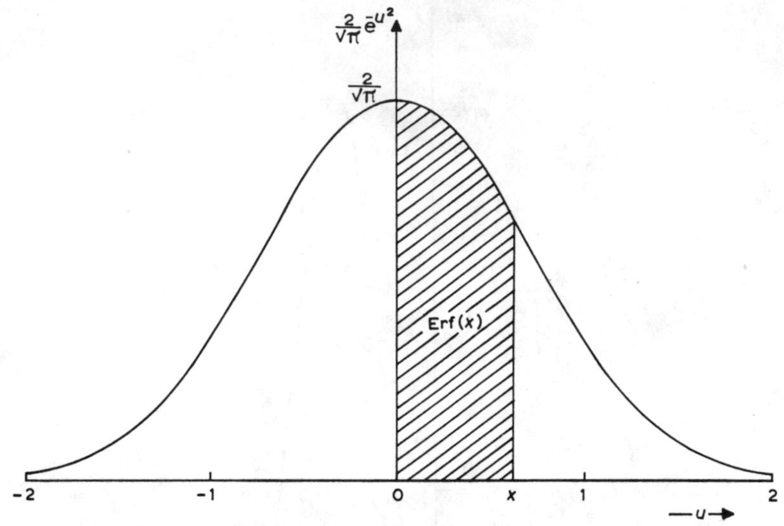

Fig. 92. Interpretation of the error function as an area under the curve of equation (19).

It will also be observed that as $x \to \infty$, Erf (x) tends to the value

$$\frac{2}{\sqrt{\pi}}\int_0^\infty e^{-u^2}du = \frac{1}{\sqrt{\pi}}\int_0^\infty v^{-\frac{1}{2}}e^{-v}\,dv = \frac{\Gamma(\frac{1}{2})}{\sqrt{\pi}} = 1 \tag{21}$$

(cf. §12, Problem 3). The variation of Erf(x) with x is shown in Fig. 93.

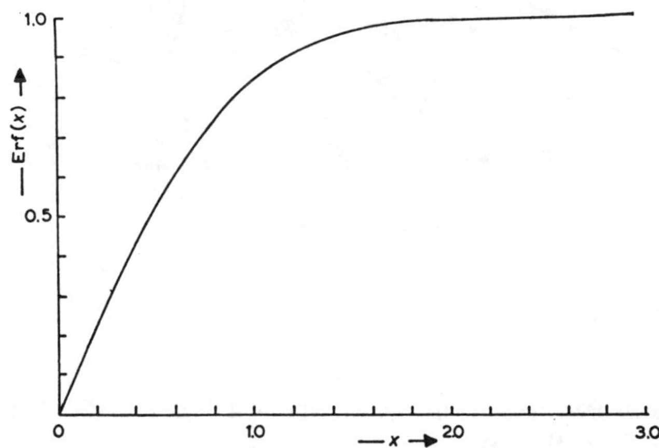

Fig. 93. Graph of the error function

$$\text{Erf}\,(x) = \frac{2}{\sqrt{\pi}}\int_0^x e^{-u^2}\,du.$$

Returning now to equation (17), we put $\xi = u\sqrt{2}$. Then

$$F(x) = \frac{1}{\sqrt{\pi}} \int_{-\infty}^{(x-m)/(\sigma\sqrt{2})} e^{-u^2}\, du.$$

We see that when $x < m$ this can be split up into two terms to give

$$F(x) = \frac{1}{\sqrt{\pi}} \int_{-\infty}^{0} e^{-u^2} du - \frac{1}{\sqrt{\pi}} \int_{-(m-x)/\sqrt{2}\sigma}^{0} e^{-u^2} du. \tag{22}$$

Similarly, if $x > m$,

$$F(x) = \frac{1}{\sqrt{\pi}} \int_{-\infty}^{0} e^{-u^2} du + \frac{1}{\sqrt{\pi}} \int_{0}^{(m-x)/\sqrt{2}\sigma} e^{-u^2} du. \tag{23}$$

If, therefore, we substitute from equations (20) and (21) into equations (22) and (23) we find that the distribution function is given by the equations

$$F(x) = \begin{cases} \dfrac{1}{2}\left\{1 - \text{Erf}\left(\dfrac{m-x}{\sqrt{2}\sigma}\right)\right\}, & x < m \\[2ex] \dfrac{1}{2}\left\{1 + \text{Erf}\left(\dfrac{x-m}{\sqrt{2}\sigma}\right)\right\}, & x > m. \end{cases} \tag{24}$$

The form of $F(x)$ is shown in Fig. 94 for the three values of σ considered in Fig. 91.

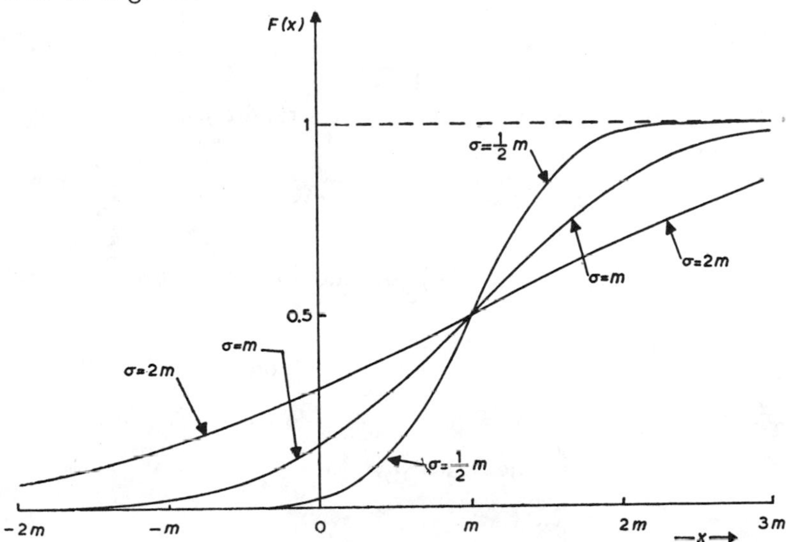

Fig. 94. Distribution functions corresponding to the frequency functions of Fig. 91.

The arithmetic mean is given by the integral

$$\mu_1' = \frac{1}{\sigma\sqrt{2\pi}} \int_{-\infty}^{\infty} x e^{-(x-m)^2/2\sigma^2} dx.$$

Making the substitution (16) we see that this equation becomes

$$\mu_1' = \frac{1}{\sqrt{2\pi}} \int_{-\infty}^{\infty} (m + \sigma\xi) e^{-\frac{1}{2}\xi^2} d\xi.$$

Or, putting $\xi = u\sqrt{2}$

$$\mu_1' = \frac{1}{\sqrt{\pi}} \int_{-\infty}^{\infty} (m + \sigma\sqrt{2}) e^{-u^2} du.$$

Now

$$\int_{-\infty}^{\infty} u e^{-u^2} du = \int_{-\infty}^{0} u e^{-u^2} du + \int_{0}^{\infty} u e^{-u^2} du = -\int_{0}^{\infty} v e^{-v^2} dv + \int_{0}^{\infty} u e^{-u^2} du = 0$$

where we have made the substitution $v = -u$ in the first integral. Using this result and the fact that

$$\int_{-\infty}^{\infty} e^{-u^2} du = 2\int_{0}^{\infty} e^{-u^2} du = \sqrt{\pi}$$

we find finally that

$$\mu_1' = m. \tag{25}$$

The moments μ_r may be evaluated straight away. We have, by definition,

$$\mu_r = \frac{1}{\sigma\sqrt{2\pi}} \int_{-\infty}^{\infty} (x - m)^r e^{-(x-m)^2/2\sigma^2} dx.$$

Making the substitution (16) in the integral, and then putting $\xi = u\sqrt{2}$, we have

$$\mu_r = \frac{2^{\frac{1}{2}r} \sigma^r}{\sqrt{\pi}} \int_{-\infty}^{\infty} u^r e^{-u^2} du.$$

Now

$$\int_{-\infty}^{\infty} u^{2n+1} e^{-u^2} du = 0,$$

$$\int_{-\infty}^{\infty} u^{2n} e^{-u^2} du = 2\int_{0}^{\infty} u^{2n} e^{-u^2} du$$

$$= \int_0^\infty v^{n-\frac{1}{2}} e^{-v} dv = \Gamma(n + \tfrac{1}{2})$$

so that

$$\mu_{2n} = \frac{2^n \sigma^{2n} \Gamma(n + \frac{1}{2})}{\Gamma(\frac{1}{2})}, \qquad \mu_{2n+1} = 0. \tag{26}$$

From the former of these two equations we get the simple result

$$\frac{\mu_{2n}}{\mu_{2n-2}} = 2\sigma^2 \frac{\Gamma(n + \frac{1}{2})}{\Gamma(n - \frac{1}{2})} = (2n - 1)\sigma^2$$

i.e.

$$\mu_{2n} = (2n - 1)\sigma^2 \mu_{2n-2} \,. \tag{27}$$

From the fact that

$$\mu_0 = \int_{-\infty}^{\infty} f(x) dx = 1$$

we see that equation (27) gives the set of values

$$\mu_2 = \sigma^2 \quad \mu_4 = 3\sigma^4, \quad \mu_6 = 15\sigma^6, \quad \ldots .$$

The moment generating function, defined by equation (11), is in this case

$$m(t) = \frac{1}{\sigma\sqrt{2\pi}} \int_{-\infty}^{\infty} e^{xt-(x-m)^2/2\sigma^2} dx. \tag{28}$$

Now

$$-\frac{(x-m)^2}{2\sigma^2} + xt = -\frac{x^2}{2\sigma^2} + \frac{x}{\sigma^2}(m + \sigma^2 t) - \frac{m^2}{2\sigma^2}$$

$$= -\frac{(x - m - \sigma^2 t)^2}{2\sigma^2} + mt + \tfrac{1}{2}\sigma^2 t^2.$$

If we make the change of variable

$$\xi = \frac{x - m - \sigma^2 t}{\sqrt{2}\sigma}$$

in the integral in equation (28), we find that

$$m(t) = \frac{1}{\sqrt{\pi}} e^{mt + \frac{1}{2}\sigma^2 t^2} \int_{-\infty}^{\infty} e^{-\xi^2} d\xi$$

showing that the moment generating function is given by the equation

$$m(t) = e^{mt + \frac{1}{2}\sigma^2 t^2}. \tag{29}$$

Expanding the exponential function we have

$$\begin{aligned} m(t) &= 1 + (mt + \tfrac{1}{2}\sigma^2t^2) + \tfrac{1}{2}(mt + \tfrac{1}{2}\sigma^2t^2)^2 + \tfrac{1}{6}(mt + \tfrac{1}{2}\sigma^2t^2)^2 + \ldots \\ &= 1 + mt + \tfrac{1}{2}t^2(\sigma^2 + m^2) + \tfrac{1}{6}t^3(m^3 + m\sigma^2) + \ldots \end{aligned}$$

so that

$$\mu_2' = \sigma^2 + m^2, \quad \mu_3' = m(\sigma^2 + m^2), \quad \ldots.$$

(*c*) THE GAMMA DISTRIBUTION

The gamma distribution is defined by the frequency function

$$f(x) = \begin{cases} 0, & x < 0; \\ \dfrac{x^\alpha\, e^{-x/\beta}}{\Gamma(\alpha+1)\beta^{\alpha+1}}, & x > 0. \end{cases} \tag{30}$$

This defines a two-parameter family of distributions, the parameters being α and β. In order that the condition (2) should be satisfied, the integral

$$\int_0^\infty x^\alpha\, e^{-x/\beta}\, dx$$

must be finite. For this to be so we need $\alpha > -1$. If we differentiate the expression for $f(x)$ we find that

$$f'(x) = \frac{1}{\Gamma(\alpha+1)\beta^{\alpha+1}}\left(\alpha - \frac{x}{\beta}\right)x^{\alpha-1}\, e^{-x/\beta}, \quad (x > 0)$$

so that $f'(x) = 0$ when $x = \alpha\beta$.

The forms of $f(x)$ for four values of α are shown in Fig. 95. When $\alpha = 0$, $f(0) = 1$ and if $\alpha > 0$, $f(0) = 0$, while if $\alpha < 0$, $f(x) \to \infty$ as $x \to 0$.

The distribution function is defined by the equation

$$F(x) = \begin{cases} 0, & x < 0; \\ \dfrac{1}{\Gamma(\alpha+1)\beta^{\alpha+1}}\displaystyle\int_0^x t^\alpha\, e^{-t/\beta}\, dt & x > 0. \end{cases}$$

Making the substitution $t = \beta u$ we see that when $x > 0$

$$F(x) = \frac{1}{\Gamma(\alpha+1)}\int_0^{x/\beta} u^\alpha e^{-u}\, du. \tag{31}$$

If α is an integer we can evaluate the integral easily. For example, if $\alpha = 1$, then, integrating by parts, we have

$$\int u e^{-u}\, du = -ue^{-u} + \int e^{-u}\, du = -(1 + u)e^{-u}$$

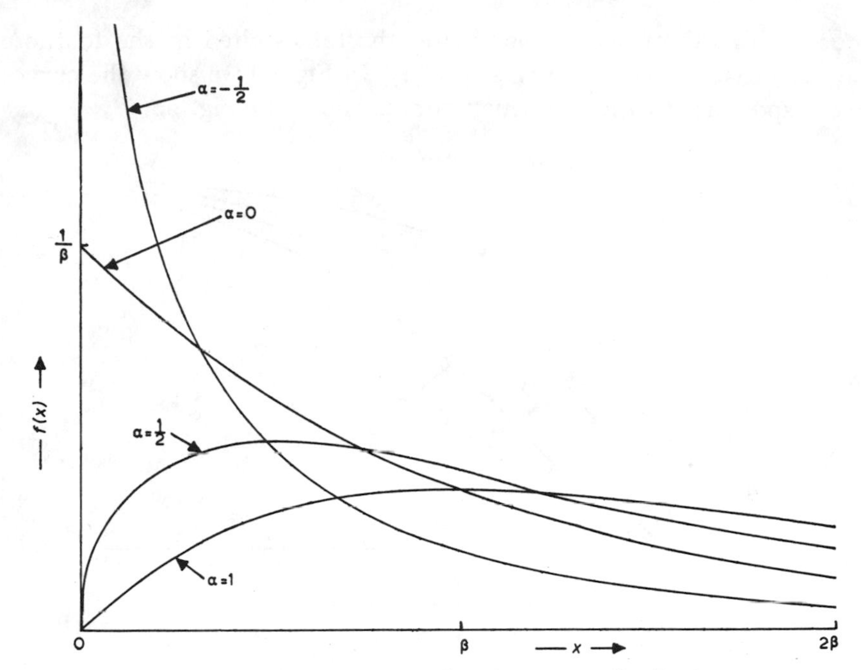

Fig. 95. Frequency curves for the gamma distribution.

and hence when $\alpha = 1$,

$$F(x) = \left[1 - \left(1 + \frac{x}{\beta}\right) e^{-x/\beta}\right].$$

When α is not a positive integer, the integral

$$\Gamma_\xi(\alpha + 1) = \int_0^\xi e^{-u} u^\alpha \, du$$

has to be evaluated numerically. The function $\Gamma_\xi(\alpha + 1)$ is called the *incomplete gamma function*. Tables of the ratio

$$I(\xi, \alpha) = \frac{\Gamma_\xi(\alpha + 1)}{\Gamma(\alpha + 1)} = \frac{1}{\Gamma(\alpha + 1)} \int_0^\xi e^{-u} u^\alpha \, du \tag{32}$$

are available.* In this notation, equation (31) gives the expression

$$F(x) = I(x/\beta, \alpha)$$

* K. Pearson, edit., "Tables of the Incomplete Γ-function" (London, H.M.S.O., 1922). Tables for integral values of 2α are much more widely available, namely as tables of the χ^2 distribution, which in turn, is closely related to the Poisson distribution (Chapter 10, § 12). See e.g. Table 7 in E.S. Pearson and H.O. Hartley, ed., Biometrika tables for statisticians, Vol. 1, Cambridge University Press 1954, where the mutual relationships are clearly set out.

for the distribution function. Using the tables cited in the footnote we can easily draw the curves of $F(x)$. In Fig. 96 we show the curves corresponding to the frequency curves shown in Fig. 95.

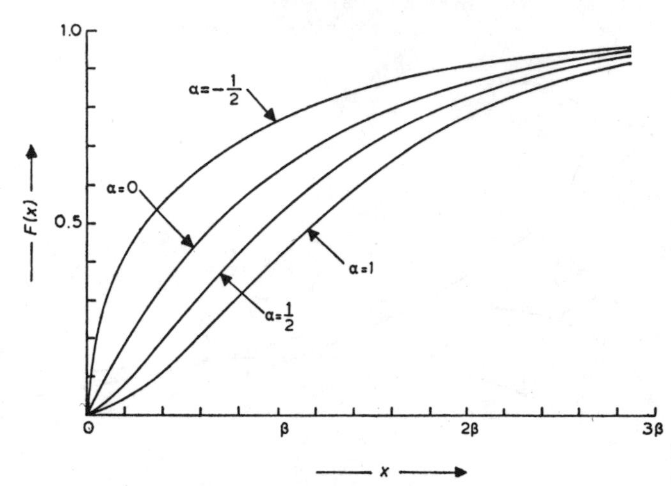

Fig. 96. Distribution curves for the gamma distribution.

The moments about the origin are, by definition,

$$\mu_r' = \frac{1}{\Gamma(\alpha+1)\beta^{\alpha+1}} \int_0^\infty x^{\alpha+r} e^{-x/\beta}\, dx.$$

Changing the variable of integration from x to t where $x = \beta t$ we see that

$$\int_0^\infty x^{\alpha+r} e^{-x/\beta}\, dx = \beta^{\alpha+1+r} \int_0^\infty t^{\alpha+r} e^{-t}\, dt = \beta^{\alpha+1+r}\Gamma(\alpha+r+1)$$

so that

$$\mu_r' = \frac{\Gamma(\alpha+r+1)}{\Gamma(\alpha+1)}\beta^r. \tag{33}$$

Using the fact (equation (4) of § 10) that $\Gamma(n) = (n-1)\Gamma(n-1)$ we see that

$$\mu_r' = (\alpha+r)(\alpha+r-1)\cdots(\alpha+1)\beta^r. \tag{34}$$

In particular

$$\mu_1' = (\alpha+1)\beta, \qquad \mu_2' = (\alpha+1)(\alpha+2)\beta^2. \tag{35}$$

From the relation (10) we find that

$$\mu_2 = \beta^2\{(\alpha+1)(\alpha+2)-(\alpha+1)^2\} = \alpha(\alpha+1)\beta^2. \tag{36}$$

The moment generating function is, by definition,

$$m(t) = \frac{1}{\Gamma(\alpha+1)\beta^{\alpha+1}} \int_0^\infty x^\alpha e^{-x(1/\beta - t)}\,dx. \tag{37}$$

Making the substitution

$$u = x(1/\beta - t) = x/\beta(1-\beta t)$$

we see that

$$\int_0^\infty x^\alpha e^{-x(1/\beta-t)}\,dx = \frac{\beta^{\alpha+1}}{(1-\beta t)^{\alpha+1}} \int_0^\infty u^\alpha e^{-u}\,du = \frac{\beta^{\alpha+1}\Gamma(\alpha+1)}{(1-\beta t)^{\alpha+1}}$$

and it follows from equation (37) that

$$m(t) = \frac{1}{(1-\beta t)^{\alpha+1}}. \tag{38}$$

(*d*) THE BETA DISTRIBUTION

The frequency function of the beta distribution is

$$f(x) = \begin{cases} 0, & x < 0; \\ \dfrac{\Gamma(\alpha+\beta+2)}{\Gamma(\alpha+1)\Gamma(\beta+1)} x^\alpha (1-x)^\beta, & 0 < x < 1; \\ 0, & x > 1; \end{cases} \tag{39}$$

where α and β are constants. To ensure that the integral in (2) exists we need to have $\alpha > -1$, $\beta > -1$. Differentiating this function, we find that

$$f'(x) = \frac{\Gamma(\alpha+\beta+2)}{\Gamma(\alpha+1)\Gamma(\beta+1)} x^{\alpha-1}(1-x)^{\beta-1}[\alpha-(\alpha+\beta)x]$$

so that $f'(x) = 0$ when

$$x = \frac{\alpha}{\alpha+\beta}.$$

If $\alpha < 0$ the curve is asymptotic to the y-axis and if $\beta < 0$ it is asymptotic to the ordinate $x = 1$. Some typical frequency curves are shown in Fig. 97.

The distribution function is given by

$$F(x) = \frac{\Gamma(\alpha+\beta+2)}{\Gamma(\alpha+1)\Gamma(\beta+1)} \int_0^x u^\alpha (1-u)^\beta\,du, \quad 0 < x < 1. \tag{40}$$

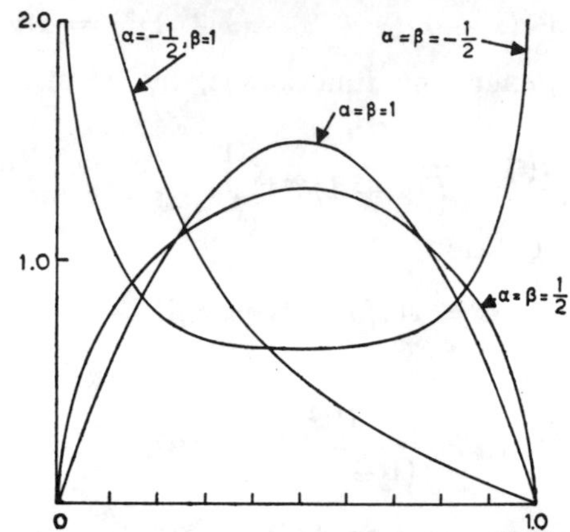

Fig. 97. Frequency curves for the beta distribution.

This can be evaluated in certain cases. For example, if $\alpha = \beta = 1$

$$f(x) = 6x(1 - x)$$

so that

$$F(x) = x^2(3 - 2x)$$

and if $\alpha = -\frac{1}{2}$, $\beta = 1$,

$$f(x) = \tfrac{3}{4}(x^{-\frac{1}{2}} - x^{\frac{1}{2}})$$

so that

$$F(x) = \tfrac{1}{2}x^{\frac{1}{2}}(3 - x).$$

In the general case the integral can be evaluated in terms of the incomplete beta function defined by the equation

$$B_x(\alpha, \beta) = \int_0^x u^{\alpha-1}(1 - u)^{\beta-1}\,du. \tag{41}$$

From equation (40) we see that

$$F(x) = I_x(\alpha + 1, \beta + 1) \tag{42}$$

where

$$I_x(p, q) = \frac{B_x(p, q)}{B(p, q)}. \tag{43}$$

Tables giving the values of $I_x(p, q)$ are available* to facilitate the numerical work.

* K. Pearson, ed., "Tables of the Incomplete Beta Function" (Cambridge University Press, London, 1932).

The r-th moment about the origin is

$$\mu_r' = \frac{\Gamma(\alpha+\beta+2)}{\Gamma(\alpha+1)\Gamma(\beta+1)} \int_0^1 x^{\alpha+r}(1-x)^\beta\,dx$$
$$= \frac{\Gamma(\alpha+\beta+2)}{\Gamma(\alpha+1)\Gamma(\beta+1)} \cdot \frac{\Gamma(\alpha+r+1)\Gamma(\beta+1)}{\Gamma(\alpha+\beta+r+2)} \tag{44}$$
$$= \frac{\Gamma(\alpha+r+1)}{\Gamma(\alpha+1)} \cdot \frac{\Gamma(\alpha+\beta+2)}{\Gamma(\alpha+\beta+r+2)}.$$

Putting $r = 1$ we find that the arithmetic mean is

$$\mu_1' = \frac{\alpha+1}{\alpha+\beta+2} \tag{45}$$

and putting $r = 2$, that

$$\mu_2' = \frac{(\alpha+1)(\alpha+2)}{(\alpha+\beta+2)(\alpha+\beta+3)}. \tag{46}$$

Hence the second moment about the arithmetic mean is (by equation (10) with $a = 0$)

$$\mu_2 = \frac{(\alpha+1)(\alpha+2)}{(\alpha+\beta+2)(\alpha+\beta+3)} - \frac{(\alpha+1)^2}{(\alpha+\beta+2)^2}$$

and a little algebraic manipulation shows that

$$\mu_2 = \frac{(\alpha+1)(\beta+1)}{(\alpha+\beta+2)^2(\alpha+\beta+3)}. \tag{47}$$

There is no simple expression for the moment generating function of this distribution.

14. Cumulants

For both theoretical and practical statistics, cumulants (the old name is seminvariants) are even more important than moments.

Suppose we have two statistical distributions, of x_1 and x_2, and we multiply the moment generating functions (of their frequency functions) $m_1(t)$ and $m_2(t)$, together. It can be show from (11) or (12) that $m_1(t)m_2(t)$ is the moment generating function for the distribution of $x_1 + x_2$. (This is equivalent mathematically to the convolution theorem in § 16, equation (20).)

The logarithm of the moment generating function is called the *cumulant generating function* $k(t)$, and the cumulants $k_1, k_2, k_3, \ldots k_n$ are defined to be the coefficients of $t^n/n!$ in $\log m(t)$.

Hence the cumulants for the distribution of $x_1 + x_2$ are generated by

$$\log\{m_1(t)m_2(t)\} = \log m_1(t) + \log m_2(t) = k_1(t) + k_2(t). \qquad (1)$$

The cumulants for the distribution of $x_1 + x_2$ are simply the sums of those for the separate distributions.*

Suppose the second distribution consists of just one constant a. Its moments are then $a, a^2, a^3, \ldots$; its moment generating function is

$$1 + at + a^2t^2/2! + \ldots = e^{at},$$

hence its cumulant generating function is at. So the cumulants of the distribution of $x_1 + a$ are: $k_1 + a$; k_2, k_3, $k_4, \ldots$ etc. Adding this constant has changed the mean and left the other cumulants unchanged.

It follows that we can obtain all the cumulants except the first one from (1) by putting the mean equal to 0, and expanding

$$\log(1 + \mu_2 t^2/2! + \mu_3 t^3/3! + \ldots). \qquad (2)$$

Hence the first four cumulants are

$$k_1 = \mu_1'; \quad k_2 = \mu_2; \quad k_3 = \mu_3; \quad k_4 = \mu_4 - 3\mu_2^2. \qquad (3)$$

The second cumulant is better known as the variance (the mean of the square of the deviations about the mean). Its additive properties give the name to one of the most important techniques of practical statistical analysis, the analysis of variance, which is described in any text book on the subject. If equation (1) refers to the *second* moments and cumulants the so-called "components of variance" are simply the two terms on the right.

For most important probability distributions, the cumulant generating function is very simple, and so the cumulants form a simple pattern. For the Poisson distribution (Chapter 10, § 12) they are all equal to the mean m. For the normal distribution

$$k(t) = mt + \tfrac{1}{2}\sigma^2 t^2. \qquad (4)$$

* Strictly we should add a second suffix throughout e.g. $k_{1,r}(t) + k_{2,r}(t)$, to indicate that (1) refers to one particular cumulant (the rth).

Hence all the cumulants beyond the second are zero. In the gamma distribution, from (38)

$$k(t) = -(\alpha + 1)\log(1 - \beta t) = (\alpha + 1)(\beta + \beta^2 t^2/2! + 2\beta^3 t^3/3! + \ldots) \quad (5)$$

So the first 4 cumulants are $\alpha + 1$ times β, β^2, $2\beta^3$ and $6\beta^4$ respectively.

15. Some Skew Probability Curves; Distributions of Time Intervals in Biology

We conclude this introduction to the mathematics of probability and statistics with some examples of skew probability curves and frequency distributions. These are observed especially often for time intervals in biology, e.g. between an event and a response to it, which cannot be negative.

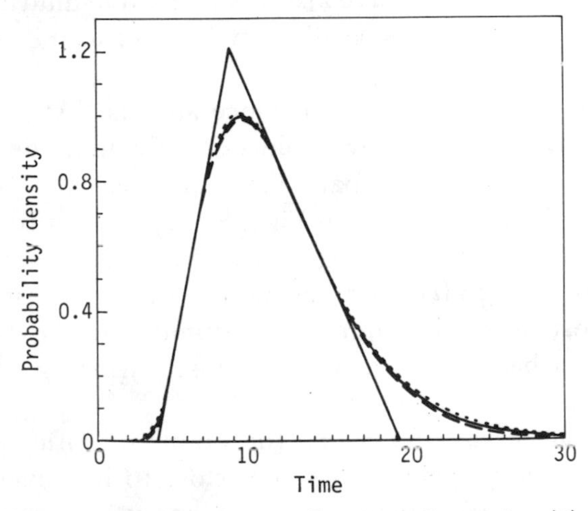

Fig. 98. Solid curve: the random walk curve of equation (6); dotted, the log normal curve defined by (2) and (3); dashed, the gamma curve defined by (4) and (5), with numerical parameters given in the text. All the curves have the same two tangents at their points of inflexion. From Wise, Acta Physiol. Pharmacol. Neerl. **14** (1966) 175–204.

Fig. 98 gives three of these distributions in one figure, and shows at the same time how surprisingly alike they can be, except for very long time intervals. The bottom curve is for a gamma distribution, as in § 13, eq. (30) but with a change of origin and scale. We shall explain in a moment why this particular curve was chosen.

The middle curve is for the distribution

$$y(t) = at^{-\frac{1}{2}} \exp[-\varphi(t/\mu + \mu/t)] \tag{1}$$

with $\mu = 1$, $\varphi = 3$. The area under the curve is unity when $a = e^{2\varphi}\varphi^{\frac{1}{2}}(\pi\mu)^{-\frac{1}{2}}$.

Hence the curve is the top one of Fig. 95, multiplied by a constant c and by $\exp(-\varphi\mu/t) = \exp(-3/t)$ in this case. Instead of going to infinity as $t \to 0$ from above it goes to zero as do all its derivatives. (This is a simple example of the limit theorem in Chapter 6, §3, eq. (6).) It arises in Einstein's classic model for Brownian movement and it has been accurately observed. If a cloud of particles start together, and move from left to right in a stream, spreading out by Brownian movement as they move, $y(t)$ is the density–time curve at a fixed point. This curve also fits many dye-dilution curves, (see Chapter 6, §12), and we shall meet it again in the section on diffusion, Chapter 9, §24c).

The curve has a single maximum. Since also its first derivative is zero at 0 and infinity, it has two points of inflexion, one in the rising part and one in the falling part, like a normal curve but unlike those shown in Figs. 95 and 97. In Fig. 98 the tangents to the curve at these points are shown (cf. Fig. 58, in particular the tangent at $x = 0$, $y = -10$). Where $y(t)$ is increasing, it is extremely close to this tangent for most of the way up to the maximum. In the descending limb $y(t)$ approaches this tangent more closely near the maximum, but departs from it sooner in the long tail.

In the normal curves of Fig. 91 the points of inflexion are at $x = m \pm \sigma$. Here too each curve is very close to its tangent at the corresponding point of inflexion. These tangents form a triangle with base extending from $m-2\sigma$ to $m + 2\sigma$, and with vertex exactly twice as high as the points of inflexion (Chapter 6, §4d, Problem 2). The other two curves in Fig. 98 have been chosen to have the same tangents at their points of inflexion, although those points are not the same. Through this little known device we see how we can find parameters for the top and bottom curves such that these are extremely close to the middle one, although their mathematical forms look rather different. We shall not give any of the derivations here. These sometimes involve tricky mathematical manipulations but the mathematics is not more advanced than anything in this book.

The top curve in fact is a well known one, the log normal curve, and this result helps to explain why log normal distributions so very often fit to biological data*. The true mechanism may lead to one of the other formulae, or to a number of others, but they all yield distribution curves that are nearly the same except for the very long time intervals.

The log normal curve is

$$y = y_m \, e^{-\frac{1}{2}\xi^2} \tag{2}$$

where

$$\xi = \{\log(t - t_0) - \log t_L\} / \sigma$$

or

$$t = t_0 + t_L \, e^{\sigma\xi}, \tag{3}$$

and for the top curve in Fig. 98 $t_0 = 0.04112$, $t_L = 0.87746$, $y_m = 1.20801$ and $\sigma = 0.42241$. The area under the curve is just greater than 1.

The gamma curve is

$$y = A e^{-\frac{1}{2}x} x^{\alpha} \tag{4}$$

where

$$x = x_0 + ct, \tag{5}$$

with $\alpha = 2.87609$, $x_0 = -2.41835$, $C = 8.82577$, $A = 0.11402$. Its area is just less than 1.

If φ in (1) is made much smaller, and σ in (2) much larger, the peaks of the curves become much sharper and the tails longer. Then the (even more steeply) rising parts remain close, but the long tails become more separate.

We shall mention one more skew distribution that is biologically important. This too, can be produced by a cloud of Brownian particles that all start together; it is the distribution of their first arrivals at a fixed point, whilst (1) includes particles that have been past this point and have returned. The distribution is

$$z(t) = At^{-3/2} \{\exp - \varphi(t/m + m/t)\}, \tag{6}$$

* i.e. besides when $\log t$ behaves like a normal variable, see any textbook explanation on how the normal error curve arises. See also A. Koch, J. Theor. Biol. 23 (1969) 251–268.

and the area under the curve is unity if

$$A = e^{2\varphi}(\lambda\mu)^{\frac{1}{2}}\pi^{-\frac{1}{2}}. \tag{7}$$

Apart from this constant, it is the same as (1) except for the power of t, $t^{-3/2}$ instead of $t^{-\frac{1}{2}}$.

The cumulant generating function is simple, namely

$$k(u) = 2\varphi\{1-(1-m/\varphi)^{\frac{1}{2}}\}, \tag{8}$$

and the first 4 cumulants are m, $\frac{1}{2}m^2/\varphi$, $3m^3/(4\varphi^2)$ and $15m^4/(8\varphi^3)$.

For (1) the cumulant generating function is

$$-\tfrac{1}{2}\log\left(1-\frac{u\mu}{\varphi}\right)+2\varphi-2\varphi(1-u\mu/\varphi)^{\frac{1}{2}}, \tag{9}$$

and the first 4 cumulants are:

$$\mu(1+1/2\varphi),\quad \mu^2\{1/(2\varphi)+1/(2\varphi^2)\},\quad \mu^3\left(\frac{1\cdot 3}{4\varphi^2}+\frac{1\cdot 2}{2\varphi^3}\right)$$

and

$$\mu^4\left(\frac{1\cdot 3\cdot 5}{8\varphi^3}+\frac{1\cdot 2\cdot 3}{2\varphi^4}\right).$$

As explained, we do not give anything on statistical *analyses* in this introduction. But all the distributions introduced are biologically important in themselves.

16. The Laplace Transform

The Laplace transform of a function (of t) is mathematically very like its moment generating function, at any rate when t is positive. However it arises from different problems and the notation is slightly different.

If $y(t)$ is a continuous function of the variable t, and if there exists a real constant c such that

$$\lim_{t\to\infty} e^{-ct}y(t)$$

is finite, then it can be shown that the integral

$$\bar{y}(p) = \int_0^\infty e^{-pt}y(t)\mathrm{d}t \tag{1}$$

exists for all values of the parameter p whose real part is greater than c. When the integral (1) does exist, we call $\bar{y}(p)$ the *Laplace transform* of the function $y(t)$, and we write

$$\bar{y}(p) = \mathscr{L}\{y(t)\}. \tag{2}$$

In the case in which $y(t)$ is a continuous function of t, it was shown by M. Lerch that there is no other continuous function which has $\bar{y}(p)$ as its Laplace transform.

LAPLACE TRANSFORMS OF SOME SIMPLE FUNCTIONS

We have collected together in Table 15 the Laplace transforms of some simple functions. We shall now indicate how these results are arrived at.

1. This follows from the fact that

$$\int_0^\infty e^{-pt} dt = \left[-\frac{1}{p} e^{-pt}\right]_0^\infty = \frac{1}{p}.$$

2. If $p > a$, then

$$\int_0^\infty e^{-pt} \cdot e^{at} dt = \int_0^\infty e^{-(p-a)t} dt = \frac{1}{p-a},$$

the condition $p > a$ being necessary for the exponential function to vanish at the upper limit.

3. This follows from the fact that

$$\int_0^\infty t^n e^{-pt} dt = p^{-n-1} \int_0^\infty u^n e^{-u} du = \Gamma(n+1) p^{-n-1},$$

where, in the second integral, we have put $u = pt$. The condition $n > -1$ is necessary if the gamma function is to exist.

4. This is an immediate consequence of 3. When n is an integer

$$\Gamma(n+1) = n!.$$

5, 6. These results have already been established in example 26 (§ 6) of this chapter.

7. We have

$$\int_0^\infty \cosh (at)\, e^{-pt} dt = \tfrac{1}{2} \int_0^\infty (e^{at} + e^{-at}) e^{-pt} dt$$

$$= \frac{1}{2}\left\{\frac{1}{p-a} + \frac{1}{p+a}\right\}$$

$$= \frac{p}{p^2 - a^2}.$$

TABLE 15

A short table of Laplace transforms

$$\bar{y}(p) = \mathscr{L}\{f(t)\} = \int_0^\infty e^{-pt} y(t) dt$$

	Function $y(t)$	Transform $\bar{y}(p) = \mathscr{L}\{y(t)\}$
1	1	$\dfrac{1}{p}$
2	e^{at}	$\dfrac{1}{p-a} \quad (p > a)$
3	$t^n \quad (n > -1)$	$p^{-n-1}\Gamma(n+1)$.
4	$t^n \quad (n = 1, 2, 3, \ldots)$	$p^{-n-1}\, n!$
5	$\cos at$	$\dfrac{p}{p^2+a^2}$
6	$\sin at$	$\dfrac{a}{p^2+a^2}$
7	$\cosh at$	$\dfrac{p}{p^2-a^2}$
8	$\sinh at$	$\dfrac{a}{p^2-a^2}$
9	$t \cos at$	$\dfrac{p^2-a^2}{(p^2+a^2)^2}$
10	$t \sin at$	$\dfrac{2ap}{(p^2+a^2)^2}$
11	$\cos at - \frac{1}{2}at \sin at$	$\dfrac{p^3}{(p^2+a^2)^2}$
12	$\frac{1}{2}\sin at - \frac{1}{2}at \cos at$	$\dfrac{a^3}{(p^2+a^2)^2}$
13	$H(t-k) = \begin{cases} 0 \text{ for } t < k; \\ 1 \text{ for } t > k > 0 \end{cases}$	$\dfrac{e^{-kp}}{p}$
14	$e^{-a/t}\sqrt{(a/\pi t^3)}$	$e^{-2\sqrt{ap}}$

8. Similarly

$$\int_0^\infty \sinh(at) e^{-pt}\, dt = \tfrac{1}{2}\int_0^\infty (e^{at} - e^{-at}) e^{-pt}\, dt$$

$$= \frac{1}{2}\left\{\frac{1}{p-a} - \frac{1}{p+a}\right\}$$

$$= \frac{a}{p^2-a^2}.$$

9. If we differentiate both sides of the equation

$$\int_0^\infty \cos(at)\, e^{-pt}\, dt = \frac{p}{p^2 + a^2} \tag{3}$$

with respect to p we find that

$$\int_0^\infty -t\cos(at)\, e^{-pt}\, dt = -\frac{p^2 - a^2}{(a^2 + p^2)^2}$$

showing that the entry 9 is correct.

Remark: The technique of differentiating under the integral sign with respect to a parameter (as here where p is considered as a variable) needs justification and this can usually be supplied when it is employed in deducing new Laplace transforms from known ones.

10. If we differentiate both sides of equation (3) with respect to a, we get the entry 10.

11, 12 follow from 9 and 10 by simple algebraic operations.

13. In entry 13 the function $H(t)$ is *Heaviside's unit function* defined by the equations

$$H(t) = \begin{cases} 0, & t < 0; \\ \frac{1}{2}, & t = 0; \\ 1, & t > 0. \end{cases} \tag{4}$$

It follows from these equations that

$$H(t-k) = \begin{cases} 0, & t < k; \\ \frac{1}{2}, & t = k; \\ 1, & t > k \end{cases} \tag{5}$$

so that the function $H(t-k)$ has the graph shown in Fig. 99.

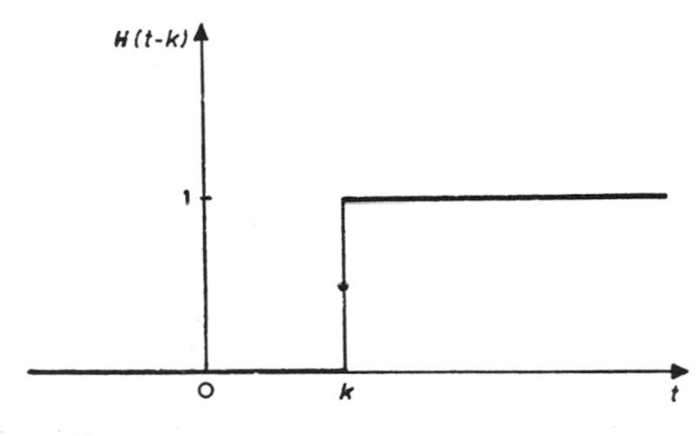

Fig. 99. Graph of the Heaviside function $H(t-k)$.

It follows immediately from this definition that

$$\mathscr{L}\{H(t-k)\} = \int_0^\infty e^{-pt} H(t-k)\mathrm{d}t = \int_k^\infty e^{-pt}\,\mathrm{d}t = \frac{e^{-kp}}{p}.$$

RULES FOR MANIPULATING LAPLACE TRANSFORMS

We shall now derive a few rules which are useful in the handling of Laplace transforms. It is useful to introduce the following notation: if $\bar{y}(p)$ is the Laplace transform of $y(t)$, we say that $y(t)$ is the *inverse transform* of $\bar{y}(p)$ and write

$$\mathscr{L}^{-1}\bar{y}(p) = y(t).$$

The symbol $\mathscr{L}^{-1}$ may therefore be taken to read "the function whose Laplace transform is". It is readily shown from the definition of a Laplace transform that

$$\mathscr{L}\{ay_1(t) + by_2(t)\} = a\bar{y}_1(p) + b\bar{y}_2(p).$$

Similarly from the definition we have that

$$\mathscr{L}\{y(at)\} = \int_0^\infty e^{-pt} y(at)\mathrm{d}t = \frac{1}{a}\int_0^\infty e^{-(p/a)u} y(u)\mathrm{d}u$$

showing that

$$y(at) = \frac{1}{a}\bar{y}\left(\frac{p}{a}\right), \tag{6}$$

which can also be written in the form

$$\mathscr{L}^{-1}\{\bar{y}(p/a)\} = ay(at). \tag{7}$$

If we replace $y(t)$ by $e^{-bt}y(t)$ in the definition (1) we find that

$$\mathscr{L}\{e^{-bt}y(t)\} = \int_0^\infty e^{-(p+b)t} y(t)\mathrm{d}t$$

and this integral is just the definition of a Laplace transform with p replaced by $p + b$. In other words we have shown that

$$\mathscr{L}\{e^{-bt}y(t)\} = \bar{y}(p + b), \tag{8}$$

or

$$\mathscr{L}^{-1}\{\bar{y}(p + b)\} = e^{-bt}y(t). \tag{9}$$

Using these simple rules we can extend the range of our Table 15. For instance if we apply the rule (9) to entry 1 of Table 15 we find that

$$\mathscr{L}^{-1}\left\{\frac{1}{(p+b)^{n+1}}\right\} = \frac{1}{\Gamma(n+1)} t^n e^{-bt}. \tag{10}$$

Similarly from entries 5 and 6 of Table 15 and equation (9) we can deduce that

$$\mathscr{L}^{-1}\left\{\frac{p+b}{(p+b)^2+a^2}\right\}=\mathrm{e}^{-bt}\cos at,\quad \mathscr{L}^{-1}\left\{\frac{a}{(p+b)^2+a^2}\right\}=\mathrm{e}^{-bt}\sin at. \qquad (11)$$

We can make use of these results with the method of partial fractions to find the inverse transforms of more complicated functions. To show how this can be achieved we have:-

EXAMPLE 36. *Find the functions of which*

$$\frac{p}{p^3+p^2-2} \quad \textit{and} \quad \frac{1}{p^3+p^2-2}$$

are the Laplace transforms.

We can factorize the denominator of these expressions as follows:-

$$p^3+p^2-2=(p-1)(p^2+2p+2)=(p-1)[(p+1)^2+1].$$

Now by the method of partial fractions we can split the complicated function up into two simpler ones. We have

$$\frac{A}{p-1}+\frac{B(p+1)+C}{p^2+2p+2}=\frac{(A+B)p^2+(2A+C)p+(2A-B-C)}{p^2+2p+2}.$$

The numerator will be equal to p if we take

$$A+B=0,\quad 2A+C=1,\quad 2A-B-C=0,$$

i.e. if we take

$$A=\tfrac{1}{5},\quad B=-\tfrac{1}{5},\quad C=\tfrac{3}{5}.$$

Hence we have

$$\frac{p}{p^3+p^2-2}=\frac{1}{5}\left\{\frac{1}{p-1}-\frac{p+1}{(p+1)^2+1}+\frac{3}{(p+1)^2+1}\right\}.$$

From entry 2 of Table 15 and the equations (11) we deduce that

$$\mathscr{L}^{-1}\left\{\frac{p}{p^3+p^2-2}\right\}=\tfrac{1}{5}\{\mathrm{e}^t-(\cos t-3\sin t)\mathrm{e}^{-t}\}.$$

If, on the other hand, we put

$$A+B=0,\quad 2A+C=0,\quad 2A-B-C=1,$$

that is,

$$A=\tfrac{1}{5},\quad B=-\tfrac{1}{5},\quad C=-\tfrac{2}{5},$$

we find that

$$\frac{1}{p^3 + p^2 - 2} = \frac{1}{5}\left\{\frac{1}{p-1} - \frac{p+1}{(p+1)^2+1} - \frac{2}{(p+1)^2+1}\right\}$$

from which we deduce that

$$\mathscr{L}^{-1}\left\{\frac{1}{p^3+p^2-2}\right\} = \tfrac{1}{5}\{e^t - (\cos t + 2\sin t)e^{-t}\}.$$

We next consider a set of results which are of great use in the application of the theory of Laplace transforms to the solution of differential equations (See § 12 of Chapter 9). We have, in fact, established the basic results in Problem 2 of § 6 above, but they are so important that we shall derive them again here. For convenience, we shall denote dy/dt, d^2y/dt^2, d^3y/dt^3 and d^4y/dt^4 by $\dot{y}$, $\ddot{y}$, $\dddot{y}$ and y^{iv} respectively. If we apply the theorem for integrating by parts we find that

$$\mathscr{L}\{\dot{y}\} = \int_0^\infty e^{-pt}\dot{y}(t)dt = [ye^{-pt}]_0^\infty + p\int_0^\infty e^{-pt}y(t)dt.$$

If we assume that $y(t)$ is such that $e^{-pt}y(t) \to 0$ as $t \to \infty$, then we may write this equation in the form

$$\mathscr{L}\{\dot{y}\} = -y(0) + p\bar{y}(p). \tag{12}$$

Now, if we use the fact that $\ddot{y}$ is the derivative of $\dot{y}$, we see that equation (12) implies that

$$\mathscr{L}\{\ddot{y}\} = -\dot{y}(0) + p\mathscr{L}\{\dot{y}\}. \tag{13}$$

Substituting from equation (12) into equation (13) we find that

$$\mathscr{L}\{\ddot{y}\} = -py(0) - \dot{y}(0) + p^2\bar{y}(p). \tag{14}$$

Corresponding to the equation (13) we have

$$\mathscr{L}\{\dddot{y}\} = -\ddot{y}(0) + p\mathscr{L}\{\ddot{y}\}. \tag{15}$$

By virtue of this equation and (14) we find that

$$\mathscr{L}\{\dddot{y}\} = -p^2y(0) - p\dot{y}(0) - \ddot{y}(0) + p^3\bar{y}(p). \tag{16}$$

In a similar way we can show that

$$\mathscr{L}\{y^{iv}\} = -p^3y(0) - p^2\dot{y}(0) - p\ddot{y}(0) - \dddot{y}(0) + p^4\bar{y}(p). \tag{17}$$

We shall now state formally a rule which we have already used to derive entry 9 of Table 15 from entry 5. If we differentiate both sides of equation (1) with respect to p, we have

$$\frac{d\bar{y}}{dp} = \int_0^\infty (-t)e^{-pt}y(t)dt,$$

a result which can be written in the form

$$\mathscr{L}\{ty(t)\} = -\frac{d\bar{y}}{dp}. \tag{18}$$

This process can be repeated any (finite) number of times to give

$$\mathscr{L}\{t^n y(t)\} = (-1)^n \frac{d^n \bar{y}}{dp^n}. \tag{19}$$

Finally, we shall state a result which we shall not attempt to prove. This is the *convolution theorem* which states that, if $\bar{f}(p)$ and $\bar{g}(p)$ are, respectively, the Laplace transforms of $f(t)$ and $g(t)$, then

$$\mathscr{L}^{-1}\{\bar{f}(p)\,\bar{g}(p)\} = \int_0^t f(u)\,g(t-u)du. \tag{20}$$

For example, if we make use of equation (11), we find that

$$\mathscr{L}^{-1}\left\{\frac{\bar{f}(p)}{(p+b)^2+a^2}\right\} = \frac{1}{a}\int_0^t f(u)e^{-b(t-u)}\sin a(t-u)du. \tag{21}$$

PROBLEMS

1. Show that if

$$\bar{y}(p) = \frac{1}{(p-a)(p-b)}, \text{ then } y(t) = \frac{e^{at}-e^{bt}}{a-b}.$$

2. If α, β, and γ are real constants such that $\alpha\gamma > \beta^2$, show that

$$\mathscr{L}^{-1}\left\{\frac{1}{\alpha p^2+2\beta p+\gamma}\right\} = \frac{1}{\alpha\omega}e^{-\beta t/\alpha}\sin \omega t,$$

$$\mathscr{L}^{-1}\left\{\frac{p}{\alpha p^2+2\beta p+\gamma}\right\} = \left\{\frac{1}{\alpha}\cos \omega t - \frac{\beta}{\alpha^2\omega}\sin \omega t\right\}e^{-\beta t/\alpha}$$

where $\omega^2 = (\alpha\gamma-\beta^2)/\alpha^2$.

3. Show that

$$\mathscr{L}^{-1}\left\{\frac{p^2+2c^2}{p^4+4p^2c^2+3c^4}\right\} = \frac{1}{2c}\left\{\sin(ct) + \frac{1}{\sqrt{3}}\sin(\sqrt{3}ct)\right\}$$

$$\mathscr{L}^{-1}\left\{\frac{1}{p^4+4p^2c^2+3c^4}\right\} = \frac{1}{2c^3}\left\{\sin(ct) - \frac{1}{\sqrt{3}}\sin(\sqrt{3}ct)\right\}.$$

4. Show that

$$\mathscr{L}^{-1}\left\{\frac{5+2p}{2p^2[(p+1)^2+4]}\right\} = \tfrac{1}{2}t - \tfrac{1}{4}\sin(2t)e^{-t}.$$

5. Taking $f(t) = t^{m-1}$, $g(t) = t^{n-1}$ in equation (20) show that

$$\int_0^t u^{m-1}(t-u)^{n-1}du = \Gamma(m)\Gamma(n)\mathscr{L}^{-1}(p^{-m-n}) = \frac{\Gamma(m)\Gamma(n)}{\Gamma(m+n)}t^{m+n-1}.$$

Deduce that*

$$B(m, n) = \frac{\Gamma(m)\Gamma(n)}{\Gamma(m+n)}.$$

17. The Use of Tables of Integrals

There are in existence several tables of integrals which facilitate the evaluation of indefinite and definite integrals. It is not necessary for a student to know how all the entries in such tables have been arrived at — just as, at a more elementary level, a student may use tables of logarithms without knowing how the values in these tables were calculated. All that he must learn is how to use such tables correctly and effectively.

A useful set of tables of integrals is contained in:-

H. B. Dwight, "Tables of Integrals and Other Mathematical Data", 2nd. edition, (Macmillan, New York, 1947).

Any reader who is contemplating doing a lot of routine integrations is recommended to buy this book. The only way to become proficient in the use of tables of integrals is to do a great many calculations with their use. Nothing can take the place of these very necessary exercises. The following two examples are offered merely as a guide to the procedure involved.

EXAMPLE 37. *Find the value of*

$$\int_0^{\frac{1}{2}} (1-4x^2)^{\frac{5}{2}}\,dx.$$

If we look up Dwight's "Tables of Integrals" we find that, in the section on irrational algebraic functions, there are listed integrals

* We said in § 12 that the proof of this result depended on double integration. There is no contradiction here. The proof of (2) also involves the evaluation of a double integral.

involving the quantity

$$t = (a^2 - x^2)^{\frac{1}{2}}.$$

Now we can write

$$(1 - 4x^2)^{\frac{5}{2}} = (4)^{\frac{5}{2}}(\tfrac{1}{4} - x^2)^{\frac{5}{2}} = 32(a^2 - x^2)^{\frac{5}{2}} = 32t^5,$$

where $a^2 = \frac{1}{4}$, $t = (\frac{1}{4} - x^2)^{\frac{1}{2}}$. Now the integral 350.05 listed by Dwight (on p. 67 of his book) is

$$\int t^5 \,dx = \tfrac{1}{6}xt^5 + \tfrac{5}{24}a^2xt^3 + \tfrac{5}{16}a^4xt + \tfrac{5}{16}a^6 \sin^{-1}(x/a),$$

so that

$$\int_0^{x_1} t^5 \,dx = \tfrac{1}{6}x_1t_1^5 + \tfrac{5}{24}a^2x_1t_1^3 + \tfrac{5}{16}a^4x_1t_1 + \tfrac{5}{16}a^6 \sin^{-1}(x_1/a),$$

where $t_1 = (a^2 - x_1^2)^{\frac{1}{2}}$. In our case $a = \frac{1}{2}$, $x_1 = \frac{1}{4}$, $t_1 = \frac{1}{4}\sqrt{3}$ so that we have

$$\int_0^{\frac{1}{4}} t^5 \,dx = \frac{1}{6}\cdot\frac{1}{4}\cdot\frac{9\sqrt{3}}{1024} + \frac{5}{24}\cdot\frac{1}{4}\cdot\frac{1}{4}\cdot\frac{3\sqrt{3}}{64} + \frac{5}{16}\cdot\frac{1}{16}\cdot\frac{1}{4}\cdot\frac{\sqrt{3}}{4} + \frac{5}{16}\cdot\frac{1}{64}\sin^{-1}(\tfrac{1}{2}).$$

Using the fact that $\sin^{-1}(\frac{1}{2}) = \frac{1}{6}\pi$, we find that

$$\int_0^{\frac{1}{4}} (1 - 4x^2)^{\frac{5}{2}} \,dx = \frac{3\sqrt{3}}{256} + \frac{5\sqrt{3}}{256} + \frac{5\sqrt{3}}{128} + \frac{5}{32}\frac{\pi}{6},$$

giving finally

$$\int_0^{\frac{1}{4}} (1 - 4x^2)^{\frac{5}{2}} \,dx = \frac{9\sqrt{3}}{128} + \frac{5\pi}{192}.$$

In many cases it is not possible to find an analytical expression for the integral in question, but often tables of numerical values of a few standard forms of integral are available and our problem reduces to showing that the integral we have to evaluate can be put into one of these forms. We have already encountered this procedure in dealing with integrals which can be reduced to a gamma function. To illustrate it further we have:-

EXAMPLE 38. *Find the value of the integral*

$$\int_0^{\frac{1}{2}\pi} \frac{d\phi}{\sqrt{(4 - \sin^2\phi)}}.$$

On p. 171 of Dwight's tables, we find that integrals of the type

$$K = \int_0^{\frac{1}{2}\pi} \frac{d\phi}{\sqrt{(1 - \sin^2\theta \sin^2\phi)}}, \tag{1}$$

where θ is a constant, can be evaluated only in the form of infinite series of powers of $\sin\theta$, but that numerical tables are available (Dwight, pp. 234—235) for their evaluation. Now we can write

$$\int_0^{\frac{1}{2}\pi} \frac{d\phi}{\sqrt{(4 - \sin^2\phi)}} = \frac{1}{2}\int_0^{\frac{1}{2}\pi} \frac{d\phi}{(1 - \frac{1}{4}\sin^2\phi)} = \frac{1}{2}\int_0^{\frac{1}{2}\pi} \frac{d\phi}{(1 - \sin^2 30^\circ \sin^2\phi)}$$

From Table 1040 on p. 234 of Dwight's book, we find that when $\phi = 30^\circ$, K has the value 1.686, so that we have, finally,

$$\int_0^{\frac{1}{2}\pi} \frac{d\phi}{\sqrt{(4 - \sin^2\phi)}} = 0.843.$$

Integrals of the type K defined by equation (1) are known as *complete elliptic integrals of the first kind*. Similarly integrals of the type

$$E = \int_0^{\frac{1}{2}\pi} \sqrt{(1 - \sin^2\theta \sin^2\phi)}d\phi \tag{2}$$

(values of which are given in Table 1041 of Dwight's book) are known as *complete elliptic integrals of the second kind.*

18. Applications

An equation involving one or more derivatives is called a differential equation.

Although differential equations are discussed in Chapter 9, we shall now illustrate the use of some simple types of differential equations.

(a) DECAY OF A RADIOACTIVE ISOTOPE

We shall consider the decay of radioactive iodine as a function of time.

The rate at which ^{131}I (and any other radioactive substance) decomposes at any instant, is proportional to the total amount of this substance that is present at that instant. This statement may be written symbolically

$$dm/dt = -km \tag{1}$$

where m = quantity of ^{131}I at any instant, k = (positive) constant. The negative sign denotes the decrease of m; t = time. The solution of this differential equation is found by integration, after rearranging as follows

$$\mathrm{d}m/m = -k\mathrm{d}t. \tag{2}$$

This rearrangement is called "*separating the variables*". Integrating (2) yields:

$$\log m = -kt + C \tag{3}$$

or $\log m - C = -kt$.

Let us put $C = \log c$ (this simplifies the formula), then

$$\log m - \log c = -kt \quad \text{or} \quad \log m/c = -kt,$$

or $m/c = \mathrm{e}^{-kt}$ or,

$$m = c\mathrm{e}^{-kt}.$$

The physical meaning of c is found by noting that at $t = 0$ $m_0 = c$, hence c is the amount m_0 at $t = 0$.

Hence we write

$$m = m_0\,\mathrm{e}^{-kt}. \tag{4}$$

This result means that the substance decreases exponentially with time.

The term "half-value period" will be familiar to many readers and its exact significance will now be discussed. Briefly this *half-value period* means the *time* after which the preparation will have diminished to half its original amount. Let us denote this time by $\tau_{1/2}$ then from (4) we have

$$\tfrac{1}{2}m_0 = m_0\,\mathrm{e}^{-k\tau_{1/2}} \text{ or } \log\tfrac{1}{2} = -k\tau_{1/2} \text{ or } \frac{-\log\frac{1}{2}}{k} = \tau_{1/2} \text{ or } \tau_{1/2} = \frac{\log 2}{k}.$$

The treatment of this physical problem is justified by the fact that radioactive substances have become an important tool both in medicine and biology.

The activity of the radioactive isotope is often denoted by N, so that the above formula may be written

$$N = N_0\,\mathrm{e}^{-0.693t/\tau_{1/2}}$$

where
N = activity at time t
N_0 = activity at $t = 0$
$\tau_{1/2}$ = half-value period.

Starting from N_0 radioactive nuclei, the number of radioactive nuclei at 1 half-value period will be $\frac{1}{2}N_0$, after 2 half-value periods $\frac{1}{2} \cdot \frac{1}{2}N_0 = \frac{1}{4}N_0$ etc.

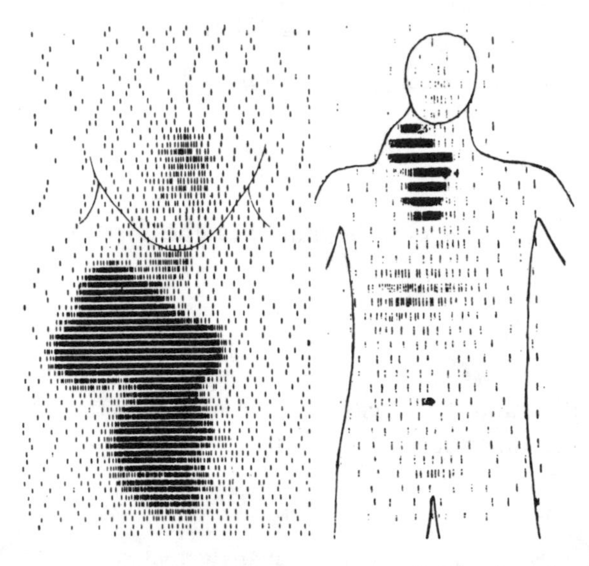

Fig. 100. Whole body scan (right) and area scan (left) of patient with intrathoracic goitre made after administration of ^{131}I. Uptake in liver due to metabolization of endogenously labelled thyroid hormone. (Tracer log no. 95.)

In Plate I, facing p. 432, an apparatus (scintillation counter) is shown which automatically measures the ^{131}I distribution in the thyroid gland. In Figs. 100 and 101 diagrams obtained with this method are shown: the method is of great diagnostic value

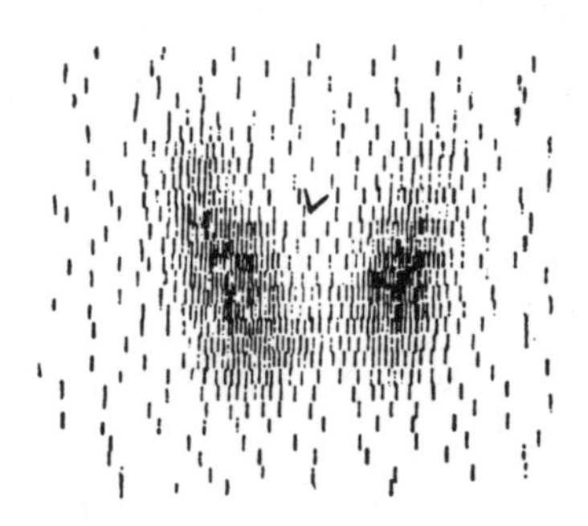

Fig. 101. Area scan of thyroid gland after administration of ^{131}I. (Tracer log no. 96.)

Another important clinical method involving the use of radioisotopes is the radioactive ^{51}Cr method for the determination of total blood volume.*

(b) BALLISTOCARDIOGRAPHY

When a subject lies quietly on a "swing", or on a table mounted on flat springs in such a manner that only motion in the long axis of the table is possible, small vibrations of the system may be observed.

These motions are associated with the "ballistic recoil" of the body when the blood is ejected from the heart (most of the blood, of course, then goes in the direction of the legs).

Ballistocardiography, the recording of these oscillatory motions, is of considerable clinical interest, since it yields qualitative information concerning the mechanical behaviour of the circulation. In a first approximation the phenomenon is described by a linear second order differential equation with constant coefficients (see Chapter 9), or

$$M\frac{\mathrm{d}^2x}{\mathrm{d}t^2}+\beta\frac{\mathrm{d}x}{\mathrm{d}t}+Dx=F \tag{1}$$

where M = mass of patient+"swing", x is displacement of the system relative to the surroundings, $\beta\mathrm{d}x/\mathrm{d}t$ is the frictional force (caused by a damping), and Dx is the restoring force driving the swing towards its equilibrium position, F is the driving force of the system.**

$M\mathrm{d}^2x/\mathrm{d}t^2$ is the Newtonian force.

The restoring force Dx is associated with the springs or the suspensions of the swing. The friction force $\beta\mathrm{d}x/\mathrm{d}t$ is caused by air resistance etc., or some additional damping device.

The driving force F (when a boy is pushing a swing, the boy's action constitutes a driving force) is caused by the propulsion of blood in the body: when m is mass and y is displacement of the centre of gravity of the blood ejected, $m\mathrm{d}^2y/\mathrm{d}t^2$ is the internal reaction force of that

* Their use has increased explosively since our first edition appeared. See, e.g. references 1 and 2 for Chapter 10, § 16, page 631. These books of symposia contain about 500 and 900 pages respectively; no. 1 contains more theory and no. 2 mainly clinical applications, but almost all the contributions involve biomathematics and models.

** The physical meaning of eq. (1) will be better understood after reading § 1 in Chapter 10, where the concept "forcing term" is discussed.

blood on the body: i.e. $F = m\mathrm{d}^2y/\mathrm{d}t^2$ so that we may write

$$M\frac{\mathrm{d}^2x}{\mathrm{d}t^2} + \beta\frac{\mathrm{d}x}{\mathrm{d}t} + Dx = m\frac{\mathrm{d}^2y}{\mathrm{d}t^2}. \tag{2}$$

What precisely is measured, when the displacement x of the table is recorded, depends on the relative magnitudes of M, β and D.

a) When the binding force Dx of the system to its surroundings is very strong (i.e. M and β are negligible relative to D) equation (2) reduces to

$$Dx = m\frac{\mathrm{d}^2y}{\mathrm{d}t^2} \tag{3}$$

so that the recorded deviation x gives a picture of the internal force: the internal force $m\mathrm{d}^2y/\mathrm{d}t^2$ is proportional to the displacement x.

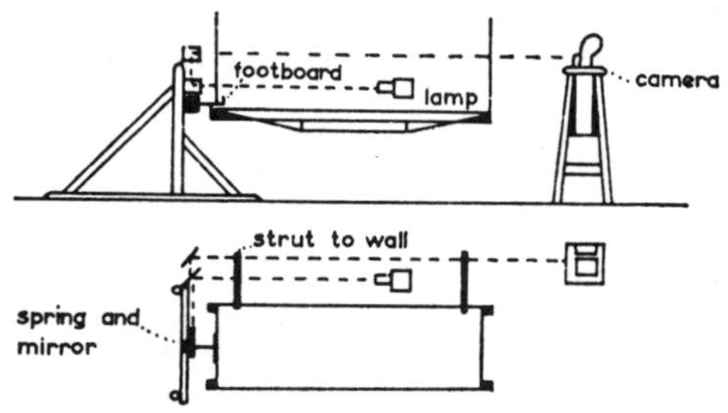

Fig. 102. The original high-frequency apparatus of Starr from the side and from above.

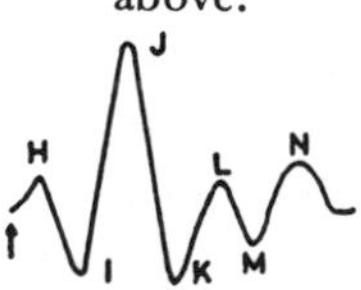

Fig. 103. Diagram of the commonest form of the normal ballistocardiogram obtained from Starr's high-frequency table. The arrow indicates the beginning of ventricular systole. The letters indicate the peaks.

b) When the damping and restoring forces are both small (2) reduces to

$$M\frac{\mathrm{d}^2x}{\mathrm{d}t^2} = m\frac{\mathrm{d}^2y}{\mathrm{d}t^2}. \tag{4}$$

Integrating equation (4) twice we obtain

$$Mx = my \quad \text{or} \quad x = my/M. \tag{5}$$

In this case the constants of integration are zero for physical reasons.

From this it follows that the recorded displacement is proportional to the displacement of the center of gravity of the blood, y, within the body.

For a thorough mathematical analysis of this problem the reader is referred to the papers cited at the end of this chapter, page 363.

In Fig. 102 the original high-frequency ballistocardiograph of Starr is shown. In Fig. 103, a record is shown.

(*c*) TRYPSIN DIGESTION

We are all more or less familiar with the law of Guldberg and Waage for chemical reactions, which states that the reaction velocity v is proportional to the product of the concentrations of the reacting substances. For example, if we have the reaction

$$\mathrm{HCl} + \mathrm{NaOH} \rightarrow \mathrm{NaCl} + \mathrm{H_2O},$$

the reaction velocity of the reaction from the left to the right is

$$v = k[\mathrm{HCl}][\mathrm{NaOH}]$$

where the square brackets denote concentrations. In the case of a so-called mono-molecular reaction, for example,

$$\mathrm{PCl_5} \rightarrow \mathrm{PCl_3} + \mathrm{Cl_2},$$

is obviously

$$v = k[\mathrm{PCl_5}].$$

It is equally clear that in this case

$$v = -\frac{\mathrm{d}[\mathrm{PCl_5}]}{\mathrm{d}t};$$

in words: v is equal to the decrease in concentration per unit time.

We shall now treat the case of hydrolysis of a protein by the trypsin. The reaction takes place between the enzyme, the water and the protein (e.g. caseine), or,

$$S + E + \mathrm{H_2O} \rightarrow \text{reaction products},$$

where S is the substrate (in our case protein), E is the enzyme. Applying the law of mass action, we have

$$v = k[S][E][\mathrm{H_2O}] \tag{1}$$

where v is the reaction velocity and k is a constant. Since water is present in abundance its concentration remains virtually constant and hence $[H_2O]$ may be included in the constant. In the living organism, the enzyme concentration does not remain constant during the reaction but decreases as a result of inhibition by combination with the reaction products and temperature effects.

In an experimental set-up, however, the enzyme concentration may be kept "constant" by using a large amount of enzyme and by working at a low temperature. In this case, equation (1) simplifies to

$$v = K[S], \tag{2}$$

where $K = k[E][H_2O]$. This case clearly resembles a monomolecular reaction, and is called a *pseudo-monomolecular reaction*. Since the velocity may in this case be defined as the decrease of substrate per unit time, we write

$$v = -\frac{\mathrm{d}[S]}{\mathrm{d}t}. \tag{3}$$

Hence from (2), we have

$$-\frac{\mathrm{d}[S]}{\mathrm{d}t} = K[S]. \tag{4}$$

Separating the variables and integrating we find that

$$\log[S] = -Kt + C, \tag{5}$$

where C is a constant of integration. It is convenient to write $C = \log c$. Substituting this form into equation (5) and rearranging we find that

$$\log\{[S]/c\} = -Kt,$$

or

$$[S] = c\mathrm{e}^{-Kt}. \tag{6}$$

The physical meaning of c becomes clear by noting that, when $t = 0$, $[S] = c$. Thus c is the concentration at time $t = 0$, which we may write as $[S]_0$. Hence equation (6) may be written in the form

$$[S] = [S]_0\,\mathrm{e}^{-Kt}. \tag{7}$$

The time course of the hydrolysis of protein under the (above) experimental conditions is predicted by equation (7). How well does this formula agree with the experimental findings? Northrop[1] has used the following approach. We may write equation (7) in the form

$$\frac{1}{t}\log\frac{[S]_0}{[S]} = -K, \tag{8}$$

from which it follows that, if we measure $[S]$ and $[t]$ simultaneously, then the calculated quantity $t^{-1}\log [S]_0/[S]$ should be constant whatever the value of t. Table 16, which shows some of Northrop's results, indicates that this in fact is the case.

TABLE 16

The hydrolysis of casein*

$[S]_0 = 1$ % casein		$[S]_0 = 3$ % casein	
t (in hours)	$t^{-1}\log [S]_0/[S]$	t (in hours)	$t^{-1}\log [S]_0/[S]$
0.11	0.235	0.20	0.39
0.215	0.236	0.40	0.39
0.53	0.242	0.80	0.36
1.00	0.255	1.60	0.37
2.51	0.238	3.40	0.35
4.00	0.238		

References for Chapter 7, § 18a, b, c

Section 18a

See also the beginning of § 16 in Chapter 10, and its references page 631, especially no. 2; nos. 7 and 8 involve ^{131}I.

Section 18b

1. J.L. Nickerson and H.J. Curtis, Amer. J. Physiol. **142** (1944) 1.
2. J.L. Nickerson and A.J.L. Mathers, Amer. Heart J. **47** (1954) 1.
3. A. Noordergraaf, Physical basis of ballistocardiography, Thesis (Utrecht, Holland, 1956).
4. H.C. Burger and A. Noordergraaf, Amer. Heart J. **56** (1956) 179.

Section 18c

1. J.H. Northrop, J. General Physiol. **6** (1924) 417.

* Modified and abbreviated from Northrop, Op. cit.

CHAPTER 8

FUNCTIONS OF MORE THAN ONE VARIABLE *

1. Functions of Several Variables

In the previous chapters of this book we have been considering functions of a single variable. In a great many physical situations however, the quantity in which we are interested depends not on one independent variable but on several.

The simplest case of such functional dependence occurs in elementary mathematics: we know that the area A of a rectangle depends both on the length x of the rectangle and on its breadth y. In fact

$$A = xy \tag{1}$$

so that, if x or y varies, A will vary too. Also, by the nature of things, it is possible to vary x and y independently. We say then that A is a *function* of the two independent variables x and y.

Another simple example, which arises in medicine, occurs in the direct Fick method of determining the cardiac output. If we denote the cardiac output by u, the CO_2 output by x, the arteriovenous CO_2 difference in volume per cent by y, then it is well known that

$$u = 100x/y. \tag{2}$$

The quantity u therefore depends upon both x and y which may vary independently.

In general, if a quantity f is defined for values of the independent variables x and y in a prescribed set we say that f is a *function* of x and y and we write it in the form $f(x, y)$. If we denote the pair of values (x, y) by a point in the xy-plane, then, as x and y assume values within the specified set, the point will fall within a certain part of the xy-plane called the *region* or *domain* of definition of the function.

For example, if the independent variables are restricted by the inequalities $a \leqq x \leqq b$, $c \leqq y \leqq d$, the domain D will be a rectangle

* Since the ideas of this chapter are used only slightly in the theory of ordinary differential equations, the reader may postpone reading it until he is familiar with the contents of Chapter 9 (except Sections 5, 6, 16 and further).

(cf. Fig. 104), while if they satisfy $x \geqq 0$, $y \geqq 0$, the corresponding domain will be the positive quadrant of the xy-plane.

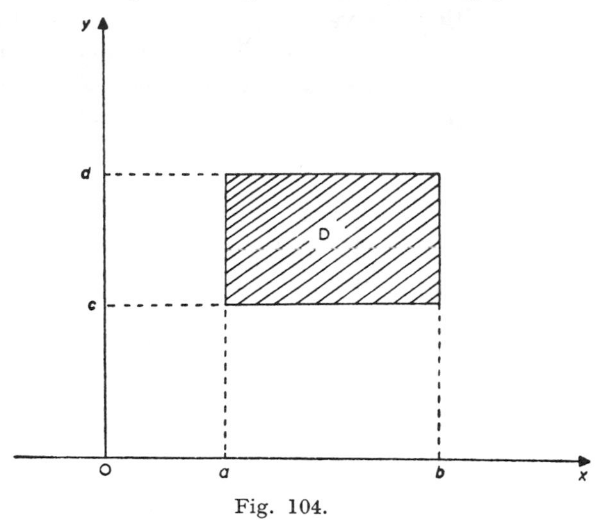

Fig. 104.

We can then define a function formally:

If D is a domain in which x and y may vary independently, and if, according to some law, a unique value f is assigned to each point of this region, then f is said to be a single-valued function of the two variables x and y. It will be observed that in this definition a functional relation associates a *unique* value of f with each pair of values (x, y) of the independent variables.

There are physical quantities which depend on more than two independent variables. For instance, if the edges of a rectangular box have lengths x, y and z, the volume of the box, V, is defined by the formula

$$V = xyz$$

and its surface area, S, by

$$S = 2(yz + zx + xy),$$

and, by their nature, the lengths x, y and z can be varied independently of one another. A trio of values (x, y, z) of the independent variables can be represented by a point P in the three-dimensional xyz-space by the simple device of fixing three mutually perpendicular axes and constructing (by analogy with the plane case) a point P by taking

$OP_1 = x$, $OP_2 = y$, and so constructing P_3 in the xy-plane; then taking P_3P perpendicular to that plane and of length z, we arrive at the point P representing the values (x, y, z). The limitations on the values of x, y and z define a domain which in this case is a solid. For example, if x, y, z are restricted by the inequalities $0 \leqq x \leqq a$, $0 \leqq y \leqq a$, $0 \leqq z \leqq a$, the domain is a cube of edge a.

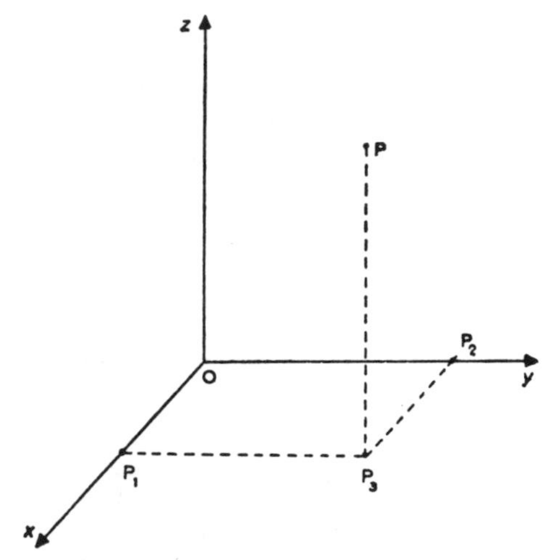

Fig. 105. Cartesian axes in three dimensions.

These concepts can be generalized readily to the case of n independent variables which we may denote by $x_1, x_2, \ldots, x_n$. In that case we do not have a simple geometrical analogue to guide us but we still refer to the set of values $(x_1, x_2, \ldots, x_n)$ of the independent variables as a 'point' and we talk of the 'domain' of the function when we mean the whole set of values which are allowed to the independent variables. Just as in the case of two independent variables we can then define a function of n independent variables as follows:

If D is a domain in which $x_1, x_2, \ldots, x_n$ may vary independently and if a unique value of f is assigned to each point of this region, we say that f is a single-valued function of the n independent variables $x_1, x_2, \ldots, x_n$ and we denote it by the symbol $f(x_1, x_2, \ldots, x_n)$.

In this chapter we shall be concerned mainly with the properties of functions of two independent variables but we shall indicate also the corresponding results for functions of an arbitrary number, n, of independent variables.

2. A Note on Continuity

The whole concept of continuity is much more difficult in the case of functions of more than one independent variable. For instance, if we consider the function $f(x, y)$, we can say that f is *continuous with respect to* x at the point (x_0, y_0) if the value $f(x_0, y_0)$ is clearly specified and if $f(x, y_0)$ tends to the value $f(x_0, y_0)$ as x tends to x_0. Similarly we can say that f is *continuous with respect to* y at the point (x_0, y_0) if $f(x_0, y)$ tends to the value $f(x_0, y_0)$ as y tends to y_0. In the theory of functions of two independent variables it is necessary for theoretical purposes, to define continuity of another kind. We say that the function $f(x, y)$ is *continuous with respect to the pair of variables* (x, y) at (x_0, y_0) if the function $f(x, y)$ tends to the value $f(x_0, y_0)$ as the point (x, y) tends to the point (x_0, y_0).

It is readily shown* that if $f(x, y)$ is continuous at the point (x_0, y_0) with respect to the pair of variables (x, y) it is continuous at that point with respect to each of the variables x and y separately, but that if $f(x, y)$ is continuous with respect to each of x and y separately it does not follow that it is continuous with respect to the pair (x, y). As an example of this latter fact, Gillespie cites the function $f(x, y)$ defined to be $xy/(x^2 + y^2)$ when x and y are not both zero and to be zero when $x = y = 0$; this function is not continuous with respect to the pair of variables (x, y) since it possesses no limit as (x, y) tends to $(0, 0)$.

A function $f(x, y)$ is said to be continuous with respect to the pair of variables (x, y) in a domain D if it is continuous with respect to these variables at each point of the domain.

In a proper treatment of the theory of functions of several variables it is necessary to treat the concept of continuity in some detail in order to establish the necessary theorems in a rigorous fashion. We shall not attempt such a treatment here but will content ourselves with developing the subject heuristically.

3. Partial Derivatives

Suppose that

$$z = f(x, y)$$

is a single-valued function of two independent variables x and y.

* See, for instance, R. P. Gillespie, "Partial Differentiation" (Oliver & Boyd, Edinburgh, 1951) p. 5.

If we assign a fixed value, b say, to the variable y, we obtain a function $f(x, b)$ of the single variable x. If this function is differentiable, we can calculate its derivative with respect to x, namely

$$\left[\frac{\mathrm{d}}{\mathrm{d}x} f(x, b)\right]_{x=a} \tag{1}$$

at the point $x = a$. This derivative is called the *partial derivative of z with respect to x* at the point (a, b) and it is denoted by the symbol $z_x(a, b)$ or $f_x(a, b)$.

By the definition of the derivative (1) — see p. 118 above — we see that

$$f_x(a, b) = \lim_{h \to 0} \frac{f(a + h, b) - f(a, b)}{h}. \tag{2}$$

In many theoretical discussions it is convenient to consider the value of the partial derivative f_x at a general point (x, y). Just as in the case of functions of a single variable we often wish to discuss the form of $f'(x)$. In these cases we write for the partial derivative

$$\left[\frac{\mathrm{d}}{\mathrm{d}x} f(x, y)\right]_{y \text{ kept constant}}$$

not $f_x(x, y)$ but more usually

$$\frac{\partial f}{\partial x} \quad \text{or} \quad \frac{\partial z}{\partial x},$$

just as in the one-variable case we usually write $\mathrm{d}f/\mathrm{d}x$ instead of $f'(x)$.

Similarly the symbols

$$f_y(a, b), \; z_y(a, b), \; \frac{\partial z}{\partial y}, \; \frac{\partial f}{\partial y}$$

denote the partial derivative of the function with respect to y.

In calculating the value of the partial derivative f_x at a point we can either follow the procedure suggested by the expression (1) above or we can evaluate the partial derivative $\partial f/\partial x$ for general values of x and y and then substitute the relevant particular values of x and y.*

* This is a simple extension of the two methods available in the one-variable case. For example, if we wish to calculate the derivative with respect to x at the point $x = 1$ of the function $f(x) = x^2$ we can put *either* $f'(1) = \lim_{h\to 0}\{(h + 1)^2 - 1\}/h = 2$, *or* $f'(x) = 2x$, so that $f'(1) = 2$.

To illustrate both procedures let us calculate the partial derivative with respect to x of the function

$$f(x, y) = x^3y^2 + x^2y^3 \tag{3}$$

at the point $x = 2$, $y = 1$. For this function

$$f(x, 1) = x^3 + x^2,$$

so that

$$f_x(2, 1) = [3x^2 + 2x]_{x=2} = 16.$$

On the other hand if we differentiate (3) with respect to x keeping y constant we find that

$$\frac{\partial f}{\partial x} = 3x^2y^2 + 2xy^3,$$

and if we put $x = 2$ and $y = 1$ in this equation we find, as before, that $f_x(2, 1)$ has the value 16.

It should be noted that a modification of our notation is used in some practical applications — for example in thermodynamics. In thermodynamical problems it is often possible to choose a variety of pairs of variables as the independent variables.* If when we write down the partial derivative of a function f with respect to an independent variable x we wish to draw attention to the fact that the other independent variable is y, we write

$$\left(\frac{\partial f}{\partial x}\right)_y.$$

In this notation the suffix y denotes that y is the second independent variable and that in this differentiation it is kept constant.

The calculation of partial derivatives obviously requires no new techniques. In the case of functions of two independent variables x and y we keep y (or x) fixed and differentiate with respect to x (or y). The following will serve as illustrations:-

EXAMPLE 1. For the function A defined by equation (1) of Section 1

$$\frac{\partial A}{\partial x} = y, \qquad \frac{\partial A}{\partial y} = x.$$

* See, for example, Section 13 of this chapter.

2. For the function u defined by equation (2) of Section 1,

$$\frac{\partial u}{\partial x} = \frac{100}{y}, \quad \frac{\partial u}{\partial y} = -\frac{100x}{y^2}.$$

3. If $z = 3yx^3 + 2y$, then

$$\frac{\partial z}{\partial x} = 9x^2y, \quad \frac{\partial z}{\partial y} = 3x^3 + 2.$$

4. If $z = y + y^3x$, then

$$\frac{\partial z}{\partial x} = y^3, \quad \frac{\partial z}{\partial y} = 1 + 3xy^2.$$

5. If $r = \sqrt{(x^2 + y^2)}$,

$$\frac{\partial r}{\partial x} = \frac{x}{\sqrt{(x^2 + y^2)}} = \frac{x}{r}, \quad \frac{\partial r}{\partial y} = \frac{y}{\sqrt{(x^2 + y^2)}} = \frac{y}{r}.$$

6. If $\theta = \tan^{-1}(y/x)$,

$$\frac{\partial \theta}{\partial x} = \frac{1}{1 + (y/x)^2}\left(\frac{-y}{x^2}\right) = -\frac{y}{x^2 + y^2} = -\frac{y}{r^2},$$

$$\frac{\partial \theta}{\partial y} = \frac{1}{1 + (y/x)^2}\left(\frac{1}{x}\right) = \frac{x}{x^2 + y^2} = \frac{x}{r^2}.$$

4. The Geometrical Interpretation of Partial Derivatives

In the case of a function of two independent variables we can give a simple geometrical interpretation to the partial derivatives. If $z = f(x, y)$ is such a function, we can give a geometrical form to the functional relation in the following way. We choose a point with coordinates (x, y) in the xy-plane and erect at that point a line of length $f(x, y)$ perpendicular to the xy-plane. The end of that line is then a point which can be taken to represent $f(x, y)$. As x and y vary throughout the domain of definition of the function this point will trace out a surface of the kind shown in Fig. 106(a). Suppose that P is a typical point on this surface corresponding to the point $P_0(x, y)$ in the xy-plane. Through this point P we cut a section of the surface by a plane parallel to the xz-plane — this is the plane BPA shown in Fig. 106(a). This section yields the curve BPA shown in Fig. 106(b).

On this curve the value of y is fixed so that the gradient of the tangent $T'PT$ is

$$\left[\frac{\mathrm{d}}{\mathrm{d}x}f(x,\ y)\right]_{y=\text{constant}}$$

i.e. the gradient of the tangent $T'PT$ is a measure of the partial derivative

$$\frac{\partial f}{\partial x}$$

evaluated at the point $(x,\ y)$. Similarly, if we take a section QPR of the surface by a plane through P parallel to the yz-plane we get a

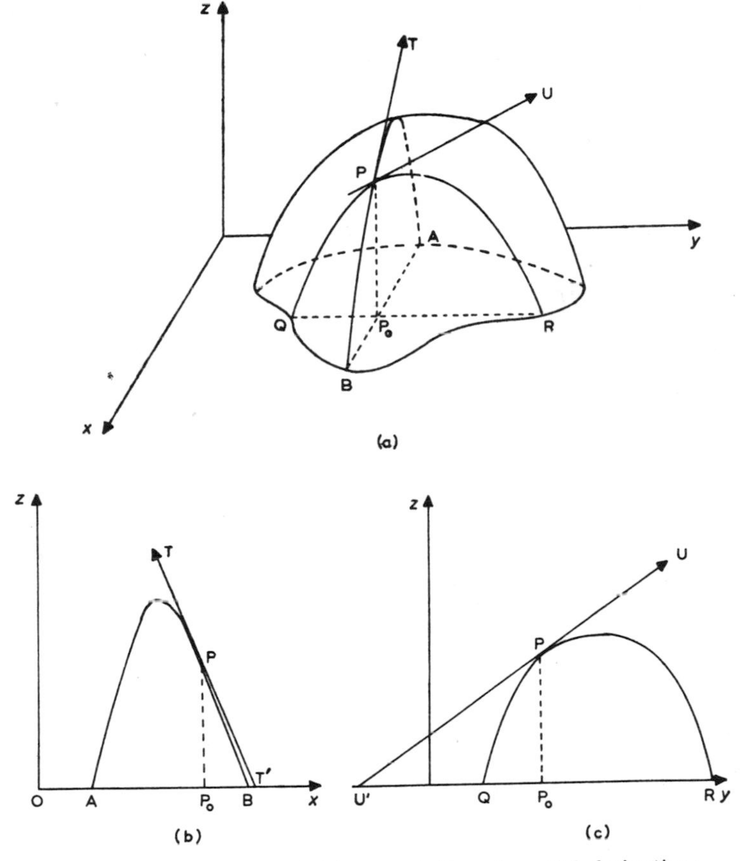

Fig. 106. The geometrical interpretation of partial derivatives.

curve QPR along which x is constant so that the gradient of the tangent $U'PU$ to that curve is a measure of the partial derivative $\partial f/\partial y$ at the point P. (Cf. Fig. 106(c).)

5. Change of Variable in Partial Differentiations

It often happens that a quantity u is a function of two variables x and y each of which is itself a function of another pair of variables ξ and η. It follows that, in these circumstances, u is a function of ξ and η and that it is meaningful to talk of the partial derivatives $\partial u/\partial \xi$ and $\partial u/\partial \eta$. We shall now show how these partial derivatives may be calculated in terms of the partial derivatives $\partial u/\partial x$, $\partial u/\partial y$, $\partial x/\partial \xi$, $\partial x/\partial \eta$, $\partial y/\partial \xi$, $\partial y/\partial \eta$.

Suppose that we give ξ a small increment h but keep η fixed, then $x(\xi, \eta)$ will have a resulting increment

$$h_1 = x(\xi + h, \eta) - x(\xi, \eta),$$

and $y(\xi, \eta)$ will have an increment

$$h_2 = y(\xi + h, \eta) - y(\xi, \eta),$$

and these two increments will in turn result in $u(x, y)$ taking an increment

$$H = u(x + h_1, y + h_2) - u(x, y).$$

By adding $u(x, y + h_2)$ to this expression, then subtracting it and regrouping the terms we see that we can write

$$H = \{u(x+h_1, y+h_2) - u(x, y+h_2)\} + \{u(x, y+h_2) - u(x, y)\}.$$

Now, by the definition of a partial derivative, we see that

$$\begin{aligned}\frac{\partial u}{\partial \xi} &= \lim_{h\to 0} \frac{H}{h} \\ &= \lim_{h_1, h_2 \to 0} \frac{u(x + h_1, y + h_2) - u(x, y + h_2)}{h_1} \lim_{h\to 0} \frac{h_1}{h} \\ &+ \lim_{h_2\to 0} \frac{u(x, y + h_2) - u(x, y)}{h_2} \lim_{h\to 0} \frac{h_2}{h}.\end{aligned}$$

By the definition of a partial derivative

$$\lim_{h\to 0} \frac{h_1}{h} = \lim \frac{x(\xi + h, \eta) - x(\xi, \eta)}{h} = \frac{\partial x}{\partial \xi},$$

and, similarly,

$$\lim_{h \to 0} \frac{h_2}{h} = \frac{\partial y}{\partial \xi}.$$

Also

$$\lim_{h_1, h_2 \to 0} \frac{u(x + h_1,\ y + h_2) - u(x,\ y + h_2)}{h_1} = \frac{\partial u}{\partial x},$$

$$\lim_{h_2 \to 0} \frac{u(x,\ y + h_2) - u(x,\ y)}{h_2} = \frac{\partial u}{\partial y}.$$

Substituting these limit results in the above expression for $\partial u/\partial \xi$, we obtain the result

$$\frac{\partial u}{\partial \xi} = \frac{\partial u}{\partial x}\frac{\partial x}{\partial \xi} + \frac{\partial u}{\partial y}\frac{\partial y}{\partial \xi}. \tag{1}$$

Similarly, if we had given η a small increment and studied the consequent changes in x, y and u we would obtain the result

$$\frac{\partial u}{\partial \eta} = \frac{\partial u}{\partial x}\frac{\partial x}{\partial \eta} + \frac{\partial u}{\partial y}\frac{\partial y}{\partial \eta}. \tag{2}$$

We shall now illustrate these results by some typical examples.

EXAMPLE 7. If $u = u(x, y)$ and $x = r \cos \theta$, $y = r \sin \theta$, then, since

$$\frac{\partial x}{\partial r} = \cos \theta = \frac{x}{r}, \qquad \frac{\partial y}{\partial r} = \sin \theta = \frac{y}{r},$$

it follows that

$$\frac{\partial u}{\partial r} = \frac{1}{r}\left(x \frac{\partial u}{\partial x} + y \frac{\partial u}{\partial y}\right).$$

Since

$$\frac{\partial x}{\partial \theta} = -r \sin \theta = -y, \qquad \frac{\partial y}{\partial \theta} = r \cos \theta = x,$$

we have

$$\frac{\partial u}{\partial \theta} = -y \frac{\partial u}{\partial x} + x \frac{\partial u}{\partial y}.$$

EXAMPLE 8. If $u = u(r, \theta)$ and $x = r \cos \theta$, $y = r \sin \theta$, then $r = \sqrt{(x^2 + y^2)}$, $\theta = \tan^{-1}(y/x)$, and it follows from examples 5 and 6 (of Section 3) that

$$\frac{\partial u}{\partial x} = \cos \theta \frac{\partial u}{\partial r} - \frac{\sin \theta}{r}\frac{\partial u}{\partial \theta}, \qquad \frac{\partial u}{\partial y} = \sin \theta \frac{\partial u}{\partial r} + \frac{\cos \theta}{r}\frac{\partial u}{\partial \theta}.$$

EXAMPLE 9. If $u = u(x, y)$ and $x = \xi + \eta$, $y = \sin \xi + \sin \eta$, then

$$\frac{\partial u}{\partial \xi} = \frac{\partial u}{\partial x} + \cos \eta \frac{\partial u}{\partial y}, \qquad \frac{\partial u}{\partial \eta} = \frac{\partial u}{\partial x} + \cos \eta \frac{\partial u}{\partial y}.$$

EXAMPLE 10. If $u = u(x, y)$ and $x = 1/\xi$, $y = \xi\eta - 1/\xi$, then

$$\frac{\partial u}{\partial \xi} = -\frac{1}{\xi^2}\frac{\partial u}{\partial x} + \left(\eta + \frac{1}{\xi^2}\right)\frac{\partial u}{\partial y}, \qquad \frac{\partial u}{\partial \eta} = \xi \frac{\partial u}{\partial y},$$

so that, since $\xi = 1/x$, $\eta = x(x + y)$, we have

$$\frac{\partial u}{\partial \xi} = -x^2 \frac{\partial u}{\partial x} + (2x + y)x \frac{\partial u}{\partial y}, \qquad \frac{\partial u}{\partial \eta} = \frac{1}{x}\frac{\partial u}{\partial y}.$$

Two special forms of equations (1) and (2) are used frequently. If x and y are functions of a *single* variable, t say, then $u(x, y)$ will also be a function of t only and equations (1) and (2) are replaced by the single equation

$$\frac{\mathrm{d}u}{\mathrm{d}t} = \frac{\partial u}{\partial x}\frac{\mathrm{d}x}{\mathrm{d}t} + \frac{\partial u}{\partial y}\frac{\mathrm{d}y}{\mathrm{d}t}. \tag{3}$$

In the case where u is a function of x and y and y is a function of x then, putting $t = x$ in equation (3), we obtain the formula

$$\frac{\mathrm{d}u}{\mathrm{d}x} = \frac{\partial u}{\partial x} + \frac{\partial u}{\partial y}\frac{\mathrm{d}y}{\mathrm{d}x}. \tag{4}$$

In particular, if y is defined as a function of x through the equation

$$u(x, y) = 0, \tag{5}$$

it follows from equation (4) that

$$0 = \frac{\partial u}{\partial x} + \frac{\partial u}{\partial y}\frac{\mathrm{d}y}{\mathrm{d}x},$$

so that

$$\frac{\mathrm{d}y}{\mathrm{d}x} = -\frac{\dfrac{\partial u}{\partial x}}{\dfrac{\partial u}{\partial y}}. \tag{6}$$

EXAMPLE 11. If $u = x^4 + y^4$ and $x = \mathrm{e}^{-t}$, $y = \sin t$, then from equation (3), we have

$$\frac{du}{dt} = (4x^3)(-e^{-t}) + (4y^3)(\cos t) = -4e^{-4t} + 4\sin^3 t \cos t,$$

as is easily verified by substituting for x and y into the expression for u and then differentiating with respect to t.

EXAMPLE 12. If $u = u(x, y)$ and $y = \sin \omega x$ then

$$\frac{du}{dx} = \frac{\partial u}{\partial x} + \omega \cos \omega x \cdot \frac{\partial u}{\partial y}.$$

EXAMPLE 13. If y is defined as a function of x by the relation

$$x^5 + y^5 - 5a^3 xy = 0,$$

then, writing u for the left-hand side of this equation, we see that

$$\frac{\partial u}{\partial x} = 5x^4 - 5a^3 y, \qquad \frac{\partial u}{\partial y} = 5y^4 - 5a^3 x,$$

so we obtain the formula

$$\frac{dy}{dx} = -\frac{x^4 - a^3 y}{y^4 - a^3 x}$$

for the derivative of y with respect to x.

PROBLEMS

1. If z is a function of x and y and

$$\xi = x - ay, \qquad \eta = x + ay$$

show that

$$\frac{\partial z}{\partial x} = \frac{\partial z}{\partial \xi} + \frac{\partial z}{\partial \eta}, \qquad \frac{1}{a}\frac{\partial z}{\partial y} = -\frac{\partial z}{\partial \xi} + \frac{\partial z}{\partial \eta}$$

and deduce that

$$\frac{\partial z}{\partial x} - \frac{1}{a}\frac{\partial z}{\partial y} = 2\frac{\partial z}{\partial \xi}, \qquad \frac{\partial z}{\partial x} + \frac{1}{a}\frac{\partial z}{\partial y} = 2\frac{\partial z}{\partial \eta}.$$

2. If z is a function of x and y and

$$u = x^2 + y^2, \quad v = 2xy$$

show that

$$\frac{\partial z}{\partial x} = 2\left(x\frac{\partial z}{\partial u} + y\frac{\partial z}{\partial v}\right), \qquad \frac{\partial z}{\partial y} = 2\left(y\frac{\partial z}{\partial u} + x\frac{\partial z}{\partial v}\right)$$

and deduce that

$$x\frac{\partial z}{\partial x} + y\frac{\partial z}{\partial y} = 2\left(u\frac{\partial z}{\partial u} + v\frac{\partial z}{\partial v}\right).$$

3. If $\xi = x^n + y$ and $\eta = x^n - y$ show that for any function u of x and y,

$$x\frac{\partial u}{\partial x} = n(\xi + \eta)\left(\frac{\partial u}{\partial \xi} + \frac{\partial u}{\partial \eta}\right), \qquad \frac{\partial u}{\partial y} = \frac{\partial u}{\partial \xi} - \frac{\partial u}{\partial \eta}.$$

4. If ϕ is a function of x and y and if ξ and η are new variables given by

$$\xi = e^x \cos y, \qquad \eta = e^x \sin y$$

prove that

$$\left(\frac{\partial \phi}{\partial x}\right)^2 + \left(\frac{\partial \phi}{\partial y}\right)^2 = (\xi^2 + \eta^2)\left[\left(\frac{\partial \phi}{\partial \xi}\right)^2 + \left(\frac{\partial \phi}{\partial \eta}\right)^2\right].$$

6. Higher Partial Derivatives

Higher partial derivatives may be obtained by differentiating the first order partial derivatives. For instance, the partial derivative $\partial f/\partial x$ will be a function of x and y so that we can form its partial derivatives with respect to both x and y. The derivatives so obtained are denoted as follows:-

$$\frac{\partial}{\partial x}\left(\frac{\partial f}{\partial x}\right) \equiv \frac{\partial^2 f}{\partial x^2} \equiv f_{xx},$$

$$\frac{\partial}{\partial y}\left(\frac{\partial f}{\partial x}\right) \equiv \frac{\partial^2 f}{\partial y \partial x} \equiv f_{yx}.$$

Similarly from the partial derivative $\partial f/\partial y$ we can form the second derivatives

$$\frac{\partial}{\partial x}\left(\frac{\partial f}{\partial y}\right) \equiv \frac{\partial^2 f}{\partial x \partial y} \equiv f_{xy},$$

$$\frac{\partial}{\partial y}\left(\frac{\partial f}{\partial y}\right) \equiv \frac{\partial^2 f}{\partial y^2} \equiv f_{yy}.$$

For example, for the function $z(x, y)$ of example 3, we have

$$\frac{\partial^2 z}{\partial x^2} = \frac{\partial}{\partial x}(9x^2 y) = 18xy, \qquad \frac{\partial^2 z}{\partial y \partial x} = \frac{\partial}{\partial y}(9x^2 y) = 9x^2,$$

$$\frac{\partial^2 z}{\partial x \partial y} = \frac{\partial}{\partial x}(3x^3 + 2) = 9x^2, \qquad \frac{\partial^2 z}{\partial y^2} = \frac{\partial}{\partial y}(3x^3 + 2) = 0.$$

This example shows that the "mixed" partial derivatives z_{yx} and z_{xy} have the same value for all values of x and y. This is no coincidence but a particular example of a general result called Schwartz's theorem which states that: *if* f_{xy} *and* f_{yx} *of a function* $f(x, y)$ *are continuous in a domain D then* $f_{xy} = f_{yx}$ *at every point within the domain D.*

Schwartz's theorem states that the order of differentiation in a mixed partial derivative is (in general) immaterial. We shall not give a proof of this theorem here; any reader who is interested in seeing one is referred to part II pp. 55 and 56 of R. Courant, "Differential and Integral Calculus" (Blackie, Glasgow, 1957).

We shall illustrate the calculation of partial derivatives of higher order by two examples:-

EXAMPLE 14. *If* $z = e^{nx} \sin (x - y)$, *where n is a constant, show that*

$$\frac{\partial^2 z}{\partial x^2} = \frac{\partial^2 z}{\partial y^2} - 2n \frac{\partial z}{\partial y} + n^2 z.$$

For this function we have

$$\frac{\partial z}{\partial x} = n e^{nx} \sin (x - y) + e^{nx} \cos (x - y), \tag{1}$$

$$\frac{\partial z}{\partial y} = -e^{nx} \cos (x - y). \tag{2}$$

Differentiating the first of these equations with respect to x we have

$$\frac{\partial^2 z}{\partial x^2} = n^2 e^{nx} \sin (x - y) + 2n e^{nx} \cos (x - y) - e^{nx} \sin (x - y), \tag{3}$$

and differentiating the second with respect to y we find that

$$\frac{\partial^2 z}{\partial y^2} = -e^{nx} \sin (x - y). \tag{4}$$

Substituting from equations (2) and (4) into equation (3) we see that this latter equation can be written in the form

$$\frac{\partial^2 z}{\partial x^2} = n^2 z - 2n \frac{\partial z}{\partial y} + \frac{\partial^2 z}{\partial y^2}$$

proving the result stated.

EXAMPLE 15. *If $z = f(x + u)$ where u is a function of y only, prove that*

$$\frac{\partial z}{\partial x}\frac{\partial^2 z}{\partial x \partial y} = \frac{\partial z}{\partial y}\frac{\partial^2 z}{\partial x^2}. \tag{5}$$

We can write

$$z = f(\xi)$$

where

$$\xi = x + u$$

so that

$$\frac{\partial \xi}{\partial x} = 1, \qquad \frac{\partial \xi}{\partial y} = \dot{u}$$

where $\dot{u}$ denotes $\mathrm{d}u/\mathrm{d}y$.

$$\frac{\partial z}{\partial x} = \frac{\mathrm{d}z}{\mathrm{d}\xi}\frac{\partial \xi}{\partial x} = f'(\xi), \tag{6}$$

$f'(\xi)$ denoting $\mathrm{d}f/\mathrm{d}\xi$. Similarly

$$\frac{\partial z}{\partial y} = \frac{\mathrm{d}z}{\mathrm{d}\xi}\frac{\partial \xi}{\partial y} = f'(\xi)\dot{u}. \tag{7}$$

Differentiating (6) again with respect to x we have

$$\frac{\partial^2 z}{\partial x^2} = \frac{\mathrm{d}}{\mathrm{d}\xi}[f'(\xi)]\frac{\partial \xi}{\partial x} = f''(\xi). \tag{8}$$

Similarly if we differentiate (6) with respect to y we have

$$\frac{\partial^2 z}{\partial x \partial y} = \frac{\mathrm{d}}{\mathrm{d}\xi}[f'(\xi)]\frac{\partial \xi}{\partial y} = f''(\xi)\dot{u}. \tag{9}$$

From equation (6), (9) we find that

$$\frac{\partial z}{\partial x}\frac{\partial^2 z}{\partial x \partial y} = f'(\xi)f''(\xi)\dot{u},$$

and from (7) and (8) that

$$\frac{\partial z}{\partial y}.\frac{\partial^2 z}{\partial x^2} = f''(\xi)f'(\xi)\dot{u},$$

showing that equation (5) is satisfied.

A result of some importance is the formula for the second order derivatives of a function $u(x, y)$ when the independent variables x

and y are changed to ξ and η, where

$$\xi = \xi(x, y), \quad \eta = \eta(x, y).$$

From equation (1) of § 5, we know that

$$\frac{\partial u}{\partial x} = \frac{\partial u}{\partial \xi}\frac{\partial \xi}{\partial x} + \frac{\partial u}{\partial \eta}\frac{\partial \eta}{\partial x}. \tag{10}$$

Differentiating both sides of this equation with respect to x and using the formula for differentiating a product we have the equation

$$\frac{\partial^2 u}{\partial x^2} = \left\{\frac{\partial}{\partial x}\left(\frac{\partial u}{\partial \xi}\right)\right\}\frac{\partial \xi}{\partial x} + \frac{\partial u}{\partial \xi}\cdot\frac{\partial^2 \xi}{\partial x^2} + \left\{\frac{\partial}{\partial x}\left(\frac{\partial u}{\partial \eta}\right)\right\}\frac{\partial \eta}{\partial x} + \frac{\partial u}{\partial \eta}\cdot\frac{\partial^2 \eta}{\partial x^2}. \tag{11}$$

If, in equation (10), we replace u by $\partial u/\partial \xi$, we obtain the relation

$$\frac{\partial}{\partial x}\left(\frac{\partial u}{\partial \xi}\right) = \frac{\partial^2 u}{\partial \xi^2}\frac{\partial \xi}{\partial x} + \frac{\partial^2 u}{\partial \xi \partial \eta}\frac{\partial \eta}{\partial x}. \tag{12}$$

Similarly, if we replace u by $\partial u/\partial \eta$ we obtain the relation

$$\frac{\partial}{\partial x}\left(\frac{\partial u}{\partial \eta}\right) = \frac{\partial^2 u}{\partial \xi \partial \eta}\frac{\partial \xi}{\partial x} + \frac{\partial^2 u}{\partial \eta^2}\frac{\partial \eta}{\partial x}. \tag{13}$$

Substituting from equations (12) and (13) into equation (11) we obtain the formula

$$\frac{\partial^2 u}{\partial x^2} = \frac{\partial^2 u}{\partial \xi^2}\left(\frac{\partial \xi}{\partial x}\right)^2 + 2\frac{\partial^2 u}{\partial \xi \partial \eta}\frac{\partial \xi}{\partial x}\frac{\partial \eta}{\partial x} + \frac{\partial^2 u}{\partial \eta^2}\left(\frac{\partial \eta}{\partial x}\right)^2 + \frac{\partial u}{\partial \xi}\frac{\partial^2 \xi}{\partial x^2} + \frac{\partial u}{\partial \eta}\frac{\partial^2 \eta}{\partial x^2}. \tag{14}$$

The corresponding formula for $\partial^2 u/\partial y^2$. is obtained merely by replacing x by y in this formula. We find that

$$\frac{\partial^2 u}{\partial y^2} = \frac{\partial^2 u}{\partial \xi^2}\left(\frac{\partial \xi}{\partial y}\right)^2 + 2\frac{\partial^2 u}{\partial \xi \partial \eta}\frac{\partial \xi}{\partial y}\frac{\partial \eta}{\partial y} + \frac{\partial^2 u}{\partial \eta^2}\left(\frac{\partial \eta}{\partial y}\right)^2 + \frac{\partial u}{\partial \xi}\frac{\partial^2 \xi}{\partial y^2} + \frac{\partial u}{\partial \eta}\frac{\partial^2 \eta}{\partial y^2}. \tag{15}$$

If we differentiate both sides of equation (10) with respect to y we find that

$$\frac{\partial^2 u}{\partial x \partial y} = \left\{\frac{\partial}{\partial y}\left(\frac{\partial u}{\partial \xi}\right)\right\}\frac{\partial \xi}{\partial y} + \frac{\partial u}{\partial \xi}\frac{\partial^2 \xi}{\partial x \partial y} + \left\{\frac{\partial}{\partial y}\left(\frac{\partial u}{\partial \eta}\right)\right\}\frac{\partial \eta}{\partial x} + \frac{\partial u}{\partial \eta}\frac{\partial^2 \eta}{\partial x \partial y}. \tag{16}$$

Now, from equation (1) of § 5,

$$\frac{\partial}{\partial y}\left(\frac{\partial u}{\partial \xi}\right) = \frac{\partial^2 u}{\partial \xi^2}\frac{\partial \xi}{\partial y} + \frac{\partial^2 u}{\partial \xi \partial \eta}\frac{\partial \eta}{\partial y} \tag{17}$$

$$\frac{\partial}{\partial y}\left(\frac{\partial u}{\partial \eta}\right) = \frac{\partial^2 u}{\partial \xi \, \partial \eta}\,\frac{\partial \xi}{\partial y} + \frac{\partial^2 u}{\partial \eta^2}\,\frac{\partial \eta}{\partial y} \tag{18}$$

so that, substituting from equations (17) and (18) into equation (16) we find that

$$\begin{aligned}\frac{\partial^2 u}{\partial x \, \partial y} = \frac{\partial^2 u}{\partial \xi^2}\,\frac{\partial \xi}{\partial x}\,\frac{\partial \xi}{\partial y} + \frac{\partial^2 u}{\partial \xi \, \partial \eta}\left(\frac{\partial \xi}{\partial x}\,\frac{\partial \eta}{\partial y} + \frac{\partial \xi}{\partial y}\,\frac{\partial \eta}{\partial x}\right) + \frac{\partial^2 u}{\partial \eta^2}\,\frac{\partial \eta}{\partial x}\,\frac{\partial \eta}{\partial y} \\ + \frac{\partial u}{\partial \xi}\,\frac{\partial^2 \xi}{\partial x \, \partial y} + \frac{\partial u}{\partial \eta}\,\frac{\partial^2 \eta}{\partial x \, \partial y}.\end{aligned} \tag{19}$$

The reader should not attempt to memorize these formulae, but only to familiarize himself with the procedure involved. To illustrate how the method is used in a specific problem we have:-

EXAMPLE 16. *If* $\xi = x^n + y$ *and* $\eta = x^n - y$, *find expressions for* $\partial u/\partial x$, $\partial u/\partial y$, $\partial^2 u/\partial x^2$, $\partial^2 u/\partial y^2$ *in terms of* $\partial u/\partial \xi$, $\partial u/\partial \eta$, $\partial^2 u/\partial \xi^2$, $\partial^2 u/\partial \xi \partial \eta$, $\partial^2 u/\partial \eta^2$.

Find also the equation to which the equation

$$x^2 \frac{\partial^2 u}{\partial x^2} - (n-1)x\frac{\partial u}{\partial x} = n^2 x^{2n} \frac{\partial^2 u}{\partial y^2}$$

is transformed.

$$\frac{\partial u}{\partial x} = \frac{\partial u}{\partial \xi}\,\frac{\partial \xi}{\partial x} + \frac{\partial u}{\partial \eta}\,\frac{\partial \eta}{\partial x} = nx^{n-1}\frac{\partial u}{\partial \xi} + nx^{n-1}\frac{\partial u}{\partial \eta} = nx^{n-1}\left(\frac{\partial u}{\partial \xi} + \frac{\partial u}{\partial \eta}\right) \tag{20}$$

$$\frac{\partial u}{\partial y} = \frac{\partial u}{\partial \xi}\,\frac{\partial \xi}{\partial y} + \frac{\partial u}{\partial \eta}\,\frac{\partial \eta}{\partial y} = \frac{\partial u}{\partial \xi} - \frac{\partial u}{\partial \eta} \tag{21}$$

since

$$\frac{\partial \xi}{\partial x} = nx^{n-1}, \qquad \frac{\partial \xi}{\partial y} = 1, \qquad \frac{\partial \xi}{\partial x} = nx^{n-1}, \qquad \frac{\partial \xi}{\partial y} = -1.$$

Differentiating (20) again with respect to x we find that

$$\frac{\partial^2 u}{\partial x^2} = n(n-1)x^{n-2}\left(\frac{\partial u}{\partial \xi} + \frac{\partial u}{\partial \eta}\right) + nx^{n-1}\frac{\partial}{\partial x}\left(\frac{\partial u}{\partial \xi} + \frac{\partial u}{\partial \eta}\right).$$

We can evaluate

$$\frac{\partial}{\partial x}\left(\frac{\partial u}{\partial \xi} + \frac{\partial u}{\partial \eta}\right)$$

from equation (20) by replacing u by $\partial u/\partial \xi + \partial u/\partial \eta$. In this way we find that

$$\frac{\partial^2 u}{\partial x^2} = n(n-1)x^{n-2}\left(\frac{\partial u}{\partial \xi} + \frac{\partial u}{\partial \eta}\right) + n^2 x^{2n-2}\left(\frac{\partial^2 u}{\partial \xi^2} + 2\frac{\partial^2 u}{\partial \xi \partial \eta} + \frac{\partial^2 u}{\partial \eta^2}\right). \quad (22)$$

If we differentiate both sides of (20) with respect to y we have

$$\frac{\partial^2 u}{\partial x \partial y} = nx^{n-1}\frac{\partial}{\partial y}\left(\frac{\partial u}{\partial \xi} + \frac{\partial u}{\partial \eta}\right)$$

and by equation (21)

$$\frac{\partial}{\partial y}\left(\frac{\partial u}{\partial \xi} + \frac{\partial u}{\partial \eta}\right) = \frac{\partial}{\partial \xi}\left(\frac{\partial u}{\partial \xi} + \frac{\partial u}{\partial \eta}\right) - \frac{\partial}{\partial \eta}\left(\frac{\partial u}{\partial \xi} + \frac{\partial u}{\partial \eta}\right) = \frac{\partial^2 u}{\partial \xi^2} - \frac{\partial^2 u}{\partial \eta^2}$$

so that

$$\frac{\partial^2 u}{\partial x \partial y} = nx^{n-1}\left(\frac{\partial^2 u}{\partial \xi^2} - \frac{\partial^2 u}{\partial \eta^2}\right). \quad (23)$$

Finally if we differentiate both sides of (21) with respect to y we find that

$$\frac{\partial^2 u}{\partial y^2} = \frac{\partial}{\partial y}\left(\frac{\partial u}{\partial \xi} - \frac{\partial u}{\partial \eta}\right) = \frac{\partial}{\partial \xi}\left(\frac{\partial u}{\partial \xi} - \frac{\partial u}{\partial \eta}\right) - \frac{\partial}{\partial \eta}\left(\frac{\partial u}{\partial \xi} - \frac{\partial u}{\partial \eta}\right)$$

so that

$$\frac{\partial^2 u}{\partial y^2} = \frac{\partial^2 u}{\partial \xi^2} - 2\frac{\partial^2 u}{\partial \xi \partial \eta} + \frac{\partial^2 u}{\partial \eta^2}. \quad (24)$$

If we substitute from equations (20), (22) and (24) into the equation

$$x^2\frac{\partial^2 u}{\partial x^2} - (n-1)x\frac{\partial u}{\partial x} = n^2 x^{2n}\frac{\partial^2 u}{\partial y^2} \quad (25)$$

we find that it transforms to

$$\frac{\partial^2 u}{\partial \xi \partial \eta} = 0. \quad (26)$$

We shall get further examples of this procedure in § 21 of Chapter 9.

PROBLEMS

1. If $\phi = \frac{1}{2}\log(x^2 + y^2)$, find the second order derivatives of ϕ with respect to x and y and show that

$$\frac{\partial^2 \phi}{\partial x^2} + \frac{\partial^2 \phi}{\partial y^2} = 0.$$

2. If $\phi(x, y) = x^m y^n$ show that

$$x^2 \frac{\partial^2 \phi}{\partial x^2} + 2xy \frac{\partial^2 \phi}{\partial x \partial y} + \frac{\partial^2 \phi}{\partial y^2} = (m + n)(m + n - 1)\phi.$$

3. If

$$v = \frac{x - y}{x + y}$$

show that

$$\frac{\partial^2 v}{\partial x \partial y} = \frac{1}{2}\left(\frac{\partial^2 v}{\partial x^2} + \frac{\partial^2 v}{\partial y^2}\right).$$

4. If $\psi(r, \theta) = r^n \cos (n\theta + \alpha)$ where n and α are constants, find

$$\frac{\partial^2 \psi}{\partial r^2}, \quad \frac{\partial^2 \psi}{\partial \theta^2}, \quad \frac{\partial^2 \psi}{\partial r \partial \theta}$$

and verify that

$$\frac{\partial^2 \psi}{\partial r^2} + \frac{1}{r}\frac{\partial \psi}{\partial r} + \frac{1}{r^2}\frac{\partial^2 \psi}{\partial \theta^2} = 0.$$

5. If

$$\phi(x, y) = (x + y)^2 \, e^{ax} + (x - y)^2 \, e^{-ax}$$

prove that

$$\frac{\partial^2 \phi}{\partial x^2} = \frac{\partial^2 \phi}{\partial y^2} + 2a \frac{\partial \phi}{\partial y} + a^2.$$

7. Functions of n Independent Variables

All that we have said of functions of two variables can be generalized to a function $f(x_1, x_2, \ldots, x_n)$ of n independent variables.

We define

$$\frac{\partial f}{\partial x_r},$$

the partial derivative of f with respect to x_r, to be the derivative of f with respect to x_r when the remaining $n - 1$ variables are kept constant and we write

$$\frac{\partial^2 f}{\partial x_r \partial x_s} = \frac{\partial}{\partial x_r}\left(\frac{\partial f}{\partial x_s}\right)$$

etc.

EXAMPLE 17. If $r = \sqrt{(x^2 + y^2 + z^2)}$, then

$$\frac{\partial r}{\partial x} = \frac{x}{\sqrt{(x^2 + y^2 + z^2)}} = \frac{x}{r}, \qquad \frac{\partial r}{\partial y} = \frac{y}{r}, \qquad \frac{\partial r}{\partial z} = \frac{z}{r}.$$

EXAMPLE 18. *If* $f = x^m y^n z^q$

$$\frac{\partial f}{\partial x} = m x^{m-1} y^n z^q = \frac{mf}{x}, \qquad \frac{\partial f}{\partial y} = \frac{nf}{y}, \qquad \frac{\partial f}{\partial z} = \frac{qf}{z}.$$

EXAMPLE 19. If $\theta = \tan^{-1}\{\sqrt{(x^2 + y^2)}/z\}$,

$$\frac{\partial \theta}{\partial x} = \frac{1}{1 + (x^2 + y^2)/z^2}\left\{\frac{1}{z}\frac{x}{\sqrt{(x^2 + y^2)}}\right\} = \frac{xz}{(x^2 + y^2 + z^2)\sqrt{(x^2 + y^2)}},$$

$$\frac{\partial \theta}{\partial y} = \frac{yz}{(x^2 + y^2 + z^2)\sqrt{(x^2 + y^2)}},$$

$$\frac{\partial \theta}{\partial z} = \frac{1}{1 + (x^2 + y^2)/z^2}\left\{-\frac{1}{z^2}\sqrt{(x^2 + y^2)}\right\} = -\frac{\sqrt{(x^2 + y^2)}}{x^2 + y^2 + z^2}.$$

The rules for change of variable follow the pattern of § 5. If ξ_1, $\xi_2, \ldots, \xi_n$ are independent variables which are themselves functions of the n independent variables $x_1, x_2, \ldots, x_n$, then a function $f(x_1, \ldots, x_n)$ can be thought of as a function of the n variables $\xi_1, \xi_2, \ldots, \xi_n$ and it can be shown that

$$\frac{\partial f}{\partial \xi_r} = \frac{\partial f}{\partial x_1}\frac{\partial x_1}{\partial \xi_r} + \frac{\partial f}{\partial x_2}\frac{\partial x_2}{\partial \xi_r} + \ldots + \frac{\partial f}{\partial x_n}\frac{\partial x_n}{\partial \xi_r}. \tag{1}$$

For example, if *

$$x = r \sin\theta \cos\phi, \qquad y = r \sin\theta \sin\phi, \qquad z = r\cos\theta$$

then

$$\frac{\partial x}{\partial r} = \sin\theta\cos\phi, \qquad \frac{\partial y}{\partial r} = \sin\theta\sin\phi, \qquad \frac{\partial z}{\partial r} = \cos\theta$$

so that the formula

$$\frac{\partial f}{\partial r} = \frac{\partial f}{\partial x}\frac{\partial x}{\partial r} + \frac{\partial f}{\partial y}\frac{\partial y}{\partial r} + \frac{\partial f}{\partial z}\frac{\partial z}{\partial r}$$

* *Remark:* The substitution $x = r\sin\theta\cos\phi$, $y = r\sin\theta\sin\phi$, $z = r\cos\theta$ has a simple geometrical meaning, for $x^2+y^2+z^2=r^2$, so the point (x, y, z) lies on the sphere with center O and radius r.
The coordinates (r, θ, ϕ) are called "spherical polar coordinates".

gives the equation

$$\frac{\partial f}{\partial r} = \sin\theta\cos\phi\frac{\partial f}{\partial x} + \sin\theta\sin\phi\frac{\partial f}{\partial y} + \cos\theta\frac{\partial f}{\partial z}.$$

Similarly the results

$$\frac{\partial x}{\partial\theta} = r\cos\theta\cos\phi, \qquad \frac{\partial y}{\partial\theta} = r\cos\theta\sin\phi, \qquad \frac{\partial z}{\partial\theta} = -r\sin\theta,$$

$$\frac{\partial x}{\partial\phi} = -r\sin\theta\sin\phi, \qquad \frac{\partial y}{\partial\phi} = r\sin\theta\cos\phi, \qquad \frac{\partial z}{\partial\phi} = 0$$

give the formulae

$$\frac{\partial f}{\partial\theta} = r\cos\theta\cos\phi\frac{\partial f}{\partial x} + r\cos\theta\sin\phi\frac{\partial f}{\partial y} - r\sin\theta\frac{\partial f}{\partial z}$$

$$\frac{\partial f}{\partial\phi} = -r\sin\theta\sin\phi\frac{\partial f}{\partial x} + r\sin\theta\cos\phi\frac{\partial f}{\partial y}.$$

8. Differentials *

Suppose that $z = f(x, y)$ is a unique function of the two independent variables x and y. If x and y receive increments δx and δy, which need not be small, we may form the expression

$$\frac{\partial z}{\partial x}\delta x + \frac{\partial z}{\partial y}\delta y \tag{1}$$

which is called the *differential* of z corresponding to the increments δx, δy. It is denoted by $\mathrm{d}z$ so that we have

$$\mathrm{d}z = \frac{\partial z}{\partial x}\delta x + \frac{\partial z}{\partial y}\delta y. \tag{2}$$

We may think of x as a function of x and y with the partial derivatives

$$\frac{\partial x}{\partial x} = 1, \qquad \frac{\partial x}{\partial y} = 0.$$

If we substitute these values in equation (2) we obtain the result $\mathrm{d}x = \delta x$, i.e. the differential of x for the increments δx, δy of x, y is simply δx. Similarly we can show that $\mathrm{d}y = \delta y$ and then we can

* Before reading this section the reader should revise § 6, Chapter 4 above.

write equation (2) as a relation between differentials:

$$dz = \frac{\partial z}{\partial x} dx + \frac{\partial z}{\partial y} dy. \tag{3}$$

Suppose, now, that x and y are functions of the single variable t. Then z will be a function of t only. Corresponding to an increment dt in t we shall then obtain the differentials

$$dz = \frac{dz}{dt} dt, \qquad dx = \frac{dx}{dt} dt, \qquad dy = \frac{dv}{dt} dt. \tag{4}$$

(Cf. p. 174 above.) Now by equation (3) of § 5 above we have

$$dz = \frac{dz}{dt} dt = \left(\frac{\partial z}{\partial x}\frac{dx}{dt} + \frac{\partial z}{\partial y}\frac{dy}{dt}\right) dt. \tag{5}$$

Combining equations (4) and (5) we obtain the simple relation

$$dz = \frac{\partial z}{\partial x} dx + \frac{\partial z}{\partial y} dy \tag{6}$$

connecting the differentials dx, dy, dz. It will be noted that equations (3) and (6) are formally equivalent; we have shown that the two different results (1) and (3) of § 5 assume the same form when written in terms of differentials, i.e. equation (3) is valid when x and y are independent and also when they are dependent.

In the general case in which $z = f(x_1, \ldots, x_n)$ is a function of n variables we find that the differentials dz, $dx_1, \ldots, dx_n$ are connected by the simple relation

$$dz = \frac{\partial z}{\partial x_1} dx_1 + \frac{\partial z}{\partial x_2} dx_2 + \ldots + \frac{\partial z}{\partial x_n} dx_n. \tag{7}$$

9. The Calculation of Small Errors

We saw in § 7 of Chapter 4 how the theory of differentials of functions could be used to estimate the error in a quantity $y = f(x)$ when the error in the quantity x on which it depends is known. We shall now consider the more general case in which the quantity in which we are interested depends not on the measurement of a single quantity, but on those of several.

A simple illustration is provided by the calculation by the direct

Fick method of the cardiac output by the formula

$$u = 100\frac{x}{y} \tag{1}$$

referred to above. If we make a percentage error ε_1 in x and a percentage error ε_2 in y then the false value of u will be

$$u + \Delta u = 100\frac{x\left(1 + \dfrac{\varepsilon_1}{100}\right)}{y\left(1 + \dfrac{\varepsilon_2}{100}\right)}$$

and the percentage error in the value of u will be

$$\frac{\Delta u}{u} \times 100 = \left\{\frac{1 + \dfrac{\varepsilon_1}{100}}{1 + \dfrac{\varepsilon_2}{100}} - 1\right\} \times 100.$$

Using the binomial theorem and retaining only the first two terms we can write

$$\frac{1}{1 + \dfrac{\varepsilon_2}{100}} = 1 - \frac{\varepsilon_2}{100}$$

so that the percentage error in the value of u is $\varepsilon_1 - \varepsilon_2$. Now if $p(x)$, $p(y)$ are the greatest values of the absolute errors of x and y, ε_1 may be as much as $p(x)$ and ε_2 may take the negative value $-p(y)$. Hence the percentage error in u can have an absolute value as great as $p(x) + p(y)$. Denoting by $p(u)$ the possible percentage error in the estimate of the value of u we have

$$p(u) = p(x) + p(y). \tag{2}$$

We shall now generalize this result by making the same assumption as in the simple case considered previously. The basic assumption is that the differential $\mathrm{d}f$ of a function $f(x_1, x_2, \ldots, x_n)$ is a good approximation to the increment (i.e. error) of f resulting from the increments (i.e. errors) $\mathrm{d}x_1, \ldots, \mathrm{d}x_n$ of $x_1, \ldots, x_n$. It follows from equation (7) of Section 8 that

$$\mathrm{d}f = \frac{\partial f}{\partial x_1}\mathrm{d}x_1 + \frac{\partial f}{\partial x_2}\mathrm{d}x_2 + \ldots + \frac{\partial f}{\partial x_n}\mathrm{d}x_n.$$

If we write this result in the form

$$\frac{\mathrm{d}f}{f} = \frac{x_1}{f}\frac{\partial f}{\partial x_1}\frac{\mathrm{d}x_1}{x_1} + \ldots + \frac{x_n}{f}\frac{\partial f}{\partial x_n}\frac{\mathrm{d}x_n}{x_n}$$

and write $p(x_r)$ for the numerical value of the possible percentage error in x_r then taking numerical values we have

$$p(f) \leqq \left|\frac{x_1}{f}\frac{\partial f}{\partial x_1}\right| p(x_1) + \ldots + \left|\frac{x_n}{f}\frac{\partial f}{\partial x_n}\right| p(x_n), \tag{3}$$

the symbol $|A|$ denoting the *numerical value* of the quantity A.

For example for the function defined by equation (1)

$$\frac{\partial u}{\partial x} = 100\frac{1}{y}, \qquad \frac{\partial u}{\partial y} = -100\frac{x}{y^2}$$

so that

$$\left|\frac{x}{u}\frac{\partial u}{\partial x}\right| = 1, \qquad \left|\frac{y}{u}\frac{\partial u}{\partial y}\right| = 1.$$

It follows from equation (3) that $p(u) = p(x) + p(y)$ in agreement with equation (2).

As a further example consider the problem of determining the density, ρ, of a substance by measuring the mass M and the radius r of a sphere of the substance. We then have

$$\rho = \frac{3M}{4\pi r^3}$$

and it is readily shown that

$$\frac{\partial \rho}{\partial M} = \frac{3}{4\pi r^3}, \qquad \frac{\partial \rho}{\partial r} = -\frac{9M}{4\pi r^4}$$

from which it follows that

$$\left|\frac{M}{\rho}\frac{\partial \rho}{\partial M}\right| = 1, \qquad \left|\frac{r}{\rho}\frac{\partial \rho}{\partial r}\right| = 3,$$

so that if $p(M)$ and $p(r)$ denote the possible percentage errors in the measurements of M and r, the estimate of ρ may have a percentage error as large as

$$p(\rho) = p(M) + 3p(r). \tag{4}$$

For example, if the error in the measurement of the mass is 1 % and the error in the measurement of the radius is 2 %, the estimate of the density may be out by as much as 7 % (since $p(M) = 1$, $p(r) = 2$, $p(\rho) = 7$).

Equations (2) and (4) are special cases of a more general result. If

$$f = kx_1^q x_2^r \cdots x_n^s \tag{5}$$

where $k, q, r, \ldots, s$ are constants, then

$$\frac{\partial f}{\partial x_1} = qkx_1^{q-1} x_2 \cdots x_n^s$$

so that

$$\left|\frac{x_1}{f} \frac{\partial f}{\partial x_1}\right| = |q|$$

and it follows from equation (3) that

$$p(f) = |q|p(x_1) + |r|p(x_2) + \ldots + |s|\, p(x_n). \tag{6}$$

If $x_1, x_2, \ldots, x_n$ are all liable to the same percentage error ε, it follows that f is liable to a percentage error of as much as

$$(|q| + |r| + \ldots + |s|)\varepsilon.$$

Gauss' formula

We saw that formula (1) above implied that if the percentage errors in x and y were ε_1 and ε_2 respectively then the percentage error in u would be $\varepsilon_1 - \varepsilon_2$. If ε_1 can vary between $\pm$ 2 % and ε_2 can vary between $\pm$ 1 % then the percentage error in u may be as much as $3\frac{1}{2}$ % or as little as $\frac{1}{2}$ %. In view of uncertainties of this kind it is usual to calculate the absolute error in $f(x, y)$ due to errors $\mathrm{d}x$, $\mathrm{d}y$ in x, y by Gauss' formula

$$\Delta f = \left[\left(\frac{\partial f}{\partial x}\mathrm{d}x\right)^2 + \left(\frac{\partial f}{\partial y}\mathrm{d}y\right)^2\right]^{\frac{1}{2}}$$

so that the percentage error is estimated to be

$$\begin{aligned} P(f) &= \frac{\Delta f}{f} \times 100 \\ &= \left[\left(\frac{x}{f}\frac{\partial f}{\partial x}\frac{\mathrm{d}x}{x} \times 100\right)^2 + \left(\frac{y}{f}\frac{\partial f}{\partial y}\frac{\mathrm{d}y}{y} \times 100\right)^2\right]^{\frac{1}{2}} \end{aligned}$$

so that if the percentage errors in x and y are $p(x)$, $p(y)$ then

$$P(f) = \left[\left(\frac{x}{f}\frac{\partial f}{\partial x}p(x)\right)^2 + \left(\frac{y}{f}\frac{\partial f}{\partial y}p(y)\right)^2\right]^{\frac{1}{2}}. \tag{7}$$

Whereas equation (3) gives the *maximum error* in f, equation (7) gives the *average error*. In the case considered this would give

$$P(u) = (\varepsilon_1^2 + \varepsilon_2^2)^{\frac{1}{2}}.$$

If $\varepsilon_1 = 2$ and $\varepsilon_2 = 1.5$ then

$$P(u) = (4 + \tfrac{9}{4})^{\frac{1}{2}}$$

so that the error in u is estimated by Gauss' formula to be $2\frac{1}{2}$ %.

In the case of a function $f(x_1, \ldots, x_n)$ when each x_r is liable to a percentage error $p(x_r)$ Gauss' estimate of the percentage error in f is given by the equation

$$P^2(f) = \left\{\frac{x_1}{f}\frac{\partial f}{\partial x_1}p(x_1)\right\}^2 + \ldots + \left\{\frac{x_n}{f}\frac{\partial f}{\partial x_n}p(x_n)\right\}^2. \tag{8}$$

For the function defined by equation (5)

$$P^2(f) = q^2p^2(x_1) + \ldots + s^2p^2(x_n) \tag{9}$$

we see that if $x_1, x_2, \ldots, x_n$ are all liable to the same percentage error ε, then the percentage error in f is

$$\{q^2 + r^2 + \ldots + s^2\}^{\frac{1}{2}}\varepsilon.$$

EXAMPLE 20. If $u = ax^2/y^{\frac{1}{3}}$, and the percentage errors of x and y are $\pm$ 2 % and $\pm$ 10 % respectively, then in equation (5) $q = 2$, $r = -\frac{1}{3}$, $p(x) = 2$, $p(y) = 10$ and the error in u may be as great as

$$p(u) = 2 \cdot 2 + \tfrac{1}{3} \cdot 10 = 7\tfrac{1}{3}\ \%.$$

The Gauss formula (7) gives

$$P(u) = [4 \cdot 4 + \tfrac{1}{9} \cdot 100]^{\frac{1}{2}} = [\tfrac{244}{9}]^{\frac{1}{2}} = 5.2\ \%.$$

EXAMPLE 21. In an Atwood's machine the acceleration due to gravity, g, is determined by measuring a distance s and a time t and then using the formula

$$g = \frac{2s}{t^2}.$$

If the error in measuring s is 1 % and the error in measuring t is

$\frac{1}{2}$% then putting $q = 1$, $r = -2$ in equation (5) we have

$$p(g) = p(s) + 2p(t)$$

and putting $p(s) = 1$, $p(t) = \frac{1}{2}$ we see that g may be in error by 2 %. The average error as given by equation (9) is

$$P(g) = \sqrt{1 + 4 \cdot \tfrac{1}{4}} = \sqrt{2}$$

so that the average error is approximately 1.4%.

EXAMPLE 22. The area Δ of an isosceles triangle of equal sides b and included angle A is determined by the formula

$$\Delta = \tfrac{1}{2}b^2 \sin A.$$

If the measurements in b, A are in error by δb, δA then

$$\delta\Delta = b\delta b \sin A + \tfrac{1}{2}b^2 \cos A \cdot \delta A$$

so

$$\frac{\delta\Delta}{\Delta} = \frac{2\delta b}{b} + \cot A \cdot \delta A.$$

where δA must be expressed in *radian measure*. If the error in A is ε (in minutes of arc) then

$$\delta A = \frac{\varepsilon}{60} \frac{\pi}{180} = \frac{\varepsilon\pi}{10800}.$$

PROBLEMS

1. The density ρ of dry air in g/cm³ at t°C, and pressure h mm of mercury is given by the formula

$$\rho = \frac{0.001293}{1 + 0.00367t} \times \frac{h}{760}.$$

Obtain an expression giving the approximate change in the value of ρ for small changes δt, δh in the values of t and h respectively, and use it to find the percentage change in ρ when t increases from 25 to 26 and h from 760 to 762.

2. The heat Q (in calories) liberated on mixing x gmol of sulphuric acid with y gmol of water is given by

$$Q = \frac{17860\,xy}{1.798\,x + y}.$$

Find to the nearest calorie the additional heat generated if a mixture of 5 gmol of acid and 4 gmol of water has its acid content increased by 2 % and its water content by 1 %.

3. If the period T of a pendulum of length L oscillating at a height H above the surface of the earth is given by $T = 2\pi\sqrt{(L/g)}$, where $g = g_0(1 + H/r)^{-2}$, g_0 and r being constants, find the error in T due to errors of l per cent in L, and h per cent in H.

4. The refractive index μ of a prism is given by the formula

$$\mu = \sin \tfrac{1}{2}(\delta + A)/\sin \tfrac{1}{2}A$$

where A is the angle of the prism and δ is the angle of minimum deviation. Find approximately the percentage error in μ if errors of θ, ϕ minutes are made in the measurements of A, δ respectively.

If $A = 60°$ and if the measurements of A, δ can be out by 1 minute, show that the error in the calculated value of μ will be less than 0.03 % of the correct value.

10. Exact Differentials

In applications of mathematics, for example in thermodynamics, we often construct forms of the type

$$P(x, y)\mathrm{d}x + Q(x, y)\mathrm{d}y. \tag{1}$$

The problem that then arises is whether there exists a function $\phi(x, y)$ which has this form as its differential. When such a function exists the form (1) is said to be an *exact differential.*

It is a simple matter to derive a necessary condition for the form to be exact. For, if it is, there will exist a function $\phi(x, y)$ whose differential

$$\mathrm{d}\phi = \frac{\partial \phi}{\partial x}\mathrm{d}x + \frac{\partial \phi}{\partial y}\mathrm{d}y$$

is identical with the form (1). In other words, $P(x, y)$ and $Q(x, y)$ must be such that, for some function $\phi(x, y)$

$$P(x, y) = \frac{\partial \phi}{\partial x}, \qquad Q(x, y) = \frac{\partial \phi}{\partial y}. \tag{2}$$

If we differentiate both sides of the first of these equations with respect to y and the second with respect to x we find that

$$\frac{\partial P}{\partial y} = \frac{\partial^2 \phi}{\partial y \partial x}, \qquad \frac{\partial Q}{\partial x} = \frac{\partial^2 \phi}{\partial x \partial y}$$

so that, by Schwartz's theorem we have as a necessary condition that

$$\frac{\partial P}{\partial y} = \frac{\partial Q}{\partial x}. \tag{3}$$

The question now arises as to whether the fulfillment of this condition alone is sufficient for the form (1) to be exact. We shall now prove that it is by actually finding a function $\phi(x, y)$ with the property (2). If we write

$$\phi(x, y) = \int P(x, y) \mathrm{d}x + \psi(y), \tag{4}$$

where the integration of $P(x, y)$ with respect to x is done by keeping y constant, and $\psi(y)$ is an arbitrary function of y, then $\phi(x, y)$ satisfies the first of equations (2). In order that the second one should be satisfied we must have

$$Q = \frac{\partial}{\partial y} \int P(x, y) \mathrm{d}x + \psi'(y)$$

i.e. $\psi(y)$ must be such that its derivative $\psi'(y)$ is given by

$$\psi'(y) = Q - \frac{\partial}{\partial y} \int P(x, y) \mathrm{d}x. \tag{5}$$

This will be a function of y alone only if its partial derivative with respect to x is zero, i.e. if

$$\frac{\partial Q}{\partial x} - \frac{\partial}{\partial x} \frac{\partial}{\partial y} \int P(x, y) \mathrm{d}x = 0$$

i.e. if

$$\frac{\partial Q}{\partial x} - \frac{\partial}{\partial x} \int \frac{\partial P}{\partial y} \mathrm{d}x = 0$$

or

$$\frac{\partial Q}{\partial x} - \frac{\partial P}{\partial y} = 0.$$

Since this condition is satisfied it follows that $\psi(y)$ *is* a function of

y alone. Integrating both sides of (5) with respect to y we find that

$$\psi(y) = \int \left\{ Q - \frac{\partial}{\partial y} \int P \mathrm{d}x \right\} \mathrm{d}y$$

and hence from equation (4) that

$$\phi(x, y) = \int P \mathrm{d}x + \int \left\{ Q - \frac{\partial}{\partial y} \int P \mathrm{d}x \right\} \mathrm{d}y. \tag{6}$$

We obtained equation (6) by integrating the first of the equations (2). If we had started with the second we would have obtained an expression of the form

$$\phi(x, y) = \int Q \mathrm{d}y + \int \left\{ P - \frac{\partial}{\partial x} \int Q \mathrm{d}y \right\} \mathrm{d}x. \tag{7}$$

To illustrate the procedure consider the form

$$\frac{x+y}{xy} \mathrm{d}x - \frac{x+y}{y^2} \mathrm{d}y \tag{8}$$

which has

$$P(x, y) = \frac{1}{y} + \frac{1}{x}, \qquad Q(x, y) = -\frac{x}{y^2} - \frac{1}{y}.$$

It is obvious that equation (3) is satisfied so that the form is an exact differential.

Since

$$\int P(x, y) \mathrm{d}x = \frac{x}{y} + \log x,$$

it follows that

$$\frac{\partial}{\partial y} \int P(x, y) \mathrm{d}x = -\frac{x}{y^2}$$

and that

$$\int \left\{ Q - \frac{\partial}{\partial y} \int P(x, y) \mathrm{d}x \right\} \mathrm{d}y = \int \left(-\frac{1}{y} \right) \mathrm{d}y = -\log y.$$

Hence the form (8) is an exact differential $\mathrm{d}\phi$ where

$$\phi(x, y) = \frac{x}{y} + \log x - \log y. \tag{9}$$

It is obvious from this result that not every differential form of the type (1) is an exact differential. If we wish to show that a differential form is not exact we denote it by a symbol of the type $đ\phi$ with a bar through the d or by $\Delta\phi$ or $\delta\phi$.

In certain cases in which the form

$$P\mathrm{d}x + Q\mathrm{d}y \tag{10}$$

is not exact, it sometimes happens that we can find a function $\mu(x, y)$ such that

$$\mu P\mathrm{d}x + \mu Q\mathrm{d}y \tag{11}$$

is exact. When this happens we say that $\mu(x, y)$ is an *integrating factor* of the form (10). If the form (11) is exact then $\mu(x, y)$ must be such that

$$\frac{\partial}{\partial y}(\mu P) = \frac{\partial}{\partial x}(\mu Q). \tag{12}$$

This concept is illustrated by Problems 2, 3 below. We shall return to it in § 6 of Chapter 9 below.

PROBLEMS

1. Find which of the following forms are exact

(i) $(x^3 + 3xy^2)\mathrm{d}x + (y^3 + 3x^2y)\mathrm{d}y$;

(ii) $(x^2 + y^2)\mathrm{d}x - 2xy\mathrm{d}y = 0$;

(iii) $\left(1 + \frac{y^2}{x^2}\right)\mathrm{d}x - \frac{2y}{x}\mathrm{d}y = 0$;

(iv) $(x^2 - 4xy - 2y^2)\mathrm{d}x + (y^2 - 4xy - 2x^2)\mathrm{d}y = 0$;

(v) $(x + 2y)\mathrm{d}x - x\mathrm{d}y$;

(vi) $\frac{x + 2y}{x^3}\mathrm{d}x - \frac{1}{x^2}\mathrm{d}y$.

In the cases in which the form is exact find the function of which it is the exact differential.

2. Show that the form

$$f = 2xy\mathrm{d}x - x^2\mathrm{d}y$$

is not exact but that f/y^2 is exact.

3. Find an integrating factor of

$$(4xy + y^2)\mathrm{d}x + (x^2 + 4xy)\mathrm{d}y$$

of the type $(x + y)^n$.

11. The Integral of an Exact Differential

An exact differential has an important property which is used frequently in applications. The integral of the differential form

$$P(x, y)\mathrm{d}x + Q(x, y)\mathrm{d}y \tag{1}$$

along the curve Γ joining the two points $A(x_1, y_1)$ and $B(x_2, y_2)$ (cf. Fig. 107) may be defined to be

$$I = \int_{x_1}^{x_2} \{P[x, \psi(x)] + Q[x, \psi(x)]\psi'(x)\}\mathrm{d}x \tag{2}$$

where

$$y = \psi(x)$$

is the equation of the curve Γ. In general, the value of this integral will depend on the specific curve Γ joining A and B. However, if the differential form (1) is an exact differential of a function $\phi(x, y)$ then we may write equation (2) in the form

$$I = \int_{x_1}^{x_2} \frac{\mathrm{d}\phi}{\mathrm{d}x}\mathrm{d}x = \phi(x_2, y_2) - \phi(x_1, y_1) \tag{3}$$

showing that the value of the integral I depends only upon the coordinates of the points A and B and not upon the nature of the curve Γ

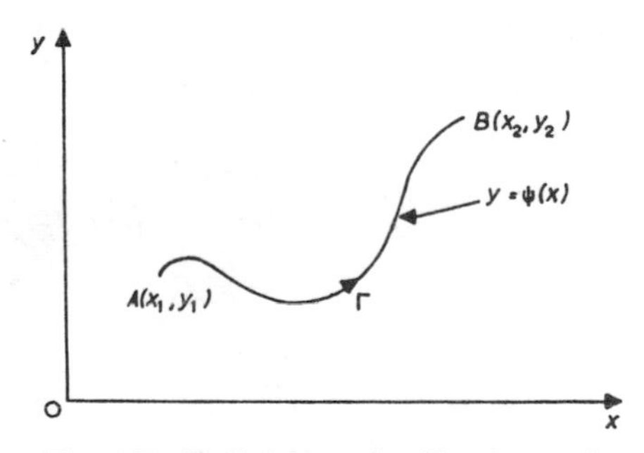

Fig. 107. Definition of a line integral.

joining them. For this reason the function $\phi(x, y)$ is sometimes called a *point function*.

Integration along a closed curve is important in applications of mathematics to cyclic processes (such as occur, for instance, in thermodynamics). If we have a closed curve Γ and are integrating an exact differential

$$\mathrm{d}\phi = P\mathrm{d}x + Q\mathrm{d}y$$

round it then taking any two points $A(x_1, y_1)$ and $B(x_2, y_2)$ on Γ (cf. Fig. 108) we see that

$$\int_\Gamma (P\mathrm{d}x + Q\mathrm{d}y) = \int_{\Gamma_1} (P\mathrm{d}x + Q\mathrm{d}y) + \int_{\Gamma_2} (P\mathrm{d}x + Q\mathrm{d}y)$$
$$= [\phi(x_2, y_2) - \phi(x_1, y_1)] + [\phi(x_1, y_1) - \phi(x_2, y_2)]$$

which shows that

$$\int_\Gamma (P\mathrm{d}x + Q\mathrm{d}y) = 0. \tag{4}$$

Thus the integral of an exact differential around a closed curve is zero.

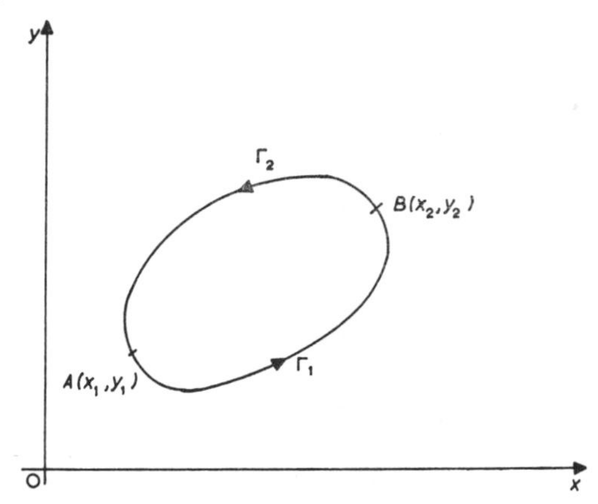

Fig. 108. Integration along a closed curve.

EXAMPLE 23. *By considering a straight line and a parabola through the points $A(1, 1)$, $B(2, 4)$ show by direct integration that the value of*

$$\int_{AB} (x\mathrm{d}y - y\mathrm{d}x)$$

depends on the path chosen, but that

$$\int_{AB} \frac{1}{x^2} (x\mathrm{d}y - y\mathrm{d}x)$$

is independent of the path.

If we take Γ_1 to be the straight line with equation $y = 3x - 2$ then

$$\int_{\Gamma_1} (x\mathrm{d}y - y\mathrm{d}x) = \int_1^2 [x \cdot 3 - (3x - 2)]\mathrm{d}x = 2\int_1^2 \mathrm{d}x = 2,$$

and if we take Γ_2 to be the parabola with equation $y = x^2$ then

$$\int_{\Gamma_2} (x\mathrm{d}y - y\mathrm{d}x) = \int_1^2 [x \cdot 2x - x^2]\mathrm{d}x = \int_1^2 x^2\,\mathrm{d}x = \tfrac{7}{3},$$

showing that the value of the integral depends on the path. On the other hand

$$\int_{\Gamma_1} \frac{1}{x^2}(x\mathrm{d}y - y\mathrm{d}x) = 2\int_1^2 \frac{\mathrm{d}x}{x^2} = 1,$$

and

$$\int_{\Gamma_2} \frac{1}{x^2}(x\mathrm{d}y - y\mathrm{d}x) = \int_1^2 \mathrm{d}x = 1.$$

In this case

$$\frac{1}{x^2}(x\mathrm{d}y - y\mathrm{d}x) = \mathrm{d}\phi$$

where

$$\phi(x, y) = \frac{y}{x}.$$

Hence $\phi(2, 4) = 2$, $\phi(1, 1) = 1$ so that the value of the integral along any curve joining A to B is $\phi(2, 4) - \phi(1, 1) = 1$.

PROBLEMS

1. Find the value of

$$\int_\Gamma [(x + y^2)\mathrm{d}x + 2xy^2\mathrm{d}y]$$

where Γ is the part of the parabola $y^2 = 4x$ between the origin and the point (1, 2).

2. Show that the value of the integral

$$\int_\Gamma [3x(x + 2y)\mathrm{d}x + 3x^2 - y^2)\mathrm{d}y],$$

where Γ is a curve joining the origin to the point (1, 2), is independent of the form of Γ, and find its value.

3. Show that the expression

$$\Delta f = (x^2y^2 - y)\mathrm{d}x + (2x^3y + x)\mathrm{d}y$$

is not an exact differential but that $x^{\ 2}\Delta f$ is exact.

By considering a straight line and a parabola through the points $A(1, 1)$, $B(2, 4)$ show by direct integration that the value of

$$\int_{AB} \Delta f$$

depends on the path chosen, but that

$$\int_{AB} x^{-2}\Delta f$$

is independent of the path.

12. Implicit Functions

We saw in § 5 that if we have a relation of the kind

$$u(x, y) = 0$$

defining y in terms of x we can calculate the derivative of y with respect to x by means of the formula

$$\frac{\mathrm{d}y}{\mathrm{d}x} = -\frac{\dfrac{\partial u}{\partial x}}{\dfrac{\partial u}{\partial y}}$$

(cf. equation (6) of § 5). When the functional relationship between x and y is of this type we say that y is an *implicit function* of x. Naturally we could equally well regard x as an implicit function of y. We now generalize this idea to relations which define one quantity as an implicit function of two other variables.

If we are given a relation

$$F(x, y, z) = 0 \tag{1}$$

connecting the three variables x, y and z, then we can regard this relation as defining any one of them in terms of the other two. In other words any one of them can be regarded as a function of the other two. For instance, we may think of z as a function of x and y, or alternatively, of x as a function of y and z.

Now we may write equation (1) in the differential form

$$\frac{\partial F}{\partial x}\,\mathrm{d}x + \frac{\partial F}{\partial y}\,\mathrm{d}y + \frac{\partial F}{\partial z}\,\mathrm{d}z = 0. \tag{2}$$

Suppose that we choose x and y to be the independent variables. Then we have*

$$\mathrm{d}z = \left(\frac{\partial z}{\partial x}\right)_y \mathrm{d}x + \left(\frac{\partial z}{\partial y}\right)_x \mathrm{d}y. \tag{3}$$

If we substitute from (3) into (2)

$$\left\{\frac{\partial F}{\partial x} + \frac{\partial F}{\partial z}\left(\frac{\partial z}{\partial x}\right)_y\right\}\mathrm{d}x + \left\{\frac{\partial F}{\partial y} + \frac{\partial F}{\partial z}\left(\frac{\partial z}{\partial y}\right)_x\right\}\mathrm{d}y = 0.$$

* We add the suffixes y and x to the partial derivatives here in order to emphasize which are the independent variables.

Since the variables x and y are independent variables we may equate the coefficients of their differentials to zero. In this way we find that

$$\left(\frac{\partial z}{\partial x}\right)_y = -\frac{\dfrac{\partial F}{\partial x}}{\dfrac{\partial F}{\partial z}} \tag{4}$$

$$\left(\frac{\partial z}{\partial y}\right)_x = -\frac{\dfrac{\partial F}{\partial y}}{\dfrac{\partial F}{\partial z}}. \tag{5}$$

Similarly if we had taken x to be a function of y and z we should have found that

$$\left(\frac{\partial x}{\partial y}\right)_z = -\frac{\dfrac{\partial F}{\partial y}}{\dfrac{\partial F}{\partial x}} \tag{6}$$

$$\left(\frac{\partial x}{\partial z}\right)_y = -\frac{\dfrac{\partial F}{\partial z}}{\dfrac{\partial F}{\partial x}}. \tag{7}$$

Finally we could have supposed that y was a function of x and z and obtained the relations

$$\left(\frac{\partial y}{\partial x}\right)_z = -\frac{\dfrac{\partial F}{\partial x}}{\dfrac{\partial F}{\partial y}} \tag{8}$$

$$\left(\frac{\partial y}{\partial z}\right)_x = -\frac{\dfrac{\partial F}{\partial z}}{\dfrac{\partial F}{\partial y}}. \tag{9}$$

Relations of three kinds can be derived from the last six equations. From equations (6) and (8) we have the relation

$$\left(\frac{\partial x}{\partial y}\right)_z = \frac{1}{\left(\frac{\partial y}{\partial x}\right)_z} \tag{10}$$

and there are two others of the same type. The equation (6), (9) and (7) show that

$$\left(\frac{\partial x}{\partial y}\right)_z \left(\frac{\partial y}{\partial z}\right)_x = -\left(\frac{\partial x}{\partial z}\right)_y \tag{11}$$

and there are two more relations of the same kind. Finally, if we multiply equations (6), (9) and (4) we find that

$$\left(\frac{\partial x}{\partial y}\right)_z \left(\frac{\partial y}{\partial z}\right)_x \left(\frac{\partial z}{\partial x}\right)_y = -1. \tag{12}$$

13. Applications to Thermodynamics

In this section we shall consider some applications of the theorems of partial differentiation to thermodynamics, a subject of some importance to students of the biological sciences.

The simplest thermodynamic systems are homogeneous fluids which are free from any external stress other than a constant hydrostatic pressure. It is found experimentally that in such a system the pressure p, the volume v and the temperature T of unit mass of the fluid cannot be independently specified. The three physical variables are related by an equation of the type

$$F(p, v, T) = 0 \tag{1}$$

which is called the equation of state of the fluid. Mathematically equation (1) is identical with equation (1) of the last section, so that equations (10), (11) and (12) of that section are applicable. Identifying p with x, v with y and T with z, we find that these equations give the results

$$\left(\frac{\partial p}{\partial v}\right)_T = 1 \Big/ \left(\frac{\partial v}{\partial p}\right)_T, \tag{2}$$

$$\left(\frac{\partial p}{\partial v}\right)_T \left(\frac{\partial v}{\partial T}\right)_p = -\left(\frac{\partial p}{\partial T}\right)_v, \tag{3}$$

$$\left(\frac{\partial p}{\partial v}\right)_T \left(\frac{\partial v}{\partial T}\right)_p \left(\frac{\partial T}{\partial p}\right)_v = -1, \tag{4}$$

which are of great importance in thermodynamics since they are capable of direct experimental verification. The partial derivatives occurring in them can be identified with physical quantities such as the coefficient of thermal expansion, the bulk modulus of elasticity etc., (see below).

If we consider unit mass of a substance, then the principle of the conservation of energy — in the form of the first law of thermodynamics — states that in a small change

$$\Delta e = \Delta q + \Delta w, \tag{5}$$

where Δe is the increase in the internal energy, Δq is the heat (in dynamical units, i.e. in joules, not calories) transferred to the system during the change, and Δw is the work done *on* the system. The change Δe depends only on the initial and final states of the system and not on how the work is done on the system. It follows from the considerations of § 11 that Δe is an exact differential so that we may write $\Delta e = \mathrm{d}e$.

The second law of thermodynamics states that Δq is not an exact differential, but that, if T is the absolute temperature of the system, $\Delta q/T$ is the exact differential of a function s called the entropy. We therefore, have

$$\frac{\Delta q}{T} = \mathrm{d}s. \tag{6}$$

For a simple system Δw takes a simple form. If it is a gas of unit mass occupying volume v at pressure p then

$$\Delta w = -p\mathrm{d}v. \tag{7}$$

If it is a wire held in tension by a force f

$$\Delta w = -f\mathrm{d}l$$

where $\mathrm{d}l$ is the change in length of the wire, and if it is a soap film of area $\mathscr{A}$ and surface tension t,

$$\Delta w = t\mathrm{d}\mathscr{A}.$$

For more complicated systems Δw can be written in the form

$$\Delta w = \sum_{i=1}^{n} X_i \mathrm{d}x_i.$$

If we substitute from equations (7) and (6) into equation (5) we obtain the equation

$$\mathrm{d}e = T\mathrm{d}s - p\mathrm{d}v. \tag{8}$$

We shall now make use of differentials to derive results of physical significance from this equation. If we regard v and s as the independent variables, then we can write

$$\mathrm{d}e = \left(\frac{\partial e}{\partial s}\right)_v \mathrm{d}s + \left(\frac{\partial e}{\partial v}\right)_s \mathrm{d}v.$$

Substituting this form into equation (8) we have

$$\left\{\left(\frac{\partial e}{\partial s}\right)_v - T\right\}\mathrm{d}s + \left\{\left(\frac{\partial e}{\partial v}\right)_s + p\right\}\mathrm{d}v = 0.$$

Since s and v are independent variables we may equate the coefficients of $\mathrm{d}s$ and $\mathrm{d}v$ to zero. In this way we have the pair of equations

$$T = \left(\frac{\partial e}{\partial s}\right)_v, \qquad p = -\left(\frac{\partial e}{\partial v}\right)_s.$$

Forming the second derivative

$$\frac{\partial^2 e}{\partial v \partial s} = \frac{\partial^2 e}{\partial s \partial v}$$

from each of these equations we find that

$$\left(\frac{\partial T}{\partial v}\right)_s = -\left(\frac{\partial p}{\partial s}\right)_v. \tag{9}$$

This is known as Maxwell's first thermodynamic relation.

If we introduce the *enthalpy* (or *heat content*) h, through the equation

$$h = e + pv \tag{10}$$

we have the differential relation

$$\mathrm{d}h = \mathrm{d}e + p\mathrm{d}v + v\mathrm{d}p.$$

From equation (8) we know that $\mathrm{d}e + p\mathrm{d}v$ is $T\mathrm{d}s$ so that

$$\mathrm{d}h = T\mathrm{d}s + v\mathrm{d}p. \tag{11}$$

If we regard p and s as the independent variables then since

$$\mathrm{d}h = \left(\frac{\partial h}{\partial s}\right)_p \mathrm{d}s + \left(\frac{\partial h}{\partial p}\right)_s \mathrm{d}p$$

we find that

$$T = \left(\frac{\partial h}{\partial s}\right)_p, \quad v = \left(\frac{\partial h}{\partial p}\right)_s$$

and, as a consequence, that

$$\left(\frac{\partial T}{\partial p}\right)_s = \left(\frac{\partial v}{\partial s}\right)_p. \tag{12}$$

Equation (12) is Maxwell's second relation.

The *Gibbs' function* (or *thermodynamic potential*) is defined by the equation

$$g = h - Ts. \tag{13}$$

The differential form of this equation is

$$\mathrm{d}g = \mathrm{d}h - T\mathrm{d}s - s\mathrm{d}T$$

which, with the aid of equation (11), can be written in the form

$$\mathrm{d}g = v\mathrm{d}p - s\mathrm{d}T.$$

Now, by definition, if p and T are the independent variables

$$\mathrm{d}g = \left(\frac{\partial g}{\partial p}\right)_T \mathrm{d}p + \left(\frac{\partial g}{\partial T}\right)_p \mathrm{d}T$$

so that

$$v = \left(\frac{\partial g}{\partial p}\right)_T, \quad s = -\left(\frac{\partial g}{\partial T}\right)_p.$$

Forming $\partial^2 g/\partial p \partial T$ from these equations we obtain the equation

$$\left(\frac{\partial v}{\partial T}\right)_p = -\left(\frac{\partial s}{\partial p}\right)_T, \tag{14}$$

which is the third of Maxwell's thermodynamic relations.

Similarly, we define the *Helmholtz free energy* f by the equation

$$f = e - Ts, \tag{15}$$

the differential form of which is

$$\mathrm{d}f = \mathrm{d}e - T\mathrm{d}s - s\mathrm{d}T.$$

From equation (8) we see that we may replace $\mathrm{d}e - T\mathrm{d}s$ by $-p\mathrm{d}v$ so that

$$\mathrm{d}f = -p\mathrm{d}v - s\mathrm{d}T.$$

Writing

$$\mathrm{d}f = \left(\frac{\partial f}{\partial v}\right)_T \mathrm{d}v + \left(\frac{\partial f}{\partial T}\right)_v \mathrm{d}T$$

and equating coefficients of $\mathrm{d}v$ and $\mathrm{d}t$, we see that

$$\left(\frac{\partial f}{\partial v}\right)_T = -p, \qquad \left(\frac{\partial f}{\partial T}\right)_v = -s.$$

Forming $\partial^2 f/\partial v \partial T$ in two ways from these equations we obtain the fourth of the Maxwell relations

$$\left(\frac{\partial p}{\partial T}\right)_v = \left(\frac{\partial s}{\partial v}\right)_T. \tag{16}$$

From thermodynamics we get simple illustrations of the physical meaning of some partial derivatives. We define the specific heat at constant pressure c_p, as the heat absorbed per unit increase of temperature of a unit mass of the substance the pressure being kept constant. We therefore have

$$c_p = \left(\frac{\Delta q}{\Delta T}\right)_p$$

where ΔT is the increase in temperature produced by the absorption of an amount of heat Δq. Writing $\Delta q = T\Delta s$ and letting $\Delta T \to 0$ we find that

$$c_p = T\left(\frac{\partial s}{\partial T}\right)_p. \tag{17}$$

Similarly the specific heat at constant volume, c_v, is defined by the equation

$$c_v = \left(\frac{\Delta q}{\Delta T}\right)_v = T\left(\frac{\partial s}{\partial T}\right)_v. \tag{18}$$

The latent heat of expansion is defined to be

$$l_v = \left(\frac{\Delta q}{\Delta v}\right)_T = T\left(\frac{\partial s}{\partial v}\right)_T \tag{19}$$

and the latent heat of change of pressure to be

$$l_p = \left(\frac{\Delta q}{\Delta p}\right)_T = T\left(\frac{\partial s}{\partial p}\right)_T. \tag{20}$$

Now we regard p and T as the independent variables of a simple system

$$\mathrm{d}s = \left(\frac{\partial s}{\partial T}\right)_p \mathrm{d}T + \left(\frac{\partial s}{\partial p}\right)_T \mathrm{d}p.$$

Multiplying both sides of this equation by T and using equations (17) and (20) we find that

$$\Delta q = c_p \mathrm{d}T + l_p \mathrm{d}p. \tag{21}$$

On the other hand, if we take v and T to be the independent variables and substitute from equations (18) and (19) into the definition

$$\mathrm{d}s = \left(\frac{\partial s}{\partial T}\right)_v \mathrm{d}T + \left(\frac{\partial s}{\partial v}\right)_T \mathrm{d}v$$

we obtain the equation

$$\Delta q = c_v \mathrm{d}T + l_v \mathrm{d}v. \tag{22}$$

If we regard T as a function of v and p, then, equating the right hand sides of equations (21) and (22) we find that

$$(c_p - c_v)\mathrm{d}T = -l_p \mathrm{d}p + l_v \mathrm{d}v \tag{23}$$

from which we deduce that

$$c_p - c_v = -l_p \left(\frac{\partial p}{\partial T}\right)_v \tag{24}$$

$$c_p - c_v = l_v \left(\frac{\partial v}{\partial T}\right)_p \tag{25}$$

The coefficient of expansion α of a substance at constant pressure is the proportionate change in v for unit change in T, i.e.

$$\alpha = \frac{1}{v}\left(\frac{\partial v}{\partial T}\right)_p. \tag{26}$$

The bulk modulus of elasticity K, is defined to be

$$\frac{\Delta p}{-(\Delta v/v)}$$

at constant temperature, so

$$K = -v\left(\frac{\partial p}{\partial v}\right)_T. \tag{27}$$

PROBLEMS

1. Show that for any gas

$$\left(\frac{\partial e}{\partial v}\right)_T = T\left(\frac{\partial p}{\partial T}\right)_v - p.$$

Deduce that for a perfect gas, for which, by Joule's law, $(\partial e/\partial v)_T = 0$,

$$\left[\frac{\partial}{\partial T}\left(\frac{p}{T}\right)\right]_v = 0$$

so that

$$p = T\phi(v)$$

where $\phi(v)$ is an arbitrary function of v.

2. Show that

$$e = f - T\left(\frac{\partial f}{\partial T}\right)_v = -T^2\left[\frac{\partial}{\partial T}\left(\frac{f}{T}\right)\right]_v.$$

Prove also that

$$g = -v^2\left[\frac{\partial}{\partial v}\left(\frac{f}{v}\right)\right]_T, \qquad h = -T^2\left[\frac{\partial}{\partial T}\left(\frac{g}{T}\right)\right]_p.$$

3. Using Maxwell's third relation show that

$$l_p = -T\left(\frac{\partial v}{\partial T}\right)_p, \qquad \left(\frac{\partial c_p}{\partial p}\right)_T = -T\left(\frac{\partial^2 v}{\partial T^2}\right)_p$$

and using Maxwell's fourth relation show that

$$l_v = T\left(\frac{\partial p}{\partial T}\right)_v, \qquad \left(\frac{\partial c_v}{\partial v}\right)_T = T\left(\frac{\partial^2 p}{\partial T^2}\right)_v.$$

Deduce that

$$c_p - c_v = \left(\frac{\partial v}{\partial T}\right)_p \left(\frac{\partial p}{\partial T}\right)_v.$$

For a perfect gas, the equation of state is

$$pv = rT,$$

where r is a constant. Show that

$$c_p - c_v = r.$$

4. Show that

$$l_p = l_v\left(\frac{\partial v}{\partial p}\right)_T.$$

5. Prove that

$$l_v = p + \left(\frac{\partial e}{\partial v}\right)_T, \qquad l_p = \left(\frac{\partial e}{\partial p}\right)_T + p\left(\frac{\partial v}{\partial p}\right)_T.$$

6. Deduce from the expression for $c_p - c_v$ in **3** that

$$c_p - c_v = K\alpha^2 vT.$$

14. Taylor Series for a Function of Two Variables

In this section we shall derive the analogue for functions of two independent variables of the Taylor series expansion for functions of a single variable.

If we make the assumption that it is possible to expand a function $f(x, y)$ in the form of a double series

$$\begin{aligned} f(x, y) = c_{0,0} &+ c_{1,0}(x-a) + c_{1,1}(y-b) + c_{2,0}(x-a)^2 \\ &+ c_{2,1}(x-a)(y-b) + c_{2,2}(y-b)^2 \\ &+ c_{3,0}(x-a)^3 + c_{3,1}(x-a)^2(y-b) \\ &+ c_{3,2}(x-a)(y-b)^2 + c_{3,3}(y-b)^3 + \ldots \end{aligned} \tag{1}$$

then, if we differentiate both sides of this equation with respect to x, we find that

$$\begin{aligned} \frac{\partial f}{\partial x} = c_{1,0} &+ 2c_{2,0}(x-a) + c_{2,1}(y-b) + 3c_{3,0}(x-a)^2 \\ &+ 2c_{3,1}(x-a)(y-b) + c_{3,2}(y-b)^2 + \ldots \end{aligned} \tag{2}$$

$$\frac{\partial^2 f}{\partial x^2} = 2c_{2,0} + 3 \cdot 2c_{3,0}(x-a) + 2c_{3,1}(y-b) + \ldots \tag{3}$$

and, similarly, if we differentiate with respect to y we find that

$$\begin{aligned} \frac{\partial f}{\partial y} = c_{1,1} &+ c_{2,1}(x-a) + 2c_{2,2}(y-b) + c_{3,1}(x-a)^2 \\ &+ 2c_{3,2}(x-a)(y-b) + 3c_{3,3}(y-b)^2 + \ldots \end{aligned} \tag{4}$$

$$\frac{\partial^2 f}{\partial y^2} = 2c_{2,2} + 2c_{3,2}(x-a) + 3 \cdot 2c_{3,3}(y-b) + \ldots. \tag{5}$$

If we differentiate (2) with respect to y (or (4) with respect to x) we obtain the mixed derivative

$$\frac{\partial^2 f}{\partial x \partial y} = c_{2,1} + 2c_{3,1}(x-a) + 2c_{3,2}(y-b) + \ldots. \tag{6}$$

Now if we put $x = a$, $y = b$ in equations (1) to (6) and write

$$f_0,\ f_{10},\ f_{01},\ f_{20},\ f_{11},\ f_{02}$$

respectively for the values of

$$f,\ \frac{\partial f}{\partial x},\ \frac{\partial f}{\partial y},\ \frac{\partial^2 f}{\partial x^2},\ \frac{\partial^2 f}{\partial x \partial y},\ \frac{\partial^2 f}{\partial y^2}$$

when $x = a$, $y = b$ we find that

$$\begin{aligned} c_{0,0} &= f_0 \\ c_{1,0} &= f_{10}, \qquad c_{0,1} = f_{01} \\ 2c_{2,0} &= f_{20}, \qquad c_{2,1} = f_{11}, \qquad 2c_{0,2} = f_{02}. \end{aligned}$$

Substituting these values in equation (1) we obtain the Taylor expansion

$$\begin{aligned} f(x, y) &= f_0 + [(x-a)f_{10} + (y-b)f_{01}] \\ &\quad + \tfrac{1}{2}[(x-a)^2 f_{20} + 2(x-a)(y-b)f_{11} + (y-b)^2 f_{02}] + \dots \end{aligned} \tag{7}$$

We can establish Taylor's theorem for functions of two variables from the Maclaurin expansion (§ 9 of Chapter 7; p. 312 above)

$$F(t) = F(0) + \frac{t}{1!}F'(0) + \frac{t^2}{2!}F''(0) + \dots \frac{t^{n-1}}{(n-1)!}F^{(n-1)}(0) + R_n(t) \tag{8}$$

where

$$R_n(t) = \frac{t^n}{n!}F^{(n)}(\theta t) \tag{9}$$

denotes the remainder term.

If we let

$$F(t) = f(a + ht,\ b + kt)$$

i.e. if we let $F(t) = f(x, y)$ where $x = a + ht$, $y = b + kt$, then

$$F'(t) = \frac{\partial f}{\partial x}h + \frac{\partial f}{\partial y}k$$

$$F''(t) = \frac{\partial^2 f}{\partial x^2}h^2 + 2\frac{\partial^2 f}{\partial x \partial y}hk + \frac{\partial^2 f}{\partial y^2}k^2.$$

We may write these expressions in the form

$$F'(t) = \left(h\frac{\partial}{\partial x} + k\frac{\partial}{\partial y}\right)f(x, y),$$

$$F''(t) = \left(h\frac{\partial}{\partial x} + k\frac{\partial}{\partial y}\right)^2 f(x, y),$$

and it follows, by induction, that, if r is a positive integer

$$F^{(r)}(t) = \left(h\frac{\partial}{\partial x} + k\frac{\partial}{\partial y}\right)^r f(x, y).$$

Putting $t = 0$, (when $x = a$, $y = b$) we find that

$$F^{(r)}(0) = \left(h\frac{\partial}{\partial a} + k\frac{\partial}{\partial b}\right)^r f(a, b)$$

and hence that

$$t^r F^{(r)}(0) = \left(ht\frac{\partial}{\partial a} + kt\frac{\partial}{\partial b}\right)^r f(a, b) = \left\{(x-a)\frac{\partial}{\partial a} + (y-b)\frac{\partial}{\partial b}\right\}^r f(a, b).$$

Substituting this expression in equation (8) we obtain Taylor's theorem

$$f(x, y) = f(a, b) + \sum_{r=1}^{n-1}\left\{(x-a)\frac{\partial}{\partial a} + (y-b)\frac{\partial}{\partial b}\right\}^r f(a, b) + R_n \quad (10)$$

where

$$R_n = \left\{(x-a)\frac{\partial}{\partial a} + (y-b)\frac{\partial}{\partial b}\right\}^n f[a(1-\theta) + \theta x,\ b(1-\theta) + \theta y]$$
$$(0 < \theta < 1). \quad (11)$$

15. Maxima and Minima of Functions of Two Variables

If the function $f(x, y)$ has a turning value at the point $x = a$, $y = b$ then the function $f(x, b)$ must have a turning value at $x = a$ so that

$$f_a \equiv \left[\frac{\partial f}{\partial x}\right]_{\substack{x=a\\y=b}} = 0. \quad (1)$$

Similarly $f(a, y)$ must have a turning value at $y = b$ so that

$$f_b \equiv \left[\frac{\partial f}{\partial y}\right]_{\substack{x=a\\y=b}} = 0. \quad (2)$$

The values $x = a$, $y = b$ will yield a *maximum* value to the function $f(x, y)$ if

$$f(a + h, b + k) < f(a, b) \quad (3)$$

for the range of values of h and k giving all points in the neighbourhood of (a, b). (Cf. Fig. 109.)

Using the Taylor expansion of $f(a + h, b + k)$ with $f_a = f_b = 0$ we see that the condition (3) can be written in the form

$$\tfrac{1}{2}(h^2 f_{aa} + 2hk f_{ab} + k^2 f_{bb}) < 0 \tag{4}$$

for small values of h and k. If we write u for the ratio h/k we see that (4) can be written in the form

$$f_{aa} u^2 + 2u f_{ab} + f_{bb} < 0. \tag{5}$$

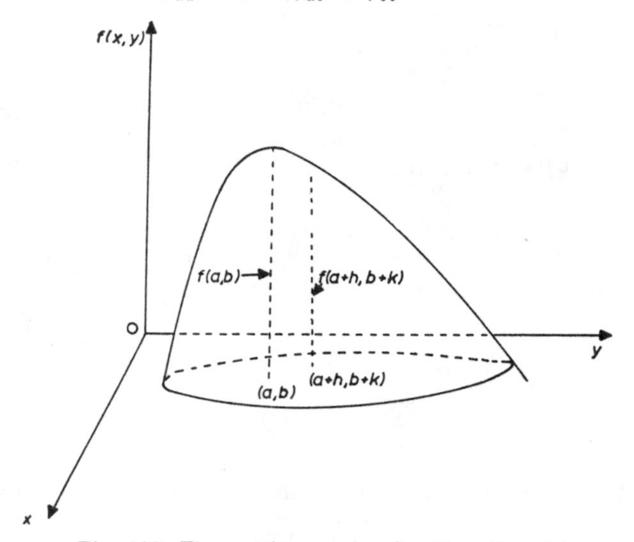

Fig. 109. The maximum of a function $f(x, y)$.

We saw in § 6 of Chapter 2 (pp. 89–90) that we can write the functional relationship

$$v = f_{aa} u^2 + 2 f_{ab} u + f_{bb}$$

in the form

$$v = f_{aa}(u - \alpha)^2 + \beta \tag{6}$$

where

$$\alpha = -\frac{f_{ab}}{f_{aa}}, \qquad \beta = f_{bb} - \frac{f_{ab}^2}{f_{aa}}.$$

From the graph of the function (6) (cf. Fig. 110) we see that if $v < 0$ for all values of u we must have $\beta < 0$, $f_{aa} < 0$. Hence for a maximum we must have

$$f_{aa} < 0, \qquad f_{ab}^2 < f_{aa} f_{bb}. \tag{7}$$

EXAMPLE 24. *Show that the function* $(x+y)(x^2+y^2-6)$ *has a maximum value at the point* $(-1, -1)$.

For this function

$$\frac{\partial f}{\partial x} = 3x^2 + 2xy + y^2 - 6, \qquad \frac{\partial f}{\partial y} = x^2 + 2xy + 3y^2 - 6,$$

$$\frac{\partial^2 f}{\partial x^2} = 6x + 2y, \qquad \frac{\partial^2 f}{\partial x \partial y} = 2(x+y), \qquad \frac{\partial^2 f}{\partial y^2} = 2x + 6y$$

so that if we put $a = -1$, $b = -1$,

$$f_a = 0, \quad f_b = 0 \quad f_{aa} = -8, \quad f_{bb} = -8, \quad f_{ab} = -4$$

showing that $f_{aa} < 0$ and $f_{ab}^2 < f_{aa}f_{bb}$. Hence at the point $(-1, -1)$ the function has the maximum value 8.

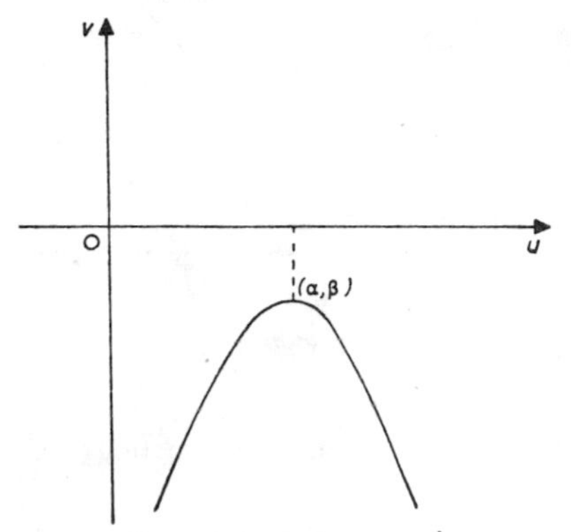

Fig. 110. The graph of the function $v = f_{aa}u^2 + 2f_{ab}u + f_{bb}$, in the case $f_{aa} < 0$, $f_{aa}f_{bb} - f^2_{ab} > 0$.

On the other hand, the values $x = a$, $y = b$ yield a *minimum* value to the function $f(x, y)$ if

$$f(a+h, b+k) - f(a, b) > 0 \tag{8}$$

for small values of h and k. (Cf. Fig. 111.) From consideration of the function defined by equation (6) we see that condition (8) is satisfied if $\beta > 0$, $f_{aa} > 0$. (Cf. Fig. 112.) Hence for a minimum we must have

$$f_{aa} > 0, \qquad f_{ab}^2 < f_{aa}f_{bb}. \tag{9}$$

EXAMPLE 25. *Show that the function* $x^4 - 2x^2 + y^2 - 2y$ *has minimum values at the points* $(-1, 1)$ *and* $(1, 1)$.

For this function we have

$$\frac{\partial f}{\partial x} = 4x^3 - 4x, \qquad \frac{\partial f}{\partial y} = 2y - 2$$

$$\frac{\partial^2 f}{\partial x^2} = 12x^2 - 4, \qquad \frac{\partial^2 f}{\partial x \partial y} = 0, \qquad \frac{\partial^2 f}{\partial y^2} = 2$$

so that when $a = -1$, $b = 1$ and when $a = 1$, $b = 1$,

$$f_a = 0, \quad f_b = 0, \quad f_{aa} > 0, \quad f_{ab}^2 < f_{aa} f_{bb}$$

showing that both points give minimum values.

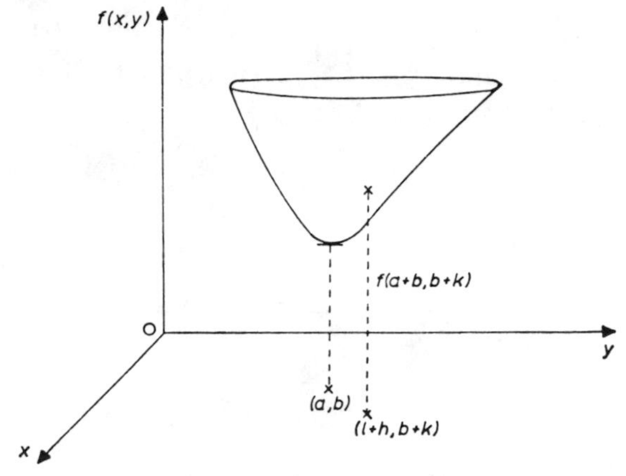

Fig. 111. The minimum of a function $f(x, y)$.

So far we have not considered what happens at a point (a, b) for

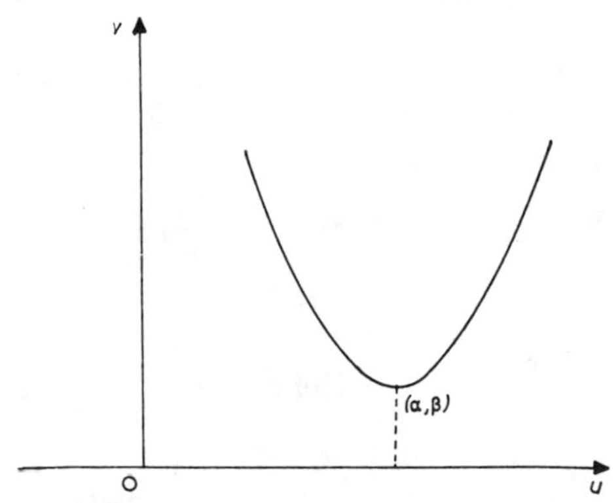

Fig. 112. The variation of the function $v = f_{aa}u^2 + 2f_{ab}u + f_{bb}$, in the case $f_{aa} > 0$, $f_{aa}f_{bb} - f^2_{ab} < 0$.

which $f_a = f_b = 0$ and

$$f_{ab}^2 > f_{aa}f_{bb}.$$

To illustrate what happens in this case we consider the function

$$f = y^2 - x^2 \tag{10}$$

at the point (0, 0). For this function

$$\frac{\partial f}{\partial x} = -2x, \quad \frac{\partial f}{\partial y} = +2y, \quad \frac{\partial^2 f}{\partial x^2} = -1, \quad \frac{\partial^2 f}{\partial x \partial y} = 0, \quad \frac{\partial^2 f}{\partial y^2} = +2$$

so that if we put $a = 0$, $b = 0$, we have

$$f_a = f_b = 0 \quad \text{and} \quad f_{ab} - f_{aa}f_{bb} = 4.$$

If we draw the surface in this case (cf. Fig. 113) we find that (0, 0) is a maximum point of the curve obtained by cutting the surface by

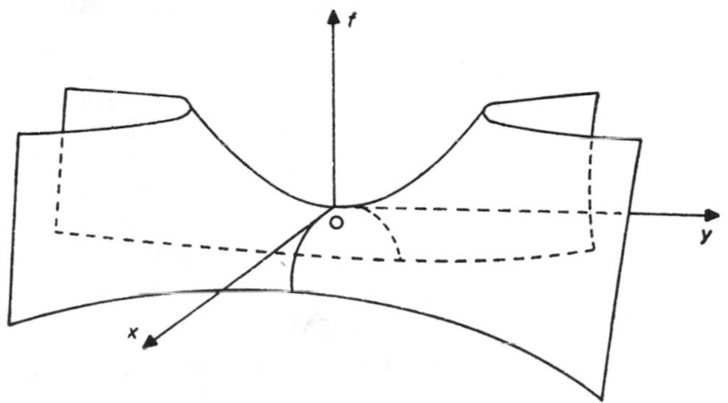

Fig. 113. Saddle point on the surface $f = y^2 - x^2$.

the plane $y = 0$, but is a minimum point of the curve obtained by cutting the surface by the plane $x = 0$. Such a point is called a *saddle point* of the surface.

In discussing particular problems it is useful to keep in mind:-

Summary of procedure

1. Values a, b of possible turning values are found by solving the equations

$$f_x = 0, \qquad f_y = 0.$$

2. If $f_{ab}^2 < f_{aa}f_{bb}$ then f is a *maximum* if $f_{aa} < 0$ and a *minimum* if $f_{aa} > 0$.
3. If $f_{ab}^2 > f_{aa}f_{bb}$ the point (a, b) is a *saddle-point*.
4. If $f_{ab}^2 = f_{aa}f_{bb}$ the case is doubtful.

As an illustration of how to set the work out we have:-

EXAMPLE 26. *Find the maximum and minimum values of the function* $x^4 + y^4 - 2x^2 - 2y^2$.

$$f = x^4 + y^4 - 2x^2 - 2y^2$$

$$\frac{\partial f}{\partial x} = 4x^3 - 4x, \qquad \frac{\partial f}{\partial y} = 4y^3 - 4y$$

$$\frac{\partial^2 f}{\partial x^2} = 4(3x^2 - 1), \qquad \frac{\partial^2 f}{\partial y^2} = 4(3y^2 - 1), \qquad \frac{\partial^2 f}{\partial x \partial y} = 0$$

$$\frac{\partial f}{\partial x} = 0 \text{ when } x = 0, \pm 1; \qquad \frac{\partial f}{\partial y} = 0 \text{ when } y = 0, \pm 1.$$

a	b	f_{aa}	f_{bb}	$f_{ab}^2 - f_{aa}f_{bb}$	Nature of turning value	$f(a, b)$
0	0	−4	−4	−16	Maximum	0
1	1	+8	+8	−64	Minimum	−2
−1	−1	+8	+8	−64	Minimum	−2
1	0	+8	−4	+32	Saddle-point	−1
0	1	−4	+8	+32	Saddle-point	−1
+1	−1	+8	+8	−64	Minimum	−2
−1	+1	+8	+8	−64	Minimum	−2

PROBLEMS

1. Find the turning values of the following functions and examine their nature

(i) $xy^2(3x + 6y - 2)$;

(ii) $x^4 + y^4 - 2(x - y)^2$;

(iii) $\cos x \cos y \cos(x + y)$ and x, y lie between $-\frac{1}{2}\pi$ and $\frac{1}{2}\pi$.

2. Find the turning values of the function

$$xy\left(1 - \frac{x}{a} - \frac{y}{b}\right),$$

where a, b are positive constants, and show that there is only one true maximum and no minimum, and find the value of this maximum.

16. Double Integrals

In this section we shall describe briefly what we mean by a definite integral of the function $f(x, y)$. Unlike the one-variable case, we cannot assign a meaning to an indefinite integral.

If $f(x, y)$ is a single-valued function of x and y, then the equation

$$z = f(x, y) \tag{1}$$

is the equation of a surface. If, for values of (x, y) lying in a region A of the xy-plane, $f(x, y) > 0$, we get a surface of the kind shown in Fig. 114. Through each point of the boundary of A we draw a line

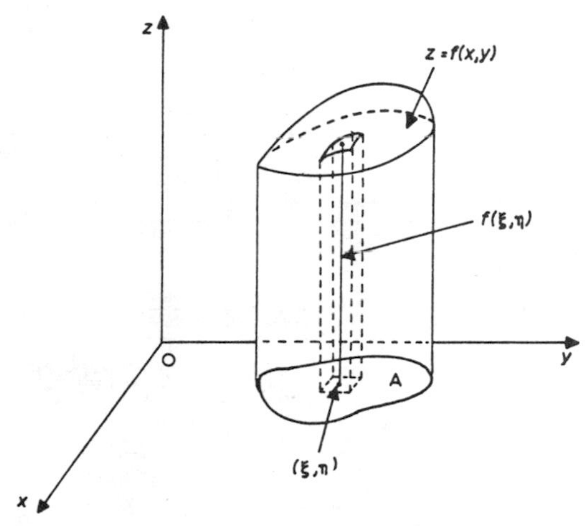

Fig. 114.

perpendicular to the xy-plane and thus generate a cylindrical surface. We *define* the volume enclosed by this surface, the cylinder and the xy-plane to be the double integral

$$\iint_A f(x, y)\mathrm{d}x\mathrm{d}y. \tag{2}$$

The notation

$$\int_A f\mathrm{d}A \tag{3}$$

is sometimes also used for this integral.

Now, just as in the case of functions of one variable (cf. § 2 of Chapter 5), we can obtain an arithmetical expression for this volume. Suppose that the area A can be wholly enclosed within a rectangle whose sides are parallel to the x- and y-axes, and that the equations of these sides are

$$x = a, \quad x = b, \quad y = c, \quad y = d,$$

where $a < b$, $c < d$. The rectangle is now divided into mn rectangles by lines $x = x_r$, $(r = 0, 1, \ldots, m)$, $y = y_s$, $(s = 0, 1, \ldots, n)$ where

$$a = x_0 < x_1 < x_2 \ldots < x_m = b,$$
$$c = y_0 < y_1 < y_2 \ldots < y_n = d.$$

In each of the mn rectangles any point (ξ_{rs}, η_{rs}) is chosen. We now

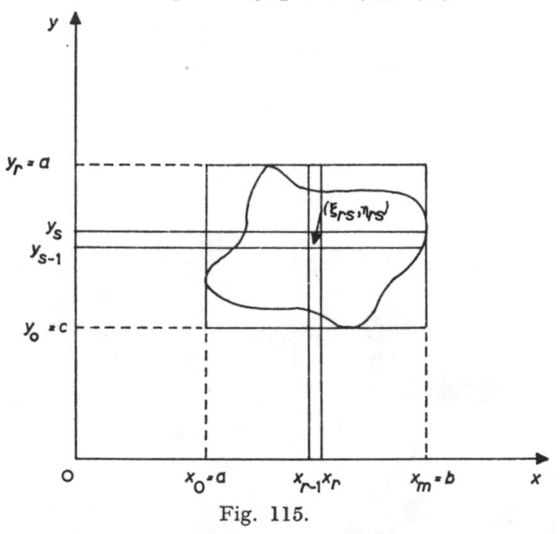

Fig. 115.

define a function $F(x, y)$ to be equal to $f(x, y)$ if the point (x, y) lies within the region A or on its boundary, and to be equal to zero if the point (x, y) lies outside that region. The expression

$$F(\xi_{rs}, \eta_{rs})\Delta x_r \Delta y_s,$$

where $\Delta x_r = x_r - x_{r-1}$ and $\Delta y_s = y_s - y_{s-1}$, is the volume of the cuboid on the small element of area $\Delta x_r \Delta y_s$ as base and with height $F(\xi_{rs}, \eta_{rs})$. If we sum all such expressions over all the rectangles we have a sum of volumes of cuboids which is approximately equal to the volume of the solid. As m and n get larger in such a way that each of the elements of area gets smaller, this sum approaches closer and closer to the volume of the solid. We say that the volume of the solid is the limit of the sum of such terms as m and n tend to infinity in such a way that the area of each of the rectangular elements tends to zero; that is,

$$\iint_A f(x, y)\mathrm{d}x\mathrm{d}y = \lim_{m, n \to \infty} \sum_{r=1}^{m} \sum_{s=1}^{n} F(\xi_{rs}, \eta_{rs})\Delta x_r \Delta y_s. \qquad (4)$$

We sometimes refer to A as *the field of integration.*

If $f(x, y)$ is taken to be identically equal to 1, the double integral

$$\iint_A dxdy$$

is the volume of the cylinder of unit height standing upon the region A. It is therefore the area of the region A.

Just as in the one-variable case we find that the definition (4) is not very useful in the actual evaluation of double integrals. In this case we do not have the analogue of the fundamental theorem of the

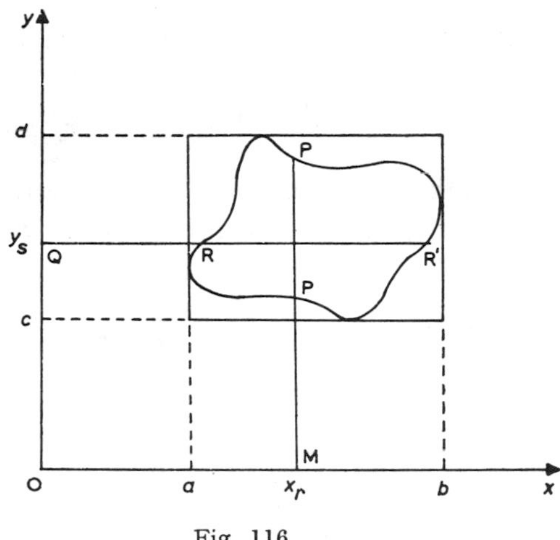

Fig. 116.

integral calculus, but we shall now show that the double integration can be carried out in two stages, each stage being an ordinary integration.

We choose the smallest rectangle which contains the region A (cf. Fig. 116), and yet has its sides parallel to the axes. We now form the sum

$$\sum_{r=1}^{m} \sum_{s=1}^{n} F(\xi_{rs}, \eta_{rs}) \Delta x_r \Delta y_s$$

in the following way. We keep r at a fixed value, and form the sum from $s = 1$ to $s = n$; then we complete the summation by letting r range over the values 1 to m. The first sum is

$$\Delta x_r \sum_{s=1}^{n} F(x_r, \eta_{rs}) \Delta y_s$$

which tends to

$$\Delta x_r \int_c^d F(x_r, y)\mathrm{d}y = \Delta x_r \int_{MP}^{MP'} f(x_r, y)\mathrm{d}y = \phi(x_r)\Delta x_r, \text{ say,}$$

as $n \to \infty$. Completing the sum we have

$$\sum_{r=1}^{m} \phi(x_r)\Delta x_r$$

which tends to

$$\int_a^b \phi(x)\mathrm{d}x,$$

as m tends to infinity in the prescribed way. Hence we have shown that

$$\iint_A f(x, y)\mathrm{d}x\mathrm{d}y = \int_a^b \mathrm{d}x \int_{MP}^{MP'} f(x, y)\mathrm{d}y. \tag{5}$$

(It will be observed that MP' and MP are, in general, dependent on x.)

We could, of course, have let s have a fixed value and make the

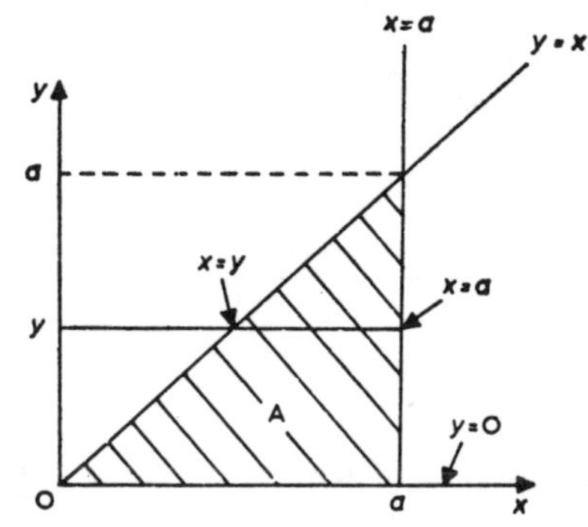

Fig. 117. Field of integration for example 27.

sum from $r = 1$ to $r = m$ first, and then completed the summation by letting s range over the values 1 to n. In that case we should have found the relation

$$\iint_A f(x, y)\mathrm{d}x\mathrm{d}y = \int_c^d \mathrm{d}y \int_{QR}^{QR'} f(x, y)\mathrm{d}x. \tag{6}$$

We shall illustrate the procedure to be followed by two examples:-

EXAMPLE 27. *Evaluate the integral*

$$\iint_A \frac{f'(y)}{\sqrt{(a-x)(x-y)}}\mathrm{d}x\mathrm{d}y,$$

where A is the region bounded by the lines $x = a$, $y = x$, $y = 0$.

Referring to Fig. 117, we see that this double integral can be written as the repeated integral

$$\int_0^a f'(y)\,\mathrm{d}y \int_y^a \frac{\mathrm{d}x}{\sqrt{(a-x)(x-y)}}.$$

If we make the substitution $x = a\sin^2\theta + y\cos^2\theta$, we can show that the integral with respect to x has the value π. The double integral is therefore equal to

$$\pi\int_0^a f'(y)\mathrm{d}y = \pi[f(a) - f(0)].$$

This is a case in which it is possible to evaluate the integral by doing the x-integration first, but impossible by doing the y-integration first.

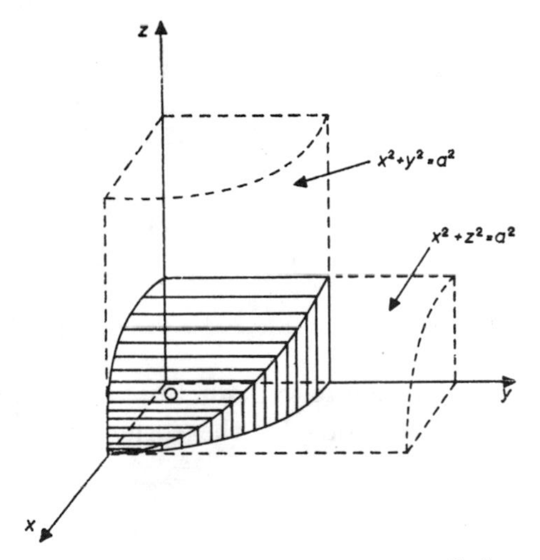

Fig. 118. Volume common to two cylinders.

EXAMPLE 28. *Find the volume common to the two cylinders* $x^2+y^2=a^2$, $x^2+z^2=a^2$.

If we denote the required volume by V, then we see from Fig. 118 that

$$\tfrac{1}{8}V = \iint_A (a^2 - x^2)^{\frac{1}{2}}\,\mathrm{d}x,$$

where A is the region in the first quadrant bounded by the circle $x^2 + y^2 = a^2$ (cf. Fig. 119). Hence we have

$$\begin{aligned}\tfrac{1}{8}V &= \int_0^a \mathrm{d}x \int_0^{(a^2-x^2)^{\frac{1}{2}}} (a^2 - x^2)^{\frac{1}{2}}\,\mathrm{d}y \\ &= \int_0^a \mathrm{d}x(a^2 - x^2) \\ &= \frac{2a^3}{3},\end{aligned}$$

showing that

$$V = \frac{16a^3}{3}.$$

We shall not discuss the general method by which the independent variables in a double integral can be changed. One special case is

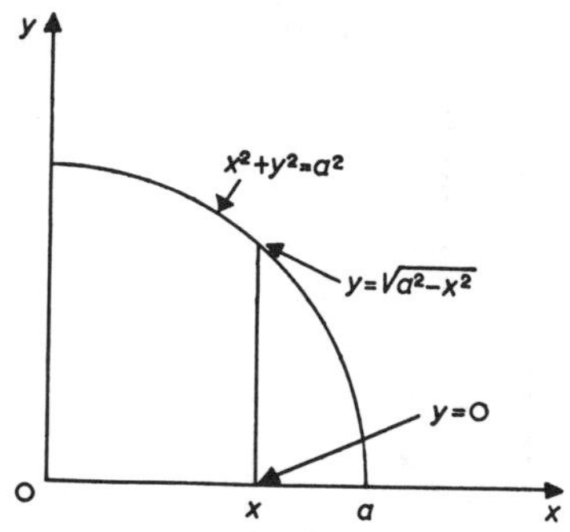

Fig. 119. The field of integration for example 28.

not only very useful, but easy to explain. Suppose that we had formed the volume of the solid bounded by the surface (1), the xy-plane and

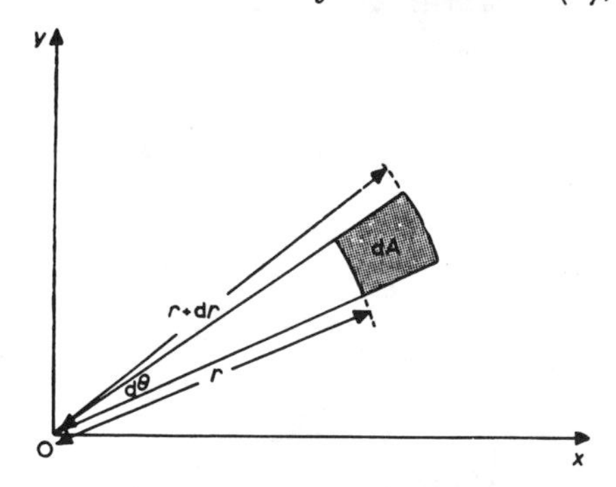

Fig. 120. Element of area in polar coordinates.

the cylinder generated by the boundary of the region A, by using polar coordinates r and θ. Then from Fig. 120 we see that the element of area $\mathrm{d}A$ is given by

$$\mathrm{d}A = \tfrac{1}{2}(r + \mathrm{d}r)^2 \mathrm{d}\theta - \tfrac{1}{2}r^2 \mathrm{d}\theta,$$

(the area of a circular sector of radius R and angle α is $\frac{1}{2}R^2\alpha$), that is that $\mathrm{d}A = r\mathrm{d}r\mathrm{d}\theta$ to the first order in $\mathrm{d}r$. If we remember that $x = r\cos\theta$, $y = r\sin\theta$, so that $f(x, y) = f(r\cos\theta, r\sin\theta)$ we see that

$$\iint_A f\mathrm{d}A = \iint_A f(x, y)\mathrm{d}x\mathrm{d}y = \iint_A f(r\cos\theta, r\sin\theta) r\mathrm{d}r\mathrm{d}\theta. \tag{7}$$

It is also easily seen that the double integral in the variables r and θ

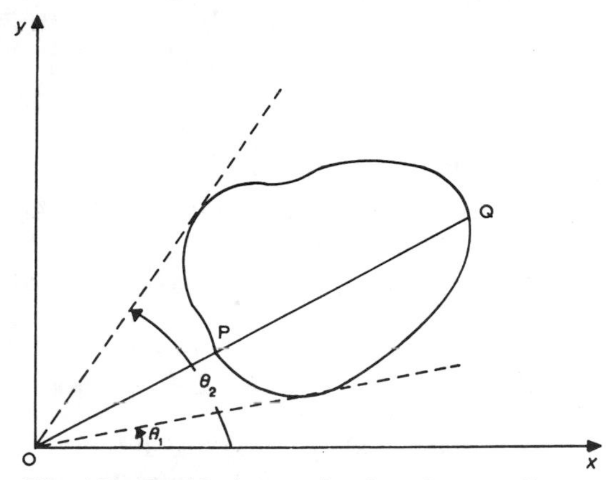

Fig. 121. Double integration in polar coordinates.

can be written as the repeated integral

$$\int_{\theta_1}^{\theta_2} \mathrm{d}\theta \int_{OP}^{OQ} f(r\cos\theta, r\sin\theta)\mathrm{d}r,$$

where the symbols have the meanings shown in Fig. 121.

EXAMPLE 29. *Find the value of the integral*

$$\iint_A \frac{\sqrt{(1 - x^2 - y^2)}}{\sqrt{(1 + x^2 + y^2)}} \mathrm{d}x\mathrm{d}y$$

where A is the region in the first quadrant bounded by the circle $x^2 + y^2 = 1$.

In this case

$$f(x, y) = \sqrt{\frac{1 - r^2}{1 + r^2}},$$

and, in the notation of Fig. 121, $\theta_1 = 0$, $\theta_2 = \frac{1}{2}\pi$, $OP = 0$, $OQ = 1$, so that the value of the integral is

$$I = \int_0^{\frac{1}{2}\pi} d\theta \int_0^1 \sqrt{\frac{1-r^2}{1+r^2}}\, r dr.$$

Performing the θ-integration and changing the variable in the r-integration to $u = r^2$, we find that

$$\begin{aligned} I &= \tfrac{1}{2}\pi \int_0^1 \sqrt{\frac{1-u}{1+u}}\,\frac{du}{2} \\ &= \tfrac{1}{4}\pi \int_0^1 \frac{1-u}{\sqrt{(1-u^2)}}\, du \\ &= \tfrac{1}{4}\pi\left[\sin^{-1} u + \sqrt{(1-u^2)}\right]_0^1 \\ &= \tfrac{1}{8}\pi(\pi - 2). \end{aligned}$$

Corresponding to double integration of a function $f(x, y)$ over an area, we can define triple integration of a function $f(x, y, z)$ throughout a volume V. In this case we cannot give a simple geometrical definition of the integral but, by analogy with equation (4) above, we can define

$$\iiint_V f(x, y, z)\, dx dy dz = \lim_{m,n,p\to\infty} \sum_{r=1}^{m} \sum_{s=1}^{n} \sum_{t=1}^{p} F(\xi_{rst}, \eta_{rst}, \zeta_{rst}) \Delta x_r \Delta y_s \Delta z_t.$$

The evaluation of triple integrals as repeated integrals can be carried out by methods similar to those outlined above. For details the reader is referred to Chapter III of R. P. Gillespie, "Integration" (Oliver and Boyd, Edinburgh, 1939).

PROBLEMS

1. Show that

$$\iint_A y\, dx dy = \tfrac{54}{5},$$

where A is the region in the first quadrant bounded by the straight line $y = x$, and the parabola $y = 4x - x^2$.

2. Find the volume of the portion of the cylinder $x^2 + y^2 = 2ax$ which lies between the planes $x + y - z + a = 0$, and $3x + 2y - z + 4a = 0$.

3. Prove that

$$\iint_A (x + y + a)^2 = \tfrac{3}{2}\pi a^4,$$

where $a > 0$, and the field of integration A is the circular area $x^2 + y^2 = a^2$.

4. Evaluate

$$\iint_A (x^2 + y^2)\mathrm{d}x\mathrm{d}y,$$

where A is the area bounded by the circle $(x-a)^2 + (y-b)^2 = a^2 + b^2$.

17. The Accuracy of Spectrophotometric and Colorimetric Measurements[1]

Photometry is based on the law of Lambert-Beer, which states

$$I = I_0\, \mathrm{e}^{\alpha c} \tag{1}$$

or

$$\log (I/I_0) = \alpha c, \tag{2}$$

where I_0 is the intensity of incident light, I is the intensity of the light leaving the solution, c is the concentration of the coloured substance and α is a negative constant whose physical meaning is immaterial here.

The ratio I/I_0 is called the transmission T, so that $T \equiv I/I_0$.

Generally a photometer measures the transmission T, even when the scale of the instrument is calibrated in so-called extinction units. Let us study the effect of a fractional error a in the measurement of T, on the error in the concentration c. We may obviously write (2)

$$\log T = \alpha c \tag{3}$$

or

$$c = \frac{\log T}{\alpha}. \tag{4}$$

Hence $\mathrm{d}c/\mathrm{d}T = 1/(\alpha T)$, so that Gauss' formula yields

$$\Delta c = \sqrt{\left(\frac{1}{\alpha T} \cdot aT\right)^2} = a/\alpha. \tag{5}$$

Hence, the procentual error in c is

$$\frac{\Delta c}{c} \times 100\,\% = \frac{a/\alpha}{\log T/\alpha} \times 100 = \frac{100a}{\log T}. \tag{6}$$

It easily follows from (6) that with a constant error in T, the error in the concentration c increases with increasing transmission. This

[1] For references see page 438.

may be illustrated by a numerical example. Let the procentual error in T be 2%.

Then, at $T = 0.2$, the procentual error in c is about 1.2 %, as may be computed with the aid of (6).

At $T = 0.9$, the procentual error in c is about 17.0 %.

From this it follows that measurements at high transmissions should be avoided.

It should, however, not be inferred from this simplified discussion that measurements should be made at extremely low transmission: at low transmission the error in T becomes so great that the accuracy of c again suffers.

In practice it is best to measure near $T = 0.5$, although in certain instruments different optimal T values obtain.

In Plate II a modern photometer is shown.

18. Discussion of Errors in the Calculation of Cardiac and Pulmonary Shunts

We shall now briefly discuss the problem of finding the accuracy of the computed value of blood distribution.

The example is taken from a paper of Defares *et al.*[1]

Although at first sight their study does not deal with shunts in the clinical sense of the term, the problem is essentially the same from the point of view of error discussion.

Since we are only concerned with the discussion of errors, we shall state the problem only in oversimplified form.

Briefly we wish to know the percentage of total bloodflow (= cardiac output) that flows through each lung. Because of the presence of unilateral hypoxia the conventional O_2 consumption method with bronchospirometry could not be used.

This information can be obtained by applying the Fick principle explained in Chapter 6, Section 11, to each lung.

For special reasons CO_2 was used in the Fick equation. The mixed venous blood CO_2 concentration may be used for both lungs as the CO_2 concentration in the incoming blood.

The CO_2 output of each lung was determined separately by using special catheters (bronchospirometry).

The CO_2 concentration in the outflowing blood was determined by an indirect method.

For present purposes we may state that the CO_2 concentration in the mixed venous blood is "absolutely" known, so that any error in the arterio-venous difference results from the error in the determination of the CO_2 in the blood leaving each lung (mixed end-capillary CO_2).

Let

$$Q_r = 100\, a/b, \tag{1}$$

$$Q_l = 100\, c/g \tag{2}$$

where Q_r = bloodflow through right lung, Q_l = bloodflow through left lung, a and c represent CO_2 output of right and left lung respectively; b and g are arterio-venous CO_2 content differences in right and left lungs respectively.

We wish to know the accuracy of the bloodflow through the right lung expressed as percentage P of total bloodflow (cardiac output).

Obviously

$$P = \frac{Q_r}{Q_r + Q_l}\, 100. \tag{3}$$

Substituting (1) and (2) in (3) and rearranging

$$P = \frac{100\, ag}{ag + bc}. \tag{4}$$

The partial derivatives P_a, P_b, P_c, and P_d are easily found to be

$$P_a = \frac{100\, bcg}{(ag + bc)^2}, \qquad P_b = \frac{-100\, agc}{(ag + bc)^2},$$

$$P_c = \frac{-100\, agb}{(ag + bc)^2}, \qquad P_d = \frac{100\, bca}{(ag + bc)^2},$$

so that, using Gauss' treatment of the total differential,

$$\mathrm{d}P = \left[\left\{\frac{100\, bcg}{(ag + bc)^2}\,\frac{\gamma a}{100}\right\}^2 + \left\{\frac{100\, bca}{(ag + bc)^2}\,\frac{\beta g}{100}\right\}^2 + \left\{\frac{-100\, agc}{(ag + bc)^2}\,\frac{\psi b}{100}\right\}^2 + \left\{\frac{-100\, agb}{(ag + bc)^2}\,\frac{\sigma c}{100}\right\}^2\right]^{\frac{1}{2}}$$

where $\gamma/100$, $\beta/100$, $\psi/100$, $\sigma/100$ are fractional errors in the subunits a, g, b and c respectively.

The procentual error $100\, \mathrm{d}P/P$ is, after some simplifications,

$$100\,\mathrm{d}P/P = \frac{bc}{ag + bc}\sqrt{\gamma^2 + \beta^2 + \psi^2 + \sigma^2} \tag{5}$$

where γ, β, etc. are of course percentage errors in the subunits.

Let us apply equation (5) to our figures.

The percentage error in the determination of CO_2 output (of each lung) was $\pm$ 2 %.

It should further be realized that the right and left arteriovenous differences b and g are *not* measured.

We only measure the CO_2 content of the mixed venous blood (here with "absolute" accuracy: see original paper) and the CO_2 content of the blood leaving each lung (end-capillary CO_2). The error in the determination of this end-capillary CO_2 content is $\pm$ 2 %.

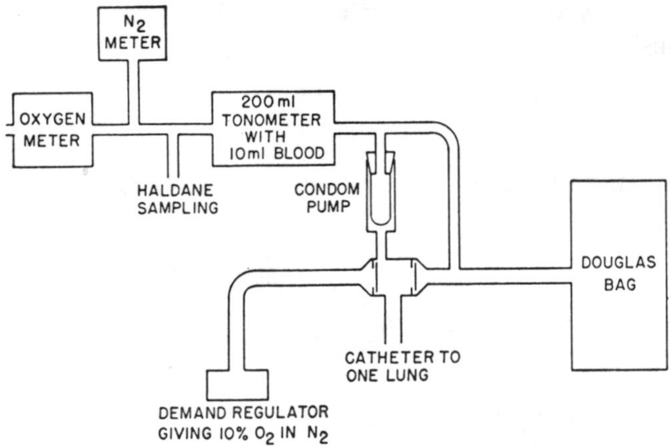

Fig. 122. The CO_2 content of arterialized blood of a single lung determined by equilibrating a blood sample with end-tidal (alveolar) air.

In a particular experiment the following values were found.

The CO_2 output of right lung is 130cc, or $a = 130$.

The CO_2 output of left lung is 167cc, or $c = 167$.

The CO_2 content of the mixed venous blood is 41.000 vols %.

The CO_2 content of the end-capillary blood on the right is 37.4 vols %.

The CO_2 content of the end-capillary blood on the left is 39.0 vols %.

Our question is: what is the accuracy of P?

With this information the percentage error in b is easily found to be $\frac{7}{36} \times 100\% = 20\%$, while the percentage error in g is $\frac{8}{20} \times 100\% = 40\%$.

Inserting the numerical values of a, b, c and g in equation (5) and noting that $\gamma = 2\%$, $\sigma = 2\%$, $\psi = 20\%$ and $\beta = 40\%$, we obtain for the percentage error in P (percentage of total bloodflow on right)

$$\frac{\mathrm{d}P}{P} \times 100\% = \frac{3.6 \times 167}{2.0 \times 130 + 3.6 \times 167} \sqrt{20^2 + 40^2 + 2^2 + 2^2} = 31.5\%.$$

The calculated value of P is, by using (1), (2) and (3), easily shown to be 33 % of total bloodflow. But the percentage error in P is 31.5% which means that the real value of P is 33 ± 10, which again means that the actual value of P very probably lies somewhere between 22 % and 43 % of (total) cardiac output.

N.B. The percentage value $P = 33\%$ should not be confused with the percentage error; P is a *percentage of total bloodflow*. In Plate III, the experimental set up for the determination of end-capillary CO_2 content and CO_2 output is shown: the patient's blood placed in a tonometer is equilibrated with the end-tidal (alveolar) air whose CO_2 tension is in equilibrium with the end-capillary CO_2 tension.

The above calculation shows that while the "naive" observer would believe that the real value of P is precisely 33%, all we really can say is that P lies somewhere between 23 % and 43 %, apparently a poor accuracy, which however happened to be sufficient for the purpose of the investigation.

19. Cybernetics and Entropy

The concept of entropy has been briefly discussed in the section on thermodynamics. The same concept, which may be expressed mathematically in different ways, plays a fundamental rôle in "Cybernetics", a term coined by Wiener.

Below we shall briefly discuss the meaning of entropy as employed in cybernetics.

Norbert Wiener of the M.I.T. who, with Shannon (1949) and Hartley (1928), may be considered as the founder of cybernetics (1942—1949), has made important contributions to the application of cybernetics to physiology. Especially his close association with the distinguished physiologist Rosenblueth resulted in important studies on such subjects as clonus, ataxia and conduction in the heart.

Cybernetics was defined by Wiener as "the science of control and communication, in the animal and the machine".

Coordination, regulation, control, feed-back, homeostasis, etc. are the central themes of cybernetics, and it may safely be stated that the theory of cybernetics constitutes a most powerful tool for the study of the animal organism from the viewpoint of homeostasis and regulation.

For the study of cybernetics three books may be recommended to the biologist (these books should be studied conjunctly).

a) An introduction to cybernetics, W. Ross Ashby (Chapman and Hall), 1957 (only elementary mathematics are used in this book).
b) The mathematical theory of communication, C. E. Shannon and W. Weaver (University of Illinois Press: Urbana), 1949.
c) Mathematical foundations of information theory, A. I. Khinchin, 1957 (Dover, translated from Russian.)

Although Khinchin's book is quite difficult, it is indispensable for a good theoretical understanding of Shannon's treatment. The mathematics given in our book are adequate to follow most of the material given in these books, although additional knowledge of such subjects as the Lebesgue integral, Markov chains (a topic of probability theory) etc. is sometimes required. Wiener's book "Cybernetics" is of great interest but cannot be recommended to anyone beginning the study of cybernetics.

The concept of entropy in cybernetics

Entropy, a term which will be defined below, is employed in cybernetics as a measure of the "degree of uncertainty".

Although this use of the entropy is apparently new, von Neumann has pointed out that in 1894 Boltzmann observed that entropy is related to "missing information". In certain formulations of statistical mechanics the entropy H is defined as

$$H = -\sum p_i \log p_i$$

where p_i is the probability of the system being in cell i of its phase space. All this will probably be quite obscure to the reader, but the above statement merely serves to show the formal similarity between the cybernetical concept of entropy and the classical entropy.

The concept of entropy, as used in cybernetics can be understood without any knowledge of statistical mechanics whatsoever. The discussion that follows is based on the treatment given by Khinchin.

We shall start with a brief discussion of "probability".

Definition: A complete system of events $A_1, A_2, \ldots, A_n$ is defined as a set of events such that one and only one of these events must occur at each trial.

We can think of a die, where 1, 2, 3, 4, 5 and 6 points are the events. At any trial only one of these events can occur. The events $A_1, A_2, \ldots A_n$ of a complete system together with their probabilities $p_1, p_2, \ldots, p_n$ ($p_i \geqq 0$, $\sum_{i=1}^{n} p_i = 1$), constitute, by definition, a finite scheme written as

$$A = \begin{pmatrix} A_1 & A_2 \ldots A_n \\ p_1 & p_2 \ldots p_n \end{pmatrix}. \tag{1}$$

Taking the example of a true die, we have the finite scheme

$$\begin{pmatrix} A_1 & A_2 & A_3 & A_4 & A_5 & A_6 \\ \frac{1}{6} & \frac{1}{6} & \frac{1}{6} & \frac{1}{6} & \frac{1}{6} & \frac{1}{6} \end{pmatrix}$$

since each event (number of points) has the probability $\frac{1}{6}$.

Every finite scheme implies a state of uncertainty

Generally the amount of uncertainty will differ with different schemes. Consider e.g. the three schemes

$$\begin{pmatrix} A_1 & A_2 \\ 0.5 & 0.5 \end{pmatrix} \quad \begin{pmatrix} A_1 & A_2 \\ 0.90 & 0.10 \end{pmatrix} \quad \begin{pmatrix} A_1 & A_2 \\ 0.999 & 0.001 \end{pmatrix}.$$

The outcome of the first, having a 50 % : 50 % chance, is of course quite uncertain, since A_1 and A_2 are equally likely to occur. The outcome of the second scheme is much more certain since there is a 90 % chance that A_1 will occur at a trial.

The amount of uncertainty in the third scheme is practically nil, since we are almost sure that A_1 will occur.

Now it is desirable, especially in cybernetics, to have a yardstick for measuring this amount of uncertainty associated with any scheme. By very intricate mathematical reasoning it can be shown that the quantity H, defined as (the base of the logarithm is arbitrary)

$$H = -\sum_{k=1}^{n} p_k \log p_k \tag{2}$$

serves as the only suitable measure of the uncertainty associated with the finite scheme (1).

Before proceeding we note that we are forced to define $p_k \log p_k = 0$ if $p_k = 0$ (since log 0 itself does not exist).

Earlier we have been faced with a similar situation when we defined $0! = 1$.

The way in which (2) came to be selected does not concern us here. We shall merely show that the function $H(p_1, p_2, \ldots, p_n)$ defined by (2) really possesses those properties which we might expect from a measure of uncertainty of a finite scheme.

1) Since we wish H to be a measure of the amount of uncertainty H should be zero if (and only if) the amount of uncertainty is zero, i.e. if the outcome is certain.

When the outcome is sure, one of the numbers $p_1, p_2, \ldots, p_n$ is one and all the others are zero (sum of all probabilities is always one). It is easy to check the following statement: $H = 0$, if and only if one of the numbers $p_1, p_2, \ldots, p_n$ is one (and all the others are zero).

In all other cases H is positive. Let us compute H for the finite scheme

$$\begin{pmatrix} A_1 & A_2 \\ 0.3 & 0.7 \end{pmatrix}.$$

$H = -(0.3 \log 0.3 + 0.7 \log 0.7) = +0.265.$

2) For a given n the scheme with the greatest uncertainty is obviously the one with equally likely outcomes, i.e. $p_k = 1/n$ $(k = 1, 2, \ldots, n)$ e.g. a true die has the greatest uncertainty. Because of this fact, we must expect H to attain its greatest value for equal values of p_k.

That this really happens can easily be checked by numerical examples and can be proved in a general way. We shall use an inequality which is easily shown to be applicable here.

The following is true for any continuous convex function $\varphi(x)$

$$\varphi\left(\frac{1}{n}\sum_{k=1}^{n} a_k\right) \leqq \frac{1}{n}\sum_{k=1}^{n} \varphi\,(a_k) \tag{3}$$

where $a_1, a_2, \ldots, a_n$ are any positive numbers. Do not worry about (3), just accept it as valid.

Setting $a_k = p_k$ and $\varphi(a_k) = a_k \log a_k$, and noting that

$\sum_{k=1}^{n} p_k = 1$, we obtain

$$\varphi\left(\frac{1}{n}\right) = \frac{1}{n}\log\frac{1}{n} \leqq \frac{1}{n}\sum_{k=1}^{n} p_k \log p_k = -\frac{1}{n}H(p_1, p_2, \ldots, p_n)$$

or

$$\frac{1}{n}\log\frac{1}{n} \leqq -\frac{1}{n}H(p_1, p_2, \ldots, p_n).$$

Multiplying both sides by $-n$, we have

$$-\log\frac{1}{n} \geqq H(p_1, p_2, \ldots, p_n),$$

according to the general rule that on multiplication by a negative number the sense of the inequality is reversed.
Or

$$H(p_1, p_2, \ldots, p_n) \leqq -\log\frac{1}{n} = \log n = H\left(\frac{1}{n}, \frac{1}{n}, \ldots, \frac{1}{n}\right),$$

since

$$H\left(\frac{1}{n}, \frac{1}{n}, \ldots, \frac{1}{n}\right) = -\sum_{k=1}^{n}\frac{1}{n}\log\frac{1}{n}$$

$$= -\left(\frac{1}{n}\log\frac{1}{n} + \frac{1}{n}\log\frac{1}{n} \ldots + \frac{1}{n}\log\frac{1}{n}\right) = -\log\frac{1}{n} = \log n.$$

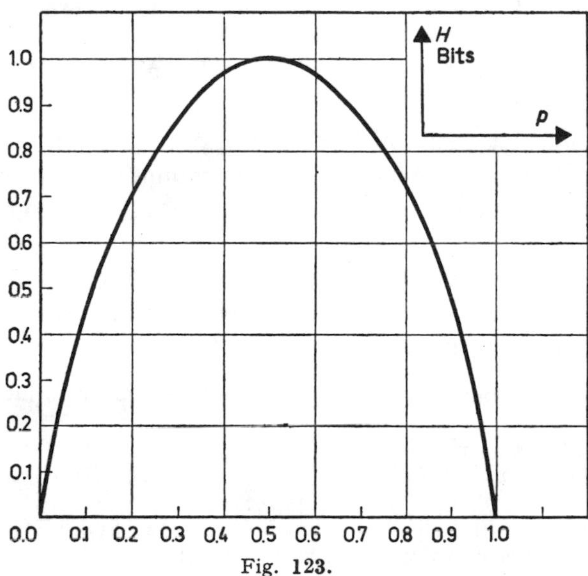

Fig. 123.

Thus the proof that H attains its greatest value when all the p_k's are equal, is complete.

The entropy in the case of two possibilities with probabilities p and $q = 1 - p$, that is $H = -(p \log p + q \log q)$, is plotted in Fig. 123 as a function of p (after Shannon and Weaver).

We cannot pursue this fascinating subject further.

A typical problem of cybernetics is presented by a communication system, schematically shown in Fig. 124.

The *information source* (e.g. the brain) selects a desired message out of a set of possible messages (Weaver). The *transmitter* (e.g. the voice mechanism including the centre of Broca etc.) transforms this message into the signal, which will be sent over the communication channel (nerve, air, telephone wire etc.) to the *receiver*, which is a sort of inverse transmitter handing the message on to its destination.

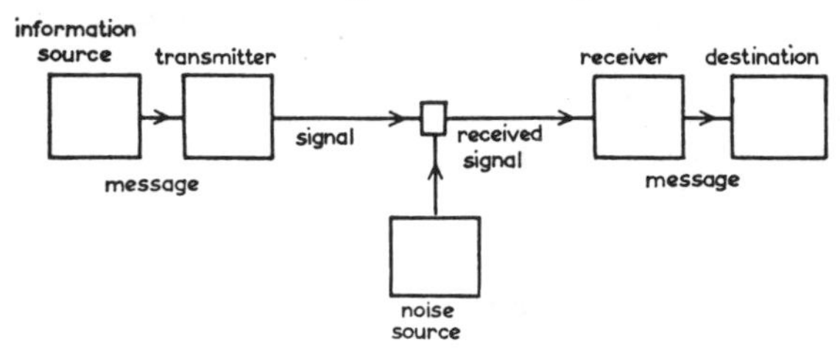

Fig. 124. Schematic diagram of a general communication system. (From Shannon and Weaver.)

Distortion of the transmitted signal by all sorts of extraneous influences (noise) may occur.

Typical questions with which communication theory is concerned are:

a) what is the best measure of the amount of information?

b) how does one measure the *capacity* of the communication channel?

c) how does one minimize the disturbing effects of noise?*

It should be fairly obvious that the entropy H forms a key to the answer to the first question.

* Among other books on cybernetics of possible interest to the physiologist may be mentioned:

C. Cherry, On Human Communication (John Wiley, 1957);

H. Quastler (editor), Information Theory in Biology (University of Illinois Press, 1953);

E. Schrödinger, What is life? (Anchor Books, 1948); Schrödinger defines life in terms of "negative entropy". See also §§ 20 and 21.

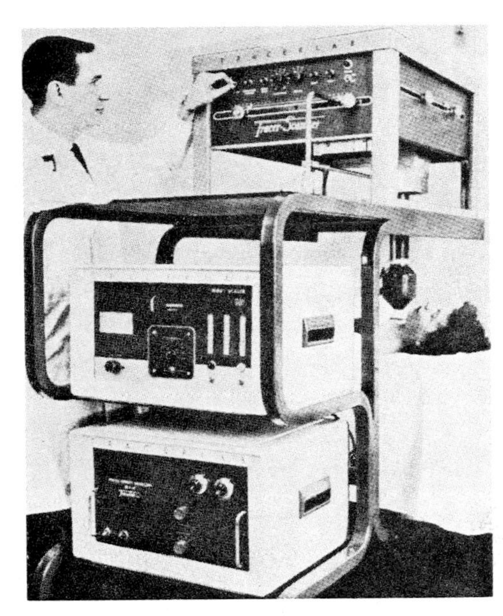

Plate I. Tracer-Scanner for scanning body areas in which a radio-isotope has localized. (Tracerlog no. 96).

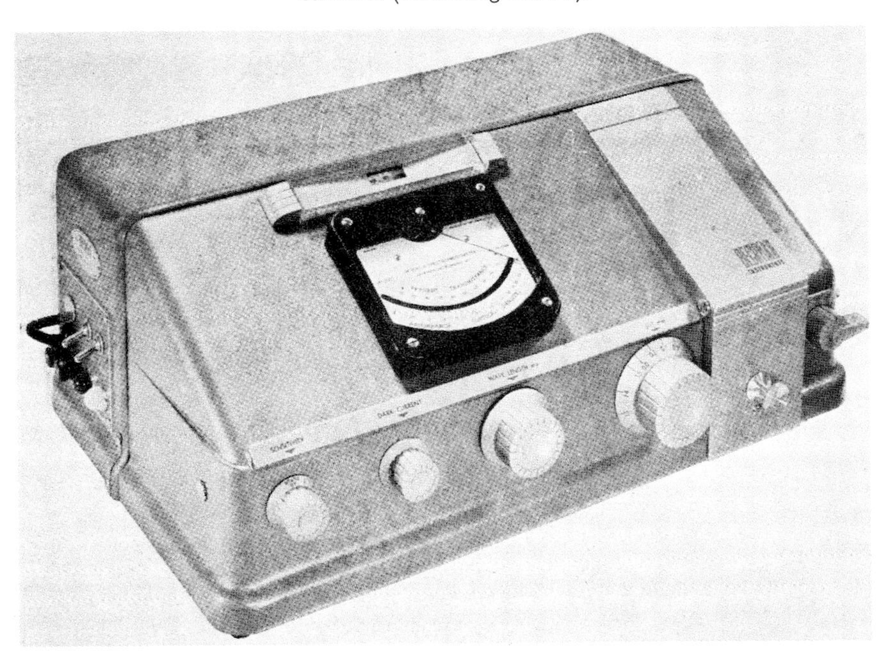

Plate II. Beckman spectrophotometer, model B.

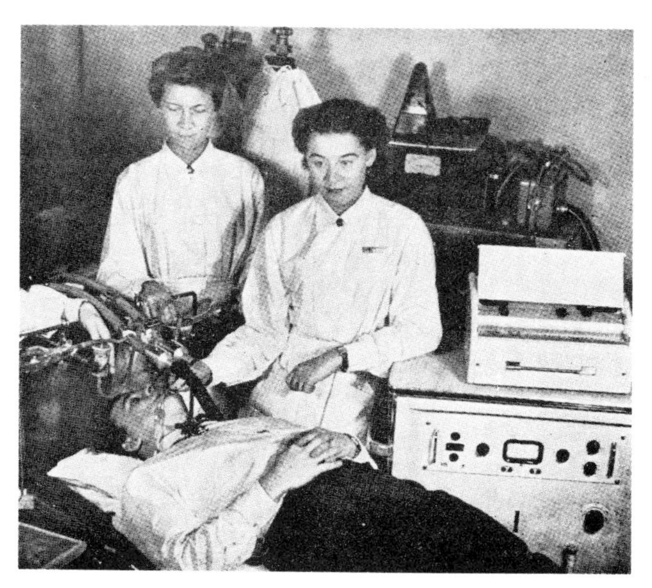

Plate III. Experimental set-up for the determination of bloodflow partition in the lungs. (Defares *et al.*). For references see page 438.

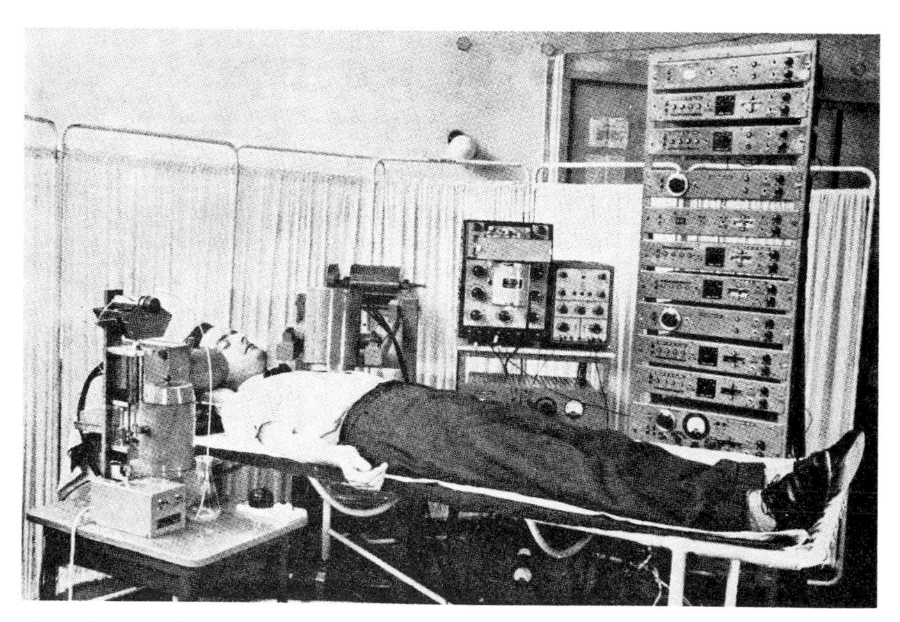

Plate IV. Cardiac output determination with the radioisotope method. [Thesis Y. van der Feer.] For references see page 438.

In Fig. 125 a communication system with a so-called "correction channel", which may be looked upon as a sort of "*feed-back loop*" to correct for noise, is shown. For its meaning the reader is referred to Shannon and Weaver.

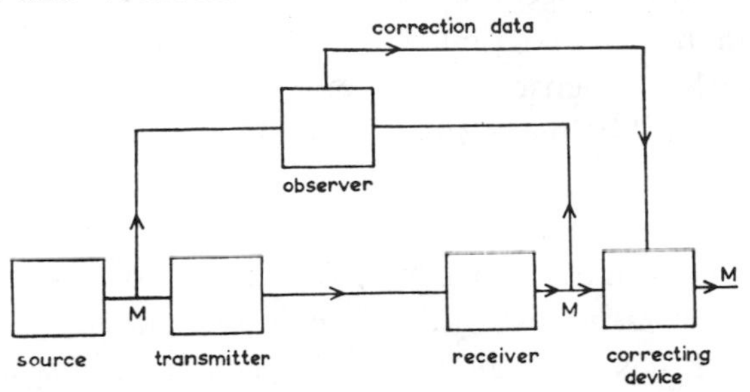

Fig. 125. Schematic diagram of a correction system. (After Shannon and Weaver.)

20. The Relevance of Thermodynamic Principles to Living Systems

The second law of thermodynamics states that *order can not originate from disorder.*

The evolution of living systems originating in some primordial soup to higher and higher levels of organization (order) seems to make a mockery of the second law. Prigogine[1] stated in 1955: "The behaviour of living organisms has always seemed so strange from the point of view of classical thermodynamics, that the applicability of thermodynamics to such systems has often been questioned". Indeed the question of the applicability of the second law was the subject of an international conference convened in Paris in 1938. The conference (Brillouin[3] reported on it) ended in total disorder (maximum entropy) and no agreement was reached.

According to the modern point of view a *living system – environment* may be regarded as a steady open *thermodynamic system.* The meaning of an *open* system is best understood by examining the meaning of a closed system. This is defined as a system that will ultimately come to *equilibrium*, i.e. ultimately reach a state of maximum entropy (zero energy, maximum disorder, i.e. disintegration). A living organism thus reaches equilibrium after death.

The living state can only be maintained as part of an *open* system. Theoretically this open system can remain indefinitely in a *steady state*, which is characterized by a steady flow of energy and matter.[4]

In the steady state the energy input equals the energy output through metabolic processes. In accord with the modern theory an organism can maintain its entropy at a low level and can even decrease it at the expense of the entropy of its environment. As Schrödinger[2] said "living matter feeds on negative entropy or negentropy".

The negentropy (properties of order) of the living system is very high (stored information in DNA, RNA, folded protein structures, etc.). The quantitative determination of negentropy content of the genetic code in DNA can be made as discussed below. The calculation of negentropy content of a whole organism is, however, at present a meaningless endeavour.[5,6]

21. Entropy, Information and DNA

It seems intuitively obvious that a close connection exists between the concepts entropy (a measure of uncertainty) and *information.* In classical information theory as first formulated by Shannon the term information is used within the context of *transmission.* If we are interested in the information content of the genetic molecule, DNA, we are concerned with the problem of information *storage* rather than transmission (Gatlin[1], 1966). In classical information theory information is equated with entropy. If we define stored information as equivalent to entropy, this would be like saying that a library would contain its maximum amount of stored information if books were cut into single letters and these randomly mixed. Obviously such a conclusion would make nonsense of the usual meaning of the concept "information".

In the usual sense information is proportional, not to the degree of disorder (randomness) but to the degree of order: the greater the ordering the greater the amount of stored information. A suitable measure of (stored) information, I, was first proposed by Brillouin.[2] This is given by defining

$$I = H^{\max} - H^{\text{obs}} \qquad (1)$$

where $H^{\max} = \log n$ as shown in § 19, equation (4) and H^{obs} is the entropy determined from the actual finite scheme under consideration.

If the disorder is maximal, i.e. if all p_i's (see again § 19) are equal, H^{obs} is a maximum, i.e. equals H^{max}, so that from eq. (1) $I = 0$ (minimal information).

At the other extreme, if the disorder is minimal, i.e. one of the p_i is unity (and all the others are zero), $H^{obs} = 0$, so that I attains its maximum value, viz. H^{max}. We thus see that the above definition is a perfectly sensible quantitative definition of information.

Before applying this definition of I to the problem of stored information in DNA we must briefly consider the "source" concept. In communication engineering a "source" is anything that emits "signals". Mathematically, a source is an abstraction whose output is any linear sequence of symbols. For example the linear sequence of bases in the DNA or RNA molecule may be regarded as the output of an (abstract) source.

The set of symbols emitted by the source is called the alphabet, A. For DNA, the alphabet is $\mathsf{A} = \{A, T, C, G\}$ where A = adenine, T = thymine, C = cytosine and G = guanine. Let the number of symbols in the alphabet be denoted by a, then $a = 4$ for DNA. In what follows Khinchin's treatment is given. Let a source emit a sequence of symbols from its alphabet,

$$(\ldots x_{t_{-2}}, x_{t_{-1}}, x_{t_0}, x_{t_1}, x_{t_2}, \ldots) \tag{2}$$

where x_{t_i} is the symbol emitted at time t_i. Let the source emit a specified symbol α_{t_i} at time t_i, $1 \leqslant i \leqslant n$ then, following Khinchin, the set of all such sequences is called a cylinder, C.

EXAMPLE 30

$$C = \left\{(\ldots x_{t_{-1}}, x_{t_0} | A_{t_1}, C_{t_2}, G_{t_3} | x_{t_4}, x_{t_5}, \ldots)\right\}. \tag{3}$$

Here the α_{t_i} are sequential symbols, A, C, and G, $n = 3$ and the x_{t_i} on both sides of the triplet may be any symbol of this alphabet. For any given source a definite probability, $p(C)$, is associated with the emission of any given cylinder. Consider a stationary source, i.e. one in which the probability distribution of its output is time invariant. Let us select the set of all a^n different sequences of length n and of the form (3) as elementary events of a complete system. By doing this we have reduced the problem to the familiar situation discussed above. The entropy, H_n, is from its definition,

$$H_n = -\sum_{i=1}^{a^n} p(C_i) \log p(C_i) \tag{4}$$

where $p(C_i)$ is the probability of emission of the ith cylinder. Clearly,

$$\sum_{i=1}^{a^n} p(C_i) = 1.$$

From equation (1)

$$I_n = H_n^{\max} - H_n^{\text{obs}} \tag{5}$$

where

$$H_n^{\max} = \log a^n.$$

Equation (5) enables us to compute the information content of DNA (RNA) provided we can determine H_n^{obs}. Since the treatment of this rather abstruse subject is sketchy, the student may be satisfied with an "intuitive" understanding of the situation. For an analytical understanding a study of Khinchin's paper[3] appears in order. I_n should be interpreted as the average information per sequence. The average information per emitted symbol is given by I_n/n, a quantity that is called the *rate of information emission* (Khinchin).

H^{obs} can be obtained by identifying $p(C_i)$ with the frequency of specified symbols in the DNA (RNA). $I_2/2$ may be calculated* from eqs. (4) and (5) when the "nearest neighbor frequencies" are given. (Josse et al.[4]). These represent the cylinder probabilities of the 16 cylinders of length 2 associated with DNA or RNA.

Example 31 (after Gatlin)

The nearest neighbor frequencies of micrococcus lysodeikticus may be written in matrix form. (See Appendix, §4.)

Clearly $A = 0.147$, $T = 0.145$, $C = 0.354$, $G = 0.354$.

* For $n = 2$ the Markov assumption is not required.

TABLE 17

3′→5′	*A*	*T*	*C*	*G*	Column sums
A	0.019	0.022	0.057	0.049	0.147
T	0.011	0.017	0.063	0.054	0.145
C	0.052	0.050	0.113	0.139	0.354
G	0.065	0.056	0.121	0.112	0.354
Row sums	0.147	0.145	0.354	0.354	

Hence from equation (4) and Table 17 H_2 is the sum of 16 terms, namely:

$$-0.019 \log_{10} 0.019 - 0.022 \log_{10} 0.022 - \ldots = 1.1227 \text{ Hartleys.}$$

The term Hartley is used for the unit of H when the base 10 is used. Since 1 Hartley is 3.32 bits, we have $H_2 = 3.727$ bits. Using eq. (5),

$$I_2 = 3.32 \log_{10} 4^2 - 3.727 = 0.271 \text{ bits.}$$

Hence

$$\tfrac{1}{2} I_2 = 0.135 \text{ bits.}$$

Biologically, the relative value (ordering) of I among different organisms is of interest. In Table 18 the ordering of three unicellular groups, protozoa, bacteria and phage is given.

TABLE 18

Organism	$\frac{1}{2} I_2$ (bits)
tetrahymena pyriformis	0.2180
M. lysodeikticus	0.1352
H. influenza	0.0518
B. subtilis	0.0197
E. coli B_b	0.0117
E. coli B_a	0.0080
λ_{dg}	0.0073
λT	0.0054

References for Chapter 8, § 17–21 and Plates I–IV

Section 17

1. D.L. Drabkin, Photometry, in: Otto Glasser, editor, Medical physics (The Year Book Publishers, Chicago, 1956).

Section 18 and Plate III

1. J.G. Defares, G. Lundin, M. Arborelius and G. Strömblatt, J. Appl. Physiol. **15** (1960) 169.

Plate IV (also for Chapter 6, § 12 and Chapter 7, § 18a)

1. Y. van der Feer, The determination of cardiac output by the injection method, Thesis (Utrecht, Holland, 1958).
2. H.C. Burger, Y. van der Feer and J.H. Douma, Acta Cardiologica **11** (1956) 1.

Section 20

1. I. Prigogine, Introduction to thermodynamics of irreversible processes (C.C. Thomas, Springfield, Ill., 1955).
2. E. Schrödinger, What is Life? (Anchor Books, 1948).
3. L. Brillouin, Amer. Scientist **37** (1949) 554.
4. L. von Bertalanffy, Human Biol. **23** (1951) 302.
5. J.P. Wesly, Currents Mod. Biol. **1** (1967) 214.
6. R.M. Friedenberg, Pioneering concepts in modern science **1** (Hafner, New York, 1968).

Section 21

1. L.L. Gatlin, J. Theor. Biol. **10** (1966) 281.
2. L. Brillouin, Science and information theory (Academic Press, New York, 1956).
3. A.I. Khinchin, Mathematical foundations of information theory (Dover, New York, 1957) 44.
4. J. Josse, A.D. Kaiser and A. Kornberg, J. Biol. Chem. **236** (1961) 864.
5. M.J. Apter and L. Wolpert, J. Theor. Biol. **8** (1965) 244.

CHAPTER 9

DIFFERENTIAL EQUATIONS

1. Differential Equations

In § 10 of Chapter 6 we became familiar with the concept of a differential equation. For instance, in example 10 of that section, we saw that a certain function $P_A(t)$ could be determined from the fact that there was a relation between it and its derivative of the form

$$\frac{\mathrm{d}P_A}{\mathrm{d}t} + KP_A = KP_V, \tag{1}$$

and that

$$P_A(0) = P_A^{(0)}, \textit{ a constant.} \tag{2}$$

The relation (1), involving the derivative of P_A as well as that function itself, is called a *differential equation*; the equation (2) is said to express the *initial conditions* of the function.

We can generalize this idea by defining a *differential equation* to be any relation which involves a dependent variable, the independent variables upon which it depends, and derivatives with respect to these variables.

If there is only one independent variable, x say, and if we denote the dependent variable by y, then such an equation would be of the form

$$f\left(\frac{\mathrm{d}^n y}{\mathrm{d}x^n}, \ldots, \frac{\mathrm{d}y}{\mathrm{d}x}, y, x\right) = 0. \tag{3}$$

An equation of the type (3) is called an *ordinary differential equation* since it involves "ordinary" (i.e. not partial) derivatives.

The *order* of a differential equation is defined to be the order of the highest derivative that occurs in the equation. If it is possible to express an ordinary differential equation so that it is of the form of a polynomial of various derivatives equated to zero then the equation is said to be in *standard form*, and the index of the power to which the highest derivative is raised is called the *degree* of the equation. For example, equation (1) is an ordinary differential equation of the first degree and the first order. The equation

$$\left(\frac{d^4y}{dx^4}\right)^3 + 5\left(\frac{dy}{dx}\right)^6 = x^2y^3, \tag{4}$$

is of order 4 and degree 3.

A certain type of ordinary differential equation arises frequently in applications of mathematics to physical and biological problems and, for that reason, it is studied closely. If a differential equation can be written in the form

$$a_0\frac{d^ny}{dx^n} + a_1\frac{d^{n-1}y}{dx^{n-1}} + \dots + a_{n-1}\frac{dy}{dx} + a_ny = f(x), \tag{5}$$

where the coefficients $a_0, a_1, \dots, a_n$ are functions of the independent variable x, it is called a *linear* equation of degree n. For example, equation (1) is a linear equation of the first degree.

To illustrate how ordinary differential equations arise in biological problems, we now consider the equation arising in a simple study of the growth of populations.* Suppose that we denote by $p(t)$ the number of inhabitants of a given area at time t. The difference in number at time $t + h$ from that at time t is $p(t+h) - p(t)$ and this will depend upon:-

(1) the number of individuals born in that time; we assume that this number is Nph, where N may depend on both p and t:
(2) the number of individuals dying in that time; we assume that this is Mph, the quantity M being in general a function of p and t:
(3) the number of individuals entering the area; this is taken to be Ih, where I depends on p and t;
(4) the number of individuals leaving the area; this is taken to be Eh, where E is a function of p and t.

If we make up the balance we find that

$$p(t+h) - p(t) = Nph - Mph + Ih - Eh$$

from which we derive the relation

$$\frac{p(t+h) - p(t)}{h} = (N - M)p + (I - E).$$

Letting h tend to zero we obtain the differential equation

* V. A. Kostitzin, "Mathematical Biology", (Translated by T. H. Savory), (Harrap, London, 1939), p. 18.

$$\frac{\mathrm{d}p}{\mathrm{d}t} = (N - M)p + (I - E), \tag{6}$$

where, in general, N, M, I and E will be functions of both p and t.

In one population model we assume that I and E are constants and that

$$N = n - \nu p, \qquad M = m + \mu p, \tag{7}$$

where m, n, μ and ν are constants. If we substitute from equations (7) into equation (6) and write

$$n - m = \varepsilon, \qquad \mu + \nu = k, \tag{8}$$

we obtain the differential equation

$$\frac{\mathrm{d}p}{\mathrm{d}t} = \varepsilon p - kp^2 + I - E. \tag{9}$$

This equation is an ordinary differential equation of first order and first degree, but (because of the term kp^2) it is *non-linear*. It is known as *Verhulst's equation*, the constant ε is called the *coefficient of increase* and k is called the *limiting coefficient.*

If the population is isolated (i.e. if there is neither immigration or emigration), $I = E = 0$ and equation (9) reduces to the simpler form

$$\frac{\mathrm{d}p}{\mathrm{d}t} = \varepsilon p - kp^2. \tag{10}$$

Let us return now to equation (1). We saw previously that the function

$$P_\mathrm{A} = P_\mathrm{v} + C\mathrm{e}^{-Kt}, \tag{11}$$

satisfies the equation (1) for all values of the arbitrary constant C. This relation involves no derivatives, and is called the *solution* or the *integral* or the *primitive* of the differential equation (1). Since it involves an arbitrary constant C, this solution is also called the *general solution.* If we give C the value $P_\mathrm{A}^{(0)} - P_\mathrm{v}$ we obtain the solution

$$P_\mathrm{A}(t) = P_\mathrm{v} + (P_\mathrm{A}^{(0)} - P_\mathrm{v})\mathrm{e}^{-Kt}. \tag{12}$$

Such a solution, satisfying the initial conditions (2) is called a *particular integral* of the equation (1). In obtaining the general solution (11) from the differential equation (1) we are said to have *solved* or *integrated* the differential equation.

It can be proved (but we shall omit the proof here) that a differential equation of order n will have a general solution involving n arbitrary constants. If, therefore, we can find a solution of equation (3) of the form

$$F(x, y, c_1, \ldots, c_n) = 0, \tag{13}$$

where $c_1, \ldots, c_n$ denote arbitrary constants, we can be sure we have found the most general solution of the equation. All the particular solutions of the equation (3) will be obtained from the general solution (13) by assigning particular values to the arbitrary constants c_1, $c_2, \ldots, c_n$.

Differential equations of another type also arise in biology. In the study of the growth of one population feeding upon another* we find that, if x is the total number or biomass of the predators and y the biomass of the prey, then the variation of x and y with the time t is determined by the pair of equations

$$\begin{aligned} \frac{\mathrm{d}y}{\mathrm{d}t} &= \varepsilon_1 y - h_{11}y^2 - h_{12}xy, \\ \frac{\mathrm{d}x}{\mathrm{d}t} &= -\varepsilon_2 x + h_{21}xy - h_{22}x^2, \end{aligned} \tag{14}$$

where ε_1, ε_2, h_{11}, h_{12}, h_{21} and h_{22} are constants. In this case x and y are determined, not by two separate equations (one for each variable), but by two differential equations which must be satisfied simultaneously. Equations of the type (14) are, for obvious reasons, called *simultaneous ordinary differential equations*.

Finally, in applications of mathematics, we encounter partial differential equations. A *partial differential equation* is one in which there occur two or more independent variables and partial derivatives with respect to them of the function sought. For example, in some studies of population dynamics** it is found that $y(x, t)$, the expected number of individuals at position x at time t is determined by one or other of the partial differential equations

* V. A. Kostitzin, *loc. cit.*, p. 80.

** See J. G. Skellam, "The Mathematical Approach to Population Dynamics" in J. B. Cragg & N. W. Pirie (editors), "The Numbers of Man and Animals" (Oliver & Boyd, Edinburgh, 1955), and the papers referred to there.

$$\frac{\partial y}{\partial t} = \frac{\partial^2}{\partial x^2}(\lambda y), \tag{15}$$

$$\frac{\partial y}{\partial t} = \frac{\partial}{\partial x}\left\{\lambda^2 \frac{\partial}{\partial x}\left(\frac{y}{\lambda}\right)\right\}, \tag{16}$$

$$\frac{\partial y}{\partial t} = \frac{\partial}{\partial x}\left(\lambda \frac{\partial y}{\partial x}\right), \tag{17}$$

the equation chosen depending on what hypothesis is made about the biological and physical considerations which affect the movement of an individual from a position a to an adjacent position b. As in the case of ordinary differential equations we define the order of a partial differential equation to be the order of the partial derivative of highest order occurring in the equation. Since we may write

$$\frac{\partial}{\partial x}\left(\lambda \frac{\partial y}{\partial x}\right) = \lambda \frac{\partial^2 y}{\partial x^2} + \frac{\partial \lambda}{\partial x}\frac{\partial y}{\partial x}, \text{ etc.,}$$

we see that each of these equations is a second order partial differential equation.

In this chapter we shall consider some simple aspects of the theory of differential equations, illustrating our procedures with some simple examples. In the next chapter we shall consider the application of these ideas to the discussion of problems arising in biology and medicine.

The sections of this chapter will be grouped as follows:-

A. First Order Ordinary Differential Equations.
B. Linear Equations of the Second and Higher Order.
C. Simultaneous Ordinary Differential Equations.
D. Partial Differential Equations.

A. FIRST ORDER ORDINARY DIFFERENTIAL EQUATIONS

2. Equations of the First Order and the First Degree

A typical equation of the first order and the first degree may be written in the form

$$\frac{dy}{dx} = f(x, y), \tag{1}$$

where the function $f(x, y)$ is continuous with respect to the pair of variables x and y and is single-valued. Corresponding to each point (x, y) in the region in which $f(x, y)$ is defined, equation (1) assigns the definite value $f(x, y)$ to the derivative dy/dx, i.e. at every point, equation (1) defines a direction in the plane. If a curve could be drawn so that at each of its points, its gradient is given by equation (1), then this curve would have an equation which was a solution of the ordinary differential equation (1). Any such curve is called an *integral curve* of the equation.

Suppose we now try to construct an integral curve, starting at an

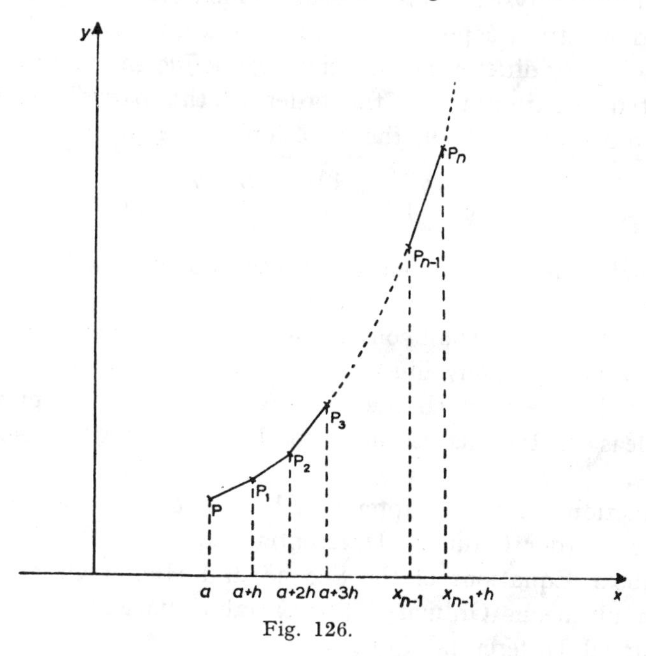

Fig. 126.

arbitrary point P with coordinates (a, b). If we take h to be a small positive number and define a point $P_1(x_1, y_1)$ by the equations

$$x_1 = a + h, \qquad y_1 = b + hf(a, b),$$

and a point $P_2(x_2, y_2)$ by the equations

$$x_2 = x_1 + h, \qquad y_2 = y_1 + hf(x_1, y_1),$$

and so on, the series of points $P_1, P_2, \ldots, P_n, \ldots$ being defined by the relations

$$x_n = x_{n-1} + h, \qquad y_n = y_{n-1} + hf(x_{n-1}, y_{n-1}),$$

then, if we join these points by straight lines (cf. Fig. 126), we obtain a polygonal line $P_1P_2P_3 \ldots$ with the property that the segment $P_{n-1}P_n$ has gradient $f(x_{n-1}, y_{n-1})$. If h is made very small and we draw a smooth curve through the points $P_1P_2P_3 \ldots$, the gradient of this curve at any point (x, y) on it will be very nearly given by $f(x, y)$. A point which starts at P and moves to the right in such a way that at every point (x, y) of its path its direction of motion has gradient $f(x, y)$ traces out a *unique* curve, since $f(x, y)$ is assumed to be single-valued. Thus we might infer that, in the limit as h tends to zero, the polygonal line tends to a unique limiting curve which is an integral curve of the differential equation (1).

This curve has an equation giving y as a function of x, such that $y = b$ when $x = a$. Since the initial point may be chosen in any way it follows that the equation to the curve must possess an arbitrary element. If this equation is

$$F(x, y) = 0,$$

then the arbitrary element is determined completely by the relation

$$F(a, b) = 0.$$

Therefore, there is only one arbitrary element in the equation and, since only one curve passes through the point (a, b), it is clear that this arbitrary element appears to the first degree. Hence the equation $F(x, y) = 0$ can be expressed in the form

$$\phi(x, y) = C, \tag{2}$$

where C is an arbitrary constant.

It should be emphasized that these remarks do not constitute a proper formal proof,* but only an indication of the validity of the result. For our purposes, the important thing about this theorem is that it asserts that if we can find a solution of the differential equation (1) involving one arbitrary constant, then that solution is the *general solution* of the equation.

Equations of the first order can be split into a variety of types which we shall now discuss.

* For this the reader is referred to: J. C. Burkill, "The Theory of Ordinary Differential Equations" (Oliver & Boyd, Edinburgh, 1956).

3. Equations with Separable Variables

If a first order equation can be written in the form

$$f_1(y)\frac{dy}{dx} + f_2(x) = 0, \tag{1}$$

we say that the variables x and y are *separable*. Integrating both sides of this equation with respect to x we obtain the general solution

$$\int f_1(y)\frac{dy}{dx}dx + \int f_2(x)dx = C, \tag{2}$$

where C is an arbitrary constant. This can be written as a form which suggests that it is convenient to write the given differential equation (1) symbolically as

$$\int f_1(y)dy + \int f_2(x)dx = 0.$$

EXAMPLE 1. *Solve the differential equation*

$$(a^2 - x^2)\frac{dy}{dx} + xy = 0.$$

If we write this equation in the form

$$\frac{dy}{y} + \frac{xdx}{a^2 - x^2} = 0,$$

we see that its solution is

$$\int \frac{dy}{y} + \int \frac{xdx}{a^2 - x^2} = C,$$

where C is an arbitrary constant. The integrations are elementary and we find that

$$\log y - \tfrac{1}{2}\log(a^2 - x^2) = \log c.$$

Here we use the trick of putting $C = \log c$ to obtain the solution in the more elegant form

$$y = c\sqrt{(a^2 - x^2)},$$

where c is an arbitrary constant.

PROBLEMS

Find the general solutions of the equations:-

1. $x\dfrac{dy}{dx} = y(1 + x);$

2. $x(2y + a)\dfrac{dy}{dx} = y(a + y)$, where a is a constant;

3. $\dfrac{y^2 + 1}{2x} = y(1 + x)\dfrac{dy}{dx};$

4. $\dfrac{dy}{dx} = \tan 2x \tan y;$

5. $dy/dx = y(3x^2y + 1)/x(x^2y + 2)$, by first changing the dependent variable from y to z, where $y = zx^{-2}$.

4. Homogeneous Equations

We say that a function $f(x, y)$ is *homogeneous of degree n* in x and y if it can be written in the form

$$f(x, y) = x^n g(y/x), \tag{1}$$

where $g(y/x)$ is a function of the single variable y/x.

If the two functions $f_1(x, y)$, $f_2(x, y)$ are homogeneous in x and y of the same degree, the differential equation

$$f_1(x, y)\frac{dy}{dx} + f_2(x, y) = 0 \tag{2}$$

is said to be *homogeneous*. If we write $f_1(x, y) = x^n g_1(y/x)$, $f_2(x, y) = x^n g_2(y/x)$, we see that equations of the type (2) can be written in the form

$$g_1(y/x)\frac{dy}{dx} + g_2(y/x) = 0. \tag{3}$$

The form of this equation suggests that we might change the dependent variable from y to z, where

$$z = y/x. \tag{4}$$

From this last equation we have $y = xz$ so that

$$\frac{dy}{dx} = x\frac{dz}{dx} + z. \tag{5}$$

Substituting from equations (4) and (5) into equation (3) we see that it is transformed to

$$g_1(z)x\frac{dz}{dx} + zg_1(z) + g_2(z) = 0. \tag{6}$$

Rewriting this equation as

$$\frac{g_1(z)}{zg_1(z) + g_2(z)}\,dz + \frac{dx}{x} = 0,$$

we see that the variables are separable and that the solution of the equation (6) is

$$\log x + \int \frac{g_1(z)}{zg_1(z) + g_2(z)}\,dz = C, \tag{7}$$

where C is an arbitrary constant.

It is not intended that the reader should memorize these formulae. The important thing for him to remember is that the substitution (4) reduces a homogeneous equation to one in which the variables are separable. The procedure is illustrated by

EXAMPLE 2. *Find the solution $y(x)$ of the differential equation*

$$x^2\frac{dy}{dx} = xy - y^2,$$

which satisfies the condition $y(1) = 1$.

We can write the differential equation as

$$\frac{dy}{dx} = \frac{xy - y^2}{x^2} = \frac{y}{x} - \left(\frac{y}{x}\right)^2,$$

showing that it is homogeneous. If we let $y = zx$ we find that

$$x\frac{dz}{dx} + z = z - z^2,$$

which can be written in the separable form

$$\frac{dz}{z^2} + \frac{dx}{x} = 0$$

with solution

$$-1/z + \log x = C,$$

where C is a constant. If $y = 1$ when $x = 1$, then $z = 1$ when $x = 1$, so that (remembering that $\log 1 = 0$) we find that $C = -1$. Hence wc have

$$-\frac{x}{y} + \log x = -1$$

so that the required solution is

$$y = \frac{x}{1 + \log x}.$$

A more general type of differential equation can be reduced to the homogeneous type. Suppose that we have to solve an equation of the type

$$\frac{dy}{dx} = f\left(\frac{y-k}{x-h}\right), \tag{8}$$

where h and k are constants. Then if we put

$$y' = y - k, \qquad x' = x - h,$$

and use the fact that

$$\frac{dy}{dx} = \frac{dy'}{dx} = \frac{dy'}{dx'}\frac{dx'}{dx} = \frac{dy'}{dx'},$$

since $dx'/dx = 1$, we see that equation (8) is reduced to the homogeneous form

$$\frac{dy'}{dx'} = f\left(\frac{y'}{x'}\right).$$

An important special case of this is when the equation is of the type

$$\frac{dy}{dx} = \frac{ax + by + c}{a'x + b'y + c'}, \tag{9}$$

where a, a', b, b', c, c' are constants. If we let

$$x = x' + h, \qquad y = y' + k, \tag{10}$$

then

$$ax + by + c = ax' + by' + (ah + bk + c),$$
$$a'x + b'y + c' = a'x' + b'y' + (a'h + b'k + c').$$

If we choose the constants h and k so that

$$ah + bk + c = 0, \qquad a'h + b'k + c' = 0, \tag{11}$$

then the substitution (10) reduces the equation (9) to the homogeneous form

$$\frac{dy'}{dx'} = \frac{ax' + by'}{a'x' + b'y'}$$

which can be solved by putting $z = y'/x'$.

Often the substitution (10) can be derived by inspection of the equation, rather than by solving the simultaneous equations (11). As an example we have:-

EXAMPLE 3. *Find the general solution of the equation*

$$\frac{dy}{dx} = \frac{x - y + 3}{x + y - 5}.$$

If we rewrite the equation as

$$\frac{dy}{dx} = \frac{(x-1) - (y-4)}{(x-1) + (y-4)}$$

we see that the substitutions $x' = x - 1$, $y' = y - 4$ transform the equation to the homogeneous equation

$$\frac{dy'}{dx'} = \frac{x' - y'}{x' + y'}.$$

Now putting $z = y'/x'$ we find that

$$z + x'\frac{dz}{dx'} = \frac{1 - z}{1 + z},$$

that is,

$$\frac{z + 1}{1 - 2z - z^2}\,dz = \frac{dx'}{x'}$$

which can be integrated immediately to give

$$-\tfrac{1}{2}\log(1 - 2z - z^2) = \log x' + C,$$

where C is an arbitrary constant. If we write $C = -\frac{1}{2}\log c$ we see that

$$x'^2(1 - 2z - z^2) = c,$$

that is,

$$x'^2 - 2x'y' - y'^2 = c,$$

or

$$(x-1)^2 - 2(x-1)(y-4) - (y-4)^2 = c.$$

Simplifying the left hand side we see that the required solution is

$$x^2 - 2xy - y^2 + 6x + 10y = A,$$

where A denotes an arbitrary constant.

PROBLEMS

Find the general solutions of the equations:-

1. $(5x - y)\dfrac{dy}{dx} + x = 5y;$

2. $2x^2 \dfrac{dy}{dx} = x^2 + y^2;$

3. $x\dfrac{dy}{dx} = \dfrac{x^2 + y^2}{x + y};$

4. $\dfrac{dy}{dx} = \dfrac{x + y - 2}{x - y};$

5. $(2x + y + 3)^2 \dfrac{dy}{dx} = (x + 2y)^2.$

5. Exact Equations

If we write $f(x, y) = -P(x, y)/Q(x, y)$ the first order equation

$$\frac{dy}{dx} = f(x, y) \tag{1}$$

may be written symbolically in the form

$$P(x, y)dx + Q(x, y)dy = 0. \tag{2}$$

If the differential form on the left hand side of this equation is an exact differential (cf. § 10 of Chapter 8) we say that the differential equation (1) is *exact*. In that case there will be a function $\phi(x, y)$ such that

$$d\phi(x, y) = 0$$

so that the solution of (1) is

$$\phi(x, y) = C, \tag{3}$$

where C is an arbitrary constant.

We saw in § 10 of the previous chapter that a necessary and sufficient condition for the equation (2) to be exact is that

$$\frac{\partial P}{\partial y} = \frac{\partial Q}{\partial x}.$$

We also saw that, when this condition is satisfied,

$$\phi(x, y) = \int P\mathrm{d}x + \int \left\{ Q - \frac{\partial}{\partial y} \int P\mathrm{d}x \right\} \mathrm{d}y. \tag{4}$$

EXAMPLE 4. *Solve the equation*

$$\left(2xy + \frac{1}{y}\right) + x\left(x - \frac{1}{y^2}\right)\frac{\mathrm{d}y}{\mathrm{d}x} = 0.$$

This is of the same form as equation (2) above with

$$P(x, y) = 2xy + 1/y, \qquad Q(x, y) = x^2 - x/y^2,$$

so that

$$\frac{\partial P}{\partial y} = 2x - \frac{1}{y^2} = \frac{\partial Q}{\partial x},$$

showing that the equation is exact. Also

$$\int P(x, y)\mathrm{d}x = x^2 y + \frac{x}{y}$$

and

$$Q - \frac{\partial}{\partial y}\int P(x, y)\mathrm{d}x = x^2 - \frac{x}{y^2} - x^2 + \frac{x}{y^2} = 0,$$

so that, by equation (4), the solution of the equation is given by equation (3) with

$$\phi(x, y) = x^2 y + x/y.$$

PROBLEMS

Show that the following equations are exact and find their general solutions:-

1. $(2y + x^2)\mathrm{d}x + (2x + 12y^3)\mathrm{d}y = 0;$
2. $(8x^3 + 4xy)\mathrm{d}x + (2x^2 + 12y^3)\mathrm{d}y = 0;$
3. $(3x + 2y)\mathrm{d}x + (2x + 3y)\mathrm{d}y = 0;$
4. $(2x + y)\mathrm{d}x + (x + 2y)\mathrm{d}y = 0;$
5. $(x - 3y)\mathrm{d}x + \dfrac{(1 - 3xy^2)\mathrm{d}y}{y^2} = 0.$

6. Integrating Factors

When the equation

$$Pdx + Qdy = 0 \tag{1}$$

is not exact we may seek to make it exact by multiplying throughout by a suitable function $\mu(x, y)$. Applying the condition for the equation

$$(\mu P)dx + (\mu Q)dy = 0$$

to be exact, we see that such a function μ must satisfy the equation

$$\frac{\partial}{\partial y}(\mu P) = \frac{\partial}{\partial x}(\mu Q). \tag{2}$$

If we can find a function μ which satisfies this equation we say that we have found an *integrating factor* of the equation (1).

It is not an easy matter to find functions which satisfy equation (2) unless we make some simplifying assumption about the nature of μ.

μ a function of x alone: If we assume that $\mu = \mu(x)$ is a function of x alone,

$$\frac{\partial \mu}{\partial y} = 0, \qquad \frac{\partial \mu}{\partial x} = \mu'(x),$$

and equation (2) reduces to

$$\mu \frac{\partial P}{\partial y} = \mu'(x)Q + \mu \frac{\partial Q}{\partial x},$$

which may be written in the form

$$\frac{\mu'(x)}{\mu(x)} = \frac{1}{Q}\left(\frac{\partial P}{\partial y} - \frac{\partial Q}{\partial x}\right). \tag{3}$$

Now the function on the left hand side of this equation contains x alone and this must therefore also be true of the function on the right hand side of this equation. Our assumption that μ is a function of x alone is justified, therefore, only if it turns out that

$$\frac{1}{Q}\left(\frac{\partial P}{\partial y} - \frac{\partial Q}{\partial x}\right)$$

is a function of x alone. If we denote this function by $\psi(x)$, equation (3) may be put in the form

$$\frac{d}{dx}\log \mu(x) = \psi(x),$$

showing that

$$\log \mu(x) = \int \psi(x) \mathrm{d}x, \tag{4}$$

an equation from which we can easily determine $\mu(x)$.

μ a function of y: If we assume that $\mu = \mu(x, y)$ is a function of y alone,

$$\frac{\partial \mu}{\partial x} = 0, \qquad \frac{\partial \mu}{\partial y} = \mu'(y),$$

and equation (2) reduces to

$$\mu \frac{\partial P}{\partial y} + \mu'(y)P = \mu \frac{\partial Q}{\partial x},$$

which may be written in the form

$$\frac{\mu'(y)}{\mu(y)} = -\frac{1}{P}\left(\frac{\partial P}{\partial y} - \frac{\partial Q}{\partial x}\right),$$

showing that the initial assumption that $\mu(y)$ is a function of y alone is justified only if it turns out that

$$\frac{1}{P}\left(\frac{\partial P}{\partial y} - \frac{\partial Q}{\partial x}\right)$$

is a function of y alone, $-\chi(y)$ say. As in the last case we then find that

$$\log \mu(y) = \int \chi(y) \mathrm{d}y, \tag{5}$$

and $\mu(y)$ is determined.

If, therefore, we are given an equation of the type (1) and form the function

$$R(x, y) = \frac{\partial P}{\partial y} - \frac{\partial Q}{\partial x}, \tag{6}$$

then:

(i) If $R(x, y) = 0$, the equation is exact.

(ii) If $R(x, y)/Q(x, y) = \psi(x)$, the equation has an integrating factor $\mu(x)$ given by equation (4).

(iii) If $R(x, y)/P(x, y) = -\chi(y)$, the equation has an integrating factor $\mu(y)$ given by equation (5).

It is, of course, perfectly possible that none of these conditions will be satisfied. An integrating factor $\mu(x, y)$ might still exist but its determination from the partial differential equation (2) will probably present serious difficulties, and it is probably better to try to solve the equation by a completely different method.

As an example of the use of integrating factors we have:-

EXAMPLE 5. *Find the general solution of the equation*

$$\frac{dy}{dx} = \frac{y(2xy+1)}{x+2y^3}.$$

We can write this equation in the form (1) by putting

$$P(x, y) = 2xy^2 + y, \qquad Q(x, y) = -x - 2y^3, \tag{7}$$

so that

$$\frac{\partial P}{\partial y} = 4xy + 1, \qquad \frac{\partial Q}{\partial x} = -1,$$

and hence

$$R(x, y) = 4xy + 2.$$

It follows that

$$\frac{R(x, y)}{P(x, y)} = \frac{4xy+2}{2xy^2+y} = \frac{2}{y},$$

and, therefore, that the integrating factor is $\mu(y)$ where

$$\log \mu(y) = -2\int \frac{dy}{y} = \log\left(\frac{1}{y^2}\right).$$

The equation therefore has an integrating factor y^{-2}. Multiplying both sides of equation (1), with P and Q given by equations (7), by y^{-2} we see that it is equivalent to the equation

$$\left(2x + \frac{1}{y}\right) dx - \left(\frac{x}{y^2} + 2y\right) dy = 0,$$

the terms of which can be grouped to give

$$2x dx - 2y dy + \left(\frac{dx}{y} - \frac{x dy}{y^2}\right) = 0,$$

from which it is obvious that the general solution is

$$x^2 - y^2 + 2\frac{x}{y} = C,$$

where C is an arbitrary constant.

PROBLEMS

1. Show that

$$\frac{1}{x^2}, \quad \frac{1}{y^2}, \quad \frac{1}{xy}, \quad \frac{1}{x^2+y^2}, \quad \frac{1}{x^2-y^2}$$

are all integrating factors of the differential equation

$$x\mathrm{d}y - y\mathrm{d}x = 0.$$

2. Solve the differential equations

(i) $x^2(1+3y) + (x^3+y)\dfrac{\mathrm{d}y}{\mathrm{d}x} = 0;$

(ii) $2y\dfrac{\mathrm{d}y}{\mathrm{d}x} + 3y^2 + 4x = 0;$

(iii) $\dfrac{\mathrm{d}y}{\mathrm{d}x} = \dfrac{2xy - y^3}{2x^2 - y^3};$

(iv) $\dfrac{x}{x^4+y^2}\dfrac{\mathrm{d}y}{\mathrm{d}x} = 2\left(1 + \dfrac{y}{x^4+y^2}\right).$

3. Find an integrating factor of the form $f(xy)$ for the differential equation

$$y(x-2y)\mathrm{d}x + x(2x-y)\mathrm{d}y = 0,$$

and hence solve the equation.

7. Linear Equations of the First Order

Any ordinary differential equation which can be put in the form

$$\frac{\mathrm{d}y}{\mathrm{d}x} + p(x)y = q(x), \tag{1}$$

where $p(x)$ and $q(x)$ are functions of x alone (i.e. do *not* involve y) is called a linear equation of the first order, in conformity with the definitions of § 1.

Writing this equation as

$$\mathrm{d}y + p(x)y\mathrm{d}x = q(x)\mathrm{d}x$$

we see that it is of the type

$$P(x, y)\mathrm{d}x + Q(x, y)\mathrm{d}y = 0, \tag{2}$$

with

$$P(x, y) = p(x)y - q(x), \qquad Q(x, y) = 1. \tag{3}$$

Since, for the functions defined by equations (3), we have

$$\frac{1}{Q}\left(\frac{\partial P}{\partial y} - \frac{\partial Q}{\partial x}\right) = p(x),$$

a function of x alone, it follows from the considerations of the previous section that equation (2) has an integrating factor $\mu(x)$ which is determined by the equation

$$\log \mu(x) = \int p(x)\mathrm{d}x,$$

and hence is

$$\mu(x) = \mathrm{e}^{\int p\mathrm{d}x} \tag{4}$$

If, therefore, we multiply both sides of equation (1) by the integrating factor (4) we get an exact equation

$$\mathrm{e}^{\int p\mathrm{d}x}\left(\frac{\mathrm{d}y}{\mathrm{d}x}\right) + p\mathrm{e}^{\int p\mathrm{d}x} y = q(x)\mathrm{e}^{\int p\mathrm{d}x},$$

which can be integrated at once to give the general solution

$$\mathrm{e}^{\int p\mathrm{d}x} y = \int q(x)\mathrm{e}^{\int p\mathrm{d}x}\,\mathrm{d}x + C, \tag{5}$$

where C is an arbitrary constant.

Example 6. *Solve the differential equation*

$$(x^2 + 1)\mathrm{d}y/\mathrm{d}x + 2xy = \sqrt{y}.$$

At first sight this is not a linear equation, but if we make the substitution

$$v = y^{\frac{1}{2}}, \qquad \frac{\mathrm{d}v}{\mathrm{d}x} = \frac{1}{2\sqrt{y}}\frac{\mathrm{d}y}{\mathrm{d}x},$$

we see that the given equation transforms to

$$2(x^2 + 1)\frac{\mathrm{d}v}{\mathrm{d}x} + 2xv = 1,$$

which is a first order linear equation in v. For this equation we have

$$\frac{\mathrm{d}v}{\mathrm{d}x} + p(x)v = q(x)$$

where $p(x) = x/(x^2 + 1)$ and $q(x) = 1/2(x^2 + 1)$. For this $p(x)$

$$\int p(x)\mathrm{d}x = \frac{1}{2}\int\frac{2x}{x^2 + 1}\,\mathrm{d}x = \tfrac{1}{2}\log(x^2 + 1),$$

so that the appropriate integrating factor is $(x^2 + 1)^{\frac{1}{2}}$. Multiplying by this factor we have

$$(x^2 + 1)^{\frac{1}{2}} \frac{dv}{dx} + x(x^2 + 1)^{-\frac{1}{2}} v = \tfrac{1}{2}(x^2 + 1)^{-\frac{1}{2}},$$

that is

$$\frac{d}{dx}[(x^2 + 1)^{\frac{1}{2}} v] = \tfrac{1}{2}(x^2 + 1)^{-\frac{1}{2}},$$

which can be integrated immediately to give

$$(x^2 + 1)^{\frac{1}{2}} v = \tfrac{1}{2}[\log x + \sqrt{(x^2 + 1)}] + C,$$

where C is an arbitrary constant. Reverting to the original variable and and putting $C = \frac{1}{2}c$, we see that the general solution of the given equation is

$$y = \frac{1}{4(x^2 + 1)} [\log x + \sqrt{(x^2 + 1)} + c]^2,$$

where c denotes an arbitrary constant.

The special case of the first order linear equation in which $p(x)$ is a constant, p say, is of special importance, since it arises frequently in applications. It follows from equation (4) that the equation

$$\frac{dy}{dx} + py = q(x) \tag{6}$$

has an integrating factor e^{px}. Hence

$$e^{px} \frac{dy}{dx} + pe^{px} y = q(x)e^{px},$$

or

$$\frac{d}{dx}(e^{px} y) = q(x)e^{px}.$$

This last equation may be integrated at once to give

$$ye^{px} = \int q(x)e^{px} dx + C,$$

where C denotes an arbitrary constant. We therefore have

$$y = e^{-px} \int q(x)e^{px} dx + Ce^{-px}, \tag{7}$$

with C arbitrary, as the general solution of the equation (6).

In the particular case in which $q(x)$ is also a constant, we find that

$$\int q(x)e^{px}\,dx = \frac{q}{p}e^{px},$$

showing that the general solution of the equation

$$\frac{dy}{dx} + py = q, \qquad (p,\ q\ \textit{constants}), \tag{8}$$

is

$$y = \frac{q}{p} + Ce^{-px}, \tag{9}$$

where C is an arbitrary constant. We have already had an example of the occurrence of an equation of this type (example 10 of Chapter 6, p. 265 above).

PROBLEMS

Find the general solutions of the equations:-

1. $(1 - x^2)\dfrac{dy}{dx} - xy = \sqrt{(1 - x^2)};$

2. $\dfrac{dy}{dx} = (x - a)e^{ax} + xy;$

3. $\dfrac{dy}{dx} + \dfrac{y}{x} = x^4;$

4. $x(x^2 - 1)\dfrac{dy}{dx} + (x^2 + 1)y + x = 0;$

5. $a^2xy\dfrac{dy}{dx} = x^4 + a^2y^2$. (Note that this equation is linear in y^2.)

8. Equations Reducible to Linear Equations

In this section we shall show how ordinary differential equations of two different types can be transformed to linear equations (whose solutions may be derived by the methods of the previous section).

The first of these equations is:-

(a) Bernoulli's equation

The name *Bernoulli's equation* is given to the differential equation

$$\frac{dy}{dx} + P(x)y = Q(x)y^n \tag{1}$$

in which the coefficients P and Q may depend on x but not on y. We may reduce this equation to linear form by the substitution

$$w = \frac{1}{y^{n-1}}. \tag{2}$$

Differentiating both sides of equation (2) with respect to x we find that

$$\frac{dw}{dx} = -\frac{n-1}{y^n}\frac{dy}{dx},$$

so that

$$\frac{dy}{dx} = -\frac{y^n}{n-1}\frac{dw}{dx}. \tag{3}$$

Substituting from equation (3) into equation (1) and multiplying throughout by $-(n-1)y^{-n}$ we find that w satisfies the linear equation

$$\frac{dw}{dx} - (n-1)Pw = -(n-1)Q, \tag{4}$$

which may be solved by the method of the last section. Once w has been determined, it follows from equation (2) that

$$y = \left(\frac{1}{w}\right)^{1/(n-1)}. \tag{5}$$

EXAMPLE 7. *Find the general solution of the equation*

$$\frac{dy}{dx} - y = xy^2.$$

This is a Bernoulli equation with $n = 2$, so that we put

$$w = 1/y, \qquad y = 1/w$$

from which it follows that

$$\frac{dy}{dx} = -\frac{1}{w^2}\frac{dw}{dx}.$$

Substituting this expression in the differential equation we find that w satisfies the equation

$$-\frac{1}{w^2}\frac{dw}{dx} - \frac{1}{w} = \frac{x}{w^2},$$

which can be simplified to

$$\frac{\mathrm{d}w}{\mathrm{d}x} + w + x = 0.$$

The integrating factor of this equation is e^x so that we have

$$\frac{\mathrm{d}}{\mathrm{d}x}(e^x w) + xe^x = 0.$$

Using the formula for integrating by parts, we find that

$$\int xe^x \,\mathrm{d}x = xe^x - \int e^x \,\mathrm{d}x = xe^x - e^x,$$

so that

$$e^x w + (x-1)e^x = C,$$

where C is an arbitrary constant. We therefore have

$$w = Ce^{-x} + 1 - x$$

from which it follows that the general solution of the original differential equation is

$$y = \frac{1}{1 - x + Ce^{-x}},$$

where C is an arbitrary constant.

(*b*) *Riccati's equation*

The differential equation

$$\frac{\mathrm{d}y}{\mathrm{d}x} = P + Qy + Ry^2, \tag{6}$$

in which P, Q and R are functions of x alone is called *Riccati's equation.*

We shall now show that the general solution of this equation can be found if we know a particular solution of it.

Suppose that $u(x)$ is a particular solution of equation (6) so that

$$\frac{\mathrm{d}u}{\mathrm{d}x} = P + Qu + Ru^2. \tag{7}$$

If we let $y = z + u$ in equation (6) we find that

$$\frac{dz}{dx} + \frac{du}{dx} = P + Qu + Ru^2 + Qz + R(2zu + z^2),$$

and using (7) we see that this becomes

$$\frac{dz}{dx} = (Q + 2Ru)z + Rz^2, \tag{8}$$

so that the function z satisfies an equation of Bernoulli type which can be solved by the method of the last subsection. Since the n in this case is 2, we solve the equation by means of the substitution $z = w^{-1}$ which produces a linear equation in w. Hence we solve a Riccati equation by changing the dependent variable from y to w where

$$y = u(x) + 1/w. \tag{9}$$

EXAMPLE 8. *Show that $y = x^2$ is a particular solution of the differential equation*

$$\frac{dy}{dx} + (2x^3 - 1)y - xy^2 = x^5 - x^2 + 2x,$$

and hence find the general solution of the equation.

If $y = x^2$, then $dy/dx = 2x$ and

$$\frac{dy}{dx} + (2x^3 - 1)y - xy^2 = 2x + (2x^3 - 1)x^2 - x^5 = x^5 - x^2 + 2x,$$

verifying that $y = x^2$ is a particular solution of the given equation. We now let $y = z + x^2$, so that

$$\frac{dy}{dx} = \frac{dz}{dx} + 2x.$$

Substituting in the original differential equation we find that z satisfies the equation

$$\frac{dz}{dx} + 2x + (2x^3 - 1)(z + x^2) - x(x^2 + z)^2 = x^5 - x^2 + 2x,$$

which is easily shown to reduce to

$$\frac{dz}{dx} - z = xz^2.$$

We saw in example 7 that the general solution of this equation is

$$z = \frac{1}{1 - x + Ce^{-x}}$$

so that the general solution of the present equation is

$$y = x^2 + \frac{1}{1 - x + Ce^{-x}}.$$

PROBLEMS

1. Solve the differential equations:

(i) $xy - \dfrac{dy}{dx} = y^3 e^{-x^2};$

(ii) $2\dfrac{dy}{dx} - y + e^{-x} y^3 = 0;$

(iii) $x^2 y - x^3 \dfrac{dy}{dx} = y^4 \cos x.$

2. Show that $y = -1$ is a particular solution of the differential equation

$$x(x-1)\frac{dy}{dx} + y^2 - (2x-3)y - 2(x-1) = 0,$$

and hence deduce the general solution.

3. Show that $y = -2x$ is a particular solution of the equation

$$(x^2+1)\left(2 + \frac{dy}{dx}\right) = 4x^2 - y^2,$$

and deduce the general solution.

4. Find the general solution of the equation

$$x^4 \frac{dy}{dx} + x^4 y^2 = 4,$$

given that it has a particular integral which is a quadratic function of $1/x$.

B. LINEAR EQUATIONS OF THE SECOND AND HIGHER ORDER

9. Linear Differential Equations

From the definition of a linear ordinary differential equation given in § 1 we see that the most general linear equation of the n-th order is

$$p_n \frac{d^n y}{dx^n} + p_{n-1} \frac{d^{n-1} y}{dx^{n-1}} + \ldots + p_1 \frac{dy}{dx} + p_0 y = f(x), \tag{1}$$

where $p_n, p_{n-1}, \ldots, p_1, p_0$ and $f(x)$ are functions of x alone. In the case when $f(x) = 0$ the equation becomes

$$p_n \frac{d^n y}{dx^n} + p_{n-1} \frac{d^{n-1} y}{dx^{n-1}} + \ldots + p_1 \frac{dy}{dx} + p_0 y = 0. \tag{2}$$

We say that (1) is an *inhomogeneous* linear equation, and that (2) is the *homogeneous* equation corresponding to equation (1).

To avoid a great deal of writing we introduce a notation which simplifies the left hand sides of these equations. We let

$$\mathscr{D}y = p_n \frac{d^n y}{dx^n} + p_{n-1} \frac{d^{n-1} y}{dx^{n-1}} + \ldots + p_1 \frac{dy}{dx} + p_0 y, \tag{3}$$

so that symbolically we may write

$$\mathscr{D} = p_n \frac{d^n}{dx^n} + p_{n-1} \frac{d^{n-1}}{dx^{n-1}} + \ldots + p_1 \frac{d}{dx} + p_0. \tag{4}$$

The symbol $\mathscr{D}$ only has a meaning when there is a function on its right upon which it can operate; in other words, $\mathscr{D}$ is a *differential operator.* With this notation we can write equation (1) in the form

$$\mathscr{D}y = f(x), \tag{5}$$

and the corresponding homogeneous equation (2) as

$$\mathscr{D}y = 0. \tag{6}$$

We shall now consider some properties of the differential operator.

THEOREM 38. *If c is a constant and y is any function of x,*

$$\mathscr{D}(cy) = c\mathscr{D}y.$$

This follows immediately from the definition of $\mathscr{D}$ and the fact that, if c is a constant

$$\frac{\mathrm{d}^r}{\mathrm{d}x^r}(cy) = c\frac{\mathrm{d}^r y}{\mathrm{d}x^r}.$$

Since $\mathscr{D}(cy)$ is a sum of terms like

$$p_r \frac{\mathrm{d}^r}{\mathrm{d}x^r}(cy),$$

the result follows.

THEOREM 39. *If c_1, c_2 are constants and y_1, y_2 are functions of x,*

$$\mathscr{D}(c_1 y_1 + c_2 y_2) = c_1 \mathscr{D}y_1 + c_2 \mathscr{D}y_2.$$

This follows at once from the fact that

$$\frac{\mathrm{d}^r}{\mathrm{d}x^r}(c_1 y_1 + c_2 y_2) = c_1 \frac{\mathrm{d}^r}{\mathrm{d}x^r} y_1 + c_2 \frac{\mathrm{d}^r}{\mathrm{d}x^r} y_2.$$

As a consequence of this result we have:-

COROLLARY. *If $c_1, c_2, \ldots, c_n$ are constants and $y_1, y_2, \ldots, y_n$ are functions of x,*

$$\mathscr{D}(c_1 y_1 + c_2 y_2 + \ldots + c_n y_n) = c_1 \mathscr{D}y_1 + c_2 \mathscr{D}y_2 + \ldots + c_n \mathscr{D}y_n.$$

THEOREM 40. *If $y_1, y_2, \ldots, y_n$ are particular solutions of the homogeneous linear equation $\mathscr{D}y = 0$, then $c_1 y_1 + c_2 y_2 + \ldots + c_n y_n$ is also a solution, where $c_1, c_2, \ldots, c_n$ are arbitrary constants.*

The proof of this theorem is quite simple. If $y_1, y_2, \ldots, y_n$ are particular solutions of the differential equation $\mathscr{D}y = 0$, then

$$\mathscr{D}y_1 = 0, \quad \mathscr{D}y_2 = 0, \quad \ldots, \quad \mathscr{D}y_n = 0.$$

Using the corollary to Theorem 39 we find that

$$\mathscr{D}(c_1 y_1 + c_2 y_2 + \ldots + c_n y_n) = c_1 \mathscr{D}y_1 + c_2 \mathscr{D}y_2 + \ldots + c_n \mathscr{D}y_n = 0,$$

showing that $c_1 y_1 + c_2 y_2 + \ldots + c_n y_n$ is also a solution of the equation.

The equation $\mathscr{D}y = 0$ is of the n-th order. By splitting this equation into n first order equations we can show that its general solution must contain n arbitrary constants.* Therefore, if $y_1, y_2, \ldots, y_n$ are n *linearly independent* particular solutions of $\mathscr{D}y = 0$, then

$$y = c_1 y_1 + c_2 y_2 + \ldots + c_n y_n \tag{7}$$

* For details of the proof of this statement see J. C. Burkill, "The Theory of Ordinary Differential Equations" (Oliver & Boyd, Edinburgh, 1956) p. 12.

is the general solution of $\mathscr{D}y = 0$. By linear independence we mean that it is not possible to find a set of constants $b_1, b_2, \ldots, b_n$, not all zero, such that there exists a relation of the form

$$b_1 y_1 + b_2 y_2 + \ldots + b_n y_n = 0, \tag{8}$$

connecting the functions $y_1, y_2, \ldots, y_n$.

To show the significance of the fact that the particular solutions must be linearly independent, let us suppose that, in fact, they are not, so that a relation of the type (8) exists. We could then regard (8) as an equation from which we could determine y_n in terms of $y_1, y_2, \ldots, y_{n-1}$ by the formula

$$y_n = -\frac{1}{b_n}(b_1 y_1 + b_2 y_2 + \ldots + b_{n-1} y_{n-1}). \tag{9}$$

Substituting from equation (9) into equation (7) we see that the solution (7) then becomes

$$y = k_1 y_1 + k_2 y_2 + \ldots + k_{n-1} y_{n-1}, \tag{10}$$

where

$$k_1 = c_1 - \frac{b_1 c_n}{b_n}, \quad k_2 = c_2 - \frac{b_2 c_n}{b_n}, \quad \ldots, \quad k_{n-1} = c_{n-1} - \frac{b_{n-1} c_n}{b_n}.$$

Since the $c_1, c_2, \ldots, c_n$ are arbitrary and the $b_1, b_2, \ldots, b_n$ are fixed constants, it follows that the constants $k_1, k_2, \ldots, k_{n-1}$ may be chosen arbitrarily. Hence the solution (10) contains only $n - 1$ arbitrary constants; it is thus not the general solution (since that must contain n arbitrary constants). Hence, in order to obtain a general solution from n particular solutions, it is necessary that these particular solutions should be linearly independent.

THEOREM 41. *If $y_1, y_2, \ldots, y_n$ are n linearly independent solutions of the corresponding homogeneous equation $\mathscr{D}y = 0$, and if Y is a particular solution of the linear equation $\mathscr{D}y = f(x)$, then the general solution of the equation*

$$\mathscr{D}y = f(x)$$

is $y = Y + c_1 y_1 + c_2 y_2 + \ldots + c_n y_n$, where $c_1, c_2, \ldots, c_n$ are arbitrary constants.

The proof of this theorem is straightforward. Since

$$\mathscr{D}y_1 = \mathscr{D}y_2 = \mathscr{D}y_3 = \ldots = \mathscr{D}y_n = 0, \qquad \mathscr{D}Y = f(x),$$

it follows from the corollary to Theorem 39 that

$$\mathscr{D}(c_1y_1 + c_2y_2 + \ldots + c_ny_n + Y)$$
$$= c_1\mathscr{D}y_1 + c_2\mathscr{D}y_2 + \ldots + c_n\mathscr{D}y_n + \mathscr{D}Y = f(x)$$

so that $y = Y + c_1y_1 + c_2y_2 + \ldots + c_ny_n$ is a solution of the inhomogeneous equation $\mathscr{D}y = f(x)$. Since it contains the correct number n of arbitrary constants, it is the general solution.

The general solution of the corresponding homogeneous equation,

$$y = c_1y_1 + c_2y_2 + \ldots + c_ny_n,$$

is called the *complementary function* of the inhomogeneous equation and the function Y is called a *particular integral* of the equation. To find the general solution of an inhomogeneous linear differential equation $\mathscr{D}y = f(x)$. we need to find a particular integral of the equation and the general solution of the corresponding homogeneous equation $\mathscr{D}y = 0$.

We shall now consider further the idea of linear independence.

THEOREM 42. *Two functions $f_1(x)$, $f_2(x)$ are linearly dependent if*

$$W(f_1, f_2) = 0,$$

where $W(f_1, f_2) = f_1(x)f_2'(x) - f_1'(x)f_2(x)$.

From the definition of linear dependence we know that, if $f_1(x)$ and $f_2(x)$ are linearly dependent then there exist constants b_1 and b_2 such that

$$b_1f_1(x) + b_2f_2(x) = 0, \tag{11}$$

for all values of x. If we differentiate both sides of equation (11) with respect to x, then we obtain the relation

$$b_1f_1'(x) + b_2f_2'(x) = 0. \tag{12}$$

Now from equation (11) we find that $b_1/b_2 = -f_2(x)/f_1(x)$. Substituting this value into equation (12) we find that $f_1(x)$ and $f_2(x)$ are linearly dependent if

$$W(f_1, f_2) = f_1(x)f_2'(x) - f_1'(x)f_2(x) = 0. \tag{13}$$

The expression $W(f_1, f_2)$ is called the *Wronskian* of the functions f_1 and f_2

EXAMPLE 9. *Show that the functions $e^{\alpha x}$, $e^{\beta x}$ are linearly independent if $\alpha \neq \beta$.*

If we let $f_1(x) = e^{\alpha x}$, $f_2(x) = e^{\beta x}$, then $f_1'(x) = \alpha e^{\alpha x}$, $f_2'(x) = \beta e^{\beta x}$, so that $W(f_1, f_2) = (\beta - \alpha)e^{(\alpha+\beta)x}$, and this vanishes if, and only if, $\alpha = \beta$.

EXAMPLE 10. *Show that the functions* 1, $\cos^2 x$, $\sin^2 x$, *are linearly dependent.*

This follows from the fact that we can write the fundamental identity of trigonometry in the form

$$1 \cdot \cos^2 x + 1 \cdot \sin^2 x + (-1) \cdot 1 = 0.$$

The discussion of the linear independence of n functions is most conveniently carried out with the aid of theory of determinants. We shall not discuss the matter at this point, but refer the interested reader to the Appendix on the properties of determinants, particularly §3, p. 642.

PROBLEMS

1. Show that the following pairs of functions are linearly independent:-

(i) $f_1(x) = x^m$, $f_2(x) = x^n$, $(m \neq n)$;

(ii) $f_1(x) = x^m$, $f_2(x) = e^{\alpha x}$;

(iii) $f_1(x) = x^m$, $f_2(x) = \cos(\alpha x)$;

(iv) $f_1(x) = x^m e^{\alpha x}$, $f_2(x) = e^{\alpha x}$, $(m \neq 0)$;

(v) $f_1(x) = x^m \sin(\alpha x + \beta)$, $f_2(x) = \sin(\alpha x + \beta)$, $(m \neq 0)$;

(vi) $f_1(x) = e^{\alpha x}$, $f_2(x) = \cos(\alpha x + \beta)$.

2. Find particular integrals of the differential equation

$$(x+1)\frac{d^2y}{dx^2} + (x-1)\frac{dy}{dx} - 2y = 0$$

of the forms e^{ax} and $px^2 + qx + r$; and hence write down the general solution.

10. Second Order Linear Equations with Constant Coefficients

In this section we shall consider the solution of linear differential equations with constant coefficients, i.e. equations of the type (1) of the last section, but with $p_n, p_{n-1}, \ldots, p_1, p_0$ constants, not functions of x. We shall discuss the second order equation

$$a\frac{d^2y}{dx^2} + 2b\frac{dy}{dx} + cy = f(x) \tag{1}$$

in which a, b, c are constants and the function $f(x)$ is prescribed. To find the complementary function of this equation we must solve the corresponding homogeneous equation

$$a\frac{d^2y}{dx^2} + 2b\frac{dy}{dx} + cy = 0,$$

which, if we introduce the differential operator $D = d/dx$ we write symbolically as

$$(aD^2 + 2bD + c)y = 0. \tag{2}$$

We now try to find solutions of equation (2) of the type

$$y = e^{mx}. \tag{3}$$

Substituting from equation (3) into equation (2) and using the fact that $D(e^{mx}) = me^{mx}$, $D^2(e^{mx}) = m^2e^{mx}$, we see that (3) will be a solution of (2), provided that, for all values of x,

$$(am^2 + 2bm + c)e^{mx} = 0.$$

Since e^{mx} is always positive we may divide both sides of this equation by it, and get the result that (3) is a solution of (2) if m is a root of the quadratic equation

$$am^2 + 2bm + c = 0. \tag{4}$$

Various possibilities arise according to the nature of the roots of this equation; we shall consider the cases separately.

Case (a): Distinct real roots

If equation (4) has two distinct real roots m_1 and m_2, say, then the two functions

$$y_1(x) = e^{m_1x}, \qquad y_2(x) = e^{m_2x}$$

will be particular solutions of equation (2), and, since their Wronskian is

$$W(y_1, y_2) = e^{m_1x}m_2e^{m_2x} - m_1e^{m_1x}e^{m_2x} = (m_2 - m_1)e^{(m_1+m_2)x} \neq 0,$$

they are linearly independent. The general solution of equation (2) is therefore

$$y = c_1 e^{m_1 x} + c_2 e^{m_2 x}, \tag{5}$$

where c_1 and c_2 denote arbitrary constants.

EXAMPLE 11. *Find the general solution of the equation*

$$\frac{d^2y}{dx^2} + \frac{dy}{dx} - 56y = 0,$$

and find the solution which satisfies the conditions $y = 1$, $dy/dx = 0$, *when* $x = 0$.

Replacing D by m in this equation, we obtain the equation

$$m^2 + m - 56 = 0,$$

which may be factorized to give

$$(m + 8)(m - 7) = 0,$$

so that its roots are 7 and -8. The general solution of the equation is therefore

$$y = c_1 e^{7x} + c_2 e^{-8x}. \tag{6}$$

If we differentiate both sides of this equation with respect to x, we find that

$$\frac{dy}{dx} = 7c_1 e^{7x} - 8c_2 e^{-8x}. \tag{7}$$

Putting $x = 0$, $y = 1$ in equation (6), $x = 0$, $dy/dx = 0$ in equation (7), we obtain the equations

$$c_1 + c_2 = 1, \qquad 7c_1 - 8c_2 = 0,$$

for the determination of the constants c_1 and c_2. These may be solved easily to give $c_1 = \frac{8}{15}$ and $c_2 = \frac{7}{15}$, and hence the expression

$$y = \tfrac{1}{15}(8e^{7x} + 7e^{-8x}),$$

for the required particular solution.

Case (b): Equal real roots

If the roots of equation (4) are equal then they are both real, and equal to $-b/a$, so that one solution of equation (2) is $e^{-bx/a}$. To find the complete solution of the equation we make the substitution

$$y = z(x)e^{-bx/a} \tag{8}$$

in equation (2). Using the results

$$\frac{dy}{dx} = \frac{dz}{dx} e^{-bx/a} - \frac{b}{a} z e^{-bx/a},$$

$$\frac{d^2y}{dx^2} = \frac{d^2z}{dx^2} e^{-bx/a} - \frac{2b}{a} \frac{dz}{dx} e^{-bx/a} + \frac{b^2}{a^2} z e^{-bx/a},$$

and remembering that, if equation (4) has equal roots, $c = b^2/a$, we find that the function $z(x)$ satisfies the equation

$$\frac{d^2z}{dx^2} = 0.$$

Integrating this equation once we find that

$$\frac{dz}{dx} = c_1,$$

where c_1 is an arbitrary constant, and integrating again, we obtain the solution

$$z = c_1 x + c_2.$$

Substituting this expression in equation (8) we find that the general solution of the equation

$$a^2 \frac{d^2y}{dx^2} + 2ab \frac{dy}{dx} + b^2 y = 0 \tag{9}$$

is

$$y = (c_1 x + c_2) e^{-bx/a}, \tag{10}$$

where c_1 and c_2 are arbitrary constants.

EXAMPLE 12. *Find the solution of the equation*

$$\frac{d^2y}{dx^2} - 2 \frac{dy}{dx} + y = 0,$$

which satisfies the conditions $y = 1$, $dy/dx = 0$, *when* $x = 0$.

If we compare this equation with equation (9) we see that it is of the form (9) with $a = 1$, $b = -1$, so that its general solution is

$$y = (c_1 x + c_2) e^x,$$

with derivative

$$\frac{dy}{dx} = c_1 e^x + (c_1 x + c_2) e^x.$$

Putting $x = 0$, $y = 1$ in the first of these two equations, and $x = 0$,

$\mathrm{d}y/\mathrm{d}x = 0$ in the second, we find that we must choose c_1 and c_2 to be such that

$$c_2 = 1, \qquad c_1 + c_2 = 0.$$

We therefore obtain the solution

$$y = (1 - x)\mathrm{e}^x.$$

Case (c): Complex roots

If the quadratic equation (4) has complex roots, they must form a pair of complex conjugates $\alpha + i\beta$, $\alpha - i\beta$. By the same argument as is employed in case (*a*) above we see that $\mathrm{e}^{(\alpha+i\beta)x}$, $\mathrm{e}^{(\alpha-i\beta)x}$ will therefore be solutions of the equation. From these two solutions we can obtain the pair of solutions [cf. p. 259, eq. (22)]

$$y_1(x) = \tfrac{1}{2}\{\mathrm{e}^{(\alpha+i\beta)x} + \mathrm{e}^{(\alpha-i\beta)x}\} = \mathrm{e}^{\alpha x}\cos\beta x,$$
$$y_2(x) = \frac{1}{2i}\{\mathrm{e}^{(\alpha+i\beta)x} - \mathrm{e}^{(\alpha-i\beta)x}\} = \mathrm{e}^{\alpha x}\sin\beta x.$$

Furthermore the Wronskian of this pair of functions is

$$W(y_1, y_2) = \mathrm{e}^{\alpha x}\cos\beta x(\alpha\mathrm{e}^{\alpha x}\sin\beta x + \beta\mathrm{e}^{\alpha x}\cos\beta x) - (\alpha\mathrm{e}^{\alpha x}\cos\beta x - \beta\mathrm{e}^{\alpha x}\sin\beta x)\mathrm{e}^{\alpha x}\sin\beta x$$

and this is equal to $\beta\mathrm{e}^{\alpha x} \neq 0$, showing that y_1 and y_2 are linearly independent. It follows therefore that the equation

$$\frac{\mathrm{d}^2y}{\mathrm{d}x^2} - 2\alpha\frac{\mathrm{d}y}{\mathrm{d}x} + (\alpha^2 + \beta^2)y = 0 \tag{11}$$

has the general solution

$$y = (c_1\cos\beta x + c_2\sin\beta x)\mathrm{e}^{\alpha x} \tag{12}$$

where c_1 and c_2 denote arbitrary constants.

EXAMPLE 13. *Find the general solution of the equation*

$$\frac{\mathrm{d}^2y}{\mathrm{d}x^2} + 4\frac{\mathrm{d}y}{\mathrm{d}x} + 5y = 0,$$

and from it deduce the particular solution which satisfies the conditions $y = 0$, $\mathrm{d}y/\mathrm{d}x = 1$, *when* $x = 0$.

Writing this differential equation in the form

$$\frac{\mathrm{d}^2y}{\mathrm{d}x^2} - 2(-2)\frac{\mathrm{d}y}{\mathrm{d}x} + \{(-2)^2 + 1\}y = 0,$$

we see that it is of the type (11) with $\alpha = -2$, $\beta = 1$. Hence, from (12), its general solution is

$$y = (c_1 \cos x + c_2 \sin x)e^{-2x}.$$

Since $y = 0$ when $x = 0$ we must have $c_1 = 0$. Differentiating the resulting solution we find that

$$\frac{dy}{dx} = c_2(\cos x - 2 \sin x)e^{-2x}.$$

Putting $x = 0$, $dy/dx = 1$ in this equation we find that $c_2 = 1$, showing that the required particular solution is

$$y = e^{-2x} \sin x.$$

So far we have concentrated on the problem of finding the complementary function of an equation of the type (1). We must now consider how we should find a particular integral of the equation, which is all we need to do to derive its general solution from the complementary function. In most cases the simplest way to proceed is to decide by inspection of the function $f(x)$ a likely form of the particular integral of the equation (1), and then to determine the values of the constants occurring in that form by direct solution into the differential equation. Later on in this section, we shall show how to derive the particular integral of (1) as an integral involving $f(x)$, but for the moment we shall concentrate on illustrating the use of the *ad hoc* procedure outlined above. The most suitable forms for the particular integral $Y(x)$ corresponding to commonly occurring forms of $f(x)$ are indicated below:-

(1) If $f(x)$ is a polynomial of degree n in x, let $Y(x) = \sum_{r=0}^{n} a_r x^r$, and determine the coefficients $a_0, a_1, \ldots, a_n$ by substituting in the given differential equation. (Cf. example 14 below.)

(2) If $f(x) = Ce^{kx}$, where k is not a root of equation (4), we let $Y(x) = Ae^{kx}$ in equation (1). Making this substitution we find that $A = C/(ak^2 + 2bk + c)$. (Cf. example 15 below.)

(3) If $f(x) = Ce^{m_1 x}$, where m_1 is a root of equation (4), we let $Y(x)$ equal $Axe^{m_1 x}$, or $Ax^2 e^{m_1 x}$, according as m_1 is a single or double root of equation (4). (Cf. example 16 below.)

(4) If $f(x) = F \sin kx + G \cos kx$, we let $Y(x) = A \sin kx + B \cos kx$. The method fails if the complementary function of the equation is $c_1 \sin kx + c_2 \cos kx$; in that case, we let $Y(x) = x(A \sin kx + B \cos kx)$. (Cf. example 17 below.)

We shall illustrate the procedure by the following examples:-

EXAMPLE 14. *Find the general solution of the equation*

$$\frac{d^2y}{dx^2} - 3\frac{dy}{dx} + 2y = x,$$

and find the particular solution corresponding to the conditions $y = dy/dx = 0$, *when* $x = 0$.

If we replace D by m we get the equation

$$m^2 - 3m + 2 = 0,$$

which has roots 1, 2 so that the complementary function is $c_1 e^x + c_2 e^{2x}$. To find the particular integral we write $Y = Ax + B$ for which

$$\frac{dY}{dx} = A, \qquad \frac{d^2Y}{dx^2} = 0.$$

Substituting these forms into the differential equation we see that we must choose the constants A and B so that

$$-3A + 2(Ax + B) = x,$$

i.e. we must take $A = \frac{1}{2}$, $B = \frac{3}{4}$ to give the particular integral $\frac{1}{2}x + \frac{3}{4}$. The general solution of the equation is therefore

$$y = c_1 e^x + c_2 e^{2x} + \tfrac{1}{2}x + \tfrac{3}{4} \tag{13}$$

for which

$$\frac{dy}{dx} = c_1 e^x + 2c_2 e^{2x} + \tfrac{1}{2}. \tag{14}$$

If we substitute the conditions $x = 0$, $y = dy/dx = 0$ into equations (13) and (14) we find that for the required solution we must choose c_1 and c_2 so that

$$c_1 + c_2 + \tfrac{3}{4} = 0, \qquad c_1 + 2c_2 + \tfrac{1}{2} = 0,$$

i.e. $c_1 = -1$, $c_2 = \frac{1}{4}$ and the required solution is

$$y = \tfrac{3}{4} + \tfrac{1}{2}x - e^x + \tfrac{1}{4}e^{2x}. \tag{15}$$

EXAMPLE 15. *Find a particular integral of the equation*

$$\frac{d^2y}{dx^2} - 5\frac{dy}{dx} + 4y = e^{2x}.$$

Since $m = 2$ is not a root of the equation $m^2 - 5m + 4 = 0$, we may take a particular integral of the equation of the form $Y = Ce^{2x}$. For this solution $Y' = 2Ce^{2x}$, $Y'' = 4Ce^{2x}$, so that the equation is satisfied if we take $4C - 10C + 4C = 1$, i.e. $C = -\frac{1}{2}$.

Hence a particular integral of the equation is $-\frac{1}{2}e^{2x}$.

EXAMPLE 16. *Find the general solution of the equation*

$$\frac{d^2y}{dx^2} - 2\frac{dy}{dx} + y = e^{x}.$$

From example 12 we know that the complementary function is $(c_1x + c_2)e^x$. For the particular integral we try $Y = Ax^2e^x$, for which $Y' = A(x^2 + 2x)e^x$, $Y'' = A(x^2 + 4x + 2)$ so that

$$Y'' - 2Y' + Y = A(x^2 + 4x + 2 - 2x^2 - 4x + x^2)e^x = 2Ae^x.$$

To get the requisite particular solution we must therefore take $A = \frac{1}{2}$. The general solution of the equation is therefore

$$y = (\tfrac{1}{2}x^2 + c_1x + c_2)e^x.$$

EXAMPLE 17. *Find a particular integral of the equation*

$$\frac{d^2y}{dx^2} + y = \cos x.$$

It is easily shown that the complementary function in this case is $c_1 \cos x + c_2 \sin x$, so that we must take a particular integral of the type $Y = x(A \cos x + B \sin x)$. For this function we have

$$Y' = x(-A \sin x + B \cos x) + (A \cos x + B \sin x),$$
$$Y'' = x(-A \cos x - B \sin x) + 2(-A \sin x + B \cos x),$$

and hence

$$Y'' + Y = -2A \sin x + 2B \cos x,$$

which reduces to $\cos x$ if we take $A = 0$, $B = \frac{1}{2}$. The particular integral is therefore $\frac{1}{2}x \sin x$.

Let us now return to the general equation (1). The methods we have already discussed show how the complementary function of the equation may be obtained. We shall now give a general method for the determination of the particular integral.* If we make the substitutions

$$y = z(x)e^{-bx/a}, \qquad f(x) = aF(x)e^{-bx/a} \tag{16}$$

in equation (1) we find that the function $z(x)$ satisfies the differential equation

$$\frac{d^2z}{dx^2} - \frac{b^2 - ac}{a^2} z = F(x). \tag{17}$$

The nature of this equation depends upon the sign of the expression $b^2 - ac$, which is, of course, related to the roots of the quadratic equation (4).

Case (a): Distinct real roots

If equation (4) has two distinct real roots, $b^2 - ac$ is positive, so that we may write

$$\omega^2 = \frac{b^2 - ac}{a^2} > 0, \tag{18}$$

so that ω is real and we may write equation (17) in the form

$$\frac{d^2z}{dx^2} - \omega^2 z = F(x). \tag{19}$$

By means of a simple differentiation we can show that

$$\frac{d}{dx}\left\{e^{\omega x}\frac{dz}{dx} - \omega e^{\omega x} z\right\} = \left\{\frac{d^2 z}{dx^2} - \omega^2 z\right\} e^{\omega x},$$

so that we can write equation (19) in the form

$$\frac{d}{dx}\left\{e^{\omega x}\frac{dz}{dx} - \omega e^{\omega x} z\right\} = F(x)e^{\omega x}.$$

Since we are looking for a particular solution we may assume that $dz/dx = 0$, $z = 0$ when $x = 0$ in which case this last equation can be integrated immediately to yield the relation

$$\frac{dz}{dx} - \omega z = e^{-\omega x}\int_0^x F(u)e^{\omega u}\, du. \tag{20}$$

On the other hand we may write equation (19) in the form

* The rest of this section may be omitted at a first reading.

$$\frac{d}{dx}\left\{e^{-\omega x}\frac{dz}{dx} + \omega e^{-\omega x} z\right\} = F(x)e^{-\omega x},$$

which can be integrated to give the relation

$$\frac{dz}{dx} + \omega z = e^{\omega x}\int_0^x F(u)e^{-\omega u}\,du. \tag{21}$$

If we subtract equations (20) and (21) and make use of the fact that

$$e^{-\omega x}e^{\omega u} - e^{\omega x}e^{-\omega u} = -\{e^{\omega(x-u)} - e^{-(x-u)\omega}\} = -2\sinh\omega(x-u),$$

we find that

$$z = \frac{1}{\omega}\int_0^x F(u)\sinh\omega(x-u)du. \tag{22}$$

Substituting from the equations (16) we can write equation (22) in terms of the original variables. We find that a particular solution of equation (1) is

$$y(x) = \frac{1}{a\omega}\int_0^x f(u)e^{-b(x-u)/a}\sinh\omega(x-u)du. \tag{23}$$

Case (b): Equal roots

If the roots of equation (4) are equal, then $b^2 = ac$, and equation (17) becomes simply

$$\frac{d^2z}{dx^2} = F(x) \tag{24}$$

which can be easily integrated.

Case (c): Complex roots

If the roots of equation (4) are complex $b^2 - ac$ is negative so that we may write $b^2 - ac = -\Omega^2 = (i\Omega)^2$. If we write $\omega = i\Omega$, we get equation (19) again. The solution is obtained from equation (23) by putting $\omega = i\Omega$, and using the relation [cf. p. 258, eq. (17)]

$$\frac{1}{i\Omega}\sinh i\Omega(x-u) = \frac{1}{\Omega}\sin\Omega(x-u).$$

We find that

$$z = \frac{1}{\Omega}\int_0^x F(u)\sin\Omega(x-u)du, \tag{25}$$

or, in the original variables,

$$y = \frac{1}{a\Omega}\int_0^x f(u)e^{-b(x-u)/a}\sin\Omega(x-u)du, \tag{26}$$

where $\Omega^2 = ac - b^2$.

EXAMPLE 18. *Find a particular integral of the equation*

$$\frac{d^2y}{dx^2} + 4y = x\sin x.$$

From equation (26) with $a = 1$, $b = 0$, $c = 4$ so that $\Omega = 2$ we see that a particular integral is

$$\tfrac{1}{2}\int_0^x u\sin u\sin(2x-2u)du = \tfrac{1}{4}\int_0^x u[\cos(2x-3u) - \cos(2x-u)]du.$$

Using the result for integrating by parts we find that this reduces to

$$\tfrac{1}{4}\Big\{[-\tfrac{1}{3}u\sin(2x-3u) + u\sin(2x-u)]_0^x + \tfrac{1}{3}\int_0^x \sin(2x-3u)du - \int_0^x \sin(2x-u)du\Big\}.$$

Carrying out the integrations we find that the particular integral is

$$\tfrac{1}{3}x\sin x - \tfrac{2}{9}\cos x.$$

PROBLEMS

Find the general solutions of the equations:-

1. $D^2y + n^2y = \cos ax$;
2. $D^2y - 7Dy + 12y = x$;
3. $6D^2y + 5Dy - 6y = 0$;
4. $D^2y - 17Dy - 60 = 0$;
5. $D^2y + 5Dy + 4y = 32x^2$;
6. $D^2y - Dy - 2y = 3e^{-x}$;
7. $2D^2y + Dy = 2x + e^{-\frac{1}{2}x}$;
8. $D^2y + Dy = x^2$;
9. $D^2y - 3Dy + 2y = 2x^2 + e^x$;
10. $D^2y + 2Dy + 10y = \cos 3x$.

11. Linear Equations of Higher Degree. The *D*-Operator

We have shown in the last section how it is possible to derive the solution of second order linear ordinary differential equations with constant coefficients. This method cannot easily be extended to equations of higher order. In this section we shall describe a method which can be used for the solution of linear equations with constant coefficients, whatever the order of the equation. This method is known as *the method of symbolic operators.*

The method of symbolic operators has its origins in the fact that if we write the equation (1) of the last section in the form

$$aD^2y + 2bDy + cy = f(x),$$

then the m's which occur in the solution (3) are the roots of the quadratic equation

$$aD^2 + 2bD + c = 0,$$

obtained by putting $f(x) = 0$ and "cancelling" y symbolically from the equation. The operator D is then treated as though it were an ordinary algebraic quantity. We shall now develop this idea to obtain a calculus for the operator D.

We shall first of all consider the n-th order homogeneous linear differential equation

$$\frac{d^n y}{dx^n} + a_{n-1}\frac{d^{n-1}y}{dx^{n-1}} + \ldots + a_1\frac{dy}{dx} + a_0 y = 0, \tag{1}$$

with constant coefficients $a_{n-1}, a_{n-2}, \ldots, a_0$. If we introduce the operator $D = d/dx$ and write symbolically

$$f(D) = D^n + a_{n-1}D^{n-1} + \ldots + a_1D + a_0, \tag{2}$$

then we can write equation (1) in the form

$$f(D)y = 0. \tag{3}$$

It will be observed that $f(D)$ is a polynomial of degree n in D. In the results which we are now going to prove, it will be assumed that $f(D)$ is a simple function of this type.

THEOREM 43. *If $f(D)$ is a polynomial in D, and p is a constant, then*

$$f(D)e^{px} = f(p)e^{px}.$$

The proof of this result is simple. If p is a constant, then

$$D^r e^{px} = p^r e^{px},$$

so that

$$f(D)e^{px} = \{\sum_{r=0}^{n} a_r D^r\}e^{px} = \sum_{r=0}^{n} a_r(D^r e^{px}) = e^{px}\sum_{r=0}^{n} a_r p^r = f(p)e^{px}.$$

THEOREM 44. *If $g(x)$ is any function of x, and $f(D)$ is any polynomial in D,*

$$f(D)[e^{px}g(x)] = e^{px}f(D+p)g(x).$$

Using Leibnitz's theorem for the r-th derivative of the product of the two functions e^{px} and $g(x)$, (equation (4), § 11, Chapter 4), we find that

$$\begin{aligned} D^r[e^{px}g(x)] &= \sum_{m=0}^{r} {}^rC_m D^{r-m} e^{px} D^m g(x) \\ &= \sum_{m=0}^{r} {}^rC_m p^{r-m} e^{px} D^m g(x) \\ &= e^{px}\{\sum_{m=0}^{r} {}^rC_m p^{r-m} D^m\} g(x) \\ &= e^{px}(D+p)^r g(x). \end{aligned}$$

We therefore have that

$$\begin{aligned} f(D)[e^{px}g(x)] &= (\sum_{r=0}^{n} a_r D^r)e^{px}g(x) \\ &= e^{px}(\sum_{r=0}^{n} a_r(D+p)^r)g(x) = e^{px}f(D+p)g(x), \end{aligned}$$

and the theorem is proved.

We now apply these results to the solution of the homogeneous equation (3). If the equation $f(p) = 0$ has n distinct roots $p_1, p_2, \ldots, p_n$ then, by Theorem 43,

$$f(D)\{c_1e^{p_1x}+c_2e^{p_2x}+\ldots+c_ne^{p_nx}\}=c_1f(p_1)e^{p_1x}+\ldots+c_nf(p_n)e^{p_nx}=0,$$

since $f(p_i) = 0$ for $i = 1, 2, \ldots, n$. Also it can be shown that the functions e^{p_ix} $(i = 1, 2, \ldots, n)$ are linearly independent, so that, if $c_1, c_2, \ldots, c_n$ are arbitrary constants,

$$y = c_1e^{p_1x} + c_2e^{p_2x} + \ldots + c_ne^{p_nx}$$

is the general solution of equation (3) when the algebraic equation $f(p) = 0$ has n distinct roots.

If the equation $f(p) = 0$ has multiple roots then we can write

$$f(p) = (p-p_1)^{q_1} \cdots (p-p_m)^{q_m}$$

and we need only consider the case

$$(D-p_i)^{q_i}y = 0 \tag{4}$$

since the solution of the equation

$$f_1(D) \cdots f_m(D)y = 0$$

may be obtained by adding the solutions of the equations

$$f_1(D)y = 0, \qquad f_2(D)y = 0, \qquad \ldots, \qquad f_m(D)y = 0.$$

If we let $y = x^s e^{p_i x}$ in equation (4) and make use of Theorem **44**, we find that s must be such that

$$(D - p_i)^{q_i}(x^s e^{p_i x}) = e^{p_i x} D^{q_i} x^s = 0,$$

implying that s must be an integer less than q_i. In this way we obtain the set of q_i particular solutions

$$e^{p_i x}, \quad xe^{p_i x}, \quad \ldots, \quad x^{q_i-1} e^{p_i x}.$$

These can be shown to be linearly independent, so that the general solution of equation (4) can be written in the form

$$y = e^{p_i x}(c_1 + c_2 x + \ldots + c_{q_i} x^{q_i-1}), \tag{5}$$

where $c_1, c_2, \ldots, c_{q_i}$ are arbitrary constants.

The above results apply whether the roots of the equation $f(p) = 0$ are real or complex but, when they are complex, it is possible to group them in pairs, since complex roots always occur in conjugate pairs. A factor

$$\{(p - a)^2 + b^2\}^r$$

if $f(p)$ will lead to a set of terms

$$(c_1 + c_2 x + \ldots + c_r x^{r-1})e^{ax} \cos bx + (c_{r+1} + c_{r+2} x + \ldots + c_{2r} x^{r-1})e^{ax} \sin bx$$

in the general solution.

Example 19. *Find the general solution of the equation*

$$\frac{d^6 y}{dx^6} + 2\frac{d^3 y}{dx^3} + y = 0.$$

This equation can be written in the form $f(D)y = 0$ where

$$\begin{aligned} f(D) &= D^6 + 2D^3 + 1 = (D^3 + 1)^2 \\ &= [(D + 1)(D^2 - D + 1)]^2 = (D + 1)^2[(D - \tfrac{1}{2})^2 + \tfrac{3}{4}]^2 \end{aligned}$$

so that the general solution is

$$y = (c_1 + c_2 x)e^{-x} + (c_3 + c_4 x)e^{\frac{1}{2}x} \cos \frac{\sqrt{3}x}{2} + (c_5 + c_6 x)e^{\frac{1}{2}x} \sin \frac{\sqrt{3}x}{2}.$$

We shall now consider the non-homogeneous linear equation

$$f(D)y = R(x). \tag{6}$$

To find the general solution of this equation we must add to the complementary function (the general solution of $f(D)y=0$) a particular integral. Now, if we agree to denote by

$$\frac{1}{f(D)} R(x)$$

a function $S(x)$ with the property that

$$f(D)S(x) = R(x),$$

we may write a particular integral of the equation (6) in the form

$$\frac{1}{f(D)} R(x).$$

As a particular example consider the expression

$$\frac{1}{D^r} (x^m).$$

Since

$$D^r \left\{ \frac{x^{m+r}}{(m+r)\cdots(m+1)} \right\} = x^m$$

it follows from the definition that

$$\frac{1}{D^r} x^m = \frac{x^{m+r}}{(m+r)\cdots(m+1)} = \frac{\Gamma(m+1)x^{m+r}}{\Gamma(m+r+1)} \tag{7}$$

and in particular that

$$\frac{1}{D^r} (1) = \frac{x^r}{r!}. \tag{8}$$

We shall now prove some simple results concerning the operator $1/f(D)$, defined in this way.

THEOREM 45. $1/f(D)\, e^{px} = 1/f(p)\, e^{px}$, *provided that* $f(p) \neq 0$.

By Theorem 43

$$f(D) \frac{1}{f(p)} e^{px} = \frac{1}{f(p)} f(D) e^{px} = e^{px},$$

so that, by the definition of the operator $1/f(D)$, we have

$$\frac{1}{f(D)}\,e^{px} = \frac{1}{f(p)}\,e^{px}, \qquad f(p) \neq 0. \tag{9}$$

In order to deal with the case in which $f(p) = 0$ we must first prove

THEOREM 46.

$$\frac{1}{f(D)}\,\{e^{px} g(x)\} = e^{px}\,\frac{1}{f(D+p)}\,g(x).$$

By Theorem 44, we have the result

$$f(D)\left[e^{px}\left\{\frac{1}{f(D+p)}\,g(x)\right\}\right] = e^{px} f(D+p)\,\frac{1}{f(D+p)}\,g(x) = e^{px} g(x),$$

and the theorem follows from the definition of the operator $1/f(D)$.

THEOREM 47. *If* $f(D) = (D-p)^r \psi(D)$, where $\psi(p) \neq 0$, then

$$\frac{1}{f(D)}\,e^{px} = \frac{x^r e^{px}}{\psi(p) r!}.$$

By Theorems 46 and 45 we have

$$\begin{aligned}
\frac{1}{f(D)}\,e^{px} &= \frac{1}{(D-p)^r}\,\frac{1}{\psi(D)}\,e^{px} \\
&= \frac{1}{(D-p)^r}\,\frac{1}{\psi(p)}\,e^{px} \\
&= \frac{1}{\psi(p)}\,e^{px}\cdot\frac{1}{D^r}\,(1) \\
&= \frac{1}{\psi(p)}\,e^{px}\cdot\frac{x^r}{r!}
\end{aligned}$$

by equation (8).

The method of symbolic operators is of use in finding particular integrals of equations of the type (6) if $R(x)$ is one of the forms

(i) $x^m e^{px}$,

(ii) $x^m \cos(qx)$ or $x^m \sin(qx)$,

(iii) $e^{px}\cos(qx)$ or $e^{px}\sin(qx)$,

or is a linear combination of them. We shall consider these three cases separately:

Case (i): $f(D)y = x^m e^{px}$, where m is an integer

By Theorem 46,

$$y = \frac{1}{f(D)} x^m e^{px} = e^{px} \frac{1}{f(D+p)} x^m,$$

so that we need only consider the means of interpreting expressions of the form

$$\frac{1}{F(D)} x^m,$$

where $F(D)$ is a polynomial in D, of degree n, say. Suppose now that $G(t)$ is the sum of the terms up to that in t^m of the expansion of $1/F(t)$ as a power series in t. Then it is readily shown that

$$F(t)G(t) = 1 + g_1 t^{m+1} + g_2 t^{m+2} + \dots$$

so that

$$F(D)G(D)x^m = x^m + g_1 D^{m+1} x^m + g_2 D^{m+2} x^m + \dots$$

where $g_1, g_2, \dots$ are constants. Now if m and r are integers $D^{m+r} x^m = 0$, so that

$$F(D)G(D)x^m = x^m,$$

a result which may be written in the form

$$\frac{1}{F(D)} x^m = G(D)x^m.$$

Example 20. *Find the general solution of the equation*

$$\frac{d^2y}{dx^2} + 2\frac{dy}{dx} + y = x^2 e^x.$$

Since the equation $D^2 + 2D + 1 = 0$ has equal roots, each being -1, the complementary function is $(c_1 + c_2 x)e^{-x}$. A particular integral is

$$y = \frac{1}{(D+1)^2} x^2 e^x = e^x \frac{1}{(D+2)^2} x^2.$$

Now, up to the term in D^2 we have (by the binomial theorem)

$$\frac{1}{(D+2)^2} = \tfrac{1}{4}(1 + \tfrac{1}{2}D)^{-2} = \tfrac{1}{4} - \tfrac{1}{4}D + \tfrac{3}{16}D^2,$$

so that

$$\frac{1}{(D+2)^2}x^2 = \tfrac{1}{4}x^2 - \tfrac{1}{2}x + \tfrac{3}{8}.$$

Hence the general solution of the equation is

$$y = (c_1 + c_2 x)e^{-x} + \tfrac{1}{8}(2x^2 - 4x + 3)e^x.$$

EXAMPLE 21. *Find the general solution of the equation*

$$\frac{d^2y}{dx^2} - 2\frac{dy}{dx} + y = x^2 e^x.$$

The roots of the equation $D^2 - 2D + 1 = 0$ are equal (both unity). A particular integral is

$$y = \frac{1}{(D-1)^2}e^x x^2 = e^x \frac{1}{D^2}x^2 = \frac{x^4 e^x}{12},$$

by means of Theorem 47 and equation (7) above. Hence the general solution of the equation is

$$y = (c_1 + c_2 x + \tfrac{1}{12}x^4)e^x.$$

Case (ii): $f(D)y = x^m \cos(qx)$, where m is a positive integer.

Suppose that we introduce a function z such that

$$f(D)z = x^m \sin(qx),$$

then

$$f(D)(y + iz) = x^m e^{iqx},$$

and the particular integral $y + iz$ can be found by the method outlined above. Equating real and imaginary parts we find the expressions for y and z.

EXAMPLE 22. *Find the general solution of the equation*

$$\frac{d^2y}{dx^2} + y = x \cos x.$$

The complementary function is easily shown to be $c_1 \cos x + c_2 \sin x$. If we let $(D^2 + 1)z = x^2 \sin x$, then

$$\begin{aligned} y + iz &= \frac{1}{D^2+1}x^2 e^{ix} \\ &= e^{ix}\frac{1}{(D+i)^2+1}x \end{aligned}$$

$$= e^{ix}\left(-\tfrac{1}{2}i\frac{1}{D} + \tfrac{1}{4} + \tfrac{1}{8}iD\right)x$$
$$= e^{ix}(-\tfrac{1}{4}ix^2 + \tfrac{1}{4}x + \tfrac{1}{8}i)$$
$$= (\cos x + i \sin x)(-\tfrac{1}{4}ix^2 + \tfrac{1}{4}x + \tfrac{1}{8}i)$$
$$= (\tfrac{1}{4}x^2 - \tfrac{1}{8}) \sin x + \tfrac{1}{4}x \cos x - i(\tfrac{1}{4}x^2 - \tfrac{1}{8})\cos x + \tfrac{1}{4}ix \sin x.$$

Equating real parts, and dropping the term $-\frac{1}{8} \sin x$ (since it can be included in the term $c_2 \sin x$ of the complementary function), we find that the particular integral we need is $\frac{1}{4}x^2 \sin x + \frac{1}{4}x \cos x$, so that the general solution of the equation is

$$y = (\tfrac{1}{4}x^2 + c_2) \sin x + (\tfrac{1}{4}x + c_1) \cos x.$$

Case (iii): $f(D)y = e^{ax} \cos (bx)$.

If we let $f(D)z = e^{ax} \sin (bx)$, then $f(D)(y + iz) = e^{(a+ib)x}$, so that

$$y + iz = \frac{1}{f(D)} e^{(a+ib)x},$$

and the expression on the right can be evaluated by the method of case (i) above. If we equate real parts we obtain the expression for

$$\frac{1}{f(D)} e^{ax} \cos (bx),$$

while if we equate imaginary parts we obtain the expression for

$$\frac{1}{f(D)} e^{ax} \sin (bx).$$

EXAMPLE 23. *Find the general solution of the equation*

$$\frac{d^2y}{dx^2} - 4\frac{dy}{dx} + 3y = e^x \cos 2x.$$

Since the roots of the equation $D^2 - 4D + 3 = 0$ are 3 and 1, the complementary function is $c_1 e^x + c_2 e^{3x}$. If we let $(D^2-4D+3)z= e^x \sin 2x$, then

$$y + iz = \frac{1}{(D-1)(D-3)} e^{(1+2i)x} = \frac{e^{(1+2i)x}}{2i(2i-2)}.$$

Now $2i(2i - 2) = -4(1 + i)$ so that $[2i(2i - 2)]^{-1} = -\frac{1}{8}(1 - i)$ and

$$y + iz = -\tfrac{1}{8}e^x(1 - i)(\cos 2x + i \sin 2x).$$

Equating real parts we obtain the particular integral

$$-\tfrac{1}{8}e^{x}(\cos 2x + \sin 2x),$$

showing that the general solution is

$$y = c_1 e^{x} + c_2 e^{3x} - \tfrac{1}{8}e^{x}(\cos 2x + \sin 2x).$$

PROBLEMS

Find the general solutions of the following differential equations:-

1. $D^4y - 4D^3y + 6D^2y - 4Dy + y = 0;$
2. $D^4y + 4D^2y + 4y = 0;$
3. $D^3y - 3D^2y + 4y = 0;$
4. $D^3y + 2D^2y - Dy - 2y = 0;$
5. $D^3y - 2D^2y + Dy - 2y = x^2;$
6. $D^2y - Dy = xe^{x};$
7. $D^3y - D^2y + Dy - y = \cos x;$
8. $D^4y - y = \cos x;$
9. $D^2y + 4Dy + 3y - e^{-x} \cos 2x;$
10. $D^2y + n^2y = x^2 \cos (nx).$

12. The Use of the Laplace Transform

When it is desired to find the solution of a linear differential equation subject to given boundary conditions, the Laplace transform may be used to determine it. To illustrate the method consider the equation

$$a\frac{d^2y}{dt^2} + 2b\frac{dy}{dt} + cy = f(t), \tag{1}$$

where a, b and c are prescribed constants and $f(t)$ is a prescribed function of t, and it is known that, initially at $t = 0$,

$$y = y_0, \qquad \frac{dy}{dt} = y_1,$$

where y_0 and y_1 are constants. Then, from equations (12) and (14) of Chapter 7, § 16, we know that

$$\mathscr{L}\{\dot{y}\} = -y_0 + p\bar{y}(p), \quad \mathscr{L}\{\ddot{y}\} = -y_1 - py_0 + p^2\bar{y}(p). \tag{2}$$

If, therefore, we multiply both sides of equation (1) by e^{-pt} and integrate with respect to t from 0 to ∞ we find that $\bar{y}(p)$, the Laplace transform of $y(t)$, satisfies the equation

$$a(-y_1 - py_0 + p^2\bar{y}) + 2b(-y_0 + p\bar{y}) + c\bar{y} = \bar{f}(p), \tag{3}$$

where $\bar{f}(p)$ denotes the Laplace transform of $y(t)$. Solving equation (3) for $\bar{y}$ we find that

$$\bar{y}(p) = \frac{(ay_1 + 2by_0) + ay_0 p + \bar{f}(p)}{ap^2 + 2bp + c}. \tag{4}$$

Once $\bar{f}(p)$ has been calculated, $\bar{y}(p)$ can be determined by means of this formula and the function $y(t)$ of which it is the Laplace transform may be determined by use of Table 15 and the rules for manipulating Laplace transforms.

EXAMPLE 24. *Find the solution of the equation*

$$\ddot{y} - 3\dot{y} + 2y = t,$$

satisfying the initial conditions $y(0) = \dot{y}(0) = 0$.

Since $y(0) = \dot{y}(0) = 0$, we have that $\mathscr{L}\{\ddot{y}\} = p^2\bar{y}$, $\mathscr{L}\{\dot{y}\} = p\bar{y}$, $\mathscr{L}\{t\} = p^{-2}$, so that taking the Laplace transform of both sides of the equation, we obtain the equation

$$(p^2 - 3p + 2)\bar{y} = \frac{1}{p^2},$$

from which we derive the relation

$$\bar{y}(p) = \frac{1}{p^2(p-1)(p-2)}.$$

In order to find the function $y(t)$ of which this is the Laplace transform, we simplify the right-hand side by resolving it into partial fractions. We can write

$$\bar{y}(p) = \frac{A}{p^2} + \frac{B}{p} + \frac{C}{p-1} + \frac{D}{p-2},$$

where A, B, C and D are chosen so that

$$A(p-1)(p-2) + Bp(p-1)(p-2) + Cp^2(p-2) + Dp^2(p-1) = 1.$$

If we let $p = 0$, we find that $2A = 1$; similarly putting $p = 1$ and $p = 2$, respectively, we find that $-C = 1$ and $4D = 1$. Finally if we equate to zero the coefficient of p^3 on the left hand side we find that $B + C + D = 0$. In this way we find the set of values

$$A = \tfrac{1}{2}, \qquad B = \tfrac{3}{4}, \qquad C = -1, \qquad D = \tfrac{1}{4}.$$

Hence we find that

$$\bar{y}(p) = \tfrac{1}{2}p^{-2} + \tfrac{3}{4}p^{-1} - (p-1)^{-1} + \tfrac{1}{4}(p-2)^{-1},$$

from which it follows immediately that

$$y(t) = \tfrac{1}{2}t + \tfrac{3}{4} - e^{t} + \tfrac{1}{4}e^{2t}.$$

We have obtained this solution by the D-method above (cf. example 14).

This is how we should proceed in any specific case in which the coefficients of the equation have numerical values. We can however easily derive the general solution of an equation of type (1).

If, for definiteness, we suppose that $ac - b^2 > 0$, and write $\beta = b/a$, $\alpha^2 = (ac - b^2)/a^2 > 0$, then we can put equation (4) into the form

$$\bar{y}(p) = \frac{y_1 + \beta y_0 + (p+\beta)y_0 + a^{-1}\bar{f}(p)}{(p+\beta)^2 + \alpha^2},$$

from which, by means of equations (11) and (14) of § 16 of Chapter 7 (pp. 351, 352), we derive the general solution

$$y(t) = y_0 e^{-\beta t}\cos\alpha t + \frac{y_1 + \beta y_0}{\alpha} e^{-\beta t}\sin\alpha t + \frac{1}{a\alpha}\int_0^t f(u)e^{-\beta(t-u)}\sin\alpha(t-u)\,du. \tag{5}$$

The Laplace transform method may be applied to the solution of equations of higher order. To illustrate this we consider:-

EXAMPLE 25. *Solve the equation*

$$\dddot{y} + \ddot{y} - 2y = 0,$$

subject to the initial conditions $y(0) = 0$, $\dot{y}(0) = 1$, $\ddot{y}(0) = 0$.

Putting $y_0 = 0$, $y_1 = 1$ in equations (2) we find that $\mathscr{L}\{\dot{y}\} = p\bar{y}$, $\mathscr{L}\{\ddot{y}\} = -1 + p^2\bar{y}$. Also $\mathscr{L}\{\dddot{y}\} = -\ddot{y}(0) + p\mathscr{L}\{\ddot{y}\} = -p + p^3\bar{y}$. Hence taking the Laplace transform of both sides of the equation, we have

$$-p + p^3\bar{y} + (-1 + p^2\bar{y}) - 2\bar{y} = 0,$$

whence

$$\bar{y} = \frac{p+1}{p^3 + p^2 - 2},$$

whence we have

$$y(t) = \mathscr{L}^{-1}\left(\frac{p+1}{p^3+p^2-2}\right).$$

From example 36 of Chapter 7 (p. 351 above) we find that

$$y = \tfrac{2}{5}e^t - \tfrac{1}{5}e^{-t}(2\cos t - \sin t).$$

PROBLEMS

1. If $y(t)$ is the solution of the equation

$$\ddot{y} + 2\dot{y} + 2y = e^{-t}\sin t,$$

satisfying $y(0) = \dot{y}(0) = 0$, show that

$$\bar{y}(p) = \frac{1}{(p^2+2p+2)^2},$$

and deduce that

$$y(t) = \tfrac{1}{2}e^{-t}(\sin t - t\cos t).$$

2. Solve the equation $\dddot{y} = y + 1$, under the conditions

$$y(0) = \dot{y}(0) = 0, \quad \ddot{y}(0) = 1.$$

3. Solve the equation

$$\dddot{y} + 2\ddot{y} + 5\dot{y} = \tfrac{5}{2},$$

subject to the conditions $y(0) = \dot{y}(0) = 0$, $\ddot{y}(0) = 1$.

4. Show that the Laplace transform of the solution of the equation

$$\dddot{y} + 2\ddot{y} - \dot{y} - 2y = e^{-t},$$

satisfying the conditions $y(0) = A$, $\dot{y}(0) = \ddot{y}(0) = 0$, is

$$\bar{y}(p) = \frac{A(p+1)(p^2+2p-1)+1}{(p+1)^2(p-1)(p+2)}.$$

Show that if $y \to 0$ as $t \to \infty$, we must take $A = -\frac{1}{4}$, and find the solution in this case.

5. Solve the equation

$$\dddot{y} + 2\ddot{y} + 3\dot{y} + 2y = 0,$$

subject to the conditions $y(0) = 1$, $\dot{y}(0) = -2$, $\ddot{y}(0) = 4$.

13. Linear Equations with Variable Coefficients

We shall end this part of the present chapter by making a few remarks about linear equations with variable coefficients. Consider, first of all, the second order equation

$$\frac{d^2y}{dx^2} + P\frac{dy}{dx} + Qy = R, \tag{1}$$

where P, Q and R are functions of x alone.

We change the *dependent* variable from y to v by the substitution

$$y = y_1 v, \tag{2}$$

where we shall choose the function y_1 a little later. If we differentiate equation (2) twice, we obtain the equations

$$\frac{dy}{dx} = y_1' v + y_1 \frac{dv}{dx},$$

$$\frac{d^2y}{dx^2} = y_1'' v + 2y_1' \frac{dv}{dx} + y_1 \frac{d^2v}{dx^2}.$$

If we substitute these expressions into equation (1) and regroup the terms we obtain the equation

$$y_1 \frac{d^2v}{dx^2} + (2y_1' + Py_1)\frac{dv}{dx} + (y_1'' + Py_1' + Qy_1)v = R. \tag{3}$$

There are two main methods of choosing the function $y_1(x)$:-

Method I: Method of variation of parameters

This is the name given to the method of solution when we choose

$$y_1'' + Py_1' + Qy_1 = 0,$$

that is, when we choose y_1 to be a particular solution of the homogeneous equation corresponding to equation (1). In this case the function

$$w = \frac{dv}{dx} \tag{4}$$

satisfies the first order equation

$$y_1 \frac{dw}{dx} + (2y_1' + Py_1)w = R. \tag{5}$$

Equation (5) is a linear equation for w and so it can be solved for w. The function v is then found from w by integrating equation (4), and the final solution y is obtained from equation (2).

Method II: *Removing the first derivative*

In this method we choose y_1 to be a solution of the equation

$$y_1' + \tfrac{1}{2}Py_1 = 0, \tag{6}$$

in which case the term in $\mathrm{d}v/\mathrm{d}x$ is removed from equation (3). Equation (6) may be integrated immediately (by separating the variables) to give the particular solution

$$y_1 = \mathrm{e}^{-\frac{1}{2}\int P\mathrm{d}x},$$

for which

$$y_1' = -\tfrac{1}{2}P\mathrm{e}^{-\frac{1}{2}\int P\mathrm{d}x}, \qquad y_1'' = (\tfrac{1}{4}P^2 - \tfrac{1}{2}P')\mathrm{e}^{-\frac{1}{2}\int P\mathrm{d}x}.$$

Substituting these expressions into equation (3) we obtain the equation

$$\frac{\mathrm{d}^2v}{\mathrm{d}x^2} + (Q - \tfrac{1}{2}P' - \tfrac{1}{4}P^2)v = R\mathrm{e}^{\frac{1}{2}\int P\mathrm{d}x}.$$

If it happens that $Q - \frac{1}{2}P' - \frac{1}{4}P^2$ turns out to be a constant or a constant divided by x^2 we may integrate the equation using constant coefficient methods.*

We shall illustrate both methods by using them to solve the same equation.

EXAMPLE 26. *Find the general solution of the equation*

$$\frac{\mathrm{d}^2y}{\mathrm{d}x^2} - \frac{1}{x}\frac{\mathrm{d}y}{\mathrm{d}x} + \frac{y}{x^2} = 1.$$

Method I: It is easily seen that $y = x$ is a particular solution of the corresponding homogeneous equation. We therefore make the substitution $y = xv$. Since $y' = v + xv'$, $y'' = 2v' + xv''$, we find that the equation is transformed to

$$xw' + w = 1,$$

where $v' = w$. This equation is easily shown to have general solution $w = 1 + c_1/x$, where c_1 is a constant. Integrating the equation

$$v' = 1 + \frac{c_1}{x}$$

we find that

* For the method of solution in the second case see Problem 1 below.

$$v = x + c_1 \log x + c_2$$

so that the required general solution is

$$y = x^2 + c_1 x \log x + c_2 x.$$

Method II: For this equation $P = -1/x$, so that

$$-\tfrac{1}{2}\int P\mathrm{d}x = \tfrac{1}{2}\log x,$$

and so

$$y_1 = x^{\frac{1}{2}}.$$

Also

$$Q - \tfrac{1}{2}P' - \tfrac{1}{4}P^2 = \frac{1}{x^2} - \tfrac{1}{2}\frac{1}{x^2} - \tfrac{1}{4}\frac{1}{x^2} = \frac{1}{4x^2}$$

so that, if we write $y = x^{\frac{1}{2}} v$, the equation for v becomes

$$x^2 v'' + \tfrac{1}{4}v = x^{\frac{3}{2}}.$$

From Problem 1 below we see that we solve this equation by writing $x = \mathrm{e}^t$, a change of variable which transforms it to

$$(D'^2 - D' + \tfrac{1}{4})v = \mathrm{e}^{\frac{3}{2}t},$$

where D' denotes here $\mathrm{d}/\mathrm{d}t$. This equation can readily be solved by the methods of § 11; we find that

$$v = (c_1 t + c_2)\mathrm{e}^{\frac{1}{2}t} + \mathrm{e}^{\frac{3}{2}t} = (c_1 \log x + c_2)x^{\frac{1}{2}} + x^{\frac{3}{2}}.$$

If we multiply this expression by $x^{\frac{1}{2}}$ to obtain y, we get the same solution as before.

We shall end our discussion of second order linear equations with variable coefficients by considering equations which are exact. We say that a second order equation is *exact* if it is the derivative of a first order equation. Consider the linear equation

$$a_2 D^2 y + a_1 Dy + a_0 y = f(x), \tag{7}$$

where a_2, a_1, a_0 and f are known functions of x. Since, from the formula for differentiating a product, we have

$$D(a_1 y) = a_1 Dy + a_1' y,$$

it follows that we may write

$$a_1 Dy = D(a_1 y) - a_1' y. \tag{8}$$

Replacing a_1 by a_2, and y by Dy in this result we find that

$$a_2 D^2 y = D(a_2 Dy) - a_2' Dy = D(a_2 Dy) - D(a_2' y) + a_2'' y. \qquad (9)$$

If we substitute from equations (8) and (9) into equation (7) and rearrange the terms, we obtain the equation

$$D[a_2 Dy - (a_2' - a_1)y] + (a_2'' - a_1' + a_0)y = f(x),$$

from which it follows that, if

$$a_2'' - a_1' + a_0 = 0, \qquad (10)$$

the equation may be written as

$$D[a_2 Dy - (a_2' - a_1)y] = f(x).$$

This equation may be integrated immediately to yield the first order linear equation

$$a_2 Dy - (a_2' - a_1)y = \int f(x)\mathrm{d}x + c_1,$$

where c_1 denotes an arbitrary constant. This equation can, in turn, be integrated by the method of § 7 above.

Equation (10) is the condition to be satisfied by the coefficients if the equation (7) is exact. We shall illustrate the method by:-

EXAMPLE 27. *Solve the equation*

$$x(x + 3)D^2 y + 3(x + 1)Dy + y = (x + 1)e^x.$$

This equation is of the type (7) with $a_2 = x^2 + 3x$, $a_1 = 3x + 3$, $a_0 = 1$. We therefore have $a_2' - a_1 = -x$, $a_2'' - a_1' + a_0 = 0$, so that the equation is exact and can be written in the form

$$D[(x^2 + 3x)Dy + xy] = (x + 1)e^x = D(xe^x).$$

Integrating this equation we obtain the equation

$$(x^2 + 3x)Dy + xy = c_1 + xe^x,$$

that is,

$$(x + 3)Dy + y = \frac{c_1}{x} + e^x$$

which can be integrated at once to give the solution

$$(x + 3)y = e^x + c_1 \log x + c_2,$$

where c_1 and c_2 are arbitrary constants.

PROBLEMS

1. If $x = e^t$, $D = d/dx$, $D' = d/dt$, show that

$$D'y = Dy \cdot e^t = xDy, \qquad D'^2 y = x^2 D^2 y + xDy,$$

and hence show that this change of the independent variable transforms the equation

$$ax^2 D^2 y + 2bxDy + cy = f(x),$$

in which a, b and c are constants, to the form

$$aD'^2 y + (2b - a)D'y + cy = f(e^t).$$

Use this method to find the general solution of the equation

$$x^2 D^2 y + xDy + y = \log x.$$

2. Solve the equation

$$\frac{d^2y}{dx^2} + 2\frac{dy}{dx} + \left(1 - \frac{2}{x^2}\right) y = 0,$$

by removing the first derivative.

3. Show that $y = x$ is a particular integral of the homogeneous equation corresponding to the equation

$$x(1 - x^2)\frac{d^2y}{dx^2} - 2x^2\frac{dy}{dx} + 2xy = 1 - 3x^2,$$

and hence find its general solution.

4. Show that the equation

$$x(x + 1)\frac{d^2y}{dx^2} + (3x + 1)\frac{dy}{dx} + y = 9x^2,$$

is exact, and hence solve it.

5. Show that the substitution

$$y = e^{\int u dx}$$

transforms the homogeneous second order linear equation

$$\frac{d^2y}{dx^2} + P\frac{dy}{dx} + Qy = 0$$

to the first order Riccati equation

$$\frac{du}{dx} + Q + Pu + u^2 = 0.$$

C. SIMULTANEOUS ORDINARY DIFFERENTIAL EQUATIONS

14. Simultaneous Linear Equations

We shall consider equations involving x, y, z, . . . as dependent variables, the independent variable being t. This notation is suggested by the fact that in most problems in which simultaneous differential equations arise, the independent variable is the time t. The derivatives involved are

$$\frac{dx}{dt}, \frac{d^2x}{dt^2}, \ldots, \frac{dy}{dt}, \frac{d^2y}{dt^2}, \ldots,$$

which are often denoted by

$$\dot{x}, \ddot{x}, \ldots, \dot{y}, \ddot{y}, \ldots$$

respectively. There are no partial derivatives involved in the equations.

We have already quoted (in § 1 of this chapter) an example of the occurrence of simultaneous differential equations in biology. A further example will be given in § 10 of Chapter 10.

We shall now illustrate the manner in which such equations arise by considering a very simple physical problem (but one which requires

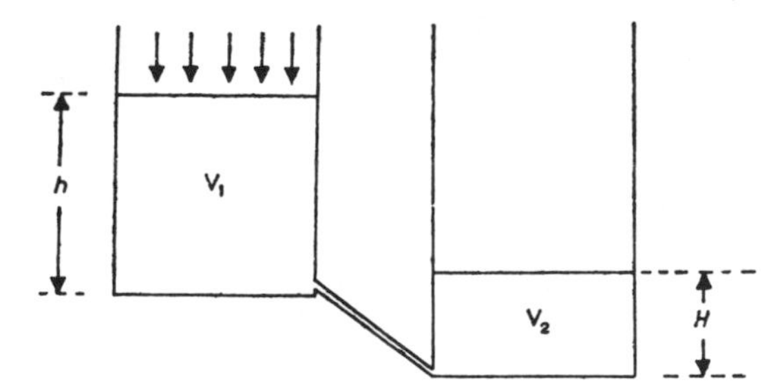

Fig. 127. Flow of water between two tanks.

no knowledge of physics!). The situation we consider is shown in Fig. 127. Two tanks of equal cross-sectional area are connected by a pipe in such a way that the rate of flow of water from the first to the second is proportional to $(h - H)^{\frac{1}{2}}$, where h is the depth of water in the first, and H the depth in the second tank, h being greater than or equal to H. We assume that the tanks are empty at time $t = 0$ and that water is pumped into the first tank at a constant rate. The problem is to find the relation between the height H and the time t.

If we denote the volume of water in the first tank by V_1 and that in the second tank by V_2, then, since both tanks have the same cross-sectional area, A say, we have $V_1 = Ah$, $V_2 = AH$. Since A is a constant,

$$\frac{\mathrm{d}V_1}{\mathrm{d}t} = A\frac{\mathrm{d}h}{\mathrm{d}t}, \qquad \frac{\mathrm{d}V_2}{\mathrm{d}t} = A\frac{\mathrm{d}H}{\mathrm{d}t}. \tag{1}$$

Since the rate of flow of water from the first tank to the second is proportional to $(h - H)^{\frac{1}{2}}$, we may write

$$\frac{\mathrm{d}V_2}{\mathrm{d}t} = \alpha(h - H)^{\frac{1}{2}}; \tag{2}$$

the water flows into the first tank at constant rate, β say, and flows out at a rate $\alpha(h - H)^{\frac{1}{2}}$. The net flow into the first tank is therefore given by the equation

$$\frac{\mathrm{d}V_1}{\mathrm{d}t} = \beta - \alpha(h - H)^{\frac{1}{2}}. \tag{3}$$

Substituting from the equations (1) into equations (2) and (3), we find that h and H satisfy the pair of simultaneous ordinary differential equations

$$A\frac{\mathrm{d}H}{\mathrm{d}t} = \alpha(h - H)^{\frac{1}{2}}, \qquad A\frac{\mathrm{d}h}{\mathrm{d}t} = \beta - \alpha(h - H)^{\frac{1}{2}}. \tag{4}$$

If we write $a = \alpha/A$, $b = \beta/A$, we see that we can write these equations in the form

$$\frac{\mathrm{d}H}{\mathrm{d}t} = a(h - H)^{\frac{1}{2}}, \qquad \frac{\mathrm{d}h}{\mathrm{d}t} = b - a(h - H)^{\frac{1}{2}}. \tag{5}$$

If we add these two equations we obtain the simple relation

$$\frac{\mathrm{d}}{\mathrm{d}t}(h + H) = b$$

which may be integrated immediately to give

$$h + H = C + bt,$$

where C is a constant. Now, since $h = H = 0$ when $t = 0$ it follows that $C = 0$ and we have

$$h + H = bt, \quad \text{or} \quad h = bt - H.$$

If we substitute this expression for h in the first of equations (5), we obtain the equation

$$\frac{\mathrm{d}H}{\mathrm{d}t} = a(bt - 2H)^{\frac{1}{2}}. \tag{6}$$

To solve equation (6) we make the substitution $bt - 2H = z^2$, i.e. $H = \frac{1}{2}(bt - z^2)$, for which

$$\frac{\mathrm{d}H}{\mathrm{d}t} = \tfrac{1}{2}b - z\frac{\mathrm{d}z}{\mathrm{d}t}$$

so that equation (6) becomes

$$\tfrac{1}{2}b - z\frac{\mathrm{d}z}{\mathrm{d}t} = az,$$

that is,

$$\frac{2z\mathrm{d}z}{b - 2az} = \mathrm{d}t.$$

This may be rearranged to the form

$$-\frac{2a}{b - 2az}\mathrm{d}z + \frac{2a}{b}\mathrm{d}z + \frac{2a^2}{b}\mathrm{d}t = 0$$

which can be integrated at once to give

$$\log(b - 2az) + \frac{2az}{b} + \frac{2a^2t}{b} = K,$$

where K is a constant. When $t = 0$, $H = 0$, so that $z = 0$ and we find that $K = \log b$. The relation between H and t is therefore

$$\log\left(1 - \frac{2a}{b}\sqrt{bt - 2H}\right) + \frac{2a}{b}\sqrt{(bt - 2H)} + \frac{2a^2t}{b} = 0.$$

It will be observed that, in arriving at this result, we introduced two constants of integration C and K (which we evaluated from the known initial conditions).

It will not have escaped the reader's notice that we were very fortunate to be able to integrate this pair of simultaneous equations.

Because of their non-linear character they would have proved pretty difficult had it not been for the fact that we were able to construct an easily integrable equation for $h + H$. The general theory of simultaneous equations might then be expected to be awkward. We shall restrict ourselves to the simplest case — in which the derivatives have constant coefficients.

15. Simultaneous Linear Equations with Constant Coefficients

We shall illustrate the procedures available by considering two examples.

EXAMPLE 28. *Solve the differential equations*

$$\frac{\mathrm{d}y}{\mathrm{d}t} + z = 1, \qquad y + \frac{\mathrm{d}z}{\mathrm{d}t} = 0,$$

given that $y = 1$, $z = A$ *when* $t = 0$. *Find the value of* A *if* $y \to 0$ as $t \to \infty$.

We shall first show how to use the D-method. We can write the equations in the form

$$Dy + z = 1, \qquad y = -Dz.$$

Substituting from the second equation into the first we find that

$$(-D^2 + 1)z = 1,$$

an equation whose general solution is

$$z = c_1 \mathrm{e}^t + c_2 \mathrm{e}^{-t} + 1,$$

where c_1 and c_2 are constants. Since $y = -Dz$, it follows that

$$y = -c_1 \mathrm{e}^t + c_2 \mathrm{e}^{-t}.$$

Substituting the conditions $y = 1$, $z = A$ when $t = 0$ we obtain the equations

$$c_1 + c_2 + 1 = A, \qquad -c_1 + c_2 = 1$$

for the determination of the arbitrary constants c_1 and c_2. We find that $c_1 = -1 + \frac{1}{2}A$, $c_2 = \frac{1}{2}A$, so that the required solution is

$$y = (1 - \tfrac{1}{2}A)\mathrm{e}^t + \tfrac{1}{2}A\mathrm{e}^{-t}, \quad z = 1 - (1 - \tfrac{1}{2}A)\mathrm{e}^t + \tfrac{1}{2}A\mathrm{e}^{-t}. \quad (1)$$

If $y \to 0$ as $t \to \infty$, then we must choose A so that $(1 - \frac{1}{2}A) = 0$, i.e. we must take $A = 2$. The solution (1) then becomes

$$y = e^{-t}, \qquad z = 1 + e^{-t}. \tag{2}$$

Next we shall indicate how equations of this kind can be solved by the use of the Laplace transform. Inserting the initial conditions we have

$$\mathscr{L}(y) = -1 + p\bar{y}, \qquad \mathscr{L}(z) = -A + p\bar{z}.$$

Using these results and the fact that $\mathscr{L}(1) = 1/p$, we find that the Laplace transforms of the given equations are

$$p\bar{y} - 1 + \bar{z} = \frac{1}{p}, \qquad \bar{y} + p\bar{z} = A.$$

Solving these equations we find that

$$\bar{y} = \frac{p + 1 - A}{p^2 - 1} = \frac{1 - \frac{1}{2}A}{p - 1} + \frac{\frac{1}{2}A}{p + 1},$$

$$\bar{z} = \frac{1}{p} + 1 - p\bar{y} = \frac{1}{p} + 1 - p\left\{\frac{1 - \frac{1}{2}A}{p - 1} + \frac{\frac{1}{2}A}{p + 1}\right\}.$$

Finding the inverse Laplace transforms of the functions on the right-hand sides of these equations we obtain the solution (1). The solution (2) may be derived in the same way as before, or, more quickly, by noting that if $y \to 0$ as $t \to \infty$ its Laplace transform must not involve a factor $p - 1$ in the denominator. In other words $p - 1$ must be a factor of the numerator, showing that $A = 2$.

Example 29. *Solve for* x *the differential equations*

$$\ddot{x} - x + \dot{y} - y = 0, \qquad -2\dot{x} - 2x + \ddot{y} - y = e^{-t}$$

subject to the conditions $x = 0$, $\dot{x} = -1$, $y = 1$, $\dot{y} = 1$ *when* $t = 0$.

In a problem like this, where we are given numerical values for the initial values of the functions and their derivatives and where we are asked to determine only one of the functions, the Laplace transform method is usually the quickest to use. We shall solve the equations by that method; the reader is advised to solve them by the D-method for himself.

Using the results

$$\mathscr{L}(\dot{x}) = p\bar{x}, \qquad \mathscr{L}(\ddot{x}) = 1 + p^2\bar{x}, \qquad \mathscr{L}(\dot{y}) = -1 + p\bar{y},$$

$$\mathscr{L}(\ddot{y}) = -1 - p + p^2\bar{y}, \qquad \mathscr{L}(e^{-t}) = \frac{1}{p + 1},$$

we see that the given equations have the transforms

$$1 + p^2\bar{x} - \bar{x} - 1 + p\bar{y} - \bar{y} = 0, \tag{3}$$

$$-2(p + 1)\bar{x} + (p^2 - 1)\bar{y} = p + 1 + \frac{1}{p+1}. \tag{4}$$

From equation (3) we find that

$$\bar{y} = -(p + 1)\bar{x}.$$

Substituting this expression in equation (4) and solving for $\bar{x}$, we find that

$$\bar{x} = -\frac{1}{p^2+1} - \frac{1}{(p+1)^2(p^2+1)}.$$

Now

$$\frac{A}{p+1} + \frac{B}{(p+1)^2} + \frac{Cp+D}{p^2+1}$$
$$= \frac{(A+C)p^3 + (A+B+2C+D)p^2 + (A+C+2D)p + A+B+D}{(p+1)^2(p^2+1)}$$

and the numerator reduces to 1 if we choose A, B, C, D such that

$$A+C=0,\ A+C+2D=0,\ A+B+2C+D=0,\ A+B+D=1, \text{ i.e.}$$
$$A=\tfrac{1}{2},\ B=\tfrac{1}{2},\ C=-\tfrac{1}{2},\ D=0.$$

We therefore have

$$\bar{x} = \frac{1}{2}\frac{p}{p^2+1} - \frac{1}{p^2+1} - \frac{1}{2}\frac{1}{p+1} - \frac{1}{2}\frac{1}{(p+1)^2}.$$

From the table of Laplace transforms we find that

$$x = \tfrac{1}{2}\cos t - \sin t - \tfrac{1}{2}e^{-t} - \tfrac{1}{2}te^{-t}.$$

In finding the *general* solution of relatively simple equations the D-method is probably easier to use. As a check on the solution we add the orders of the differential equations involved. This number is the number of arbitrary constants in the general solution. For example, the general solution of the system of equations in Problem 1 below would involve $2 + 2 = 4$ arbitrary constants. Similarly the general solution of the system in Problem 4 would involve $1 + 1 + 1 + 1 = 4$ arbitrary constants, while that in 5 would have $1 + 1 + 1 = 3$.

PROBLEMS

1. The quantities $x(t)$, $y(t)$ satisfy the simultaneous equations

$$\ddot{x} + 2n\dot{x} + n^2 x = 0, \qquad \ddot{y} + 2n\dot{y} + n^2 y = \mu\dot{x},$$

where $x(0) = 0$, $\dot{x}(0) = \lambda$; $y(0) = 0$, $\dot{y}(0) = 0$. Show that

$$y(t) = \tfrac{1}{2}\lambda\mu t^2 \mathrm{e}^{-nt}(1 - \tfrac{1}{3}nt).$$

2. Solve the simultaneous equations

$$\frac{\mathrm{d}x}{\mathrm{d}t} + ky = a \sin kt, \qquad \frac{\mathrm{d}y}{\mathrm{d}t} - kx = a \cos kt,$$

subject to the conditions that $x = 0$, $y = b$ when $t = 0$.

3. Find, for $t > 0$, the solution of the equations

$$\frac{\mathrm{d}^2x}{\mathrm{d}t^2} = c^2(y - 2x), \qquad \frac{\mathrm{d}^2y}{\mathrm{d}t^2} = c^2(x - 2y),$$

which satisfies the conditions $x = y = \mathrm{d}y/\mathrm{d}t = 0$, $\mathrm{d}x/\mathrm{d}t = V$, when $t = 0$.

4. Solve the equations

$$\dot{x}_3 = x_2 - x_3, \qquad \dot{x}_2 = x_1 - x_2, \qquad \dot{x}_1 = x_0 - x_1, \qquad \dot{x}_0 = -x_0,$$

given that, when $t = 0$, $x_0 = 1$, and $x_1 = x_2 = x_3 = 0$.

5. Solve the simultaneous equations

$$\dot{x} = 3x + y + z, \qquad \dot{y} = x + 5y + z, \qquad \dot{z} = x + y + 3z,$$

given that $x = 3$, $y = z = 0$ when $t = 0$.

D. PARTIAL DIFFERENTIAL EQUATIONS

16. Partial Differential Equations

As we have remarked in § 1, a *partial differential equation* is any relation between the partial derivatives of a function. For example, if ψ is a function of two independent variables x and y, then the equations

$$\frac{\partial\psi}{\partial x} + a\frac{\partial\psi}{\partial y} = 0 \tag{1}$$

$$\frac{\partial^2\psi}{\partial x^2} + x^2\frac{\partial^2\psi}{\partial y^2} + 2xy\frac{\partial\psi}{\partial x} = 0, \tag{2}$$

are partial differential equations for the determination of the function ψ. If ψ is a function of x and t, then

$$\frac{\partial^2 \psi}{\partial t^2} = x \frac{\partial \psi}{\partial x}$$

is a partial differential equation for the function $\psi(x, t)$.

Certain partial differential equations raise frequently in applications of mathematics. For example, the equation

$$\frac{\partial^2 \psi}{\partial x^2} = \frac{1}{k} \frac{\partial \psi}{\partial t} \tag{3}$$

describes linear diffusion, the variable x being a length, t the time, and k a constant; ψ may be a temperature or a chemical concentration. The equation (3) is called the *linear* (or *one-dimensional*) *diffusion equation*. The linear propagation of waves is likewise described by the equation

$$\frac{\partial^2 \psi}{\partial x^2} = \frac{1}{c^2} \frac{\partial^2 \psi}{\partial t^2}, \tag{4}$$

where c is a constant, the velocity of wave propagation. This equation is known as the *one-dimensional wave equation*. Two-dimensional problems in electrostatics and fluid mechanics are studied through the partial differential equation

$$\frac{\partial^2 \psi}{\partial x^2} + \frac{\partial^2 \psi}{\partial y^2} = 0 \tag{5}$$

where x and y are the coordinates of a typical point in the plane region under consideration. This equation is known as *the two-dimensional Laplace equation* or as *the plane harmonic equation*.

In the discussion of three-dimensional problems in certain physical subjects the analogues of these equations occur. The discussion of diffusion problems in solid bodies is based upon the study of the equation

$$\frac{\partial^2 \psi}{\partial x^2} + \frac{\partial^2 \psi}{\partial y^2} + \frac{\partial^2 \psi}{\partial z^2} = \frac{1}{k} \frac{\partial \psi}{\partial t} \tag{6}$$

which is accordingly known as *the diffusion equation*. Similarly the propagation of waves in three-dimensional space is governed by the partial differential equation

$$\frac{\partial^2 \psi}{\partial x^2} + \frac{\partial^2 \psi}{\partial y^2} + \frac{\partial^2 \psi}{\partial z^2} = \frac{1}{c^2} \frac{\partial^2 \psi}{\partial t^2} \tag{7}$$

where ψ is the disturbance at time t and the point with cartesian coordinates (x, y, z). This equation is known as *the wave equation.* Problems concerned with the distribution of electrostatic potential in solid bodies can be solved by consideration of the solution of the partial differential equation

$$\frac{\partial^2 \psi}{\partial x^2} + \frac{\partial^2 \psi}{\partial y^2} + \frac{\partial^2 \psi}{\partial z^2} = 0, \tag{8}$$

which is known as *Laplace's equation* or *the harmonic equation.*

It will be observed that, in the partial differential equations appropriate to these three-dimensional problems, the expression

$$\frac{\partial^2 \psi}{\partial x^2} + \frac{\partial^2 \psi}{\partial y^2} + \frac{\partial^2 \psi}{\partial z^2}$$

occurs in each case. It makes the equations simpler if we denote this quantity by the symbol

$$\nabla^2 \psi,$$

called the *Laplacian* of ψ. In terms of this symbol, the partial differential equations (6), (7) and (8) can be written in the forms

$$\nabla^2 \psi = \frac{1}{k} \frac{\partial \psi}{\partial t}, \tag{6a}$$

$$\nabla^2 \psi = \frac{1}{c^2} \frac{\partial^2 \psi}{\partial t^2}, \tag{7a}$$

$$\nabla^2 \psi = 0. \tag{8a}$$

EXAMPLE 30. *Solve the equation*

$$\frac{\partial z}{\partial \xi} = f(\xi, \eta). \tag{9}$$

The function

$$\int f(\xi, \eta) \mathrm{d}\xi$$

has the property that

$$\frac{\partial}{\partial \xi} \int f(\xi, \eta) \mathrm{d}\xi = f(\xi, \eta).$$

However, if the function $\phi(\eta)$ is an arbitrary function of η alone, then

$$\frac{\partial}{\partial \xi}\phi(\eta) = 0$$

(since $\phi(\eta)$ does not contain ξ). Hence

$$\frac{\partial}{\partial \xi}\left\{\int f(\xi, \eta)\mathrm{d}\xi + \phi(\eta)\right\} = f(\xi, \eta)$$

so that

$$z(\xi, \eta) = \int f(\xi, \eta)\mathrm{d}\xi + \phi(\eta) \tag{10}$$

satisfies the given equation.

Now it can be shown that the general solution of a first order linear differential equation contains one arbitrary function. Hence (10) is the general solution of equation (9).

EXAMPLE 31. *If z is a function of ξ and η, and*

$$\frac{\partial^2 z}{\partial \xi\, \partial \eta} = 0, \tag{11}$$

show that z can be written in the form

$$z = f(\xi) + g(\eta)$$

where the functions f and g are arbitrary.

If we integrate equation (11) with respect to ξ we find that

$$\frac{\partial z}{\partial \eta}$$

is an arbitrary function of η. To facilitate the integration involved in the next step we write this arbitrary function as $\mathrm{d}g/\mathrm{d}\eta$, thus

$$\frac{\partial z}{\partial \eta} = \frac{\mathrm{d}g}{\mathrm{d}\eta}.$$

Integrating this equation with respect to η we have

$$z = g(\eta) + f(\xi) \tag{12}$$

where $f(\xi)$ is an arbitrary function of ξ, since

$$\frac{\partial}{\partial \eta}f(\xi) = 0$$

for any $f(\xi)$.

PROBLEMS

1. Show that

$$\psi(x,\ y,\ z) = \frac{1}{\sqrt{(x^2 + y^2 + z^2)}}$$

is a solution of Laplace's equation $\nabla^2\psi = 0$.

2. Show that

$$\psi(x,\ t) = \cos \omega(x - ct)$$

is a solution of the one-dimensional wave equation

$$\frac{\partial^2\psi}{\partial x^2} = \frac{1}{c^2}\frac{\partial^2\psi}{\partial t^2}.$$

3. Show that

$$\psi(x,\ t) = \mathrm{e}^{-ax} \cos (\omega t - ax)$$

is a solution of the linear diffusion equation

$$\kappa\frac{\partial^2\psi}{\partial x^2} = \frac{\partial\psi}{\partial t}$$

if $\omega = 2\kappa a^2$.

17. Spherical Symmetry

In many problems we know that the dependent variable is a function of x, y, z in the combination $(x^2 + y^2 + z^2)^{\frac{1}{2}}$. In this section we shall derive the form of $\nabla^2\psi$ in this case.

If $\psi = \psi(r,\ t)$ is a function of r and t where

$$r = \sqrt{(x^2 + y^2 + z^2)} \tag{1}$$

is the distance of the point with cartesian coordinates $(x,\ y,\ z)$ from the origin, then

$$\frac{\partial\psi}{\partial x} = \frac{\partial\psi}{\partial r}\frac{\partial r}{\partial x}. \tag{2}$$

From equation (1) we find that

$$\frac{\partial r}{\partial x} = \frac{x}{\sqrt{(x^2 + y^2 + z^2)}} = \frac{x}{r} \tag{3}$$

so that

$$\frac{\partial\psi}{\partial x} = \frac{x}{r}\frac{\partial\psi}{\partial r}. \tag{4}$$

Differentiating both sides of this equation with respect to x, we find, on using the rule for differentiating a product, that

$$\begin{aligned}\frac{\partial^2\psi}{\partial x^2} &= \frac{\partial}{\partial x}\left(x\frac{1}{r}\frac{\partial\psi}{\partial r}\right)\\ &= \frac{1}{r}\frac{\partial\psi}{\partial r} + x\frac{\partial}{\partial x}\left(\frac{1}{r}\frac{\partial\psi}{\partial r}\right).\end{aligned} \tag{5}$$

Now if we replace ψ by $1/r\ \partial\psi/\partial r$ in equation (4) we find that

$$\frac{\partial}{\partial x}\left(\frac{1}{r}\frac{\partial\psi}{\partial r}\right) = \frac{x}{r}\frac{\partial}{\partial r}\left(\frac{1}{r}\frac{\partial\psi}{\partial r}\right) = \frac{x}{r}\left(\frac{1}{r}\frac{\partial^2\psi}{\partial r^2} - \frac{1}{r^2}\frac{\partial\psi}{\partial r}\right). \tag{6}$$

Substituting from equation (6) into equation (5) we find that

$$\frac{\partial^2\psi}{\partial x^2} = \frac{1}{r}\frac{\partial\psi}{\partial r} + \frac{x^2}{r^2}\left(\frac{\partial^2\psi}{\partial r^2} - \frac{1}{r}\frac{\partial\psi}{\partial r}\right). \tag{7}$$

If we differentiate with respect to y and z, we obtain the expressions

$$\frac{\partial^2\psi}{\partial y^2} = \frac{1}{r}\frac{\partial\psi}{\partial r} + \frac{y^2}{r^2}\left(\frac{\partial^2\psi}{\partial r^2} - \frac{1}{r}\frac{\partial\psi}{\partial r}\right) \tag{8}$$

$$\frac{\partial^2\psi}{\partial z^2} = \frac{1}{r}\frac{\partial\psi}{\partial r} + \frac{z^2}{r^2}\left(\frac{\partial^2\psi}{\partial r^2} - \frac{1}{r}\frac{\partial\psi}{\partial r}\right). \tag{9}$$

Adding equations (7), (8), (9) we find that

$$\nabla^2\psi = \frac{3}{r}\frac{\partial\psi}{\partial r} + \frac{x^2+y^2+z^2}{r^2}\left(\frac{\partial^2\psi}{\partial r^2} - \frac{1}{r}\frac{\partial\psi}{\partial r}\right)$$

and using the fact that $x^2 + y^2 + z^2 = r^2$ we see that, if ψ is a function of r only

$$\nabla^2\psi = \frac{\partial^2\psi}{\partial r^2} + \frac{2}{r}\frac{\partial\psi}{\partial r}. \tag{10}$$

PROBLEMS

1. Show that

$$\frac{\partial^2\psi}{\partial r^2} + \frac{2}{r}\frac{\partial\psi}{\partial r} = \frac{1}{r^2}\frac{\partial}{\partial r}\left(r^2\frac{\partial\psi}{\partial r}\right).$$

2. Show that if $\psi = \psi(r)$ and $\nabla^2\psi = 0$ then

$$\psi(r) = \frac{A}{r} + B$$

where A and B are arbitrary constants.

3. Prove that

(i) $r^2 \nabla^2 r^n = n(n+1)r^n$;

(ii) $\nabla^2 (\log r) = \dfrac{1}{r^2}$.

18. Solution of Partial Differential Equations by Change of the Independent Variables

We may often obtain the general solution of a partial differential equation by changing the independent variables in such a way that the equation is transformed to a simple form. We shall illustrate the procedure by three examples.

EXAMPLE 32. *Show that the general solution of the equation*

$$\frac{\partial z}{\partial x} = \frac{1}{a} \frac{\partial z}{\partial y}$$

is

$$z = f(x + ay)$$

where the function f is arbitrary.

We saw in Problem 1 of § 5 of Chapter 8 that if we change the independent variables from x and y to ξ and η where

$$\xi = x - ay, \qquad \eta = x + ay$$

then

$$\frac{\partial z}{\partial x} - \frac{1}{a} \frac{\partial z}{\partial y} = 2 \frac{\partial z}{\partial \xi}.$$

The given equation therefore transforms to

$$\frac{\partial z}{\partial \xi} = 0$$

with solution

$$z = f(\eta),$$

where f is an arbitrary function. If we substitute into this last equation the expression for η, we get the solution stated.

EXAMPLE 33. *Change the independent variables in the equation*

$$x(x+y)\frac{\partial u}{\partial x} + y^2 \frac{\partial u}{\partial y} = ax \tag{1}$$

to ξ *and* η *where*

$$\xi = x e^{y/x}, \qquad \eta = y e^{y/x},$$

and hence solve the equation.

Using formula (1) of § 5, Chapter 8, we find that

$$\frac{\partial u}{\partial x} = \frac{\partial u}{\partial \xi}\left(e^{y/x} - \frac{y}{x} e^{y/x}\right) + \frac{\partial u}{\partial \eta}\left(-\frac{y^2}{x^2} e^{y/x}\right)$$
$$= \frac{1}{x^2}\left\{x(x-y)\frac{\partial u}{\partial \xi} - y^2 \frac{\partial u}{\partial \eta}\right\} e^{y/x}$$

and that

$$\frac{\partial u}{\partial y} = \frac{\partial u}{\partial \xi} e^{y/x} + \frac{\partial u}{\partial \eta}\left\{1 + \frac{y}{x}\right\} e^{y/x}$$
$$= \frac{1}{x}\left\{x \frac{\partial u}{\partial \xi} + (x+y)\frac{\partial u}{\partial \eta}\right\} e^{y/x}.$$

Substituting these expressions into equation (1) we find that it becomes

$$x^2 e^{y/x} \frac{\partial u}{\partial \xi} = ax$$

i.e.

$$\frac{\partial u}{\partial \xi} = \frac{a}{\xi}.$$

Integrating this equation we find that

$$u = a \log \xi + f(\eta)$$

where f denotes an arbitrary function. Reverting to the original variables we see that the solution of equation (1) is

$$u = a(y/x + \log x) + f(y e^{y/x}), \tag{2}$$

the function f being arbitrary.

EXAMPLE 34. *By changing the independent variables from* x, y *to* $\xi = x^n + y$, $\eta = x^n - y$, *find the general solution of the equation*

$$x^2 \frac{\partial^2 u}{\partial x^2} - (n-1)x \frac{\partial u}{\partial x} = n^2 x^{2n} \frac{\partial^2 u}{\partial y^2}. \tag{3}$$

We saw in example 16 of § 6, Chapter 8, that with this change of the independent variables the equation becomes

$$\frac{\partial^2 u}{\partial \xi \, \partial \eta} = 0,$$

the general solution of which we know to be

$$u = f(\xi) + g(\eta)$$

where the functions f and g are arbitrary. Reverting the original variables we see that the general solution of equation (3) is

$$u = f(x^n + y) + g(x^n - y) \tag{4}$$

where the functions f and g are arbitrary.

PROBLEMS

1. Change the variables in the equation

$$x \frac{\partial z}{\partial y} - y \frac{\partial z}{\partial x} = \frac{(y^2 - x^2)^3}{xy}$$

to ξ, η where

$$\xi = x^2 + y^2, \qquad \eta = 2xy$$

and hence solve the equation. (Cf. Problem 2 of § 5, Chapter 8.)

2. By changing the independent variables from x and y to

$$\xi = \phi(x) - \psi(y), \qquad \eta = \phi(x) + \psi(y)$$

show that the general solution of the equation

$$\psi'(y) \frac{\partial u}{\partial x} + \phi'(x) \frac{\partial u}{\partial y} = 0$$

is

$$u(x, y) = f[\phi(x) - \psi(y)]$$

where the function f is arbitrary.

3. Solve the equation

$$x \frac{\partial z}{\partial x} + y \frac{\partial z}{\partial y} = x^2 y^2 \sin(xy),$$

first changing the variables to ξ, η where $\xi = xy$, $\eta = x/y$.

4. Solve the equation

$$x\frac{\partial z}{\partial x} + y\frac{\partial z}{\partial y} = 4x^2 \log (x + y),$$

by changing the variables to ξ, η where $x = \xi\eta$, $y = \xi(1 - \eta)$.

5. Change the independent variables in the equation

$$\frac{a^2}{x}\frac{\partial}{\partial x}\left(\frac{1}{x}\frac{\partial u}{\partial x}\right) = \frac{1}{c^2}\frac{\partial^2 u}{\partial t^2}$$

to ξ, η where

$$x^2 = a(\xi + \eta), \qquad 2ct = \eta - \xi,$$

and hence find a general solution of the equation.

Deduce that

$$\cos (\omega t) \cos \left(\frac{x^2\omega}{2ac}\right)$$

is a solution of the equation.

19. Method of Separation of Variables

A useful method of finding a solution of a partial differential equation with two independent variables, x and t say, is to see if a solution of the type

$$f(x, t) = X(x)T(t)$$

in which $X(x)$ is a function of x alone and $T(t)$ is a function of t alone, can be found. If this method is successful in providing simple solutions, more complex ones can sometimes be found by addition. The method is known as the method of separation of variables.

To illustrate the method we consider the first order equation

$$\frac{\alpha'(x)}{\alpha(x)}\frac{\partial f}{\partial t} = \frac{\beta'(t)}{\beta(t)}\frac{\partial f}{\partial x} \tag{1}$$

in which $\alpha(x)$ is a function of x alone and $\beta(t)$ is a function of t alone. If we let

$$f(x, t) = X(x)T(t), \tag{2}$$

where $X(x)$ is a function of x alone and $T(t)$ is a function of t alone, then

$$\frac{\partial f}{\partial t} = X\frac{\mathrm{d}T}{\mathrm{d}t}, \qquad \frac{\partial f}{\partial x} = T\frac{\mathrm{d}X}{\mathrm{d}x}$$

so that equation (1) gives

$$\frac{\alpha'(x)}{\alpha(x)} X \frac{dT}{dt} = \frac{\beta'(t)}{\beta(t)} T \frac{dX}{dx}$$

which may be written in the form

$$\frac{\beta(t)}{\beta'(t)} \frac{1}{T} \frac{dT}{dt} = \frac{\alpha(x)}{\alpha'(x)} \frac{1}{X} \frac{dX}{dx}. \tag{3}$$

The expression on the left-hand side of this equation is a function of t alone, while that on the right hand side is a function of x alone. The only way, therefore, in which they can be equal is if both of these functions reduce to a constant. If we denote by n an arbitrary constant then equation (3) will be satisfied if

$$\frac{\beta(t)}{\beta'(t)} \frac{1}{T} \frac{dT}{dt} = n, \qquad \frac{\alpha(x)}{\alpha'(x)} \frac{1}{X} \frac{dX}{dx} = n,$$

i.e. if

$$\frac{dT}{T} = \frac{n\beta'(t)dt}{\beta(t)}, \qquad \frac{dX}{X} = \frac{n\alpha'(x)dx}{\alpha(x)}.$$

These equations have solutions

$$\log T = n \log \beta(t) + \log c_1, \qquad \log X = n \log \alpha(x) + \log c_2,$$

where c_1 and c_2 are arbitrary constants. Writing these solutions in the form

$$T = c_1[\beta(t)]^n, \qquad X = c_2[\alpha(x)]^n$$

and writing $c_n = c_1 c_2$ we see from equation (2) that if we write

$$\xi = \alpha(x)\beta(t) \tag{4}$$

then $c_n \xi^n$ is a solution of equation (1) for arbitrary n and arbitrary c_n. Now from this solution we can show by the simple process of summation that

$$F(\xi) = \sum_{n=1}^{\infty} c_n \xi^n$$

is also a solution of the equation. Now if c_n is arbitrary, the function $F(\xi)$ is an arbitrary function of ξ. In other words

$$f = F(\xi), \tag{5}$$

where $F(\xi)$ is an arbitrary function of the variable ξ, given by equation

(4), is a solution of equation (1). Since the solution contains the correct number of arbitrary elements, it is the general solution.

To illustrate the use of this solution we have

EXAMPLE 35. *Show that the general solution of the equation*

$$\frac{\partial f}{\partial t} = (x - 1)(\lambda x - \mu)\frac{\partial f}{\partial x} \tag{6}$$

is

$$f = F\left(\frac{\lambda x - \mu}{x - 1} e^{-(\lambda-\mu)t}\right)$$

where the function F is arbitrary.

We can write equation (6) in the form (1) with

$$\frac{\alpha'(x)}{\alpha(x)} = \frac{\mu - \lambda}{(x - 1)(\lambda x - \mu)} = \frac{\lambda}{\lambda x - \mu} - \frac{1}{x - 1} \tag{7}$$

and

$$\frac{\beta'(t)}{\beta(t)} = \mu - \lambda. \tag{8}$$

Integrating equation (7) we get

$$\log \alpha(x) = \log (\lambda x - \mu) - \log (x - 1)$$

(the arbitrary constant of integration may be neglected). This is equivalent to

$$\alpha(x) = \frac{\lambda x - \mu}{x - 1}.$$

Integrating equation (8) we get

$$\log \beta(t) = -(\lambda - \mu)t$$

that is

$$\beta(t) = e^{-(\lambda-\mu)t}.$$

It follows that the general solution of (6) is $f = F(\xi)$ where F is an arbitrary function of the variable

$$\xi = \frac{\lambda x - \mu}{x - 1} e^{-(\lambda-\mu)t}.$$

The use of the method of separation of variables in the solution of second order equations is illustrated again in the next section.

20. The Use of Fourier Series

In this section we shall show how the method of separation of variables, introduced in the last section, may be made more effective by the use of the theory of Fourier series. We shall illustrate the method by considering the solution of the linear diffusion equation in a special case.

We take the linear diffusion equation in the form

$$D\frac{\partial^2 c}{\partial x^2} = \frac{\partial c}{\partial t}. \tag{1}$$

If we write $c(x, t)$ as the product of two functions

$$c(x, t) = X(x)T(t)$$

then

$$\frac{\partial c}{\partial x} = \frac{\mathrm{d}X}{\mathrm{d}x}T, \qquad \frac{\partial^2 c}{\partial x^2} = \frac{\mathrm{d}^2 X}{\mathrm{d}x^2}T, \qquad \frac{\partial c}{\partial t} = X\frac{\mathrm{d}T}{\mathrm{d}t}$$

so that the equation (1) becomes

$$D\frac{\mathrm{d}^2 X}{\mathrm{d}x^2}T = X\frac{\mathrm{d}T}{\mathrm{d}t}.$$

Dividing both sides of this equation by DXT we find that

$$\frac{1}{X}\frac{\mathrm{d}^2 X}{\mathrm{d}x^2} = \frac{1}{DT}\frac{\mathrm{d}T}{\mathrm{d}t}.$$

The expression on the left-hand side of this equation is a function of x alone and that on the right-hand side is a function of t alone, so that the only way in which they can be equal is if each is equal to a constant, k say. We therefore have

$$\frac{1}{DT}\frac{\mathrm{d}T}{\mathrm{d}t} = k, \qquad \frac{1}{X}\frac{\mathrm{d}^2 X}{\mathrm{d}x^2} = k.$$

From the first of these equations we have

$$\frac{\mathrm{d}T}{\mathrm{d}t} = kDT$$

with solution

$$T = T_0 \mathrm{e}^{kDt}$$

where T_0 is the value of T at time $t = 0$. If we wish solutions for which $T \to 0$ as $t \to \infty$ we must choose the constant k to be negative, i.e. we must write $k = -p^2$ where p is real. We therefore have that

$$T = T_0 \mathrm{e}^{-p^2 Dt} \tag{2}$$

and that

$$\frac{\mathrm{d}^2 X}{\mathrm{d}x^2} + p^2 X = 0. \tag{3}$$

The general solution of equation (3) is

$$X = c_1 \cos px + c_2 \sin px. \tag{4}$$

Combining equations (2) and (4) and writing

$$T_0 c_1 = a_p, \qquad T_0 c_2 = b_p$$

we see that

$$c(x,\ t) = (a_p \cos px + b_p \sin px)\mathrm{e}^{-p^2 Dt} \tag{5}$$

is a solution of equation (1) for all real values of p. We can construct a more complicated solution by adding these simple solutions, thus

$$c(x,\ t) = \sum_p (a_p \cos px + b_p \sin px)\mathrm{e}^{-p^2 Dt} \tag{6}$$

the a_p, b_p being constants which depend on p, and the summation sign indicating that we add together a number of terms of the type (5) over a range of values of p. When $t = 0$,

$$c(x,\ 0) = \sum_p (a_p \cos px + b_p \sin px). \tag{7}$$

Suppose now that we know that at time $t = 0$

$$c(x,\ 0) = \tfrac{1}{2}a_0 + \sum_{n=1}^{\infty} \left(a_n \cos \frac{2\pi nx}{l} + b_n \sin \frac{2\pi nx}{l}\right) \tag{8}$$

then it follows from equations (6), (7) and (8) that the function

$$c(x,\ t) = \tfrac{1}{2}a_0 + \sum_{n=1}^{\infty} \left(a_n \cos \frac{2\pi nx}{l} + b_n \sin \frac{2\pi nx}{l}\right) \mathrm{e}^{-4\pi^2 Dn^2 t/l^2} \tag{9}$$

(obtained by putting $p = 2\pi n/l$) is a solution of equation (1) with the property that $c(x,\ t)$ reduces to the form (7) when $t = 0$.

The series on the right-hand side of equation (8) is called *the Fourier series* of the function $c(x,\ 0)$. We shall now consider the method of deriving the values of the coefficients of the terms of a Fourier series. We *assume* that an arbitrary function $f(x)$ can be represented by a

Fourier series

$$f(x) = \tfrac{1}{2}a_0 + \sum_{n=1}^{\infty}\left(a_n \cos\frac{2\pi nx}{l} + b_n \sin\frac{2\pi nx}{l}\right). \tag{10}$$

Integrating both sides of this equation from 0 to l we find that

$$\int_0^l f(x)\mathrm{d}x = \tfrac{1}{2}la_0 + \sum_{n=1}^{\infty}\left(a_n\int_0^l \cos\frac{2\pi nx}{l}\,\mathrm{d}x + b_n\int_0^l \sin\frac{2\pi nx}{l}\,\mathrm{d}x\right). \tag{11}$$

Changing the variable of integration from x to u by the substitution

$$u = \frac{2\pi x}{l}, \quad x = \frac{ul}{2\pi}$$

for which

$$\mathrm{d}x = \frac{l}{2\pi}\,\mathrm{d}u$$

and $u = 0,\ 2\pi$ when $x = 0,\ l$ and

$$\int_0^l \cos\frac{2\pi nx}{l}\,\mathrm{d}x = \frac{l}{2\pi}\int_0^{2\pi} \cos nu\mathrm{d}u = 0$$

$$\int_0^l \sin\frac{2\pi nx}{l}\,\mathrm{d}x = \frac{l}{2\pi}\int_0^{2\pi} \sin nu\mathrm{d}u = 0,$$

we see from equation (11) that

$$a_0 = \frac{2}{l}\int_0^l f(x)\mathrm{d}x. \tag{12}$$

Multiplying both sides of equation (10) by $\cos(2\pi mx/l)$ we find that

$$\begin{aligned}\int_0^l f(x)\cos\frac{2\pi mx}{l}\,\mathrm{d}x &\\ = \tfrac{1}{2}a_0\int_0^l \cos\frac{2\pi mx}{l}\,\mathrm{d}x &+ \sum_{n=1}^{\infty} a_n\int_0^l \cos\frac{2\pi mx}{l}\cos\frac{2\pi nx}{l}\,\mathrm{d}x \\ &+ \sum_{n=1}^{\infty} b_n\int_0^l \cos\frac{2\pi mx}{l}\sin\frac{2\pi nx}{l}\,\mathrm{d}x.\end{aligned} \tag{13}$$

Making the same substitution as before and using the results of § 8, Chapter 7, p. 303, we find that

$$\int_0^l \cos\frac{2\pi mx}{l}\cos\frac{2\pi nx}{l}\,\mathrm{d}x = \frac{l}{2\pi}\int_0^{2\pi} \cos mu \cos nu\mathrm{d}u = \begin{cases} 0 & \text{if } n \neq m, \\ \tfrac{1}{2}l & \text{if } n = m; \end{cases}$$

$$\int_0^l \cos\frac{2\pi mx}{l}\sin\frac{2\pi nx}{l}\,dx = \frac{l}{2\pi}\int_0^{2\pi}\cos mu\sin nu\,du = 0;$$

$$\int_0^l \cos\frac{2\pi mx}{l}\,dx = \frac{l}{2\pi}\int_0^{2\pi}\cos mu\,du = 0.$$

The only non-vanishing term on the right-hand side of equation (13) is, therefore, the term for which $n = m$ which has the value

$$a_m \cdot \tfrac{1}{2}l$$

showing that

$$a_m = \frac{2}{l}\int_0^l f(x)\cos\frac{2\pi mx}{l}\,dx. \tag{14}$$

Similarly if we multiply both sides of equation (10) by $\sin(2\pi mx/l)$ and integrate with respect to x from 0 to l we find that

$$b_m = \frac{2}{l}\int_0^l f(x)\sin\frac{2\pi mx}{l}\,dx. \tag{15}$$

When an expansion of the type (10) is possible, the coefficients a_0, a_m, b_m are given by equations (12), (14) and (15) respectively. In fact, a proper discussion of Fourier series shows that, for a function which has only a finite number of finite discontinuities and turning values in the range $0 \leqq x \leqq l$ the series

$$\tfrac{1}{2}a_0 + \sum_{m=1}^{\infty}\left(a_m\cos\frac{2\pi mx}{l} + b_m\sin\frac{2\pi mx}{l}\right)$$

converges for each value of x to the value $f(\xi)$ if $x = \xi$ is a point at which $f(x)$ is continuous and to the value

$$\tfrac{1}{2}\{f(\xi + 0) + f(\xi - 0)\}$$

if ξ is a point of finite discontinuity of $f(x)$. (Cf. Fig. 128.) In particular when $x = 0$ or $x = l$, the series converges to the value

$$\tfrac{1}{2}[f(0) + f(l)].$$

It should be observed that a function of the type

$$\frac{1}{1 - 2x}$$

cannot be represented by a Fourier series in the range $0 \leqq x \leqq l$ since it has an infinite discontinuity at $x = \frac{1}{2}$ and that a function like

$$\sin \frac{1}{x}$$

cannot be represented by a Fourier series in any interval $0 \leqq x \leqq l$ since it has an infinite number of turning values in the vicinity of $x = 0$. A function which satisfies the conditions of having only a

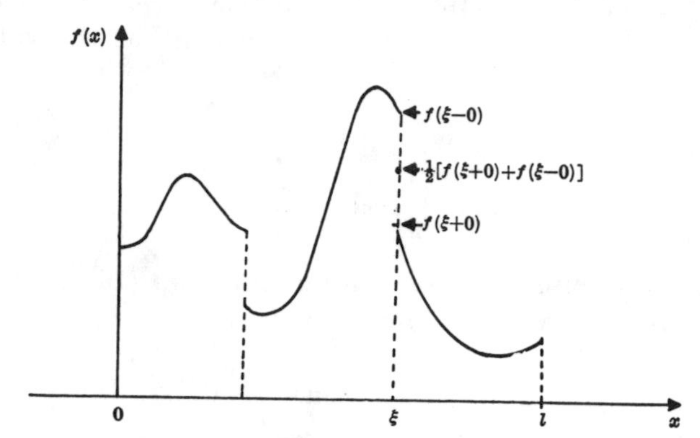

Fig. 128. Graph of a typical function which satisfies Dirichlet's conditions.

finite number of finite discontinuities and only a finite number of turning values in the range $0 \leqq x \leqq l$ is said to satisfy *Dirichlet's conditions* in that range.

The coefficients a_0, a_m, b_m defined by equations (12), (14) and (15) are called the *Fourier coefficients* of the function $f(x)$.

EXAMPLE 36. *Find the solution $c(x, t)$ of the equation*

$$D \frac{\partial^2 c}{\partial x^2} = \frac{\partial c}{\partial t}$$

for $t > 0$ and $0 \leqq x \leqq l$ satisfying the initial condition

$$c(x, 0) = c_0 x/l, \qquad (0 \leqq x \leqq l).$$

From equation (12) we have

$$a_0 = \frac{2}{l} \int_0^l \frac{c_0 x}{l} \mathrm{d}x = \frac{2c_0}{l^2} \int_0^l x \mathrm{d}x = c_0$$

and from equation (14) we have

$$a_m = \frac{2}{l} \int_0^l \frac{c_0 x}{l} \cos \frac{2\pi m x}{l} \mathrm{d}x = \frac{c_0}{\pi l} \int_0^{2\pi} u \cos mu \mathrm{d}u.$$

Using the rule for integrating by parts, we have

$$\int_0^{2\pi} u \cos mu du = \left[\frac{1}{m} u \sin mu\right]_0^{2\pi} - \frac{1}{m}\int_0^{2\pi} \sin mu du = 0,$$

so that, for all positive values of m, $a_m = 0$. Further, from equation (15)

$$b_m = \frac{2}{l}\int_0^l \frac{c_0 x}{l} \sin \frac{2\pi m x}{l} dx = \frac{c_0}{\pi l}\int_0^{2\pi} u \sin mu du$$

and the rule for integrating by parts gives

$$\int_0^{2\pi} u \sin mu du = \left[-\frac{u}{m} \cos mu\right]_0^{2\pi} + \frac{1}{m}\int_0^{2\pi} \cos mu du = -\frac{2\pi}{m}$$

so that

$$b_m = -\frac{2c_0}{lm}.$$

We therefore find that

$$c(x, 0) = \tfrac{1}{2}c_0 - \frac{2c_0}{l}\sum_{m=1}^{\infty}\frac{1}{m}\sin\frac{2m\pi x}{l}$$

so that, using equations (8) and (9) we find that the required solution is

$$c(x, t) = \tfrac{1}{2}c_0 - \frac{2c_0}{l}\sum_{m=1}^{\infty}\frac{1}{m}\sin\frac{2m\pi x}{l}\, e^{-4\pi^2 m^2 tD/l^2}.$$

PROBLEMS

1. Show that, if n is an integer,

$$y(x, t) = \sin\frac{n\pi x}{l}\cos\frac{n\pi ct}{l}$$

is a solution of the equation

$$\frac{\partial^2 y}{\partial x^2} = \frac{1}{c^2}\frac{\partial^2 y}{\partial t^2} \tag{16}$$

which has the properties

$$y(0, t) = 0, \quad y(l, t) = 0, \quad y(x, 0) = \sin\frac{n\pi x}{l}, \quad \left(\frac{\partial y}{\partial t}\right)_{t=0} = 0.$$

2. Deduce that the solution of equation (16) satisfying the conditions

$$y(0, t) = 0, \quad y(l, t) = 0, \quad y(x, 0) = f(x), \quad \left(\frac{\partial y}{\partial t}\right)_{t=0} = 0$$

is

$$y(x, t) = \sum_{n=1}^{\infty} b_n \sin\frac{n\pi x}{l} \cos\frac{n\pi ct}{l}$$

where the constants b_n are chosen so that

$$f(x) = \sum_{n=1}^{\infty} b_n \sin\frac{n\pi x}{l}. \tag{17}$$

3. Show that if a function $f(x)$ can be represented by a Fourier series of type (17),

$$b_n = \frac{2}{l}\int_0^l f(x) \sin\frac{n\pi x}{l}\,\mathrm{d}x.$$

21. Some Special Equations

In this section we shall derive some elementary solutions of the commonly occurring partial differential equations.

(a) Laplace's equation

If we take

$$\psi(x, y, z) = \frac{q(\alpha, \beta, \gamma)}{\sqrt{[(x-\alpha)^2 + (y-\beta)^2 + (z-\gamma)^2]}} \tag{1}$$

where q is a function of α, β, γ but not of x, y or z, then it is easily verified that

$$\nabla^2\psi = 0. \tag{2}$$

The function on the right-hand side of equation (1) will be a solution of Laplace's equation (2) if we integrate with respect to α, β and γ. In this way we obtain the more general solution

$$\psi(x, y, z) = \iiint \frac{q(\alpha, \beta, \gamma)\,\mathrm{d}\alpha\,\mathrm{d}\beta\,\mathrm{d}\gamma}{\sqrt{[(x-\alpha)^2 + (y-\beta)^2 + (z-\gamma)^2]}}. \tag{3}$$

For the two-dimensional equation

$$\frac{\partial^2\psi}{\partial x^2} + \frac{\partial^2\psi}{\partial y^2} = 0 \tag{4}$$

a special method of solution is available. If we write

$$\psi = f(z) \tag{5}$$

where

$$z = x + iy, \tag{6}$$

i denoting, as usual $\sqrt{-1}$, then

$$\frac{\partial \psi}{\partial x} = f'(z), \qquad \frac{\partial^2 \psi}{\partial x^2} = f''(z)$$

$$\frac{\partial \psi}{\partial y} = if'(z), \qquad \frac{\partial^2 \psi}{\partial y^2} = i^2 f''(z).$$

Since $i^2 + 1 = 0$ it follows that, if f is a twice-differentiable function of z, equation (4) is satisfied. Hence the real and imaginary parts of ψ satisfy Laplace's equation.

If we take

$$f(z) = z^4 = x^4 + 4x^3 iy + 6x^2 i^2 y^2 + 4xi^3 y^3 + i^4 y^4$$

we find that

$$x^4 - 6x^2 y^2 + y^4 + i\,\{4xy(x^2 - y^2)\}$$

is a solution of equation (4). Separating real and imaginary parts we find that

$$\psi_1 = x^4 - 6x^2 y^2 + y^4 \tag{7}$$

and

$$\psi_2 = 4xy(x^2 - y^2) \tag{8}$$

are solutions of equation (2).

Similarly if we take

$$\begin{aligned}\psi = f(z) = \sin(x + iy) &= \sin x \cos(iy) + \cos x \sin(iy) \\ &= \sin x \cosh y + i \sinh y \cos x\end{aligned}$$

we find that

$$\psi_1 = \sin x \cosh y \tag{9}$$

is a solution of equation (2).

Finally if we take

$$f(z) = \mathrm{e}^{-kz} = \mathrm{e}^{-kx}(\cos ky + i \sin ky)$$

we find that

$$\psi_1 = \mathrm{e}^{-kx} \cos ky, \qquad \psi_2 = \mathrm{e}^{-kx} \sin ky$$

are solutions of Laplace's equation (10) so that if δ is a real constant

$$\psi(x, y) = e^{-kx} \cos (ky - \delta) \tag{10}$$

is a solution.

The reader should verify by direct differentiation that the functions given by equations (7) to (10) do satisfy the two-dimensional Laplace equation.

(b) The wave equation

We begin with the linear wave equation

$$\frac{\partial^2 \psi}{\partial x^2} = \frac{1}{c^2} \frac{\partial^2 \psi}{\partial t^2} \tag{11}$$

the general solution of which we obtain by changing the variables from x and t to ξ and η, where

$$\xi = x - ct, \qquad \eta = x + ct. \tag{12}$$

Now

$$\frac{\partial \psi}{\partial x} = \frac{\partial \psi}{\partial \xi} \frac{\partial \xi}{\partial x} + \frac{\partial \psi}{\partial \eta} \frac{\partial \eta}{\partial x} = \frac{\partial \psi}{\partial \xi} + \frac{\partial \psi}{\partial \eta}$$

and

$$\frac{\partial^2 \psi}{\partial x^2} = \frac{\partial}{\partial x} \left(\frac{\partial \psi}{\partial \xi} + \frac{\partial \psi}{\partial \eta} \right) = \frac{\partial}{\partial \xi} \left(\frac{\partial \psi}{\partial \xi} + \frac{\partial \psi}{\partial \eta} \right) + \frac{\partial}{\partial \eta} \left(\frac{\partial \psi}{\partial \xi} + \frac{\partial \psi}{\partial \eta} \right)$$

$$= \frac{\partial^2 \psi}{\partial \xi^2} + 2 \frac{\partial^2 \psi}{\partial \xi \, \partial \eta} + \frac{\partial^2 \psi}{\partial \eta^2}. \tag{13}$$

Also

$$\frac{\partial \psi}{\partial t} = \frac{\partial \psi}{\partial \xi} \frac{\partial \xi}{\partial t} + \frac{\partial \psi}{\partial \eta} \frac{\partial \eta}{\partial t} = c \left(- \frac{\partial \psi}{\partial \xi} + \frac{\partial \psi}{\partial \eta} \right)$$

and

$$\frac{\partial^2 \psi}{\partial t^2} = -c \frac{\partial}{\partial \xi} c \left(- \frac{\partial \psi}{\partial \xi} + \frac{\partial \psi}{\partial \eta} \right) + c \frac{\partial}{\partial \eta} c \left(- \frac{\partial \psi}{\partial \xi} + \frac{\partial \psi}{\partial \eta} \right)$$

$$= c^2 \left(\frac{\partial^2 \psi}{\partial \xi^2} - 2 \frac{\partial^2 \psi}{\partial \xi \, \partial \eta} + \frac{\partial^2 \psi}{\partial \eta^2} \right). \tag{14}$$

It follows from equations (13) and (14) that equation (11) is transformed to

$$4 \frac{\partial^2 \psi}{\partial \xi \, \partial \eta} = 0$$

and we saw in example 31 that the general solution of this equation is

$$\psi = f(\xi) + g(\eta).$$

Reverting to the original variables we see that the general solution of equation (11) is

$$\psi = f(x - ct) + g(x + ct) \tag{15}$$

where f and g are arbitrary.

An important particular form of the solution (15) is obtained by taking $f(\xi) = A \cos(\omega\xi - \delta)$, $g(\eta) = B \cos(\omega\eta - \varepsilon)$ where ω, A, B, δ, ε are real constants. We obtain the solution

$$\psi = A \cos[\omega(x - ct) - \delta] + B \cos[\omega(x + ct) - \varepsilon]. \tag{16}$$

If we take $A = B$, $\delta = \varepsilon = 0$ we obtain the simple solution

$$\psi = 2A \cos \omega x \cos \omega ct \tag{17}$$

and if we take $B = -A$, $\delta = \varepsilon = 0$ we have

$$\psi = 2A \sin \omega x \sin \omega ct. \tag{18}$$

The case of spherical symmetry is also important. If we assume that

$$\psi = \psi(r, t) \tag{19}$$

where $r = \sqrt{(x^2 + y^2 + z^2)}$, then the wave equation

$$\nabla^2 \psi = \frac{1}{c^2} \frac{\partial^2 \psi}{\partial t^2}$$

becomes simply

$$\frac{\partial^2 \psi}{\partial r^2} + \frac{2}{r} \frac{\partial \psi}{\partial r} = \frac{1}{c^2} \frac{\partial^2 \psi}{\partial t^2}. \tag{20}$$

We now put

$$\psi = \frac{1}{r} \Psi$$

for which

$$\frac{\partial \psi}{\partial r} = \frac{1}{r} \frac{\partial \Psi}{\partial r} - \frac{1}{r^2} \Psi$$

$$\frac{\partial^2 \Psi}{\partial r^2} = \frac{1}{r} \frac{\partial^2 \Psi}{\partial r^2} - \frac{2}{r^2} \frac{\partial \Psi}{\partial r} + \frac{2}{r^3} \Psi$$

$$\frac{\partial^2 \Psi}{\partial t^2} = \frac{1}{r} \frac{\partial^2 \Psi}{\partial t^2}$$

and hence Ψ must satisfy the equation

$$\frac{\partial^2 \Psi}{\partial r^2} = \frac{1}{c^2} \frac{\partial^2 \Psi}{\partial t^2}.$$

We have just shown that the general solution of this equation is

$$\Psi = f(r - ct) + g(r + ct)$$

where the functions f and g are arbitrary. Hence the general solution of (20) is

$$\psi(r,\ t) = \frac{1}{r} f(r - ct) + \frac{1}{r} g(r + ct). \tag{21}$$

If we take

$$f(\xi) = -\tfrac{1}{2} A \sin k\xi, \qquad g(\eta) = \tfrac{1}{2} A \sin k\eta$$

where A and k are constants, we obtain the important special solution

$$\psi(r,\ t) = A \frac{\sin kr}{r} \cos (kct). \tag{22}$$

This solution reduces at the origin $r = 0$ to

$$\psi(0,\ t) = Ak \cos (kct) \tag{23}$$

since $(\sin kr)/r \to k$ as $r \to 0$.

(*c*) *The diffusion equation*

For the function

$$\psi(x,\ t) = \frac{1}{2\sqrt{(\pi \kappa t)}}\, e^{-(x-\xi)^2/4\kappa t} \tag{24}$$

we have

$$\frac{\partial \psi}{\partial x} = -\frac{x - \xi}{4\kappa t \sqrt{(\pi \kappa t)}}\, e^{-(x-\xi)^2/4\kappa t}$$

$$\frac{\partial^2 \psi}{\partial x^2} = -\frac{1}{4\kappa t \sqrt{(\pi \kappa t)}} + \frac{(x - \xi)^2}{8\kappa^2 t^2 \sqrt{(\pi \kappa t)}}\, e^{-(x-\xi)^2/4\kappa t}$$

$$\frac{\partial \psi}{\partial t} = -\frac{1}{4\pi^{\frac{1}{2}} \kappa^{\frac{1}{2}} t^{\frac{3}{2}}}\, e^{-(x-\xi)^2/4\kappa t} + \frac{1}{2\sqrt{\pi \kappa t}} \frac{(x - \xi)^2}{4\kappa t^2}\, e^{-(x-\xi)^2/4\kappa t}$$

so that the function (24) is a solution of the linear diffusion equation

$$\frac{\partial^2 \psi}{\partial x^2} = \frac{1}{\kappa} \frac{\partial \psi}{\partial t}. \tag{25}$$

If we multiply the function on the right hand side of equation (24) by an arbitrary function $\phi(\xi)$ and integrate with respect to ξ from $-\infty$ to ∞ we obtain the solution

$$\psi(x, t) = \frac{1}{2\sqrt{\pi \kappa t}} \int_{-\infty}^{\infty} \phi(\xi) \mathrm{e}^{-(x-\xi)^2/4\kappa t} \mathrm{d}\xi. \tag{26}$$

Changing the dependent variable from ξ to u where

$$u^2 = \frac{(x-\xi)^2}{4\kappa t}$$

we obtain the solution (26) in the form

$$\psi(x, t) = \frac{1}{\sqrt{\pi}} \int_{-\infty}^{\infty} \phi(x + 2u\sqrt{\kappa t}) \mathrm{e}^{-u^2} \mathrm{d}u. \tag{27}$$

It can also be shown* that for this solution

$$\psi(x, 0) = \phi(x).$$

If we separate the variables in equation (25) (see § 19 above) we find that equation (25) has solutions of the type

$$\psi(x, t) = \mathrm{e}^{-p^2 \kappa t} \sin (px - \delta_p)$$

where δ_p is a constant. From this equation we get solutions of the type

$$\psi(x, t) = \sum_p A_p \mathrm{e}^{-\kappa p^2 t} \sin (px - \delta_p). \tag{28}$$

The case of spherical symmetry can be reduced to the linear case by writing

$$\psi = \frac{\Psi}{r}$$

in the equation

$$\frac{\partial^2 \psi}{\partial r^2} + \frac{2}{r} \frac{\partial \psi}{\partial r} = \frac{1}{\kappa} \frac{\partial \psi}{\partial t}$$

to obtain (as in (*b*) above)

* See, for instance, I. N. Sneddon, "Elements of Partial Differential Equations" (McGraw-Hill, New York, 1957) p. 283.

$$\frac{\partial^2 \Psi}{\partial r^2} = \frac{1}{\kappa} \frac{\partial \Psi}{\partial t}.$$

Taking a solution of this equation in the form

$$\Psi = \sum_{n=1}^{\infty} A_n \sin \frac{n\pi r}{a} \, e^{-\kappa n^2 \pi^2 t/a^2}$$

we obtain the solution

$$\psi(r, t) = \frac{1}{r} \sum_{n=1}^{\infty} A_n \sin \frac{n\pi r}{a} \, e^{-\kappa n^2 \pi^2 t/a^2}. \tag{29}$$

This solution has the property that $\psi(0, t)$ is finite, $\psi(a, t) = 0$ and

$$\psi(r, 0) = \frac{1}{r} \sum_{n=1}^{\infty} A_n \sin \frac{n\pi r}{a}.$$

The function we started with, eq. (24), is closely related to two probability curves. As a function of x, $\psi(x, t)$ can be regarded as a normal distribution (Chapter 7, § 13, eq. (13)) with mean $m = \xi$ and $\sigma = \sqrt{2\kappa t}$. The variance σ^2 is then proportional to t; it is as if the whole distribution of values of x spreads out, starting at $\sigma = 0$, and then continuously changing into each of the three curves in Fig. 91.

Now let us suppose that besides spreading out the distribution moves to the right at a constant speed v. Then in eq. (24) we put

$$\xi = vt \tag{30}$$

and consider the variation of ψ with t for a given x. We put

$$x = \mu v \tag{31}$$

so that μ is the time at which $\xi = x$. Then we have

$$\psi(x, t) = \frac{1}{2\sqrt{(\pi\kappa t)}} \exp\left(-\frac{x^2 - 2vtx + v^2 t^2}{4\kappa t}\right)$$

$$= \frac{1}{2\sqrt{(\pi\kappa t)}} \exp\left(\frac{\mu v^2}{2\kappa}\right) \exp\left\{-\frac{\mu v^2}{4\kappa}\left(\frac{t}{\mu} + \frac{\mu}{t}\right)\right\} \tag{32}$$

which has the same form as the skew probability curve (1) in Chapter 7, § 15.

CHAPTER 10

FURTHER APPLICATIONS TO MEDICINE AND BIOLOGY

1. First Order Systems

In elementary electrical theory it is desirable to calculate the variation (with time) of the current in a simple circuit containing a resistance and either a condenser or a self-induction.* In the first instance consider a very simple system consisting of a resistance and a self-induction connected in series, an applied voltage acting across the ends of the system, as shown in Fig. 129. We denote the applied voltage by $E(t)$, to emphasize that it may depend upon the time t.

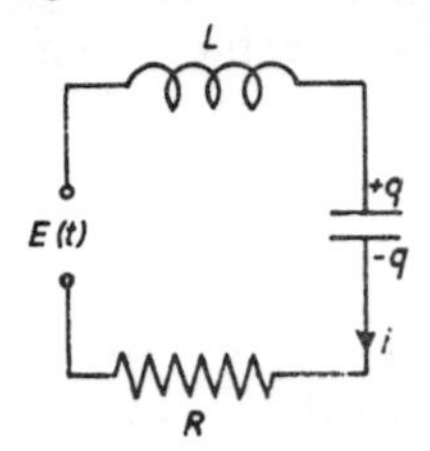

Fig. 129. A circuit containing a resistance R and a self-induction L.

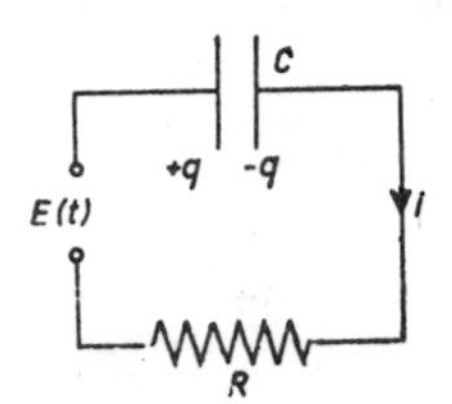

Fig. 130. A circuit containing a condenser C and a resistance R.

If the resistance is denoted by R, then, by Ohm's law, the drop of voltage across its ends is Ri, where i is the current flowing in the circuit; similarly, by the definition of self-induction, the drop in voltage across the ends of the self-inductance is $L(\mathrm{d}i/\mathrm{d}t)$. The sum of these two terms will be equal to the applied voltage, so that we obtain the differential equation

$$L\frac{\mathrm{d}i}{\mathrm{d}t} + Ri = E(t). \tag{1}$$

If now we introduce a quantity τ defined by the equation

$$\tau = L/R, \tag{2}$$

* The reader who has forgotten these terms could consult, R. Kronig (editor) "Leerboek der Natuurkunde" (Scheltema & Holkema, Amsterdam, 1954) or its English translation "Textbook of Physics" (Pergamon Press, London, 1954), p. 274.

then, dividing equation (1) throughout by L, we find that equation (1) takes the form

$$\frac{\mathrm{d}i}{\mathrm{d}t} + \frac{i}{\tau} = \frac{E(t)}{L}. \tag{3}$$

On the other hand, if we consider a circuit consisting of a condenser of capacity C in series with a resistance R, then if q is the charge on the condenser, the drop in voltage across the plates of the condenser will be q/C. As before, the drop in voltage across the ends of the resistance is Ri, and we have the basic equation

$$Ri + \frac{q}{C} = E(t). \tag{4}$$

Now the current in the circuit is equal to the rate of change of electric charge on the plates of the condenser, so that we have the relation

$$i = \frac{\mathrm{d}q}{\mathrm{d}t}. \tag{5}$$

Substituting from equation (5) into equation (4) we find that the variation with time of the charge q is determined by the differential equation

$$R\frac{\mathrm{d}q}{\mathrm{d}t} + \frac{q}{C} = E(t).$$

If we divide both sides of this equation by R and write

$$\tau = RC \tag{6}$$

we find that the equation becomes

$$\frac{\mathrm{d}q}{\mathrm{d}t} + \frac{q}{\tau} = \frac{E(t)}{R}. \tag{7}$$

It will be observed that equations (3) and (7) are very much alike. If we investigate the physical dimensions of the quantities L, R and C we find that L/R and RC both have the dimension of time, so that the τ's which occur in both of these equations are times. Both of these simple circuits are examples of first order systems.

We define a *first order system* as a physical (or physiological) system whose behaviour is governed by a differential equation which is:-

(1) ordinary;
(2) first order;
(3) linear;
(4) with constant coefficients.

In mathematical symbols, such a system is characterized by a differential equation of the type

$$\frac{\mathrm{d}x}{\mathrm{d}t} + \frac{x}{\tau} = \psi(t), \tag{8}$$

where τ is a constant and $\psi(t)$ is a prescribed function of t. We have denoted the independent variable by t (since in most problems it is the time variable) and the dependent variable by x. If we write the equation in the alternative form

$$\tau\frac{\mathrm{d}x}{\mathrm{d}t} + x = \phi(t), \tag{9}$$

(sometimes called its "physical standard form") we call the term $\phi(t)$ the *forcing term.*

The solution of the equation (8) presents no problems. (Cf. § 7 of Chapter 9.) The integrating factor is readily seen to be $\mathrm{e}^{t/\tau}$, so that we can write the equation in the form

$$\frac{\mathrm{d}}{\mathrm{d}t}(x\mathrm{e}^{t/\tau}) = \psi(t)\mathrm{e}^{t/\tau},$$

and this can be integrated immediately to give

$$x\mathrm{e}^{t/\tau} = c_1 + \int_0^t \psi(u)\mathrm{e}^{u/\tau}\,\mathrm{d}u, \tag{10}$$

where c_1 denotes a constant. If we know that, when $t = 0$, x takes the value x_0, then $c_1 = x_0$. Inserting this value and dividing the equation throughout by $\mathrm{e}^{t/\tau}$, we obtain the solution

$$x = x_0\mathrm{e}^{-t/\tau} + \int_0^t \psi(u)\mathrm{e}^{-(t-u)/\tau}\,\mathrm{d}u. \tag{11}$$

Certain special forms of this solution, corresponding to commonly occurring forms of the function $\psi(t)$ are of interest:

(i) $\psi(t) = \psi_0$, *a constant,* then

$$\int_0^t \psi(u)\mathrm{e}^{-(t-u)/\tau}\,\mathrm{d}u = \psi_0\mathrm{e}^{-t/\tau}\int_0^t \mathrm{e}^{u/\tau}\,\mathrm{d}u = \tau\psi_0(1-\mathrm{e}^{-t/\tau})$$

and the solution is

$$x = x_0 e^{-t/\tau} + \tau\psi_0(1 - e^{-t/\tau}). \tag{12}$$

(ii) $\psi(t) = \psi_0 \sin \omega t$, then, by example 26 of § 6, Chapter 7, we have

$$\int_0^t \psi(u)e^{-(t-u)/\tau}\,du = \psi_0 e^{-t/\tau}\int_0^t e^{u/\tau}\sin \omega u du$$

$$= \frac{\psi_0 \tau}{1+\omega^2\tau^2}[\sin \omega t + \omega\tau\{e^{-t/\tau} - \cos \omega t\}],$$

so that the required solution is

$$x = x_0 e^{-t/\tau} + \frac{\psi_0 \tau}{1+\omega^2\tau^2}[\sin \omega t + \omega\tau\{e^{-t/\tau} - \cos \omega t\}]. \tag{13}$$

(iii) $\psi(t) = \psi_0 \cos \omega t$, then, by example 26 of § 6, Chapter 7, we have

$$\int_0^t \psi(u)e^{-(t-u)/\tau}\,du = \psi_0 e^{-t/\tau}\int_0^t e^{u/\tau}\cos \omega u du$$

$$= \frac{\psi_0 \tau}{1+\omega^2\tau^2}\{\cos \omega t - e^{-t/\tau} + \omega\tau \sin \omega t\}$$

and hence

$$x = x_0 e^{-t/\tau} + \frac{\psi_0 \tau}{1+\omega^2\tau^2}\{\cos \omega t - e^{-t/\tau} + \omega\tau \sin \omega t\}. \tag{14}$$

For large values of t, $e^{-t/\tau} = 0$, so that

$$x = \frac{\psi_0 \tau}{1+\omega^2\tau^2}\{\cos \omega t + \omega\tau \sin \omega t\},$$

which may be written in the form

$$x = \frac{\psi_0 \tau}{\sqrt{(1+\omega^2\tau^2)}}\left\{\frac{1}{\sqrt{(1+\omega^2\tau^2)}}\cos \omega t + \frac{\omega\tau}{\sqrt{(1+\omega^2\tau^2)}}\sin \omega t\right\}.$$

If we now let

$$\cos \delta = \frac{1}{\sqrt{(1+\omega^2\tau^2)}}, \qquad \sin \delta = \frac{\omega\tau}{\sqrt{(1+\omega^2\tau^2)}} \tag{15}$$

then

$$\frac{1}{\sqrt{(1+\omega^2\tau^2)}}\cos\omega t + \frac{\omega\tau}{\sqrt{(1+\omega^2\tau^2)}}\sin\omega t = \cos\delta\cos\omega t + \sin\delta\sin\omega t,$$

a result which, by the addition formula for cosines, can be written as $\cos(\omega t - \delta)$. Hence, for values of t sufficiently large for $e^{-t/\tau}$ to be neglected, we have

$$x = \frac{\psi_0\tau}{\sqrt{(1+\omega^2\tau^2)}}\cos(\omega t - \delta), \tag{16}$$

where, because of equations (15), $\tan\delta = \omega\tau$, that is,

$$\delta = \tan^{-1}\omega\tau. \tag{17}$$

As an example of the application of these results, let us consider the variation with time of the charge on the plates of a condenser when a voltage

$$E(t) = E_0\cos\omega t \tag{18}$$

is applied to its ends through a resistance R. From equation (7) we see that this is the case (iii) with $\psi_0 = E_0/R$ and $\tau = RC$, so that, from equation (16), we see that, for values of t sufficiently great for $e^{-t/RC}$ to be negligible,

$$q = \frac{E_0 C}{\sqrt{(1+\omega^2R^2C^2)}}\cos(\omega t - \delta), \qquad (\delta = \tan^{-1}\omega RC). \tag{19}$$

This is a convenient point at which to introduce some of the definitions arising in the theory of vibrations. The function

$$x = a\cos(\omega t + \varepsilon) \qquad (a > 0) \tag{20}$$

represents a simple harmonic oscillation. If $\varepsilon = 0$, we get

$$x = a\cos\omega t,$$

while if $\varepsilon = -\frac{1}{2}\pi$, we get

$$x = a\sin\omega t.$$

Since $\cos(\omega t + \varepsilon)$ must always lie between -1 and $+1$, it follows that x must lie between $-a$ and $+a$. For this reason, the number a is known as the *amplitude* of the vibration.

Since

$$\cos\left[\omega\left(t + \frac{2\pi}{\omega}\right) + \varepsilon\right] = \cos[2\pi + (\omega t + \varepsilon)] = \cos(\omega t + \varepsilon),$$

we see that the value of x at the time $t + 2\pi/\omega$ is the same as it was at time t. Consequently,

$$T = \frac{2\pi}{\omega} \tag{21}$$

is called the *period* of the vibration. The constant ω is called the *circular frequency*; another quantity used in physics is the frequency ν defined by the equation

$$T = \frac{1}{\nu}. \tag{22}$$

Comparing equations (21) and (22) we see that the relation between the frequency ν and the circular frequency ω is

$$\omega = 2\pi\nu.$$

In equation (20) the number $\omega t + \varepsilon$ is called the *phase angle* and ε is called the *phase constant*. When ε is negative it is sometimes called the *phase lag*.

We shall now return to the discussion of equation (19). Equations

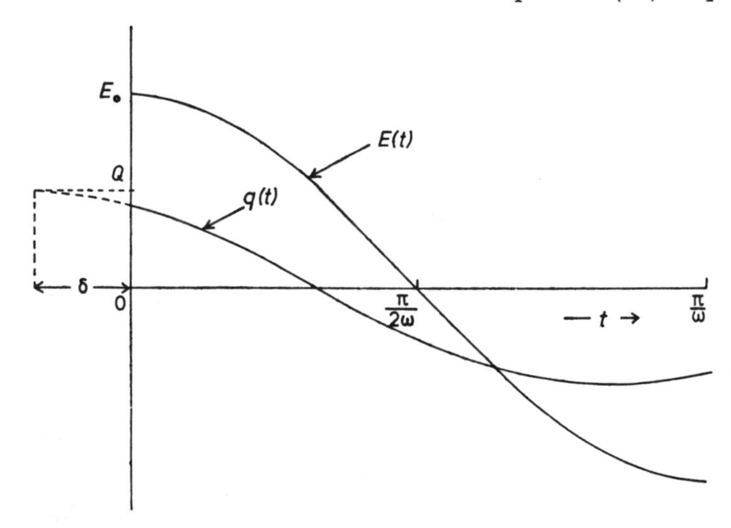

Fig. 131. The relation between the applied voltage and the charge on the plates of a condenser in a simple circuit.

(18) and (19) show that if a harmonically varying voltage of amplitude E_0 and circular frequency ω is applied to the plates of a

condenser through a resistance R, then the charge on the plates will vary harmonically with the same frequency, with amplitude $Q = E_0 C(1 + \omega^2 R^2 C^2)^{-\frac{1}{2}}$ and phase lag $\delta = \tan^{-1} \omega RC$.

The relation between the applied voltage and the electric charge on the plates of the condensor is shown in Fig. 131.

We shall now briefly discuss a practical problem. Consider the following schematic pattern:

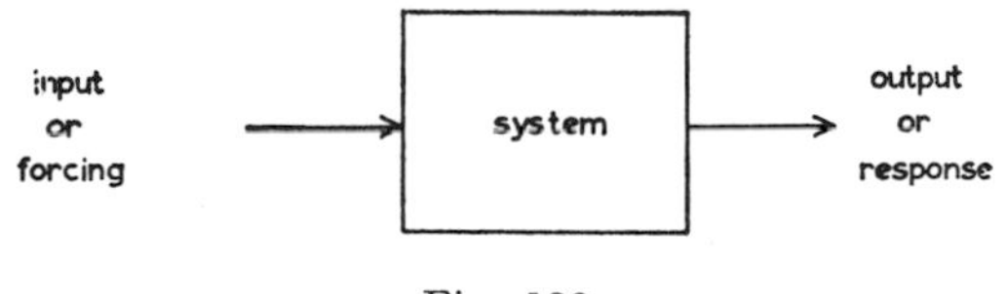

Fig. 132.

Thinking in abstract terms we may say that any given system may be submitted to an external influence or forcing resulting in a certain reaction or response of the system.

Let us take the case of the charging of a condenser (see above). The differential equation in physical standard form was found to be

$$RC\frac{\mathrm{d}q}{\mathrm{d}t} + q = CE \tag{23}$$

where q is the charge on the condenser and E is the applied voltage. Now in this case CE is the input quantity (the "cause") and q is the output, or response, or effect.

q is as it were the slave of CE. If we keep C constant and let E be some function of t, say, $E = a \cos \omega t$, we may write

$$RC\dot{q} + q = Ca \cos \omega t. \tag{24}$$

In this case the forcing quantity is $Ca \cos \omega t$.

It is easily seen that as a result of this oscillating input quantity, the output or response quantity q will also oscillate. If suddenly we let the voltage drop to zero, then we have the situation in which the *input* quantity or forcing is zero and the equation becomes

$$RC\dot{q} + q = 0. \tag{25}$$

In mathematical terms this is the homogeneous linear equation. Another example: under constant conditions the temperature of a room is a function of the temperature of the stove.

If the stove temperature oscillates, the room temperature oscillates.

If the stove temperature falls, the room temperature falls etc. Here the stove temperature is the forcing quantity and the room temperature the response.

If we write the equation in physical standard form*

$$\tau\dot{q} + q = \varphi(t) \tag{26}$$

then we define as the forcing quantity (which we shall denote by q_f) the quantity on the right-hand side of the physical standard form of the linear equation.

This means of course that we may write the physical standard form as follows

$$\tau\dot{q} + q = q_f \tag{27}$$

where q_f is the forcing quantity, $q_f = \varphi(t)$.

EXAMPLE:

$$\frac{\mathrm{d}q}{\mathrm{d}t} = 5q + 3\mathrm{e}^{at}, \textit{ find } q_f. \tag{28}$$

The only thing we have to do is to put the equation in physical standard form, i.e. to take care that $\dot{q}$ and q appear on the left-hand side and that the coefficient of q is $+1$.

We write

$$\frac{\mathrm{d}q}{\mathrm{d}t} - 5q = 3\mathrm{e}^{at}$$

whence, by division we obtain the *physical standard form*

$$-\tfrac{1}{5}\dot{q} + q = -\tfrac{3}{5}\mathrm{e}^{at}. \tag{29}$$

Here the forcing quantity q_f is $-\frac{3}{5}\mathrm{e}^{at}$.

2. The Form of the Arterial Pulse

This discussion is based upon a mathematical analysis of Otto Frank.[1,2] We shall analyse the pressure-time curve in the aorta. For this purpose we must replace the complicated conditions existing in the organism by a simplified model. In fact, we shall begin with an oversimplified model. In many biological situations it is necessary to replace the actual situation by a so-called "mathematical" model. (It is not necessary that this "model" can be visualized.) This sim-

* The essential thing is that q stands by itself.

plification not only applies to physiology but to physics just as well. Since the conditions in physiology are, in general, much more complex than in physics, it is obvious that in the former case we must simplify to a greater extent in order to make the problem accessible to mathematical treatment. In a complex problem, the first model should be as simple as possible, i.e. only the most fundamental properties of the system should be considered. At a further stage, complicating and additional factors should be taken into account. However, if the incorporation of such factors leads to great mathematical difficulties (e.g. non-linear differential equations) it is often advisable to leave these factors aside if the corresponding loss of precision is not "too large". The following will be based upon the simplest model of the pressure conditions in the aorta.

The aorta will be represented by an elastic volume container, whose capacity depends upon the pressure relations. The term "Windkessel" (well-known from textbooks) may be used for this part of the system. This "elastic container" is connected to a tube having a definite resistance to flow. This represents the "peripheral resistance".

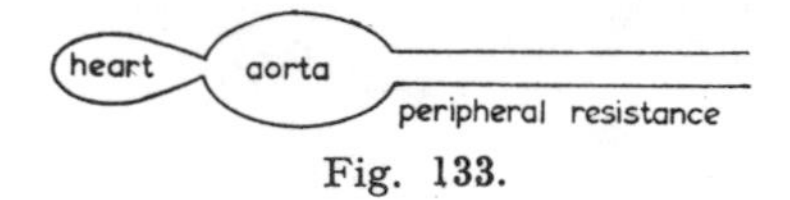

Fig. 133.

We must consider two phases:

(1) *the diastolic phase*: during this phase the rate of inflow from the heart to the aorta is zero;

(2) *the systolic phase*: during this phase blood enters the aorta from the heart. This phase will be considered later.

(1) *The diastolic phase*

We shall discuss the pressure-time course in the "Windkessel" assuming that

(a) a linear relation exists between volume and pressure in the "Windkessel";

(b) Poiseuille's law is obeyed in this system.

The linear relation between pressure and volume is expressed by

$$\frac{\mathrm{d}P}{\mathrm{d}V} = c \tag{1}$$

where c is a constant, P is the pressure in the "Windkessel" and V is the volume of blood (fluid). If we assume that the pressure at the distal end of the tube (venous-side) is zero, then Poiseuille's law may be written

$$\frac{\mathrm{d}V}{\mathrm{d}t} = -\frac{P}{\omega} \tag{2}$$

where ω is the resistance (analogue of R in Ohm's law), a constant such that in terms of the viscosity η of the blood and the radius r and the length l of the tube

$$\omega = \frac{8l\eta}{\pi r^4}.$$

The minus sign indicates that the greater the pressure P the greater the decrease of volume per unit time, in the "Windkessel".

It should be noted that equations (1) and (2) are assumptions and it should be stressed that quite different assumptions can reasonably be made. These, however, are the simplest possibilities.

Elimination of V between (1) and (2) yields the differential equation

$$\frac{\mathrm{d}P}{P} = -\frac{c}{\omega}\,\mathrm{d}t. \tag{3}$$

It is easy to find the solution in the form

$$P = P_0 \mathrm{e}^{-\frac{c}{\omega}t} \tag{4}$$

where P_0 is the value of P at $t=0$. Note that the time constant τ is ω/c.

Equation (3) tells us that during the diastolic period (closed aorta valves) the pressure in the aorta falls in a simple exponential fashion. It should be noted that the differential equation (3) is homogeneous (forcing quantity q_t is zero, i.e. no inflow from the heart.)

(2) *The systolic phase*

The essential thing about this phase is that the blood is pumped into the aorta (Windkessel) by the contraction of the heart. The rate of volume change $\mathrm{d}V/\mathrm{d}t$ in the Windkessel is now the resultant of two factors, namely the rate of inflow (from the heart) and the rate of outflow (through the peripheral resistance). The rate of outflow is given (by assumption) by Poiseuille's law, i.e.

$$\frac{\mathrm{d}V}{\mathrm{d}t} = \frac{P}{\omega}.$$

Here we omit the minus sign, since now we think of $\mathrm{d}V/\mathrm{d}t$ not as a volume decrease per unit time in the Windkessel, but as the rate of flow, i.e. as the volume of fluid passing a cross section per unit time. Let us call the volume of fluid entering the "Kessel" per unit time $i(t)$ where as indicated i may be some function of time.

It will be clear that the net change of volume per unit time in the "Kessel" is given by

$$\frac{\mathrm{d}V}{\mathrm{d}t} = i(t) - \frac{P}{\omega}. \tag{5}$$

If $i > P/\omega$, $\mathrm{d}V/\mathrm{d}t$ is positive, i.e. the volume in the "Kessel" increases, and vice versa.

Combining (5) with the assumption expressed by equation (1) we easily obtain

$$\frac{\mathrm{d}P}{\mathrm{d}t} = c\left\{i(t) - \frac{P}{\omega}\right\}$$

or

$$\frac{\mathrm{d}P}{\mathrm{d}t} + \frac{c}{\omega}P = ci(t). \tag{6}$$

Since we do not know the nature of the function $i(t)$ we must make an assumption. The simplest assumption is that i is constant during the systolic period. However, it will be clear that a better assumption would be to put $i = A \sin Bt$, i.e. to assume that the rate of outflow into the aorta is approximated by a sine function. Since normally no back flow occurs (regurgitation) we need only consider

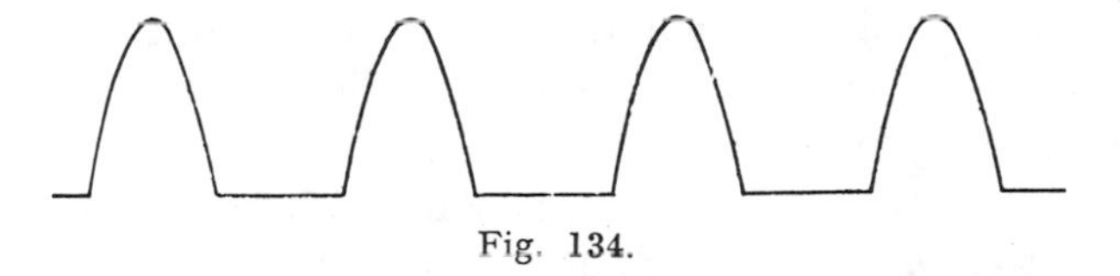

Fig. 134.

values between 0 and π/B. Since during diastole the rate of inflow is zero, the time course of i during the whole cycle (diastole + systole) may be represented by Fig. 134. Assuming the forcing quantity

$q_t = \omega i(t)$ to be a sine function, (6) becomes

$$\frac{\mathrm{d}P}{\mathrm{d}t} + \frac{c}{\omega} P = cA \sin Bt.$$

The integrating factor is clearly $\mathrm{e}^{\frac{c}{\omega}t}$. Hence, in the usual way,

$$P = C\mathrm{e}^{-\frac{c}{\omega}t} + \mathrm{e}^{-\frac{c}{\omega}t} \int cA \sin Bt\, \mathrm{e}^{\frac{c}{\omega}t}\, \mathrm{d}t.$$

We have already evaluated the integral (example 26 of Chapter 7). We then obtain

$$P = C\mathrm{e}^{-\frac{c}{\omega}t} + Ac \frac{(c/\omega) \sin Bt - B \cos Bt}{c^2/\omega^2 + B^2}. \tag{7}$$

At $t = 0$, we have $P = P_0$, hence

$$C = P_0 + \frac{cAB}{c^2/\omega^2 + B^2}$$

and, therefore,

$$P = \left\{P_0 + \frac{cAB}{c^2/\omega^2 + B^2}\right\} \mathrm{e}^{-\frac{c}{\omega}t} + \frac{Ac[(c/\omega) \sin Bt - B \cos Bt]}{c^2/\omega^2 + B^2}. \tag{8}$$

Although equation (8) exhibits some interesting features of the system, a further discussion would lead us too far astray. The interested reader is referred to the original papers (references 3 to 13, page 628).

3. The Oxygen Debt: a Heart Function Test

The problem of the oxygen debt is of acute clinical interest since measurement of oxygen debt is used as a heart-function test. The following theoretical approach to the problem is due to Prof. H. C. Burger. It should be stressed that the following is not a finished theory but merely a first approximation.

In a first approximation we must ignore all complicating factors and try to reduce the problem to its bare essentials, i.e. we must simplify as much as possible. This is the only way to obtain a clear view of the problem. At a further stage we may take the complicating factors into account. Our first problem will be to find the equation of the oxygen debt.

During (strenuous) exercise the body is able to incur an oxygen debt. This means that the muscles are able to work anaerobically, i.e. without oxygen supply, but that in the "recovery stage" oxygen is needed to replenish the energy stores (glycogen etc.) depleted during the exercise. We may say that an "oxygen debt" exists if the oxygen supply lags behind the oxygen need of the muscles. It is clear that the greater the muscular effort the greater will be the amount of oxygen needed to replenish the energy stores (glycogen) which were used up during the work.

Hence we may make the simple assumption that the oxygen debt is proportional to the work done. If we denote the oxygen debt by x and the work done by W we have

$$x = \alpha W \tag{1}$$

where α is the constant of proportionality. (This assumption is incomplete, see below.)

Let us divide equation (1) by time t, then

$$\frac{x}{t} = \alpha \frac{W}{t}. \tag{2}$$

If we choose a time unit small compared with the duration of the process, we may write (2) as difference quotients

$$\frac{\Delta x}{\Delta t} = \alpha \frac{\Delta W}{\Delta t}$$

in words: the change (increase or decrease) of oxygen debt per second is proportional to the change of work per second.

In applications we may replace the difference quotient by the derivative, hence we may write instead of the above equation

$$\frac{dx}{dt} = \alpha \frac{dW}{dt}. \tag{3}$$

On the other hand, the oxygen debt also depends upon the oxygen *supply* (amount of oxygen taken up through the lungs). If, for the moment, we ignore consumption in the muscles, we clearly may write

$$\frac{dx}{dt} = -\frac{dO}{dt}, \tag{3a}$$

where O denotes the oxygen uptake.

The minus sign is used to indicate the fact that the rate of *decrease* of the oxygen debt is proportional to the rate of *increase* of the oxygen uptake.

The net result is found by combining (3) and (3a)

$$\frac{\mathrm{d}x}{\mathrm{d}t} = \alpha \frac{\mathrm{d}W}{\mathrm{d}t} - \frac{\mathrm{d}O}{\mathrm{d}t} \tag{4}$$

where here $\mathrm{d}x/\mathrm{d}t$ is the *net* change of oxygen debt per unit time.

We now assume that the oxygen uptake per second is "regulated" by the oxygen debt, i.e. the greater the oxygen debt the greater the extra oxygen uptake per second and vice versa.

More precisely, we assume that the *extra* oxygen uptake per second (at the lungs) is proportional to the oxygen debt existing at any instant, i.e.

$$\frac{\mathrm{d}O}{\mathrm{d}t} = kx. \tag{5}$$

Substituting (5) in (4) we obtain the equation

$$\frac{\mathrm{d}x}{\mathrm{d}t} = \alpha \frac{\mathrm{d}W}{\mathrm{d}t} - kx. \tag{6}$$

In physics $\mathrm{d}W/\mathrm{d}t$ is called *power* and given a special symbol, say P. Hence we may write equation (6) in the form

$$\frac{\mathrm{d}x}{\mathrm{d}t} = \alpha P - kx \tag{7}$$

or

$$\tau \dot{x} + x = \frac{\alpha}{k} P(t) \tag{8}$$

where $\tau = 1/k$.

This is our basic equation upon which our further discussion is based. We shall study various possible functions $P(t)$.

$$P(t) = 0.$$

This case occurs, for instance, at the end of exercise. At the moment exercise is ended, the power P becomes zero.

Our differential equation then becomes

$$\tau \dot{x} + x = 0 \tag{9}$$

with solution

$$x = Ce^{-t/\tau}, \tag{10}$$

where C is an arbitrary constant whose physiological meaning is found from noting that at $t = 0$, (10) becomes

$$x = C,$$

hence C is the oxygen debt at $t = 0$, or $C = x_0$.

Hence (10) may be written

$$x = x_0 \, e^{-t/\tau}. \tag{11}$$

Physiological interpretation

Let us assume that a subject performs work. At a certain instant the oxygen debt has assumed a definite value, say x_0.

At this very instant work is ended. For our consideration this instant is $t = 0$, hence at $t = 0$ the oxygen debt has the value x_0 and from $t = 0$ onwards the forcing quantity q_t is zero, for, since $P(t) = 0$ (no work done) $q_t = \alpha P(t)/k = 0$.

(The nature of the q_t before the instant $t=0$ does not interest us here.) Equation (11) tells us that at the end of work the oxygen debt x drops to zero in an exponential fashion. (Theoretically the debt is zero only after an infinite time.)

The corresponding curve is shown in Fig. 135.

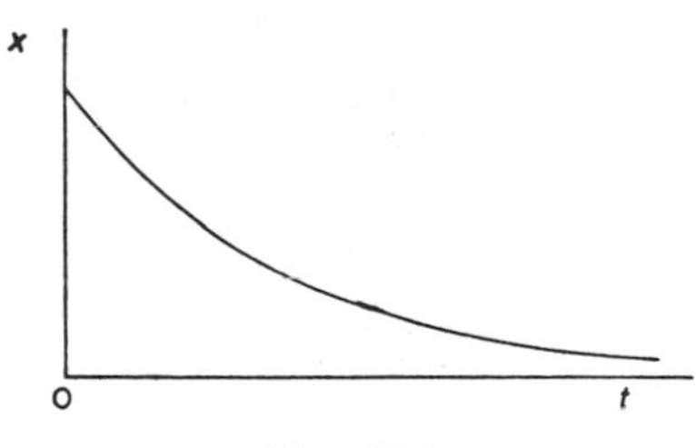

Fig. 135.

Now we are faced with a practical problem. We cannot measure the oxygen debt directly, and hence it would appear that we cannot test the equation (11).

It is obvious that if a theory cannot be tested by experiment, the theory is of little value.

But, fortunately, we can check equation (11) in an indirect manner, since in equation (5) we have a relation between x and O, the O_2-uptake, a quantity which we can measure.

Combining (5) and (11) we obtain

$$\frac{\mathrm{d}O}{\mathrm{d}t} = kx_0 \, \mathrm{e}^{-t/\tau}. \tag{12}$$

In practice, we can measure $\mathrm{d}O/\mathrm{d}t$, or the rate of extra oxygen uptake. In terms of theory, $\mathrm{d}O/\mathrm{d}t$ plotted as a function of time decreases exponentially. The curve is given in the Fig. 136 below.

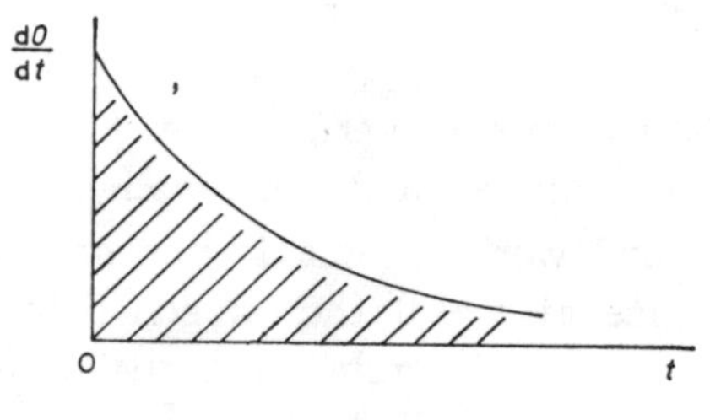

Fig. 136.

This conclusion means that if the theory is right the experimental curve of $\mathrm{d}O/\mathrm{d}t$ against t must be of the form given in Fig. 136.

Experiments show that this is indeed the case at the end of moderate work. An example will be given below.

It is interesting that the oxygen debt at $t = 0$, i.e. x_0, can be found by measuring the area under the curve in Fig. 136 with a planimeter.

This is seen as follows. Consider equation (12).

Integration of (12) between the limits 0 and ∞ gives

$$\int_0^\infty kx_0 \, \mathrm{e}^{-t/\tau} = \left[-\tau k x_0 \, \mathrm{e}^{-t/\tau}\right]_0^\infty .$$

But we have put $\tau = 1/k$, hence we have for the area under the curve

$$\left[-x_0 \, \mathrm{e}^{-t/\tau}\right]_0^\infty = x_0,$$

i.e. the oxygen debt at $t = 0$, i.e. at the instant the exercise is ended. In this way we have proved that the area under the curve of $\mathrm{d}O/\mathrm{d}t = kx_0 \, \mathrm{e}^{-t/\tau}$ between $t = 0$ and $t = \infty$ is numerically equal to the oxygen debt at $t = 0$.

It is clear that if we wish to know the oxygen debt at say, $t = 12$ we must integrate between 12 and ∞ etc.

We have seen that the quantity that we can measure is $\mathrm{d}O/\mathrm{d}t$, i.e. the rate of extra O_2 uptake (by the lungs).

In terms of theory, the experimental curve of $\mathrm{d}O/\mathrm{d}t$ against t must be a simple exponentially falling curve in the "recovery-phase", according to equation (12). How are we to test this?

Well, the obvious way would be to compare the experimental curve with the theoretical curve. But this would not be a critical test. If we obtain an experimental curve resembling the simple exponential curve, the experimental curve might just as well be a hyperbola or some other type of curve. We must reject mere inspection of the experimental curve as unreliable. The critical test runs as follows. Consider again equation (12).

Taking natural logarithms we obtain

$$\log \frac{\mathrm{d}O}{\mathrm{d}t} = \log kx_0 - \frac{1}{\tau}t. \tag{13}$$

It is easy to see from (12) by taking $t = 0$, that $kx_0 = \mathrm{d}O/\mathrm{d}t$ at $t = 0$, say $(\mathrm{d}O/\mathrm{d}t)_{t=0}$ so that (13) becomes

$$\log \frac{\mathrm{d}O}{\mathrm{d}t} = \log \left(\frac{\mathrm{d}O}{\mathrm{d}t}\right)_{t=0} - \frac{1}{\tau}t. \tag{14}$$

This is the equation of a straight line with slope $-1/\tau$ and y-intercept

$$\log \left(\frac{\mathrm{d}O}{\mathrm{d}t}\right)_{t=0}$$

Equation (14) can be used as a specific test. If the curve is exponential, then the logarithms of the experimental figures of $\mathrm{d}O/\mathrm{d}t$ plotted against time should yield a straight line.

Hence the rule for specific testing is this:

Take logarithms (it is not necessary to take natural logarithms, since decimal logarithms are proportional) of the experimental values of $\mathrm{d}O/\mathrm{d}t$ and plot these against the corresponding times.

If the resulting curve is a straight line, the curve is exponential. Instead of taking logarithms, however, it is more convenient to use semi-logarithmic paper (ordinate on logarithmic scale, abscissa (horizontal) on normal scale).

The experimental values can be directly plotted on such semi-log paper, and if the points fall on a straight line, we can conclude that the curve is exponential, i.e. that formula (12) holds, i.e. that the theory is *probably* correct.

The following results taken from a study* of H. M. Beumer illustrate the above. The oxygen consumption per time unit i.e. dO/dt, has been plotted against time (10 seconds interval).

To simplify, let us denote dO/dt by the single symbol y, i.e. $y \equiv dO/dt$. Then (12) may be written

$$y = y_0\, e^{-t/\tau} \tag{15}$$

(it is easy to show by considering $t = 0$, that $kx_0 = y_0$). It must be stressed that these results refer to the "recovery stage". The figures are shown in Table 19.

TABLE 19

Time in seconds	Rate of extra O_2 uptake in cc/sec $= y$
0	58
10	49
20	41
30	35
40	29
50	25
60	21
70	18
80	16

It will be seen that the data plotted on semi-logarithmic paper yield points that fall on a straight line (ignoring the spread, especially at the lower end of the curve).

Hence we can conclude that the experimental curve (in the "recovery stage") is exponential, and may be represented by the formula

$$y = y_0\, e^{-t/\tau} \quad (\tau = \text{time constant}). \tag{15}$$

Having shown that the experimental curve is exponential, we shall now discuss a very important problem.

How can we determine the time constant τ from the experimental curve? This may best be done as follows. Let us write (15) as follows

$$\frac{y}{y_0} = e^{-t/\tau}.$$

* The experimental method used does not concern us here.

Let us consider the instant $t = \tau$. Then

$$\frac{y}{y_0} = e^{-1} \qquad \text{or} \qquad y = \frac{1}{e} y_0 = \frac{1}{2.72} y_0 = 0.368\, y_0.$$

This result may be interpreted as follows. It is clear that y changes from y_0 (at $t = 0$) to 0 (at $t = \infty$). We may say that the total change to be made is from y_0 to zero. (y is zero means that the *extra* oxygen consumption is zero.) Hence if $y = 0.368 y_0$ we may say that $(100 - 36.8)$ or 63.2 % of the total change has been accomplished.

This is one way to obtain the value of the time constant τ from the (experimental) curve.

Consider the experimental curve on *normal* graph paper. It is seen that $y_0 = 58$ (this may also be taken from the table). Now, compute $y = 0.368\, y_0$. We obtain $y = (0.368) \times (58) = 21.344$, say 21.

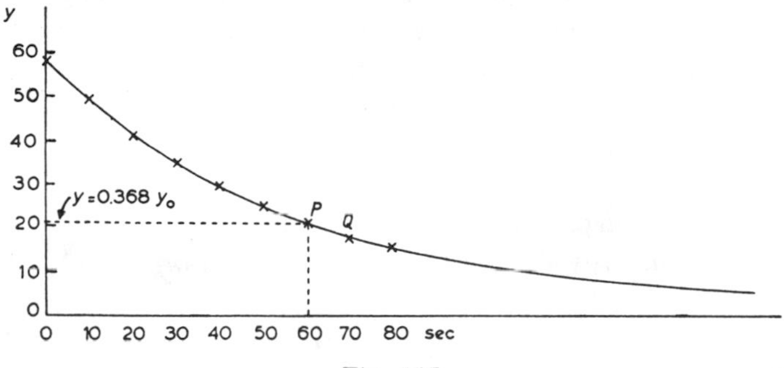

Fig. 137.

Mark off this value $y = 21$ on the vertical axis (ordinate). Then, draw a horizontal line through $y = 21$, which cuts the experimental curve at some point P. Drop, from P, a vertical line. In our case, this line cuts the abscissa at the t value $t = 60$ seconds. It follows that our desired time constant τ is 60 seconds, or $\tau = 60$. Hence, in this example the formula of the experimental curve becomes

$$y = 58\, e^{-t/60} \qquad \text{or} \qquad y = 58\, e^{-0.017t}.$$

But, suppose we could not, in practice, determine the value y_0 i.e. the y value at the instant $t = 0$. This case often occurs in practice. Fortunately this does not spoil the procedure. It is not at all necessary to start from y_0. We may take any (reasonably large) value of y as our starting point to compute τ.

We shall give an example. Consider Table 19. Let us choose the value $y = 49$ at $t = 10$. Compute $y = (0.368) \times (49) = 18$. Again draw a horizontal line through $y = 18$, which cuts the curve at some point Q. Dropping a line from Q on the abscissa we find that it cuts the abscissa at $t = 70$. Now we must take for τ the difference $70 - 10 = 60$ seconds. All this is true, because we are concerned with the "change to be made" from any point onwards, and 63.2% of this "change to be made" (in our last example from 49 to 0) is accomplished τ seconds later.

For the purpose of computing τ we may "mentally" shift the zero-point to our arbitrary starting point. In our last example, we put $t = 10$, $y_0 = 49$.

Hence $y = 49e^{-t/\tau}$ and for $t = \tau$, $y = 49e^{-1} = (0.368) \times (49) = 18$.

We have just seen how we may find τ from the experimental curve if this curve can be represented by a formula of the type $y = y_0 e^{-t/\tau}$. We shall, however, presently see that τ may be obtained in very much the same manner if we are dealing with curves that "obey" the formula

$$y = A + Be^{-t/\tau} \quad (A \text{ and } B \text{ are constants}).$$

Let us, for this purpose and others, go back to equation (8), the basic equation for the oxygen debt. This equation is

$$\tau\dot{x} + x = \frac{\alpha}{k} P(t). \tag{8}$$

Let us now consider the case where the forcing quantity $q_t = \alpha P(t)/k$ is constant. This case occurs when the work per time unit i.e. the power P is constant. Let us say $P(t) = K$ (ergs/sec), where K is a constant. Then (8) becomes

$$\tau\dot{x} + x = \frac{\alpha}{k} K \tag{16}$$

but, since α and k are also constants, we may, for the sake of convenience, replace the right-hand member by another constant, say, E, then we write (16)

$$\tau\dot{x} + x = E. \tag{17}$$

In this case $q_t = E$. If q_t is zero for negative values of t, and then, suddenly, at $t = 0$, jumps to its constant value (E in our case), we say that it is a *step-function.*

This notion is rather arbitrary and irrelevant, since it is immaterial what values q_t assumed *before* $t = 0$. It happens that in our case q_t is indeed zero (no work done) *before* $t = 0$, so that in our case the term step-function is relevant.

It is readily seen that the solution of (17) is

$$x = E + Ce^{-t/\tau}. \tag{18}$$

The meaning of C is found from the initial condition: at $t = 0$, $x = 0$ (no oxygen debt).

Hence at $t = 0$,

$$0 = E + C \quad \text{or} \quad C = -E. \tag{19}$$

Hence (18) may be written

$$x = E - Ee^{-t/\tau} = E(1 - e^{-t/\tau}). \tag{20}$$

It is readily verified that if in general $q_t = Q_2$, and if at $t = 0$, $q = Q_1$, then the solution of

$$\frac{1}{\tau}\dot{q} + q = Q_2$$

becomes

$$q = Q_2 + (Q_1 - Q_2)e^{-t/\tau}. \tag{21}$$

In our case $Q_1 = 0$, hence, from (21), $C = -E$.

From (5) and (20) we obtain

$$\frac{dQ}{dt} = kE(1 - e^{-t/\tau})$$

or putting $y = dQ/dt$, we have

$$y = kE(1 - e^{-t/\tau}). \tag{22}$$

Now (22) gives the oxygen uptake per unit of time, a quantity we can measure, as a function of time.

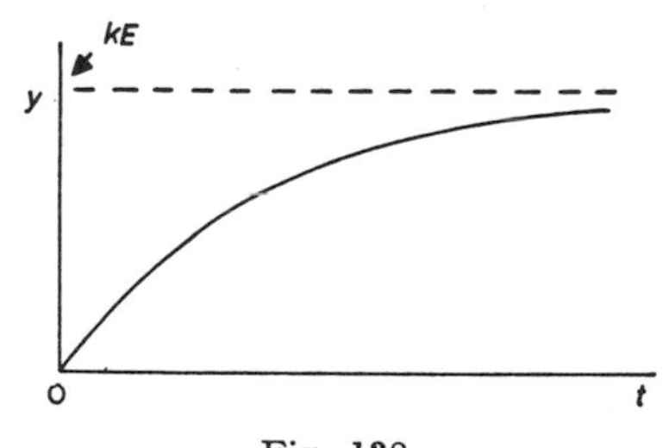

Fig. 138.

The graph of (22) is given in Fig. 138.

How are we to obtain the time constant τ from the curve in Fig. 138? Well, let us write $y = kE - kEe^{-t/\tau}$ or $kE - y = kEe^{-t/\tau}$. What happens when $t = \tau$ (seconds)? Then we have

$$kE - y = kE\mathrm{e}^{-1} = 0.368\,kE. \tag{23}$$

It should be realized that kE is the final value of y (theoretically, attained after $t = \infty$, in practice after, say, 5 minutes). We call kE the "asymptotic value" of y.

Equation (23) may be interpreted to mean that after τ seconds the difference still existing between y and its final value $kE - y$ is 36.8 % of this final value kE, or, of the total change to be made from 0 to E, 36.8 % remains to be made at $t = \tau$.

Let us now consider the general case expressed by formula (21). This can be rearranged to give

$$\frac{Q_2 - q}{Q_2 - Q_1} = \mathrm{e}^{-t/\tau}. \tag{24}$$

If we take $t = \tau$, (24) becomes

$$\frac{Q_2 - q}{Q_2 - Q_1} = 0.368, \quad \text{or} \quad Q_2 - q = 0.368\,(Q_2 - Q_1). \tag{25}$$

The "total change to be made" is from Q_1 (at $t = 0$) to Q_2 (at $t = \infty$). Equation (25) states that at $t = \tau$, 36.8 % of this total change remains to be made.

We shall illustrate the above by an example.

In order to illustrate the "general" case ($Q_1 \neq 0$) we shall not use an "oxygen-debt" curve but a curve giving the rise of CO_2 con-

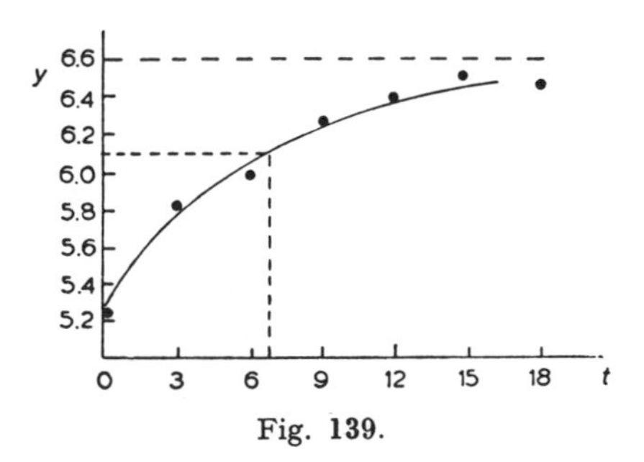

Fig. 139.

centration in the lungs during breathholding. (For reference see p. 270 above.) Consider Fig. 139.

The formula of the curve is of the type $q = Q_2 + (Q_1 - Q_2)e^{-t/\tau}$. Experimentally it was found that the value of the asymptote Q_2 was 6.6, while Q_1, i.e. q value at $t = 0$, was 5.3. At $t = \tau$, we have, according to (25)

$$6.6 - q = 0.368\ (6.6 - 5.3) = 0.48$$

or $q = 6.6 - 0.48 = 6.12$ at $t = \tau$.

Now, draw a horizontal line at $q = 6.12$, which cuts the curve at some point. The corresponding value of t must then be equal to τ. It is seen that in our case $\tau = 6.7$ seconds.

It follows that the formula describing the curve is

$$q = 6.6 - 1.3\ e^{-0.15t}.$$

(Remark: the physical meaning of the symbols is immaterial here.)

Another important problem is this: how are we to know that the experimental curve is really of the type to which equation (21) belongs?

Well, consider

$$q = Q_2 + (Q_1 - Q_2)e^{-t/\tau} \tag{21}$$

or

$$Q_2 - q = (Q_2 - Q_1)e^{-t/\tau}.$$

Taking natural logarithms, we have

$$\log(Q_2 - q) = \log(Q_2 - Q_1) - \frac{1}{\tau}t. \tag{26}$$

Now log $(Q_2 - Q_1)$ is clearly a constant, so that (26) is the equation of a straight line when log $(Q_2 - q)$ is plotted against t. The slope is $-1/\tau$ and the y-intercept is at the point log $(Q_2 - Q_1)$.

Hence we have the prescription: to test whether an (experimental) curve obeys formula (21), plot log $(Q_2 - q)$ or, what amounts to the same, $\log_{10}$ $(Q_2 - q)$ against time t, and if the points fall on a straight line, the curve may be represented by formula (21). We may, of course, use semi-logarithmic paper for this purpose.

A second method to obtain the time constant τ is the following:

Let us return to the data obtained by Beumer which may be described by the equation

$$y = y_0 e^{-t/\tau}. \tag{15}$$

The time constant τ may be obtained from the semi-log plot as follows: let us write equation (15) as

$$\log y = -\frac{t}{\tau} + \log y_0$$

or since $\log_{10} y = 0.4343 \log y$,

$$\log_{10} y = -0.4343 \frac{t}{\tau} + 0.4343 \log y_0.$$

If we use semi-log paper as commonly available, then it is just as if $\log_{10} y$ is plotted against t on normal paper. The slope is given by

$$\frac{\log_{10} y_1 - \log_{10} y_2}{t_1 - t_2} = -\frac{0.4343}{\tau}. \tag{27}$$

This is a special case of a well-known theorem from elementary analytic geometry, which states that the slope a of a straight line $y = ax + b$ is given by

$$a = \frac{y_1 - y_2}{x_1 - x_2}$$

where (x_1, y_1) and (x_2, y_2) are any two points on the curve. It is an easy exercise to obtain τ from the semi-log plot of Beumer's data (Table 19) using equation (27). As with our first method, a value of 60 seconds should be obtained.

Let us now return to the oxygen debt problem.

We have seen that during *constant* work the extra oxygen uptake per unit time ($\mathrm{d}O/\mathrm{d}t \equiv y$) is described by equation (22). When the work is stopped, y falls exponentially, according to equation (15). The combined picture clearly represents the rate of *extra* oxygen uptake *during* the *constant* work period and the "recovery" period. This is illustrated in Fig. 140.

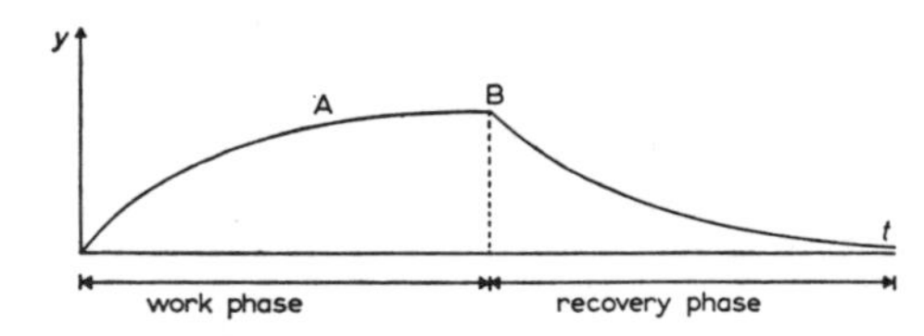

Fig. 140.

Between A and B we have the so-called steady-state. Here the "asymptote" (kE) has been reached (in practice), i.e. the rate of O_2 uptake remains constant, i.e. $dy/dt = 0$. Further, it should be noted that in terms of theory, the time constant during the work phase and the recovery phase should have the same value, i.e. τ in (22) and (15) should be the same. If we consider equation (22) we note that y consists of two terms kE and $kEe^{-t/\tau}$. The second term containing $e^{-t/\tau}$ is called the *transient* term or transient response q_t, since it dies out after a certain time (theoretically only after an infinite time) while the first term on the right (here kE) is called the forced response q_r, since q_r remains as long as the forcing quantity q_f is maintained.

In Fig. 140, we see that the transient response q_t has "died out" near A, and that between A and B only the forced response q_r operates. At B the forcing quantity is suddenly removed.

In general we call the complementary function the *transient response* and the particular integral the *forced response*.

We may write

$$q = q_r + q_t$$

where q_t = transient response, q_r = forced response.

Finally, we shall discuss the case when the power is not constant, but increases linearly with time, i.e. $P = ht$, where h = constant.

Going back to our basic equation (8) this yields

$$\tau\dot{x} + x = \frac{\alpha}{k}ht. \tag{28}$$

From (8) we remember that in our case $\tau = 1/k$ so that we may write (28) in the form

$$\tau\dot{x} + x = \tau\alpha ht. \tag{29}$$

Putting (29) into the mathematical standard form, we see that the general solution is

$$x = Ce^{-t/\tau} + e^{-t/\tau}\int \alpha hte^{-t/\tau}\,dt.$$

The integral is evaluated by integration by parts to give

$$x = \alpha h\tau t - \alpha h\tau^2 + Ce^{-t/\tau}. \tag{30}$$

The constant C is found from the initial condition: at $t = 0$, $x = 0$ (no debt).

Then,

$$0 = -\alpha h\tau^2 + C \quad \text{or} \quad C = \alpha h\tau^2.$$

Hence (30) becomes

$$x = (\alpha h\tau t - \alpha h\tau^2) + \alpha h\tau^2 \, e^{-t/\tau}. \tag{31}$$

Here the expression between brackets is the forced response q_r (particular integral), and the last term is the transient response q_t (complementary function).

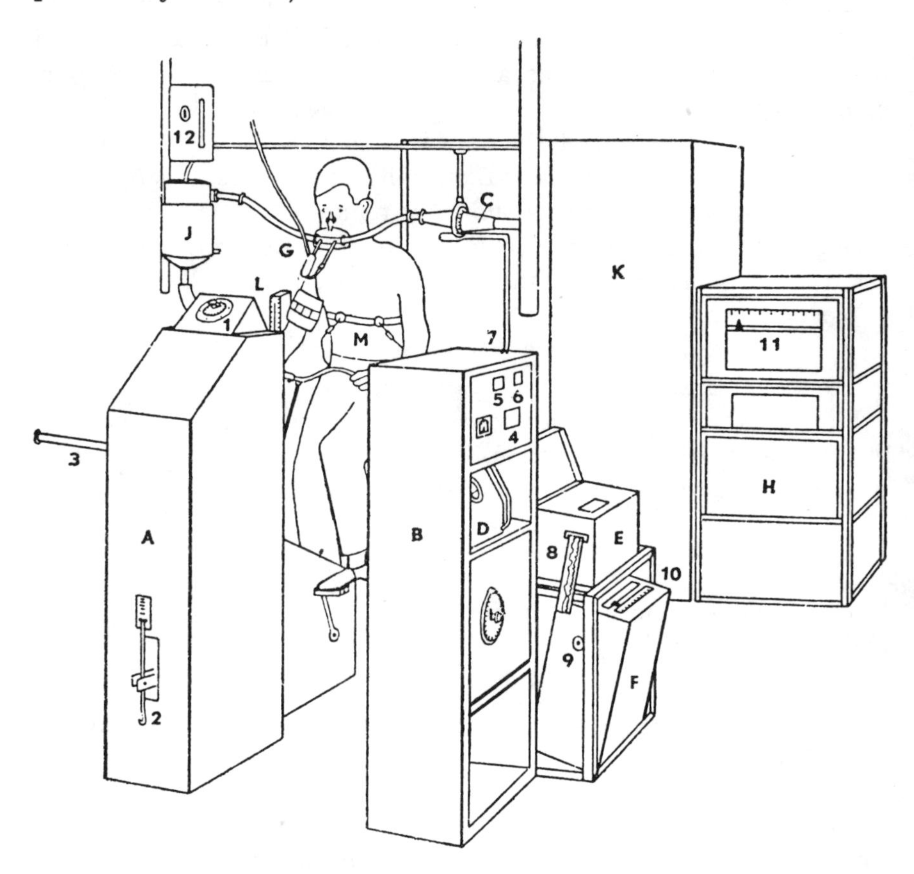

Fig. 141. A heart function test by the method of oxygen uptake during work. The set-up is used in the Institute of Preventive Medicine, Leyden. From B. Bink, "The physical working capacity of cardiac patients" (in Dutch; Thesis, Leyden, 1959).

With (5) and remembering that $\tau = 1/k$, we obtain from (31)

$$\frac{\mathrm{d}O}{\mathrm{d}t} \equiv y = (\alpha ht - \alpha h\tau) + \alpha h\tau \mathrm{e}^{-t/\tau}. \tag{32}$$

After a sufficiently long time, the transient response will have "died out", and the relation will be approximated by

$$y = \alpha ht - \alpha h\tau = \alpha h(t - \tau).$$

The reader is invited to make a sketch of (32). It may be mentioned

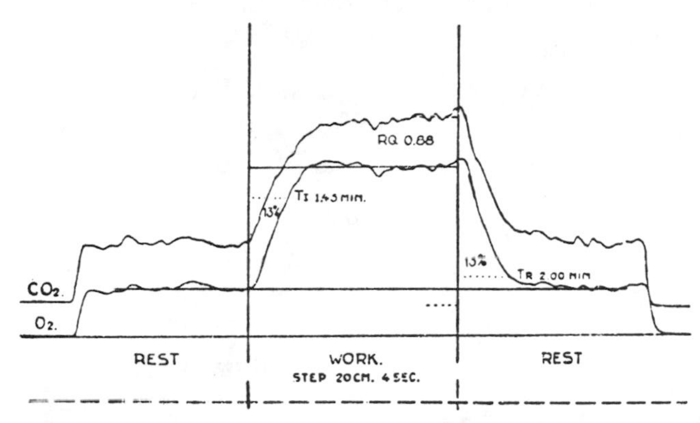

Fig. 142. Oxygen uptake during and after work in a normal subject. [Taken from J. Jongbloed, C. L. C. van Nieuwenhuizen and H. van Goor, Circulation **15** (1957) 54.]

that this linear increase of work is used in practice as a clinical heart-function test. Figs. 141 and 142 show an experimental set-up and experimental curve respectively.

4. Growth of an Isolated Population

Consider the equation

$$\frac{\mathrm{d}p}{\mathrm{d}t} = \varepsilon p - kp^2;$$

$$\text{i.e.}\quad \frac{\mathrm{d}p}{\mathrm{d}t} = kp(\omega - p),$$

where $\omega = \varepsilon/k$. This may be written as

$$\frac{\omega \mathrm{d}p}{p(\omega - p)} = \varepsilon \mathrm{d}t,$$

$$\text{i.e.} \quad \frac{\mathrm{d}p}{p} + \frac{\mathrm{d}p}{\omega - p} = \varepsilon \mathrm{d}t.$$

If $p < \omega$ then

$$\log p - \log (\omega - p) = \varepsilon t + C.$$

When $t = 0$, $p = p_0$ so

$$\log p_0 - \log (\omega - p_0) = C.$$

Therefore

$$\frac{(\omega - p)p_0}{\omega - p_0} = p\mathrm{e}^{-\varepsilon t}$$

i.e.

$$\omega p_0 = p\{p_0 + (\omega - p_0)\mathrm{e}^{-\varepsilon t}\},$$

or

$$p = \frac{\omega p_0}{p_0 + (\omega - p_0)\mathrm{e}^{-\varepsilon t}}.$$

If $p > \omega$ then

$$\frac{\mathrm{d}p}{p} - \frac{\mathrm{d}p}{p - \omega} = \varepsilon \mathrm{d}t,$$

therefore

$$\log p - \log (p - \omega) = \varepsilon t + C;$$

i.e.

$$\log p_0 - \log (p_0 - \omega) = C,$$

or

$$\log \frac{p(p_0 - \omega)}{p_0(p - \omega)} = \varepsilon t,$$

from which

$$p_0(p - \omega) = p(p_0 - \omega)\mathrm{e}^{-\varepsilon t};$$

i.e.

$$p = \frac{p_0 \omega}{p_0 - (p_0 - \omega)\mathrm{e}^{-\varepsilon t}} \quad \text{(Logistic Law).}$$

As an example of the occurrence of an equation of this type we have Verhulst's equation for an isolated population

$$\frac{\mathrm{d}p}{\mathrm{d}t} = \varepsilon p - kp^2 \tag{1}$$

(equation (10) of § 1 of Chapter 9). This equation may be written in the form

$$\frac{\mathrm{d}p}{p(\varepsilon - kp)} = \mathrm{d}t.$$

Using the fact that we may write

$$\frac{1}{p(\varepsilon - kp)} = \frac{1}{\varepsilon}\left(\frac{1}{p} + \frac{k}{\varepsilon - kp}\right)$$

we can put the equation into the form

$$\int \frac{\mathrm{d}p}{p} + \int \frac{k\mathrm{d}p}{\varepsilon - kp} = \varepsilon \int \mathrm{d}t.$$

The integrations are elementary and give

$$\log p - \log (\varepsilon - kp) = \varepsilon t - \log C.$$

(The constant of integration is written as $-\log C$ to make the final answer as simple as possible.) We therefore have

$$\frac{Cp}{\varepsilon - kp} = \mathrm{e}^{\varepsilon t}$$

from which it follows that

$$p = \frac{\varepsilon}{k + C\mathrm{e}^{-\varepsilon t}}. \tag{2}$$

If $\varepsilon > 0$, $\mathrm{e}^{-\varepsilon t} \to 0$ as $t \to \infty$ so that $p \to \varepsilon/k$ as $t \to \infty$. The variation of the function (2) has been considered in § 5 of Chapter 6.

In certain cases, the growth of a bacterial colony follows a law of the type (2). For example, Table 20 shows the comparison of the observed growth of a colony of *B. dendroides** with that calculated

TABLE 20

Growth of the bacterial colony B. dendroides

Age (days)	Area occupied in cm²	
	Observed	Calculated from equation (3)
0	0.24	0.25
2	13.53	13.08
4	47.50	47.39
5	49.40	49.02

* V. A. Kostitzin, p. 59: Population of the U.S.A., France, Belgium; p. 62: Growth of a Bacterial Colony.

by means of the formula

$$p = \frac{2.128}{0.04304 + 8.399e^{-2.128t}}, \tag{3}$$

p being estimated by measuring the area (in cm^2) occupied by the colony and t being measured in days.

5. Growth of a Non-Isolated Population

By equation (9) of p. **441** this is governed by

$$\frac{dp}{dt} = \varepsilon p(t) - kp^2(t) + I - E.$$

If
$$I - E = \nu_0 + \nu_1 p + \nu_2 p^2,$$

where ν_0, ν_1, ν_2 are positive or negative constants, then

$$\frac{dp}{dt} = \nu_0 + (\varepsilon + \nu_1)p - (k - \nu_2)p^2.$$

Putting
$$p = -\frac{\nu_0}{\varepsilon + \nu_1} + q$$

we have
$$\frac{dq}{dt} = (\varepsilon + \nu_1)q - (k - \nu_2)q^2;$$

i.e.
$$-\frac{dq}{[(\varepsilon + \nu_1) - (k - \nu_2)q]q} = dt.$$

This equation can be integrated easily (variables separable) whenever numerical values of the constants are given.

A more complicated assumption about immigration leads to the equation

$$\frac{dp}{dt} = \varepsilon p(t) - kp^2(t) - cp(t)\int_0^t k(t - \tau)p(\tau)d\tau,$$

where the function $k(t - \tau)$ is assumed to be prescribed. Let us consider the simple case in which $k(t - \tau) = 1$.

Let

$$P(t) = \int_0^t p(\tau)d\tau.$$

Then putting $P'(t) = p(t)$ in the above equation we find that

$$\frac{d^2P}{dt^2} = \varepsilon\frac{dP}{dt} - k\left(\frac{dP}{dt}\right)^2 - cP'(t)P(t).$$

Now

$$\frac{d}{dt}\left(e^{kP}\frac{dP}{dt}\right) = e^{kP}\left\{\frac{d^2P}{dt^2} + k\left(\frac{dP}{dt}\right)^2\right\} = e^{kP}\left\{\varepsilon\frac{dP}{dt} - cP'(t)P(t)\right\}$$

$$= \frac{d}{dt}\left\{e^{kP}\left(\frac{\varepsilon}{k} + \frac{c}{k^2} - \frac{c}{k}P\right)\right\}$$

Integrating we find that

$$e^{kP}\frac{dP}{dt} = e^{kP}\left(\frac{\varepsilon}{k} + \frac{c}{k^2} - \frac{c}{k}P\right) + C.$$

Suppose that when $t = 0$, $p = p_0$ then

$$t = 0, \quad \frac{dP}{dt} = p_0, \quad P = 0.$$

Hence we find that

$$p_0 - \frac{\varepsilon}{k} + \frac{c}{k^2} = C,$$

$$p_0 - \frac{\varepsilon}{k} - \frac{c}{k^2} = e^{kP}\left[\frac{dP}{dt} + \frac{c}{k}P - \frac{\varepsilon}{k} - \frac{c}{k^2}\right].$$

This is an equation with variables separable which can be solved to give $P(t)$. $p(t)$ is then found from $p(t) = dP/dt$.

For further reading see references 1 to 6, page 628.

6. Growth of Two Conflicting Populations

In Chapter 9, §1 the extended form of the Volterra equations[1] describing the growth of two conflicting populations is given (equation (14)).

Consider two variables y and x which represent respectively the number of individuals in two populations that oppose each other. Let x measure the population (A) which preys upon a second population (B) measured by y. If y is sufficiently large (A) will have an abundant food supply and x will increase. But as x increases, y, the amount of prey, will decrease. With prey scarce, x will stop increasing and, with starvation, x will actually diminish, thus allowing

y to increase again, etc. This situation may be approximated by the set of simultaneous equations

$$\frac{dy}{dt} = \xi_1 y - h_{11} yx$$
$$\frac{dx}{dt} = -\xi_2 x + h_{21} yx \tag{1}$$

where ξ_1, ξ_2, h_{11} and h_{21} are positive numbers. These differential equations, known as the Volterra equations, are derived as follows. In a closed environment the number of encounters of the members of the two species will be proportional to yx, i.e. there will be kyx encounters per unit of time. If for every encounter there is an instantaneous decrease of λ_1 members of the prey population, and an increase of λ_2 members of the other population, then the corresponding equations are

$$\frac{dy}{dt} = \xi_1 y - k\lambda_1 yx$$
$$\frac{dx}{dt} = -\xi_2 x + k\lambda_2 yx \tag{2}$$

where ξ_1 and ξ_2 are positive numbers representing the growth coefficients for species (B) and (A) respectively in the absence of the other. Note that the coefficient of x $(-\xi_2)$ is negative in agreement with the assumption that (A) depends upon (B) for its food supply. If we put $k\lambda_1 = h_{11}$ and $k\lambda_2 = h_{21}$ the system (1) is obtained. Note further that for $y = 0$, x will decrease exponentially and for $x = 0$, y will increase exponentially.

The solution of system (1) may be obtained as sketched below. By using the transformation $y = (\xi_2/h_{21})\mu$, $x = (\xi_1/h_{11})z$, system (1) is reduced to

$$\frac{dz}{dt} = \xi_1 (z - z\mu)$$
$$\frac{d\mu}{dt} = \xi_2 (\mu - z\mu). \tag{3}$$

Elimination of either variable and its derivative yields a nonlinear differential equation that cannot be integrated in terms of elementary functions. A convenient way is to write

$$\xi_2 z' - \xi_2 \xi_1 z + \xi_1 \mu' + \xi_2 \xi_1 \mu = 0, \tag{4}$$

which, as is easily verified, can be written as

$$\xi_2 z' + \xi_1 \mu' - \xi_2 \frac{z'}{z} - \xi_1 \frac{\mu'}{\mu} = 0. \tag{5}$$

Integrating, we obtain

$$\xi_2 z + \xi_1 \mu - c \log z - \xi_1 \log \mu = K \tag{6}$$

where K is an arbitrary constant. Equation (6) gives the relation between μ and z, from which the relation between y and x may be obtained.

The quadratic terms in equation (14) of Chapter 9, §1, allow for intraspecific competition of the predators for prey and of the prey for their own food. A biological weakness in the original Volterra equations (1) is the following. The presence of the xy term implies that the more prey are present for a given number of predators the more will be killed. Such gluttony is, with the exception of the glutting of seals on salmon, not found in nature. For a "surplus" of prey the number of prey killed is fairly independent of the number available, so that we should expect a relationship of the form

$$\frac{\mathrm{d}y}{\mathrm{d}t} = \xi_1 - kx \,. \tag{7}$$

For a more general discussion of this point the reader is referred to Pearce[2].

If y is plotted against x a "hysteresis" loop expressing the fact that one variable lags behind the other is obtained. More precisely, if time is eliminated in system (1) the joint values of y and x lie on a closed curve in the y, x plane. Although the deterministic model represented by system (1) is stable, the stochastic (i.e. random) variations that in reality occur around the deterministic values y and x lead to an unstable model since the stochastic variations may ultimately result in the extinction of the predator. It is of interest to note that Bartlett[3] states that whereas such extinctions generally occur after one or two cycles in laboratory experiments the theoretical estimates of the probability per cycle of an extinction based on the Lotka-Volterra model is small.

A realistic mathematical model of the "struggle for life" requires advanced mathematics such as the theory of stochastic processes, integro-differential equations (to take account of the "hereditary influences" associated with growth) and partial differential equations (e.g. to account for birth and death processes involved). The problem of birth and death processes enters in the following situation. If x, the number of predators, is not large then the population size of the predator species will be affected by probability variations that will have their greatest effect when x reaches its least value in the cycle, i.e. when the point in the x, y plane is nearest to the y-axis. As has been shown by Bailey[6] the probability of extinction of the predator at this point is critically influenced by birth and death processes. A simple example of the theory of birth and death processes is given in Chapter 10, §10. For others, see references 4, 5 and 7, page 628.

We shall end this discussion with an illustrative example.[3]

Data on *Paramecium aurelia* feeding on yeast cells, yielded, using the Volterra model, a period for the cycle of 9.4 days. The probability of extinction for any cycle was estimated to be 0.003, a very low value.

7. The Theory of the Propagated Action Potential

One of the most striking investigations in biology in the last 20 years has been that by A.F. Huxley and A.L. Hodgkin on the propagation of the action potential in nerve on the basis of empirical relationships which they discovered themselves.[1] Consider Fig. 143, where the electrical circuit represents a unit segment of the membrane of a nerve. (Consult any elementary textbook of physiology, e.g. Ganong.[8]) Let R represent a resistance, g, a conductance, then by definition, as in the figure

$$R_{\mathrm{Na}} = 1/g_{\mathrm{Na}}, \text{ etc.}$$

R_{Na} and R_{K} vary with time and membrane potential, while the other parameters are constant. Current may be carried through the membrane by: a) charging the membrane capacity and/or b) the movement of ions through the resistances in parallel with the membrane capacity. This (latter) ionic current is divided into three components: 1) a component carried by sodium ions (I_{Na}), 2) a component carried by potassium ions (I_K) and 3) a minor component carried by chlorine and other ions, called "leakage current" (I_l). In accord with Ohm's law the sodium current (I_{Na}) is equal to the dif-

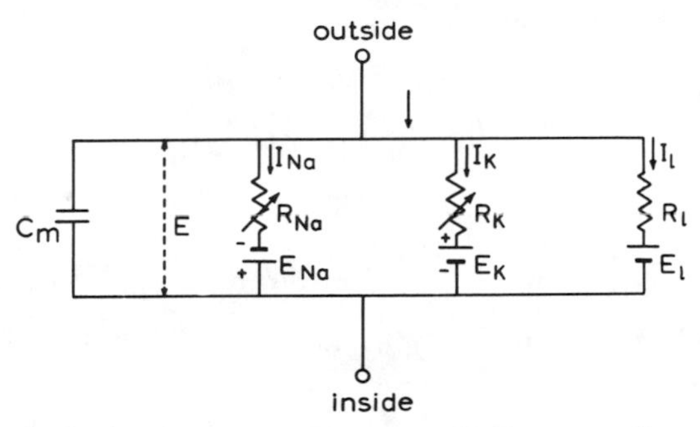

Fig. 143. Electrical circuit representing a unit segment of nerve membrane.

ference between the *membrane* potential (E) and the equilibrium potential for the sodium ion (E_{Na}) multiplied by the sodium conductance (g_{Na}). Similarly for I_K and I_l. The total membrane current is given (R and C in parallel) by

$$I = C_m \frac{dV}{dt} + I_i \tag{1}$$

where I is the total current (inward current positive), I_i is the (total) *ionic* current (inward current positive), V represents the displacement of the membrane potential from its resting value (depolarization negative). C_m is the membrane capacity per unit area. The ionic current, I_i, is given by

$$I_i = I_{Na} + I_K + I_l \tag{2}$$

with

$$I_{Na} = g_{Na} (E - E_{Na})$$

$$I_K = g_K (E - E_K)$$

$$I_l = \bar{g}_l (E - E_l)$$

where E_{Na} and E_K are the equilibrium potentials for the sodium and potassium ions, and E_l is the potential at which the leakage current,

I_l, is zero. Strictly, the bar in the notation $\bar{g}_l$ is redundant, but is retained to emphasize the constancy of g_l.

$$I_{\mathrm{Na}} = g_{\mathrm{Na}}(V - V_{\mathrm{Na}}) \tag{3}$$

$$I_{\mathrm{K}} = g_{\mathrm{K}}(V - V_{\mathrm{K}}) \tag{4}$$

$$I_l = \bar{g}_l(V - V_l) \tag{5}$$

where

$$V = E - E_r$$

$$V_{\mathrm{Na}} = E_{\mathrm{Na}} - E_r, \text{ etc.}$$

E_r is the absolute value of the resting potential.

We must focus attention on the conductances g_{K} and g_{Na}. These parameters are, as mentioned, both time dependent and membrane potential dependent. For illustration we shall consider the case of g_{K} in some detail. It is found experimentally that the "time course" of g_{K} is asymmetrical; the on-transient has an early point of inflexion, and the off-transient exponential decay, which is fast relative to the on-transient. These and other properties of g_{K} are well described by the empirical equations

$$g_{\mathrm{K}} = \bar{g}_{\mathrm{K}} n^4 \tag{6}$$

$$\frac{\mathrm{d}n}{\mathrm{d}t} = \alpha_n(1 - n) - \beta_n n \tag{7}$$

where $\bar{g}_{\mathrm{K}}$ is a constant (dimensions of conductance/cm^2), α_n and β_n are constants *that depend on voltage*, and n is a dimensionless variable (varying between 0 and 1).

As (7) implies, at rest, i.e. when $V = 0$, n has the resting value

$$n_0 = \frac{\alpha_{n_0}}{\alpha_{n_0} \mp \beta_{n_0}}. \tag{8}$$

If V is changed suddenly (step change) α_n and β_n instantly take up new steady state values. Equation (7) may be written as

$$\frac{1}{\alpha_n + \beta_n} \frac{dn}{dt} + n = \frac{\alpha_n}{\alpha_n + \beta_n},$$

from which it follows that

$$n_\infty = \frac{\alpha_n}{\alpha_n + \beta_n} \tag{9}$$

$$\tau_n = \frac{1}{\alpha_n + \beta_n}. \tag{10}$$

It follows that

$$\alpha_n = \frac{n_\infty}{\tau_n} \tag{11}$$

$$\beta_n = \frac{1 - n_\infty}{\tau_n}. \tag{12}$$

With the aid of relations (11) and (12) and on the basis of some justifiable simplifying assumptions Huxley and Hodgkin obtained, by a procedure which for lack of space will not be elaborated here, values of α_n and β_n for given levels of depolarization. By plotting of the values thus obtained of the rate constants α_n and β_n against the corresponding values V they obtained the *empirical* relationship

$$\alpha_n = \frac{0.01\,(V+10)}{e^{(V+10)/10} - 1} \tag{13}$$

$$\beta_n = 0.125\, e^{V/80}. \tag{14}$$

The sodium conductance, g_{Na}, was empirically "decomposed" along similar lines. It was assumed that g_{Na} is determined (not by one but) by two variables, each of which obeys a first order equation

$$g_{Na} = m^3 h \bar{g}_{Na} \tag{15}$$

$$\frac{dm}{dt} = \alpha_m (1 - m) - \beta_m m \tag{16}$$

$$\frac{dh}{dt} = \alpha_h (1-h) - \beta_h h \tag{17}$$

where $\bar{g}_{Na}$ is a constant and where the rate constants, α and β, are functions of V, not of t. Empirically it was found that

$$\alpha_m = \frac{0.1\,(V+25)}{e^{(V+25)/10}-1} \tag{18}$$

$$\beta_m = 4e^{V/18} \tag{19}$$

$$\alpha_h = 0.07\,e^{V/20} \tag{20}$$

$$\beta_h = \frac{1}{e^{(V+30)/10}+1}\,. \tag{21}$$

When a "sufficiently strong" local stimulus is applied to a nerve fibre a travelling "action potential" is elicited which moves with constant velocity in both directions. For this process the local circuits have to be provided by the net membrane current, a fact leading to the well-known relation

$$i = \frac{1}{r_1 + r_2}\frac{\partial^2 V}{\partial x^2} \tag{22}$$

where i is the membrane current per unit length, r_1 and r_2 are the external and internal resistances per unit length and x is the distance along the fibre.

The derivation of equation (22) involves merely Ohm's law, and differentiation, but will not be given here. Suffice it to say that it is a standard formula derived in any textbook dealing with electrical "cable theory". In most experimental situations r_1 is negligible compared to r_2 (axon surrounded by a large fluid volume), or

$$i = \frac{1}{r_2}\frac{\partial^2 V}{\partial x^2}\,.$$

Or, in terms of membrane current density, I,

$$I = \frac{a}{2R_2}\frac{\partial^2 V}{\partial x^2} \tag{23}$$

where a is the radius of the fibre and R_2 is the specific resistance of the axoplasm. Inserting expression (23) in eq. (1) and using the empirical relation for the components of I_i (see eqs. (2), (3), (4), (5), (6) and (15),

$$\frac{a}{2R_2}\frac{\partial^2 V}{\partial x^2} = C_m\frac{\partial V}{\partial t} + \bar{g}_K n^4 (V - V_K) + \bar{g}_{Na} m^3 h (V - V_{Na}) + \bar{g}_l (V - V_l). \tag{24}$$

During the steady state propagation the curve of V versus time at any location is similar to that of V against distance at any instant, or

$$\frac{\partial^2 V}{\partial x^2} = \frac{1}{\theta^2}\frac{\partial^2 V}{\partial t^2} \tag{25}$$

which is the wave equation (see Chapter 9, §21), and here θ is the velocity of conduction.

It may be seen that for dimensional reasons θ^2 must have the dimensions $(\mathrm{cm/sec})^2$, i.e. θ those of cm/sec, a velocity. Substituting eq. (25) in eq. (24) and dropping the partial d's (since only time is the independent variable) we finally obtain:

$$\frac{a}{2R_2\theta^2}\frac{d^2 V}{dt^2} = C_m\frac{dV}{dt} + \bar{g}_K n^4 (V - V_K) + \bar{g}_{Na} m^3 h (V - V_{Na}) + \bar{g}_l (V - V_l). \tag{26}$$

The numerical solution of this ordinary differential equation must be obtained by a trial and error procedure in which the (unknown) value of θ is inserted by guessing. It can be shown that only the correct value of θ brings V back to zero – the resting condition – when the action potential is over.

Of course the model only makes sense if the "correct" value of θ thus selected corresponds reasonably well with the velocity of conduction actually measured in the experiment. In one case the theoretically derived "true" value of θ was found to be 18.8 m/sec as against a measured value of 21.2 m/sec, a very good correspondence.

The main features of the actual process associated with the propagated action potential were reproduced by solving eq. (26). First (of course) the time course of V was found to correspond closely to that found experimentally. Further it was found that a) the main rise in conductance starts later than the rise of potential, b) the conductance returns to its resting value late in the so-called positive phase (undershoot of V).

For further details the reader is referred to the original papers[1–7].

8. On the Spread of the Action Current During Saltatory Conduction

The fundamental studies of Tasaki, Katz, Stämpfli, Huxley, von Muralt and others* gave rise to the so-called "theory of saltatory conduction" to explain the passage of the nerve impulse in myelinated nerve fibres. In this theory the so-called nodes of Ranvier play a predominant rôle.

The myelinated nerve fibre may be looked upon as a cable-like structure, and consists of a central conducting "wire", the axon, covered by an isolating membrane, the myelin sheath; however, over

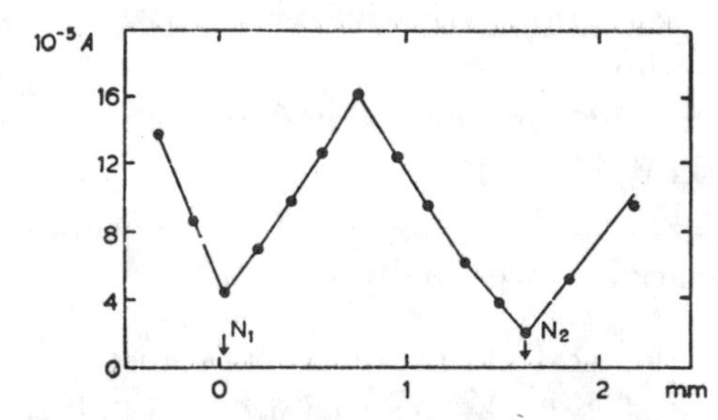

Fig. 144: Threshold strength of stimulating current plotted against distance along nerve fibre. At the nodes of Ranvier (N_1 and N_2) the threshold possesses a minimal value. (Modified from I. Tasaki, Nervous Transmission, 1953.)

short lengths the myelin sheath is broken up: these "openings" where the isolating material is lacking, are called the nodes of Ranvier. The result shown in Fig. 144 supports the view that the myelin sheath acts as an insulator and that current only passes through the nodes of Ranvier.

Now the theory of Saltatory Conduction states that the nervous impulse does not travel in a continuous manner, but "jumps from node to node", in the following manner: a node of Ranvier may be excited by, say, an external current. When this stimulating current reaches "threshold-value" the node is "transformed" into a "battery" and generates an action current. This action current spreads in both directions along the axon cylinder and this action current reaching the next node and passing outwards through this node stimulates the next node into "action", i.e. turns the next node into a "battery". This process is repeated all along the nerve, until the "nervous impulse" reaches the end-plate (in a motor nerve). Now, as a first

* See references 1 to 9 and 11, page 629.

approximation, ignoring capacity and leakage effects of the membrane, the following electrical model (after Tasaki) may be used to study the spread of the action current along the nerve. We shall thus confine our attention to a single node acting as a battery and the spread of the generated current. (See also Chapter 28 of Rashevsky's book.[10])

Let us assume that the nerve fibre is surrounded by a medium of negligible electrical resistance (e.g. a fibre lying in Ringer's solution).

Let us further assume that the membrane between the nodes acts as a perfect insulator. Let the electrical resistance of the nodes all be the same and be denoted by r, and let the internodel resistance of the axon sections be R. Further let the resistance of the neuro-muscular junction be r_e.

Then we have the following model (Fig. 145):

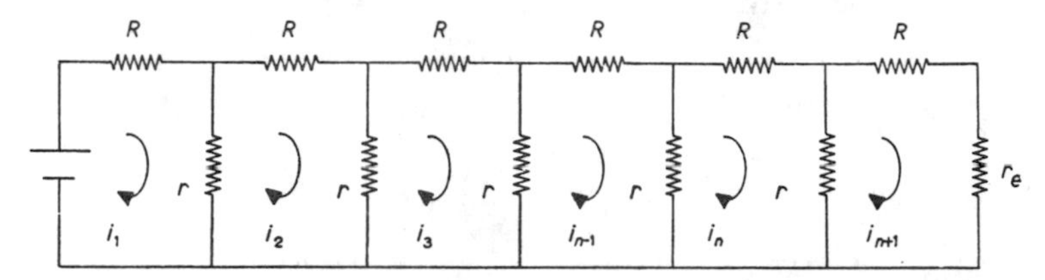

Fig. 145. Simplified electrical model of myelinated nerve fibre. (After Tasaki.)

Now from elementary circuit theory we know that we may think in terms of currents circulating around each loop, e.g. the current i_2 circulates around the second loop and thus traverses the two resistances r and the axon resistance R belonging to the second loop.

We must also make use of Kirchhoff's second law which states that the algebraic sum of the electromotive forces around a closed circuit is zero.

(Kirchhoff's law and the concept of mesh-currents can be fully grasped by a 10 minutes' study of any elementary textbook of electricity.)

Kirchhoff's second law demands that the sum of the voltage drops across all three resistances in a single loop is zero. Let us study the nth loop. The voltage drop across R is clearly $i_n R$. The voltage drop across the resistance r on the right of the loop is not simply $i_n r$, but $(i_n - i_{n+1})r$, since the net current flowing through this r is the sum of the "circular" currents i_n and i_{n+1}. But the current i_{n+1} must be given the negative sign here, since while flowing through this particular

r its direction is *opposed* to that of i_n. Hence the pressure drop here is $(i_n - i_{n+1})r$. Similarly the pressure drop across r on the left of the nth loop is $(i_n - i_{n-1})$. Hence by Kirchhoff

$$i_n R + (i_n - i_{n+1})r + (i_n - i_{n-1})r = 0 \tag{1}$$

or

$$i_n R + i_n r - i_{n+1} r + i_n r - i_{n-1} r = 0 \tag{2}$$

or

$$i_{n+1} - \left(2 + \frac{R}{r}\right) i_n + i_{n-1} = 0 \tag{3}$$

which is a homogeneous difference equation of second order with constant coefficients.

If we assume by analogy with the theory of differential equations a solution of the form

$$i_n = A\mathrm{e}^{\sigma n} \tag{4}$$

where A is an arbitrary constant, we obtain

$$A\mathrm{e}^{\sigma n}\left\{\mathrm{e}^{\sigma} - \left(2 + \frac{R}{r}\right) + \mathrm{e}^{-\sigma}\right\} = 0. \tag{5}$$

This requires for a non-trivial solution

$$2 + \frac{R}{r} = \mathrm{e}^{\sigma} + \mathrm{e}^{-\sigma} \tag{6}$$

or

$$1 + \frac{R}{2r} = \cosh \sigma. \tag{7}$$

Since $\cosh \sigma$ is an even function, this equation determines two values of σ, so that we must also have

$$\cosh(-\sigma) = 1 + \frac{R}{2r} = \cosh \sigma \tag{8}$$

(analogous to $\cos(-x) = \cos x$).

It follows that $A_1\mathrm{e}^{\sigma n}$ and $A_2\mathrm{e}^{-\sigma n}$ are solutions of (3).

The general solution may conveniently be written in terms of hyperbolic functions, so that we may write

$$i_n = M \sinh(\sigma n) + N \cosh(\sigma n) \tag{9}$$

where M and N are arbitrary constants.

With the aid of equation (9) the value of r, i.e. the resistance of the nodes of Ranvier, may be computed. This will not be discussed here. Suffice it to say, that Tasaki found the value for r to be approximately 21 megohms.

The above treatment merely serves to sketch the application of difference equations to neurophysiological problems.

The above example is given to illustrate the *use* of difference equations. The theory of linear difference equations will not be given here*. Suffice it to say, that the solution of linear difference equations with constant coefficients is very similar to that of linear differential equations with constant coefficients, as may be seen from the above.

9. Differential Equations in the Theory of Epidemics

In this section we shall discuss some of the simpler equations occurring in the mathematical theory of epidemics.[1,2]

(a) Simple epidemic with no removal

In this model it is assumed that infection spreads between members of the community, but there is no removal from circulation by death, recovery or isolation.

Suppose that we consider a community of susceptibles into which a single infective is introduced. We shall denote by x and y the numbers of susceptibles and infectives, respectively, at any time t, so that $x + y = n + 1$, the total population size. We suppose that the whole population is subject to some process of homogeneous mixing, so that the number of new infections occurring in time $\mathrm{d}t$ is $\beta xy\mathrm{d}t$, where β is the infection rate. Then

$$\frac{\mathrm{d}x}{\mathrm{d}t} = -\beta xy.$$

Substituting $y = n + 1 - x$ into this equation, we find that the time-variation of x is given by the differential equation

$$\frac{\mathrm{d}x}{\mathrm{d}t} = -\beta x(n + 1 - x),$$

with initial condition $x = n$ at time $t = 0$. If we write this equation in the form

* A clear and concise treatment of the theory of finite differences and linear difference equations is given by Pipes (L. A. Pipes, Applied Mathematics for Engineers and Physicists, McGraw-Hill, New York, 1958).

$$\frac{\mathrm{d}x}{x(n-x+1)} = -\beta \mathrm{d}t,$$

we see that the variables are separable so that the equation can be integrated immediately. Writing the left-hand side of the equation in the partial fraction form

$$\frac{1}{n+1}\left(\frac{\mathrm{d}x}{x} + \frac{\mathrm{d}x}{n-x+1}\right)$$

and multiplying both sides by $n + 1$, we see that the general solution of the equation is

$$\log x - \log(n+1-x) = -(n+1)\beta t + C,$$

where C denotes an arbitrary constant. If $x = n$ when $t = 0$, then $C = \log n$ and we find that

$$\log x - \log(n+1-x) - \log n = \mathrm{e}^{-t/\tau},$$

where

$$\tau = \frac{1}{(n+1)\beta}. \tag{1}$$

This relation is equivalent to

$$\frac{x}{n(n+1-x)} = \mathrm{e}^{-t/\tau}$$

which may be solved for x to give the solution

$$x = \frac{n(n+1)}{n+\mathrm{e}^{t/\tau}}, \tag{2}$$

where τ is defined by equation (1).

(b) Epidemic with both infection and removal

We now consider a community of total size n, comprising at time t, x susceptibles, y infectives in circulation, and z individuals who are isolated, dead or recovered and immune. Thus $x + y + z = n$. If β and γ are the infection- and removal-rate respectively, then we have the three simultaneous equations

$$\frac{\mathrm{d}x}{\mathrm{d}t} = -\beta xy, \qquad \frac{\mathrm{d}y}{\mathrm{d}t} = \beta xy - \gamma y, \qquad \frac{\mathrm{d}z}{\mathrm{d}t} = \gamma y.$$

If β and γ are constants, the first and third of these equations lead to the simple equation

$$\frac{dx}{dz} = -\frac{x}{\rho},$$

where $\rho = \gamma/\beta$ is the *relative removal rate*. The solution of this equation is elementary. If $x = x_0$ when $z = 0$, then

$$x = x_0 e^{-z/\rho} \tag{3}$$

and so

$$y = n - x - z = n - x_0 e^{-z/\rho} - z,$$

from which it follows that we can write the third equation of our set as

$$\frac{dz}{dt} = \gamma(n - x_0 e^{-z/\rho} - z).$$

If we write this equation as

$$\frac{dz}{n - x_0 e^{-z/\rho} - z} = \gamma dt \tag{4}$$

we see that the variables are separable. The integrand on the left-hand side of this equation would prove awkward to integrate as it stands. If, however, z/ρ is small we may write

$$e^{-z/\rho} = 1 - \frac{z}{\rho} + \frac{z^2}{2\rho^2},$$

so that

$$n - x_0 e^{-z/\rho} - z = n - x_0 + \left(\frac{x_0}{\rho} - 1\right) z - \frac{x_0}{2\rho^2} z^2.$$

Now, when $t = 0$, $x = x_0$, $y = y_0$, say, and $z = 0$, so that $n - x_0 = y_0$, and a little algebraic manipulation shows that

$$n - x_0 e^{-z/\rho} - z = \frac{x_0}{2\rho^2}\left\{\left(\frac{\rho^2\alpha}{x_0}\right)^2 - \zeta^2\right\},$$

where

$$\zeta = z - \frac{\rho^2}{x_0}\left(\frac{x_0}{\rho} - 1\right), \qquad \alpha^2 = \left(\frac{x_0}{\rho} - 1\right)^2 + \frac{2x_0 y_0}{\alpha^2}. \tag{5}$$

Since $d\zeta = dz$, we see that equation (4) can be put into the form

$$\frac{\dfrac{2\rho^2}{x_0}\mathrm{d}\zeta}{\left(\dfrac{\rho^2\alpha}{x_0}\right)^2 - \zeta^2} = \gamma\mathrm{d}t.$$

If we now integrate both sides of this equation and make use of the integral occurring in entry 9 of Table 11 (p. 273 above), we obtain the relation

$$\tanh^{-1}\left(\frac{x_0\zeta}{\rho^2\alpha}\right) = \tfrac{1}{2}\alpha\gamma t + C,$$

where C denotes a constant of integration. In terms of the original variable z, this becomes

$$\tanh^{-1}\left\{\frac{x_0}{\rho^2\alpha}\left[z - \frac{\rho^2}{x_0}\left(\frac{x_0}{\rho} - 1\right)\right]\right\} = \tfrac{1}{2}\alpha\gamma t + C.$$

Using the fact that $z = 0$ when $t = 0$, we find that $C = -\phi$, where

$$\phi = \tanh^{-1}\left\{\frac{1}{\alpha}\left(\frac{x_0}{\rho} - 1\right)\right\}. \tag{6}$$

Inserting this value for the constant C in the above expression for the solution, we obtain finally

$$z = \frac{\rho^2}{x_0}\left\{\frac{x_0}{\rho} - 1 + \alpha\tanh\left(\tfrac{1}{2}\alpha\gamma t - \phi\right)\right\}, \tag{7}$$

where α and ϕ are given by equations (5) and (6) respectively. The complete solution of the problem is therefore given by equations (7) and (3), the expression for y being determined by the fact that $y = n - x - z$.

(c) The simplest continuous-infection model

In this model we consider a population of n individuals made up, at time t, of x susceptibles, y infectives in circulation and z individuals who are isolated, dead, or recovered and immune. We assume that the infection- and removal-rates are β and γ respectively, so that there are $\beta xy\mathrm{d}t$ new infections and $\gamma y\mathrm{d}t$ removals in a time interval $\mathrm{d}t$. Furthermore, we introduce a birthrate parameter μ, so as to give $\mu\mathrm{d}t$ new susceptibles in time $\mathrm{d}t$. We therefore have the equations

$$\frac{dx}{dt} = -\beta xy + \mu, \qquad \frac{dy}{dt} = \beta xy - \gamma y. \tag{8}$$

Equilibrium values of x and y are reached when they yield zero values to dx/dt and dy/dt. If we denote these equilibrium values by x_0 and y_0, we obtain the relations

$$0 = -\beta x_0 y_0 + \mu, \qquad 0 = \beta x_0 y_0 - \gamma y_0,$$

from which it follows that

$$x_0 = \gamma/\beta, \qquad y_0 = \mu/\gamma. \tag{9}$$

We shall now consider small departures from these equilibrium values, i.e. we shall take

$$x = x_0(1 + u), \qquad y = y_0(1 + v), \tag{10}$$

where u and v are so small that we may write

$$xy = x_0 y_0(1 + u + v),$$

(i.e. may neglect uv in comparison with $u + v$). If we substitute from equations (10) into equations (8) and write

$$\sigma = \gamma/\beta\mu, \qquad \tau = 1/\gamma,$$

we find that the small deviations u and v satisfy the simultaneous linear equations

$$\sigma \frac{du}{dt} = -u - v, \qquad \tau \frac{dv}{dt} = u.$$

If we introduce the operator $D = d/dt$ we see that these equations may be written in the form

$$(\sigma D + 1)u + v = 0, \qquad u = \tau D v.$$

Eliminating u from these two equations we find that v satisfies the equation

$$(\sigma\tau D^2 + \tau D + 1)v = 0,$$

which is readily shown to have solution

$$v = v_0 e^{-t/2\sigma} \cos \omega t, \qquad \omega^2 = \frac{1}{\sigma\tau} - \frac{1}{4\sigma^2}, \tag{11}$$

for a suitably chosen origin of time.
Now

$$u = \tau D v = -\frac{v_0 \tau}{2\sigma} e^{-t/2\sigma} \{\cos \omega t + 2\omega\sigma \sin \omega t\},$$

so that, if we introduce an auxiliary angle δ, defined by the equations

$$\cos\delta = (1 + 4\omega^2\sigma^2)^{-\frac{1}{2}}, \qquad \sin\delta = 2\omega\sigma(1 + 4\omega^2\sigma^2)^{-\frac{1}{2}},$$

that is

$$\delta = \tan^{-1} 2\omega\sigma,$$

we find that

$$u = -\frac{v_0\tau}{2\sigma}(1 + 4\omega^2\sigma^2)^{\frac{1}{2}}\cos(\omega t - \delta)e^{-t/2\sigma}.$$

Now

$$1 + 4\omega^2\sigma^2 = 1 + \frac{4\sigma}{\tau} - 1 = \frac{4\sigma}{\tau},$$

so that, finally,

$$u = -v_0\left(\frac{\tau}{\sigma}\right)^{\frac{1}{2}} e^{-t/2\sigma}\cos(\omega t - \delta). \tag{12}$$

For further reading see references 3 and 4, page 629.

10. Bacteriology: Birth and Death Processes

First order partial differential equations arise in the theory of birth and death processes connected with bacteria. Suppose that, at time t, there are n live bacteria and that

(i) the probability of a bacterium reproducing during the time interval between t and $t + \delta t$ is $\lambda_n \delta t$;
(ii) the probability of a bacterium dying during the same interval is $\mu_n \delta t$;
(iii) the probability of the number of bacteria remaining constant in the same interval is $1 - \lambda_n\delta t - \mu_n\delta t$;
(iv) the probability of more than one birth or death occurring in a time δt is zero.

If we assume that $p_n(t)$ is the probability of there being n bacteria at time t, these assumptions lead to the equation

$$p_n(t + \delta t) = \lambda_{n-1}p_{n-1}(t)\delta t + \mu_{n+1}p_{n+1}(t)\delta t + [1 - \lambda_n\delta t - \mu_n\delta t]p_n(t)$$

which can be written in the form

$$\frac{p_n(t + \delta t) - p_n(t)}{\delta t} = \lambda_{n-1}p_{n-1}(t) + \mu_{n+1}p_{n+1}(t) - (\lambda_n + \mu_n)p_n(t).$$

Letting δt tend to zero we obtain the equation

$$\frac{\mathrm{d}p_n}{\mathrm{d}t} = \lambda_{n-1}p_{n-1}(t) + \mu_{n+1}p_{n+1}(t) - (\lambda_n + \mu_n)p_n(t). \tag{1}$$

In the general theory of such processes the parameters λ_n and μ_n depend on both n and t. However, to illustrate the manner in which the theory of first order partial differential equations enters into the discussion of these processes we consider a special case; we shall assume that both λ_n and μ_n are directly proportional to the number n of bacteria present. We shall take

$$\lambda_n = \lambda n, \qquad \mu_n = \mu n \tag{2}$$

where λ and μ are constants. Substituting from equations (2) into equation (1) we obtain the equation

$$\frac{\mathrm{d}p_n}{\mathrm{d}t} = \lambda(n-1)p_{n-1}(t) - (\lambda + \mu)np_n(t) + \mu(n+1)p_{n+1}(t). \tag{3}$$

We now introduce a generating function $f(x, t)$ defined by the equation

$$f(x, t) = \sum_{n=0}^{\infty} p_n(t)x^n. \tag{4}$$

If we multiply both sides of equation (3) by x^n and sum over n from 0 to ∞ we find that

$$\sum_{n=0}^{\infty} \frac{\mathrm{d}p_n}{\mathrm{d}t} x^n = \lambda x^2 \sum_{n=1}^{\infty} (n-1)p_{n-1}(t)x^{n-2}$$
$$- (\lambda + \mu)x \sum_{n=1}^{\infty} np_n(t)x^{n-1} + \mu \sum_{n=0}^{\infty} (n+1)p_{n+1}x^n. \tag{5}$$

Now from equation (4)

$$\frac{\partial f}{\partial t} = \sum_{n=0}^{\infty} \frac{\mathrm{d}p_n}{\mathrm{d}t} x^n, \tag{6}$$

$$\frac{\partial f}{\partial x} = \sum_{n=1}^{\infty} np_n x^{n-1}. \tag{7}$$

We can re-write the series on the right to give

$$\frac{\partial f}{\partial x} = \sum_{n=0}^{\infty} (n+1)p_{n+1}(t)x^n. \tag{8}$$

Similarly we can write

$$\frac{\partial f}{\partial x} = \sum_{n=1}^{\infty} (n-1)p_{n-1}(t)x^{n-2}. \tag{9}$$

Substituting from equations (6) to (9) into equation (5) we have

$$\frac{\partial f}{\partial t} = [\lambda x^2 - (\lambda + \mu)x + \mu]\frac{\partial f}{\partial x}.$$

Now

$$\lambda x^2 - (\lambda + \mu)x + \mu = (\lambda x - \mu)(x - 1)$$

so that the partial differential equation determining the generating function $f(x, t)$ is

$$\frac{\partial f}{\partial t} = (\lambda x - \mu)(x - 1)\frac{\partial f}{\partial x}. \tag{10}$$

We saw in example 35 of the last chapter (p. 513 above) that the general solution of this equation is

$$f = F(\xi), \qquad \xi = \frac{\lambda x - \mu}{x - 1} e^{-(\lambda-\mu)t} \tag{11}$$

where the function F is arbitrary.

If, at time $t = 0$, there are m bacteria present, then

$$p_m(0) = 1, \ p_0(0) = p_1(0) = \ldots = p_{m-1}(0) = p_{m+1}(0) = \ldots = 0,$$

so that at time $t = 0$

$$f(x, 0) = x^m.$$

But

$$f(x, 0) = F\left(\frac{\lambda x - \mu}{x - 1}\right),$$

so that the function F must be chosen so that

$$F\left(\frac{\lambda x - \mu}{x - 1}\right) = x^m. \tag{12}$$

If we put

$$\frac{\lambda x - \mu}{x - 1} = \xi \tag{13}$$

then

$$\lambda x - \mu = \xi(x - 1),$$

and so

$$x = \frac{\mu - \xi}{\lambda - \xi}. \tag{14}$$

Substituting from equation (13) and (14) into equation (12) we find that $F(\xi)$ is given by the equation

$$F(\xi) = \left(\frac{\mu - \xi}{\lambda - \xi}\right)^m.$$

Now from equation (13)

$$\frac{\mu - \xi}{\lambda - \xi} = \frac{x[\mu - \lambda e^{-(\lambda-\mu)t}] - \mu[1 - e^{-(\lambda-\mu)t}]}{\lambda x[1 - e^{-(\lambda-\mu)t}] - [\lambda - \mu e^{-(\lambda-\mu)t}]}$$

so that the solution we require is

$$f(x, t) = \left\{\frac{x[\mu - \lambda e^{-(\lambda-\mu)t}] - \mu[1 - e^{-(\lambda-\mu)t}]}{\lambda x[1 - e^{-(\lambda-\mu)t}] - [\lambda - \mu e^{-(\lambda-\mu)t}]}\right\}^m.$$

Writing this equation in the form

$$f(x, t) = \left\{\frac{\mu[1 - e^{+(\lambda-\mu)t}] - x[\lambda - \mu e^{+(\lambda-\mu)t}]}{\mu - \lambda e^{+(\lambda-\mu)t} - \lambda x[1 - e^{+(\lambda-\mu)t}]}\right\}^m$$

we see that if $\lambda < \mu$, $e^{(\lambda-\mu)t} \to 0$ as $t \to \infty$ and

$$f \to 1.$$

Ultimately, therefore $p_0 = 1$, $p_1 = p_2 = \ldots = p_n = \ldots = 0$. Since p_0 is the probability that no bacteria remain, it follows that, if $\lambda < \mu$, the ultimate extinction of the bacterial colony is certain.

11. Diffusion through Membranes

The problem of determining the concentration of a substance diffusing against a velocity arises in many biological problems.[1] We shall discuss it briefly here.

A net flow in a non-homogeneous solution caused by a pressure difference produces a change in the net rate of diffusion of a solute. If we have a concentration C of a substance which diffuses against a velocity U, then the variation of C with t and linear variable x is given by the partial differential equation

$$D\frac{\partial^2 C}{\partial x^2} - U\frac{\partial C}{\partial x} = \frac{\partial C}{\partial t} \tag{1}$$

where D is the diffusion constant. In a steady state $\partial C/\partial t = 0$ and the equation becomes

$$D\frac{\mathrm{d}C}{\mathrm{d}x} - UC = f \tag{2}$$

where f is the flux per unit area.

In equation (2) it is assumed that f is a constant, but that its numerical value is unknown. We wish to calculate f from the fact that $C = C_0$ at $x = 0$ and $C = C_\delta$ at $x = \delta$.

The equation (2) may be written in the form

$$\frac{\mathrm{d}C}{\mathrm{d}x} - \frac{U}{D}C = \frac{f}{D}$$

from which it is obvious that it has an integrating factor $\mathrm{e}^{-xU/D}$. Writing it in the form

$$\frac{\mathrm{d}}{\mathrm{d}x}(C\mathrm{e}^{-Ux/D}) = \frac{f}{D}\mathrm{e}^{-Ux/D}$$

and then integrating with respect to x we find that

$$C\mathrm{e}^{-Ux/D} = -\frac{f}{U}\mathrm{e}^{-Ux/D} + A,$$

where A is a constant of integration.

Using the fact that $C = C_0$ when $x = 0$ we obtain the relation

$$C_0 = -\frac{f}{U} + A$$

for the determination of A. We have

$$A = \frac{f}{U} + C_0$$

so that

$$C\mathrm{e}^{-Ux/D} = C_0 + \frac{f}{U}(1 - \mathrm{e}^{-Ux/D}).$$

Now using the fact that $C = C_\delta$ when $x = \delta$ we have the equation

$$C_\delta \mathrm{e}^{-U\delta/D} = C_0 + \frac{f}{U}(1 - \mathrm{e}^{-U\delta/D})$$

which may be solved for f to give

$$f = U \frac{C_\delta e^{-U\delta/D} - C_0}{e^{U\delta/D} - 1}. \tag{3}$$

In the special case in which $C_0 = 0$ this reduces to

$$f = \frac{UC_\delta}{e^{U\delta/D} - 1}. \tag{4}$$

This equation was first used by Jacobs[2].

Equation (4) applies to dissolved substances and to diffusion of water (H_2O or D_2O) in water. In the latter instance it is more convenient to work in volumes instead of concentrations. If f is in cc/cm^2 and C is in cc/cc, UA will be the net flow rate F and the absolute flux against the flow in cc/unit time will be fA which is equal to

$$\frac{FC_\delta}{e^{F\delta/AD} - 1}.$$

This expression has been used by Garby[3] in a discussion of the diffusion of heavy water against a flow through a porous membrane.

For further reading see references 4 and 5, page 629.

12. The Minimal Number of Quanta Required to Produce a Sensation of Light

AN APPLICATION OF THE POISSON DISTRIBUTION TO THRESHOLD DETERMINATION

The Poisson distribution has been used by Hecht, Schlaer, Pirenne and others to determine the minimal number of quanta absorbed by the rhodopsin molecules at threshold.[1-7]

The meaning of the Poisson distribution will be illustrated by considering the statistical behaviour of radioactive decay processes. The interval of measurement is chosen small in relation to the half-life of the radioactive isotope.

The radioactive processes of disintegration are mutually independent and randomly distributed in time.

When the same preparation is measured at different occasions under the same conditions, different values are obtained. These differences are due to the random nature of the disintegration processes.

An example will now be given to illustrate the above.

A preparation of constant strength (long half-value period) is measured 364 times, each single measurement is a count over 10 seconds.

TABLE 21

n	Observed frequency	P_n	$P_n = m^n e^{-m}/n!$ (Poisson)
0	5	0.014	0.0129
1	28	0.077	0.0561
2	42	0.116	0.1220
3	65	0.179	0.1770
4	67	0.184	0.1920
5	64	0.176	0.1670
6	42	0.116	0.1210
7	28	0.077	0.0750
8	10	0.028	0.0410
9	7	0.019	0.0200
10	5	0.014	0.0086
11	1	0.003	0.0034
12	0	0	0.0012
13	0	0	0.0004

From Hoppe-Seyler/Thierfelder, Handbuch der Physiologisch- und Pathologisch-Chemischen Analyse, Springer, 1955, p. 682.

Table 21 shows how many counts of 0, 1, 2, 3, . . . disintegrations were recorded; e.g. 5 desintegrations in the 10 second period were recorded 64 times. The graph of this frequency distribution is shown in Fig. 146. A large number of measurements (364) gave a mean value of 4.35 impulses per 10 seconds.

The asymmetrical form of the graph of Fig. 146 should be noted. When the period of measurement is extended to 60 seconds, the frequency distribution became much more symmetrical and approximated to a normal distribution, as may be seen from Fig. 147.

The frequency distribution discussed above is called the *Poisson distribution* and arises e.g. when the number of random events (disintegrations) during a single measurement is small. The Poisson distribution is given by

$$P_n = \frac{m^n}{n!} e^{-m}$$

where, in this particular case, P_n is the probability that when the mean number of impulses is m impulses/interval of measurement, the number of recorded impulses/interval is n.

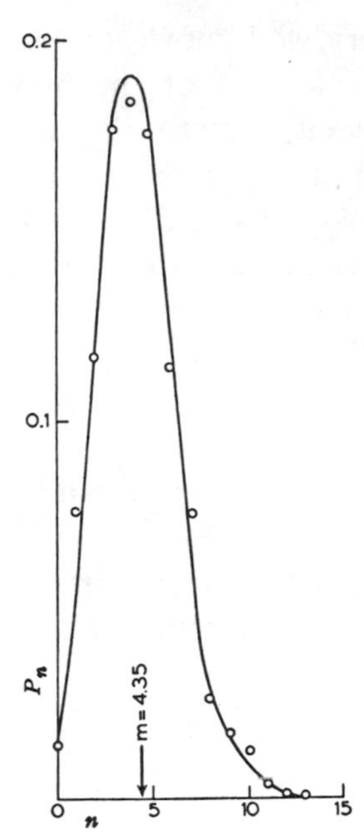

Fig. **146**. Poisson distribution, mean value = **4.35** impulses/single measurement. Dots denote experimental points. [From Hoppe-Seyler/Thierfelder, Handbuch der Physiologisch- und Pathologisch-Chemischen Analyse, **2** (1955) page 683.]

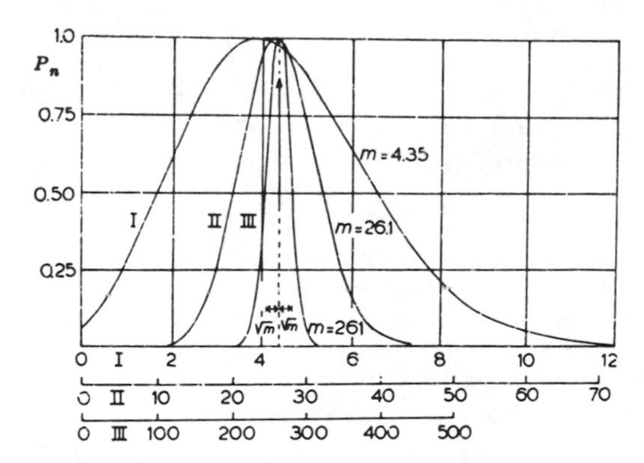

Fig. **147**. Poisson distributions. I True mean value = **4.35** impulses/single measurement. II True mean value = 26.1 impulses/single measurement. III True mean value = 261 impulses/single measurement. In the figure the mean values are made to coincide. Note the separate abscissa scales for I, II and III. [From Hoppe-Seyler/Thierfelder, Handb. der Physiologisch- und Pathologisch-Chemischen Analyse, **2** (1955) 683.]

This distribution is very well known and is usually introduced as a "law of rare events", as a limit of the binomial distribution when the probability of the event occurring in any one trial is small. The preceding example shows that the events need not necessarily be rare and that they can arise under quite different physical conditions from the binomial. We shall now show how independent events occurring at random, such as radioactive particles reaching a Geiger counter, can have a Poisson frequency distribution. The derivation given here is rarely found in textbooks. It was probably first given by Borel.

We can consider events occurring at random, but statistically uniformly, in time.

Let the mean number occurring in time t be λt; λ is a constant. Then the probability that one occurs in the interval between t and $t + \mathrm{d}t$ is $\lambda \mathrm{d}t$. It is independent of t.

0 |———————————| t |—— $\mathrm{d}t$ ——| $t + \mathrm{d}t$.

The probability that more than one occurs is of the 2nd order of smallness and so is negligible.

First we find the probability P_0 that no event occurs between 0 and t. To do this we let t increase to $t + \mathrm{d}t$, so that P_0 then increases to $P_0 + \mathrm{d}P_0$.

The events are independent. Therefore the probability that none occur between 0 and $t + \mathrm{d}t$ is the probability P_0 that none occur between 0 and t, multiplied by the probability that none occur between t and $t + \mathrm{d}t$.

The following two rules are employed here:

a) when the probability of an event occurring is P, the probability of its not occurring is $1 - P$, where P is a number such that $0 \leqq P \leqq 1$.

b) when two events a and b are independent and have the probability P_a and P_b respectively, the probability of their joint occurrence is $P_a \times P_b$; e.g. the probability of throwing a four with a dice is $\frac{1}{6}$, while the probability of throwing a five is equally $\frac{1}{6}$; the probability of throwing a four followed by a five is $\frac{1}{36}$.

Hence

$$P_0 + \mathrm{d}P_0 = P_0(1 - \lambda \mathrm{d}t) \tag{1}$$

so $\mathrm{d}P_0 = -\lambda P_0 \mathrm{d}t$. Integrating after separating the variables,

$$P_0 = ce^{-\lambda t},$$

since $P_0 = 1$, when $t = 0$, $c = 1$ and hence

$$P_0 = \mathrm{e}^{-\lambda t}. \tag{2}$$

Next we find P_1, the probability that just one event occurs between 0 and t. Again, the probability that the event occurs once (only) between 0 and $t + \mathrm{d}t$, i.e. $P_1 + \mathrm{d}P_1$ can be written down. If the event does not occur between 0 and t it must occur between t and $t + \mathrm{d}t$. The joint probability for this is $P_0 \lambda \mathrm{d}t$, P_0 for not occurring in the $0 - t$ interval and $\lambda \mathrm{d}t$ for occurring once between t and $t + \mathrm{d}t$.

We take the product because the events are independent. If the event does occur between 0 and t it must not occur between t and $t + \mathrm{d}t$. The probability of these events happening is $P_1(1 - \lambda \mathrm{d}t)$. Hence

$$P_1 + \mathrm{d}P_1 = P_1(1 - \lambda \mathrm{d}t) + P_0 \lambda \mathrm{d}t = P_1 + \lambda \mathrm{d}t(P_0 - P_1) \tag{3}$$

so

$$\frac{\mathrm{d}P_1}{\mathrm{d}t} = \lambda(P_0 - P_1) \quad \text{or} \quad \frac{\mathrm{d}P_1}{\mathrm{d}t} + \lambda P_1 = \lambda P_0$$

so that

$$P_1 = \lambda t \mathrm{e}^{-\lambda t}. \tag{4}$$

Similarly we can find for P_2

$$P_2 + \mathrm{d}P_2 = P_1 \lambda \mathrm{d}t + P_2(1 - \lambda \mathrm{d}t) \tag{5}$$

$$\mathrm{d}P_2 = (P_1 - P_2)\lambda \mathrm{d}t$$

$$\frac{\mathrm{d}P_2}{\mathrm{d}t} + \lambda P_2 = \lambda P_1 \tag{6}$$

but from (4) we know that $P_1 = \lambda t \mathrm{e}^{-\lambda t}$, so that

$$\frac{\mathrm{d}P_2}{\mathrm{d}t} + \lambda P_2 = \lambda^2 t \mathrm{e}^{-\lambda t}. \tag{7}$$

Integrating this linear first order differential equation,

$$P_2 = C\mathrm{e}^{-\lambda t} + \mathrm{e}^{-\lambda t} \frac{\lambda^2 t^2}{2}. \tag{8}$$

When $t = 0$, $P_2 = 0$, so $C = 0$, and

$$P_2 = \mathrm{e}^{-\lambda t} \left\{ \frac{\lambda^2 t^2}{2!} \right\}. \tag{9}$$

Here we write 2! instead of 2.

Similarly,

$$P_3 + \mathrm{d}P_3 = P_2 \lambda \mathrm{d}t + P_3(1 - \lambda \mathrm{d}t) \tag{10}$$

$$\frac{\mathrm{d}P_3}{\mathrm{d}t} + \lambda P_3 = \lambda P_2 \tag{11}$$

$$P_3 = \mathrm{e}^{-\lambda t} \left\{\frac{\lambda^3 t^3}{3!}\right\}. \tag{12}$$

The general result is:

$$P_n = \frac{\lambda^n t^n}{n!} \mathrm{e}^{-\lambda t}. \tag{13}$$

This is the Poisson's distribution.

It depends only upon λt, which is the mean number of events in time t. λt corresponds to m in the above example.

The events could equally well be distributed in space instead of in time. An example of this is the Poisson distribution that obtains when blood cells are counted in a counting chamber. When blood cells of the same blood are counted on different occasions, different values are obtained. Their frequency distribution may be a Poisson to a very good approximation.

THRESHOLD DETERMINATION OF VISION

The following discussion is based upon the work of Hecht, Schlaer and Pirenne.[1,2,7]

As is well known from elementary physics, light is emitted and absorbed in discrete units, called quanta.

At the threshold of vision only a small number of quanta are required to produce the sensation of light.

The problem is to find the *number of quanta* required to produce the sensation of light at threshold level.

For this purpose a small area, containing approximately 500 rods (10 minute circular visual field) was exposed to a flash of light lasting only 0.001 second. The intensity of the flash was varied. It could be shown that at threshold the number of quanta at the *cornea* ranged between 54 and 148 in a number of subjects.

However, the important thing to know is the number of quanta actually reaching the rods on the retina. On the basis of experiments it may be computed that although at threshold the number of quanta reaching the cornea varies between 54 and 148, the number of quanta actually absorbed by the rods varies between 5 and 14.

Since there are 500 rods in the exposed area and since only about 10 quanta are absorbed by the rods, it is safe to assume that each rod will only absorb 1 quantum (the probability of one rod absorbing 2 quanta is only 5 %).

We may thus conclude that in order for us to see, it is necessary that 5 to 14 rods absorb 1 quantum each.

Now, this estimate has been obtained on the basis of rather indirect evidence. By using the Poisson distribution a more direct and more precise estimate of the required number of quanta may be obtained in the following manner:

First it should be realized that the number of quanta per flash cannot be determined with precision. All we can measure, on the basis of energy density determination, is the *average number* of quanta per flash. But when a number of flashes of the same intensity and the same duration are given, then some flashes will "contain" fewer and others more quanta than this *average* number.

Since absorption of quanta by the retinal rods represents independent events occurring at random, we can apply the Poisson distribution.

Let m be the average number of quanta that any flash yields to the retina and let n be the number of quanta that must be absorbed by the rods to make "threshold vision" possible. Then we may obviously write

$$P_n = \frac{m^n}{n!} e^{-m}$$

where P_n is the probability that the flash will deliver the required n quanta.

Since the minimal number of quanta required for seeing is n, we shall also see when the rods absorb more than n quanta. It is possible to construct graphs relating the probability of absorption of n or more quanta to the average number of quanta delivered per flash.

Such "cumulated" Poisson distributions are shown in Fig. 148 for various values of n.

It may be seen that the *shape* of the distribution depends upon the value of n. From this it follows that when the distribution is

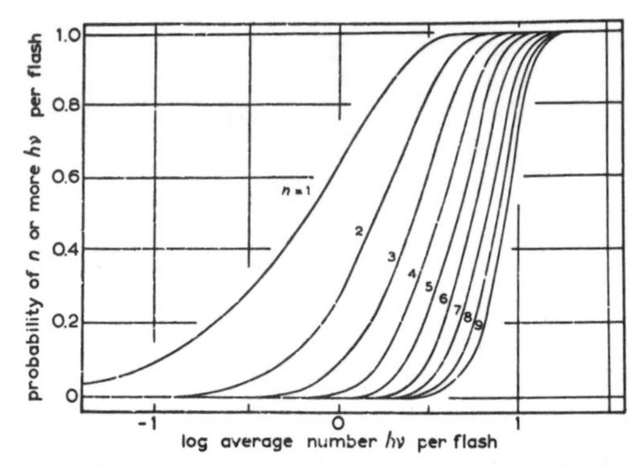

Fig. 148. Poisson distributions. For any average number of quanta ($h\nu$) per flash the ordinates give the probabilities that the flash will deliver to the retina n or more quanta, depending on the value assumed for n. [From Hecht *et al.* [2]].

determined experimentally, its shape will reveal the value of n, i.e. the number of quanta absorbed by the rods at threshold. It might be argued that only the average number of quanta at the cornea can be determined, while it is the average number of quanta delivered by the flash to the retina that should be known.

However, in Fig. 148 the *logarithm* of the average number of quanta per flash is given, which means that the absolute values of the average number of quanta per flash need not be known for comparing the experimental distributions with the theoretical ones.

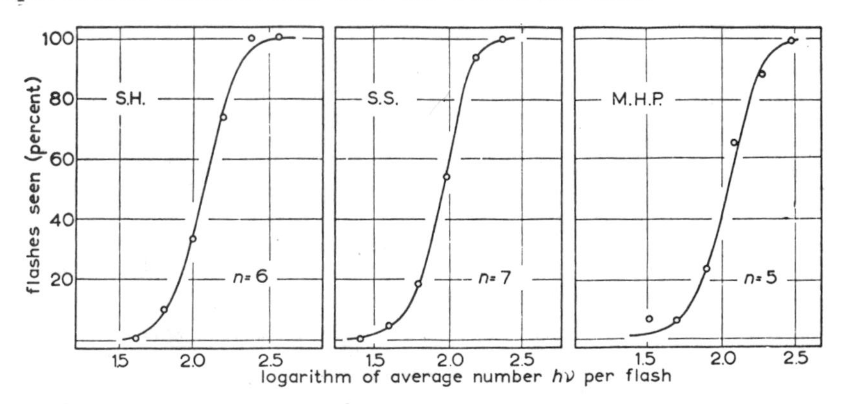

Fig. 149. Relation between the average energy content of a flash of light (in number of $h\nu$) and the frequency with which it is seen by three observers. Each point represents 50 flashes, except for S.H. where the number if 35. The curves are the Poisson distributions of Fig. 148 for n-values 5, 6 and 7. [From Hecht *et al.* [2]].

The experimental distribution is obtained as follows:

Flashes of various intensities are given; each intensity is given a large number of times, 50 times, say. The observer reports whether he has *seen* a flash or not. When, if 50 flashes of the same intensity are presented the subject *sees* only 4 flashes, then the probability of seeing a flash at that particular intensity is 8 %, etc. The data obtained from three subjects are shown in Fig. 149. Comparison with the theoretical Poisson distributions given in Fig. 148 reveals that in subject M.H.P. the value for n is 5, while in subjects S.H. and S.S. the values are $n = 6$ and $n = 7$ respectively. This result may thus be interpreted to mean that e.g. in subject M.H.P. at least 5 quanta must be absorbed by 5 rods to elicit the sensation of seeing.

13. Uses of the Poisson Distribution, especially in the Study of Bacterial Viruses

1) It is now known that the bacteriophage principle of d'Herelle is a group of viruses parasitic on bacteria. The adsorption of a virus to the host bacterium is the first step in its growth cycle.

When these phage particles are added to a bacterial culture, the distribution of phage particles among the bacteria follows the Poisson distribution when the average number n of phage particles adsorbed per bacterium is less than 2, e.g. $n = 3$ means that on the average, each bacterium adsorbs 3 phage particles.

Now we wish to know the fraction of bacteria adsorbing no phage particles, the fraction adsorbing 1 phage particle, etc. This answer may be obtained by using the Poisson formula

$$P(r) = \frac{n^r}{r!} e^{-n}$$

where $P(r)$ is the proportion of bacteria adsorbing r phage particles, and n is the average number of phage particles adsorbed per bacterium in the adsorption mixture.

Experimentally only $P(0)$, which is the proportion of bacteria which adsorbs no phage particles, can be determined, since only these uninfected bacteria can form colonies when plated on agar. Although this is certainly a limitation, we can, by determining $P(0)$, determine the total number of phage particles adsorbed. For $P(0)$, the Poisson formula degenerates into

$$P(0) = e^{-n}.$$

EXAMPLE: 5×10^7/ml phage particles are added to a bacterial culture of about 5×10^7 cells/ml. The number of uninfected bacteria was found to be 2.25×10^7 cells/ml, so that the proportion not infected $P(0) = 2.25/5.0 = 0.45 = e^{-n}$, so that $n = 0.8$.

From this it immediately follows that the number of phages adsorbed is $0.8\ (5 \times 10^7) = 4 \times 10^7$/ml.

It also follows that 80 % of the phage particles are adsorbed.

2) A more complex situation arises when two types of virus, A and B say, are added to a collection of bacteria. Let x and y respectively be the *ratio's* of the numbers of virus particles of type A and B to the number of bacteria. In each bacterium there will be present random numbers m and n of the viruses A and B respectively. Using the Poisson law it follows that the numbers m and n have independent probabilities:

$$\frac{x^m}{m!} e^{-x} \quad (m = 0, 1, \ldots..) \quad \text{and} \quad \frac{y^n}{n!} e^{-y} \quad (n = 0, 1, \ldots..).$$

Let each bacterium possess K sites for occupancy, or "participation", by a virus, where K is a fixed integer $\geqslant 1$. The bacterium is said to be "infected" if $m + n > 0$. If none of the K sites of the bacterium is occupied by virus A, we say "there is no participation by A". This may occur because $m = 0$ (no "landing" of any particle on this bacterium) or because $n \geqslant K$ and all of the K sites are, by chance, occupied by virus B. The question asked is this: what is the proportion of infected bacteria which show no participation by A? By fairly complex mathematical reasoning Franklin[1] has shown that the probability that type A virus does not participate, given that at least one virus of either type is present, is given by

$$\left(\frac{y}{x+y}\right)^K + (e^{x+y} - 1)^{-1} \sum_{k=1}^{K-1} \frac{1}{k!} y^k \left[1 - \left(\frac{y}{x+y}\right)^{K-k}\right]$$

a formula given here only for the purpose of illustration.

3) Consider DNA molecules of phage λ that superinfect *Escherichia coli* lysogenic for the phage. It has been found experimentally that

single-strand breaks caused by ^{32}P disintegrations are repaired by incubating the cells in which the damaged molecules are carried. The functions required for the repair are provided by the host bacterium. The Poisson distribution has been used in the theoretical treatment of this problem.[2]

4) Interferon is a substance produced by cells that protect the cell against virus, bacteria and cancer. Using an approach similar to that used for the threshold determination of vision, Gulford and Koch[3] have shown by mathematical analysis of their data based on the Poisson distribution that at most one molecule of interferon is required to protect one cell against viral infection. This suggests that a single molecule of interferon may protect more than one cell from infection by virus.

5) In the final example, invoking the Poisson distribution is like using a giant sledge hammer to knock in a nail! In four independent series involving a total of 1130 postmenopausal females treated with oestrogen substitution therapy the total number of the cancers found was 2 as against an expected number of 74. If the published data are based on independent random events with the given expected numbers the probability of obtaining 2 or less would be 2.05×10^{-29}. In reviewing these results reported by different authors, Defares[4] concluded that the vast majority of maturity-onset cancers in females can be prevented by maintaining optimal oestrogen levels in the body.

For further reading on the virus problems see especially Gani's papers and reviews.[5]

14. The Form(s) of Age-Incidence Curves

The next example is rather different. The mathematics mainly illustrates points in Chapter 4 and is not too difficult, but the underlying ideas are rather more difficult. In fact this is an unsolved problem well known to actuaries, who were probably the first people ever to use biomathematics professionally.

Many others too are convinced that some very important biological information is locked up somewhere in these curves. We can only select a few of the many facets of the problem, but these do serve to introduce us to some useful concepts in population studies.

They also show us examples of mathematical models as distinct from biological models, and the need to distinguish clearly between them.

It has been known for a long time that the incidence of many different diseases in a particular population (e.g. one sex and country) can be fitted by a positive power k of the age t when this is between about 35 and 70, or by a power of $t-v$ where v is a constant.[1,2,3]

Hence we put

$$y(t) = b(t-v)^k \tag{1}$$

and this is the first mathematical model that we shall consider.

The incidence $y(t)$ has to be defined carefully. It can mean the number of people aged t who get the disease, or who die from it, in a particular year, say 1970, divided by the number at risk, which would be the number born in the year $1970-t$. Then $y(t)$ would be usually what is given in an annual mortality table by age, and disease. Clearly for different ages, the tables for 1970 correspond to people born in different years – they are a *cross section*. For $y(t)$ to have biological meaning, however, it should apply to a *cohort* of people born at about the same time. If $y(t, t_b)$ is the function for people born in the year t_b, it seems often to be case that b is a function of t_b but v and k are not. In any case we shall assume that $y(t)$ refers to a cohort, or if not, that it is not affected by the year of birth. It is as if we start with a population of N people, or N mice in a laboratory at a particular time $t = 0$, and follow them *longitudinally*.

These are discrete events, but they are numerous enough for us to regard $y(t)$ as continuous. We introduce a cumulative function $F(t)$ such that $NF(t)$ – or the nearest whole number less than this – contract the disease, or die from it, as the case may be, before time t. Further we define a survival function $G(t)$, which is such that $NG(t)$ people (or, again, the next lowest integer) survive up to time t, so that a proportion $1 - G(t)$ have died from all causes, or are otherwise lost from the population.

Then strictly we should define the incidence at time t as

$$y(t) = \frac{1}{G(t)} \frac{\mathrm{d}F(t)}{\mathrm{d}t} . \tag{2}$$

If we were considering *all* causes of death, we should have $F(t) = 1 - G(t)$ and then

$$y_{\text{total}}(t) = -\frac{1}{G(t)}\frac{\mathrm{d}G(t)}{\mathrm{d}t} = -\frac{\mathrm{d}}{\mathrm{d}t}\log G(t) \qquad (3)$$

which is sometimes called the *force of mortality*.

In this case we have a complete distribution of survival times; $1-G(t)$ corresponds to the shaded area in Fig. 89 (with t replacing x of course), and corresponding to Fig. 88, we can put

$$g(t) = -\frac{\mathrm{d}G(t)}{\mathrm{d}t} \qquad (4)$$

and now $g(t)$ is the probability density as in Fig. 88; a proportion of individuals $g(t)\mathrm{d}t$ are predicted to die in the time interval t to $t + \mathrm{d}t$. It is $g(t)$ and $G(t)$ that are likely to be most directly related to fundamental biological processes.

For the particular disease in question, however, this complete distribution cannot be observed, for during any time interval individuals are being removed by death from other causes, including those who could otherwise have got that disease. We can however, invoke a standard result in the theory of *competing risks*. (We shall not try to prove it, although this is not difficult.) This says, in effect that provided the other risks work independently of the risk of getting disease 1, then the function $y(t)$, and the accumulated form $Y(t) = \int_0^t y(t)\mathrm{d}t$ describe what would happen if disease 1 was the only risk; in other words we regard $y(t)$ and $Y(t)$ as probability density and cumulative probability curves for disease no. 1 acting on its own.

At best this is an approximation, but it is certainly good enough when $y(t)$ is small and rapidly increasing, and $G(t)$ is very near to 1. In practice there are difficulties in estimating $y(t)$ from grouped data, e.g. for a 5 year interval such as $70 < t < 75$, which have recently provoked vigorous discussions (by Kimball, Chiang, and David in particular[4–8]). This is especially important when $y(t)$ is being investigated for a particular sub-group. But this need not concern us here.

From now on, then, we regard $y(t)$ and $Y(t)$ as being real distributions as in Figs. 88 and 89. For example, when Z is the yearly death rate from cancer of the stomach in men in New York State, Cook, Fellingham and Doll[1] give, for

$$Z = b(t - v)^k$$

$k = 6.0$, $v = 0$ and $b = 3.2 \times 10^{-15}$, yielding $Z = 5 \times 10^{-5}$ per year at age $t = 50$.

For cancer of the prostate in men they find that $k = 5$ and $v = 32$ in several countries; in 1961–3 in England and Wales $b = 1.4 \times 10^{-13}$ giving $Z = 38.5 \times 10^{-5}$ at age 62.5.

We shall therefore assume that there are frequency distributions, such that $Z = y(t)$, and that if N is the number in a cohort,

$$N = \int_0^\infty y(t)\mathrm{d}t = N\theta \tag{5}$$

where $N\theta$ is the number of susceptibles, so that $\theta \leqslant 1$. The value of θ is not generally known, because usually only the rising part of the distribution can be observed.

How positive integral powers can arise

Many investigators have assumed that a disease starts in the descendants of a mutated cell. This can yield a whole number for k in many different ways; such as in the following model due to Fisher.[9] In order to get a balanced view it is worth studying his full account, and the models and reviews given by Armitage, Doll, Pike and others they refer to.[10–13]

It is assumed that in a population of n_0 cells in the human body, a constant small proportion of them mutate in unit time interval, so that $n_0 \alpha dt$ mutate when $t_1 < t < t_1 + \mathrm{d}t_1$. Fisher now assumes that descendants of these cells multiply as if to produce a disc with uniformly expanding radius, so that their number is proportional to the square of the time from the mutation. Then the number of these descendants is, at time t_2, equal to

$$n_0 \beta_1 (t_2 - t_1)^2 \alpha_1 \,\mathrm{d}t_1$$

where β_1 is a constant describing the rate of growth of the "disc". So the total number n_1 of cells that are descended from cells that mutated at all times between 0 and t_1 is given by

$$n_1 = n_0 \alpha_1 \beta_1 \int_{t_1=0}^{t_2} (t_2 - t_1)^2 \, \mathrm{d}t_1 \, . \tag{6}$$

Putting $u = t_2 - t_1$, we have

$$n_1 = -n_0 \alpha_1 \beta_1 \int_{t_2}^{0} u^2 \, \mathrm{d}u = -\alpha_1 \beta_1 n_0 \left[\tfrac{1}{3}u^3\right]_{t_2}^{0} = \tfrac{1}{3}\alpha_1 \beta_1 t_2^3 n_0 . \tag{7}$$

So if the chance of getting the disease is small and proportional to n_1, for example it could be initiated by a mutation in any one of these n_1 cells, the rate of onset will be proportional to $(t-v)^3$ where t is the age and the process begins at $t = v$.

Various positive powers can be predicted depending on what the law of multiplication is assumed to be.

Mathematical models based on complete distributions

The last example was a biological model for interpreting the rapidly increasing part of $y(t)$. We now give two quite different mathematical models, which are consistent with (1) but based on complete distributions.

Doll[13] discovered that age-incidence curves for stomach cancer could be fitted to parts of normal error curves. This is a particular case of a surprising mathematical relationship which we shall now derive. Starting with a curve such as in Fig. 91, we replace x by the time variable t. Then $y = A \exp -\frac{1}{2}[(t-m)/\sigma]^2$. We write $t = \mathrm{e}^u$ (thus $u = \log t$) and $z = \log f(t)$ then we have:

$$z = \log A - \tfrac{1}{2}\left(\frac{\mathrm{e}^u - m}{\sigma}\right)^2 \tag{8}$$

where $A = 1/(\sigma\sqrt{2\pi})$. Now we consider the rising part of this curve, i.e. a plot of z as a function of u. As $u \to -\infty$, z approaches the constant value $\log A - m^2/(2\sigma^2)$ and z clearly passes through a maximum where $u = \log m$ or $t = m$. In between, by Rolle's theorem applied to $\mathrm{d}z/\mathrm{d}u$, there must be a point of inflexion, which we shall now locate. We have:

$$\frac{\mathrm{d}z}{\mathrm{d}u} = -\left(\frac{\mathrm{e}^u - m}{\sigma}\right)\frac{\mathrm{e}^u}{\sigma} = -\frac{1}{\sigma^2}(\mathrm{e}^{2u} - m\mathrm{e}^u) .$$

Hence

$$\frac{d^2 z}{du^2} = -\frac{e^u}{\sigma^2}(2e^u - m) = 0 \tag{9}$$

where $e^u = t = \frac{1}{2}m$. At this point, the tangent to the curve is extremely close to it. Its slope k (gradient) is the value of dz/du where $e^u = \frac{1}{2}m$, or $\frac{1}{4}(m^2/\sigma)$; then $z = \log A - \frac{1}{8}(m^2/\sigma^2)$. It follows that, on either side of the point of inflexion of y as function of u, a good approximation to the normal curve is

$$f(t) = Ae^{-m^2/(8\sigma^2)}(2t/m)^k \tag{10}$$

where $k = (m/2\sigma)^2$. For example, for a normal error curve with mean 100 (years), and $\sigma = 20$, $k = 6.25$.

TABLE 22

Age $= t$	Ordinate of normal curve	$4.2186 \times 10^{-13} t^{6.25}$
$35 = 100 - 3.25\sigma$	0.002029	0.001886
$40 = 100 - 3\sigma$	0.004432	0.004346
$45 = 100 - 2.75\sigma$	0.009094	0.009073
$50 = 100 - 2.5\sigma$	0.017528	0.017528
$55 = 100 - 2.25\sigma$	0.031740	0.031801
$60 = 100 - 2\sigma$	0.053991	0.054779
$65 = 100 - 1.75\sigma$	0.086277	0.090337
$70 = 100 - 1.5\sigma$	0.12952	0.14356
$75 = 100 - 1.25\sigma$	0.18265	0.21190

Table 22 shows the ordinates of the normal curve and corresponding ones for the power curve.

All this applies equally to the possible approximation $y = b(t-v)^k$. If we put $u = \log(t-v)$, then we can write (8) as

$$f(t) = A\exp - \tfrac{1}{2}[(t - v + v - m)/\sigma]^2$$

and:

$$\log f(t) = z = \log A - \frac{1}{2}\left(\frac{e^u - m_1}{\sigma}\right)^2 \tag{11}$$

where $m_1 = m - v$. So formula (10) holds good with m replaced by m_1 throughout, thus $4\sigma^2 k = m - v$ which for a given m and σ can be satisfied by continuously varying pairs of values of k and v.

This approximation makes no use of the symmetry of the normal curve. Could we get about as good a fit to a power function from a skew curve such as one of those in Fig. 98? If so, it could only be in the critical range between t = ca. 3 and 5, but this could still correspond to a large increase in incidences. It can be shown that there are no fits to $y = bt^k$ but there can be good fits to $y = b(t-v)^k$, when the random walk starts at $t = 0$. The method is the same in

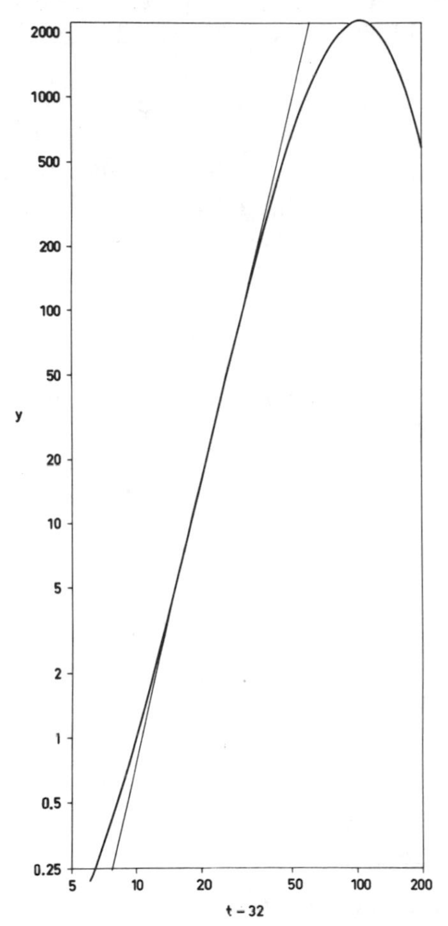

Fig. 150. The curve $y(t) = Ax^{-3/2} \exp -\varphi(x + x^{-1})$, with $x = t/\mu$, $\mu = 160$, $\varphi = 4.6052 = 2 \log_e 10$, $A = 2 \times 10^{-7}$, as for a random walk starting at $t = 0$, showing $z = \log y$ plotted against $u = \log (t-32)$, i.e. with log scales for y and for $t-32$, and the tangent to this curve at the point of inflexion where $t \approx 54.24$.

principle as for a normal error curve, but the mathematics is less simple. The main results come from equations (14) and (15) below.

Fig. 150 shows an example where $y(t)$ is the distribution of times of first arrival given in Chapter 7, §15, equation (6).

PROBLEM

Investigate the curve for log y plotted against log t when $y(t)$ takes a generalised form for a random walk starting at $t = t_0$

$$y(t) = Ax^{-W} \exp\left\{-\varphi\left(x + \frac{1}{x}\right)\right\} \tag{12}$$

(for which we have considered the cases $W = \frac{1}{2}$ and $W = \frac{3}{2}$), where $x = (t - t_0)/\mu$, and A, W, and μ are positive constants, $x \geqslant 0$, $t \geqslant 0$; t and t_0 are times. In particular show that:

1. If $z = \log y$, $v = \log t$ and $u = \log x$, then

$$\mathrm{d}z/\mathrm{d}v = -(W + 2\varphi \sinh u)\, \mathrm{d}u/\mathrm{d}v.$$

$$\frac{\mathrm{d}^2 z}{\mathrm{d}v^2} = \frac{t}{\mu^2 x^2}\left\{Wt_0 + 2\varphi\,(t_0 \sinh u - t \cosh u)\right\}. \tag{13}$$

$\left(\text{Hint: write } \frac{\mathrm{d}z}{\mathrm{d}v} = \frac{\mathrm{d}z}{\mathrm{d}u} \cdot \frac{\mathrm{d}u}{\mathrm{d}v} \text{ and } \frac{\mathrm{d}u}{\mathrm{d}v} = \frac{\mathrm{d}u}{\mathrm{d}x} \cdot \frac{\mathrm{d}x}{\mathrm{d}t} \cdot \frac{\mathrm{d}t}{\mathrm{d}v}.\right)$

2. At a point of inflexion of the log log plot, $x = x_f$, and

$$\left(\frac{Wt_0}{\varphi} - \mu\right) x_f - \mu x_f^3 = 2t_0 \tag{14}$$

$$\frac{\mathrm{d}z}{\mathrm{d}v} = -\frac{\varphi t^2}{t_0 \mu}\left(1 + \frac{1}{x_f^2}\right). \tag{15}$$

3. There is only one solution for x_f above with $x_f > 0$, on the rising part of the curve where t_0 is negative, and on the falling part where t_0 is positive.

Use these results to verify the approximation:

$$y(t) = bt^k$$

where t takes values round $t_f = t_0 + \mu x f$, e.g. $\frac{2}{3} t_f < t < \frac{3}{2} t_f$, and k is the value of $\mathrm{d}z/\mathrm{d}v$ at $x = x_f$, with the values of μ, φ and A as in the figure, and $t_0 = -32$. (Then $k = 4.316$, $x_f = 0.339$, $b = 2.0548 \times 10^{-5}$.) Hint. Use Newton's method, see Chapter 4, §11, to find x_f.

We recall that in Einstein's model for Brownian movement, a cloud of particles starts from C and makes to the right towards a threshold D, spreading out as it moves

$$C \xrightarrow{\qquad} D$$

and that (12) gives the distribution of the time taken by an individual particle to arrive for the first time at a threshold. This arrival could represent the onset of a disease, and the movement of the particle could represent small steps, mainly of deterioration of some kind during life. This could provide a simple interpretation when v is positive. However in order to interpret a fit to bt^k, i.e. $v = 0$, the process would have to begin years before birth, which makes no sense biologically. On the other hand, as Doll[2] has shown (without reference to any model) where there is a good fit with $v = 0$, there is often one with positive v and a lower value of k. Also, with this approach, k need not be an integer. As Fig. 98 shows, if one distribution of this type fits, many others may do so equally well. If there was only one set of data we could get no further; but there are very many different age incidence curves, and with the right choice of distribution, as yet unknown, there is a good chance that the fitted parameters will form a simple pattern, so that Occam's razor* will help us.

The last model we shall give is due to Burch. It combines some features of all those mentioned so far; biologically it involves dis-

* Occam's (Ockham's) razor. William of Occam was born towards the end of the 13th century at Ockham (Surrey, England) and studied in Oxford and Paris as a Franciscan. He was associated with the famous Franciscan assembly at Perugia, Italy, and led their revolt against Pope John XXII in 1322 to uphold their original doctrine of poverty. He formulated (in Latin) his principle that "no inessentials be introduced except where necessary" for his philosophy and theology. It has been debatably extended to many other fields, and it is not certain how or when the Occam's razor principle was first so named (a steel razor cuts finely and easily). Still it seems fitting that his name should live on in this way even if this is not quite right historically. The concept "scholastic-nominalism" is attributed mainly to him.

crete events, and so predicts integers for k, but it also defines a complete distribution of ages at onset of symptoms or ages at death. The mathematical model is given in the cumulative form:

$$Y(t) = \theta \{1 - e^{-K(t-l)^r}\}^n \tag{16}$$

so that the observed age incidence curve would be

$$y(t) = Y'(t) = \theta K n r (t-l)^{r-1} e^{-u} (1 - e^{-u})^{n-1} \tag{17}$$

where we write u for $K(t-l)^r$. Burch interprets the parameters biologically as follows: θ is either the proportion or the absolute number of susceptible people, since $\lim_{t\to\infty} Y(t) = \theta$; r and n are integers of which n is very often unity, and when $n > 1$, r is often unity. A disease is initiated by n distinctive "forbidden clones" of cells. The cells in any clone have a common descendant, which initiates this clone when r specific somatic gene mutations have taken place in any one cell out of a set of L growth control cells. Then if the mutation rate is η per gene at risk, per stem cell at risk, K is defined to be $L\eta^r$. It is assumed that $\eta t \ll 1$. Finally, l is a latent period, that is a time interval assumed constant between the last initiating event and the observed event, namely onset of the disease or death.[15,16]

The power function corresponding to (1) holds good when u is small. From the exponential and binomial expansions:

$$(1 - e^{-u})^{n-1} = \{1 - (1 - u + \tfrac{1}{2}u^2 \ldots)\}^{n-1}$$

$$= u^{n-1}(1 - \tfrac{1}{2}u + \ldots)^{n-1} = u^{n-1}\{1 - \tfrac{1}{2}(n-1)u + \ldots\}.$$

Hence we can put

$$(t-l)y(t) = \theta n r u^n (1 - u + \tfrac{1}{2}u^2 + \ldots)\{1 - \tfrac{1}{2}(n-1)u + \ldots\},$$

or:

$$y(t) = nrK^n(t-l)^{rn-1}\{1 - \tfrac{1}{2}(n+1)u + c_2u^2 + \ldots\}, \tag{18}$$

where c_2 is a constant. Thus the observed power K should be $rn-1$.

As t, and with it u, increases, $1 - e^{-u}$ becomes negligibly different from 1, and then

$$y(t) = \theta nru\,e^{-u}/t$$

which clearly tends to zero as u tends to infinity (cf. Chapter 6, § 3, equation (6)).

PROBLEMS

1. Deduce that, in (18)

$$c_2 = \tfrac{1}{2}n + \frac{(n-1)(3n+2)}{24} \quad .$$

2. Prove that the maximum of $y(t)$ is where $u = u_m$ and

$$u = \left(1 - \frac{1}{r}\right)\left(1 + \frac{n-1}{e^u - n}\right) \tag{19}$$

Hence show that for $n = 3$, $r = 4$, $\log_e 4 < u_m < \log_e 5$.

Burch has fitted these functions to about 200 age-incidence curves for about 150 different diseases. Where the curves differ between the sexes he often finds that the parameter K, for the same disease, is either the same for men and women or twice as large in women. There is an impressive biological discussion of the biological aspects of K and of all other parameters.

The evidence for the *mathematical* model looks most convincing where the maximum of the curve and both the rising and falling parts can be observed. Very good fits are then found for (e.g.) diabetes mellitus (data from Spiegelman and Marks)[17] with $r = 5$, $n = 1$, $K = 10^{-9}$ (yr^{-5}) and $l = 5$ years for men and women, $\theta = 6.7 \times 10^{-2}$, women, and $\theta = 4 \times 10^{-2}$, men; also for involutional psychoses (Malzberg)[18] with $n = 3$, $r = 4$, and for women and men respectively $K = 3.2 \times 10^{-7}$ and 1.6×10^{-7}, $l = 8$ and 4 years, and $\theta = 1.24 \times 10^{-2}$ and 0.76×10^{-2}. This example is shown in Fig. 151.

In a number of other fits to (17) the rising part and the maximum are observed. Good fits as far as the maximum and/or beyond it are obtained for a great variety of conditions, e.g. systemic lupus

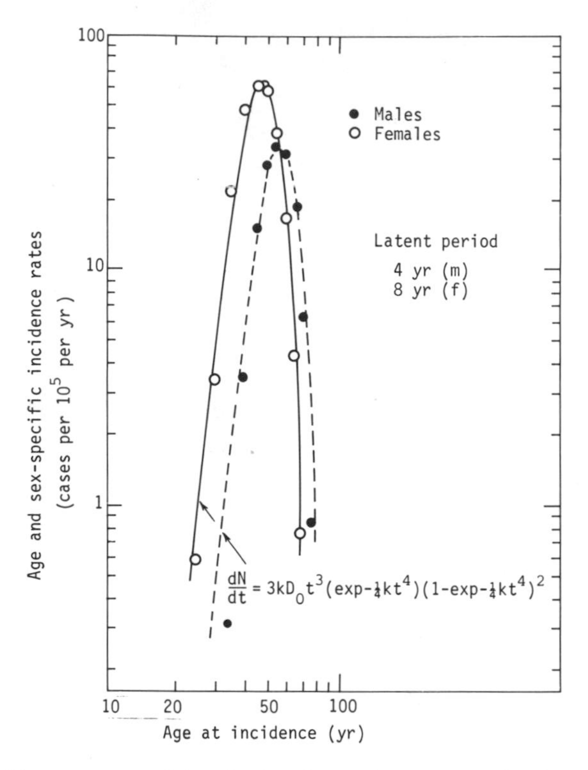

Fig. 151. Incidence rates of involutional psychoses, specific for age and sex, plotted against age on log-log scales. Derived from Malzberg's (1955) statistics of first-admission rates to hospital in New York State, 1949 to 1951. As Fig. 1 from P.R.J. Burch, Brit. J. Psychiatry 110 (1964) 827. N corresponds to our $Y(t)$ in (16), $dN/dt = y(t)$ and $D_0 = 4\theta$.

erythematosis, pityriosis rosea, ankylosing spondylitis, carpel tunnel syndrome, Wilson's disease, paralytic poliomyelitis, diverticulosis, and petit mal and grand mal; also for various cancers, such as of the ovary, corpus uteri, and for Japan, oesophagus, pancreas, stomach and liver. These too provide good evidence for (17).[15]

In many other cases only the rising part of the curve is available. For diseases of the mitral valve in England and Wales, 1961[19], $r = 4$, $n = 1$, $K = 7.5 \times 10^{-9}$ and $l = 3.5$ years, men and women, $\theta = 3.5 \times 10^{-2}$ in men and 6.4×10^{-2} in women, and in these the evidence is still consistent with (17); however one of the previous mathematical models or others might fit as well or better. When there is a good fit to the power function (1), with 3 parameters, the function (17) may fit equally well, but there is not then enough information in the data to determine each of its 5 parameters.

In other cases a complete curve *is* observed, but it has to be fitted to the sum of two functions of the form (17), and then between 6 and 10 parameters – some are the same in the two components – have to be estimated from one curve (e.g. ulcerative colitis, Figs. 4.21(a) to (g)[15]). In this and some other instances the curves can be interpreted as mixtures arising from two distinctive forms of the disease, each with its own age-distribution.

The pitfall of having possibly too many adjustable parameters in a mathematical fit is often encountered. It plays a major role in the problem treated in the next two sections.

15. Creatinine Clearances of Intravenously Injected Creatinine

We shall now study an example of a homogeneous linear second order differential equation, which is based on a study of L. A. Sapirstein *et al.*[1]

This example will be seen to be of considerable clinical interest.

Before discussing this example, we shall consider a very important problem. We shall study the semi-logarithmic graph of a second order equation and the inverse problem, i.e. the question: how are we to know that an experimental curve plotted on semi-log paper may be represented by an equation of the type

$$y = C_1 e^{-\alpha t} + C_2 e^{-\beta t}? \qquad (1)$$

We have seen that a first order equation $y = Ae^{-\gamma t}$, when plotted on semi-log paper, yields a straight line, since $\log y = -\gamma t + \log A$. It is easily seen that the semi-log plot of (1) will not yield a straight line.

Let us study an (experimental) curve $y = f(t)$, which when plotted on semi-log paper yields a curve of the type shown in Fig. 152 (continuous curve). The ordinate of this (continuous) curve is obviously $\log_{10} y$.

We observe that the part of the non-linear curve to the right of the point *a* is (practically) a straight line.

If now we extend this linear part to the left of *a* (broken line), we obtain a straight line starting at $t = 0$. Since this is a straight line on semi-log paper, we know that this straight line represents the

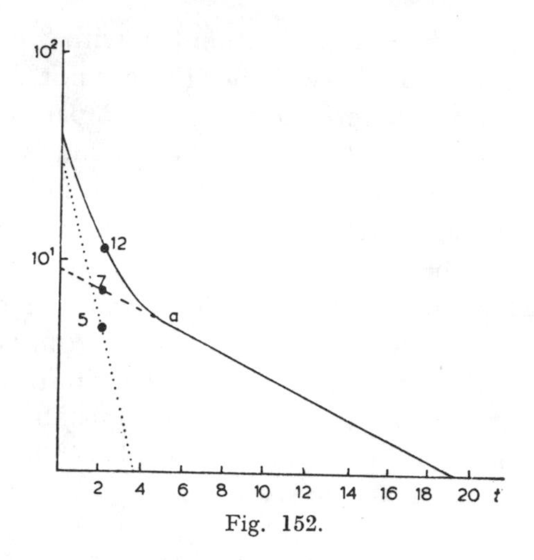

Fig. 152.

function*

$$\log_{10} z = -0.4343\ \alpha t + \log_{10} C_1 \tag{2}$$

or

$$\log z = -\alpha t + \log C_1. \tag{3}$$

Dropping logarithms this yields

$$z = C_1 e^{-\alpha t}. \tag{4}$$

Let us now direct our attention to what happens to the left of the point a. Let us choose a value of y and a value of z at, say, the point $t = 2$. We then find $y = 12$, $z = 7$. Let us now form their difference, to find $y - z = 12 - 7 = 5$.

If we plot this difference on (the same) semi-log paper, this means that we plot $\log_{10} (y - z)$ as the ordinate. If we repeat this process for a number of y and z values on the left of the point a, we notice from Figure 152, that in this particular case, another straight line (steeper than the first) is obtained. What does this imply? It means that $\log_{10} (y - z)$ is some linear function of time, say,

$$\log_{10} (y - z) = -0.4343\beta t + \log_{10} C_2 \tag{5}$$

where β is obviously a positive constant, and where the modul 0.4343 is introduced for the sake of elegance, since this enables us to write (5) as

$$\log (y - z) = -\beta t + \log C_2. \tag{6}$$

* $\log z$ is clearly the dependent variable.

Or, dropping logarithms,

$$y - z = C_2 e^{-\beta t}. \tag{7}$$

Combining (4) and (7) we thus find

$$y = C_1 e^{-\alpha t} + C_2 e^{-\beta t} \tag{1}$$

which is clearly a very interesting result with important implications.

It means that by the method just explained we can test whether a given experimental curve may be represented by the second order equation (1). The above explanation is rarely found in textbooks and the reader is urged to study it carefully, since the method is of great practical importance.*

The following points should be noted.

a) Supposing that we are really dealing with a curve representing equation (1) on semi-log paper, then that part of the curve to the right of a is theoretically never a straight line, since the second exponential term is never zero. But in practice, we may find a straight line (to the right of a point a) which means that the second exponential has become negligibly small.

b) It must be clearly realized that we are using semi-log paper. It should become very clear that if at, say, $t = 2$, the value of z (7 units) is subtracted from the value of y (12 units), and this difference (5 units) is plotted on semi-log paper, then we really have the effect of log $(y - z)$ plotted against time on normal paper. This fact forms the key to a full understanding of the method.

We shall now formulate the steps required for testing whether an experimental curve corresponds to the formula

$$y = C_1 e^{-\alpha t} + C_2 e^{-\beta t}. \tag{1}$$

1) Plot the experimental values on semi-log paper.
2) If the "latter" part of the curve is approximately a straight line, then extend this linear part to the left until it touches the y-axis.
3) Take the differences of the values of the original curve and the extrapolated part of the linear curve. Plot these differences (with the symbols used above: $y - z$) on (the same) semi-log paper.
4) If a straight line is thus obtained, this means that the curve may indeed be represented by equation (1).

* It is often called *exponential stripping.*

Remarks:

a) If we are dealing with an equation of the form

$$y = A - C_1 e^{-\alpha t} - C_2 e^{-\beta t}$$

where A = constant, then we may proceed in the same way, provided we plot $A - y$ on semi-log paper, for then we have

$$A - y = C_1 e^{-\alpha t} + C_2 e^{-\beta t}$$

where we treat $A - y$ as the dependent variable.

b) It is easy to prove, that if we do not obtain a straight line after plotting $y - z$ on semi-log paper, but a line curved at first and (practically) linear in the latter part, we may *repeat the method just described* and if then a straight line is obtained we may conclude that the curve is described by a third order equation, or

$$y = C_1 e^{-\alpha t} + C_2 e^{-\beta t} + C_3 e^{-\gamma t}.$$

Of course, repetition may be required if the curve requires, say, 5 exponential terms.

It should be stressed, however, that when an (experimental) curve can be split up in, say, more than three straight lines, then it becomes dubious whether the underlying process may be described by a sum of exponential terms.

Let us be more specific: suppose five straight lines are obtained, then we represent the experimental curve by an *empirical* equation involving five exponential terms. But such an equation is no more than an *empirical* one, i.e. it has little or no theoretical significance, which means that it can tell us nothing about the underlying mechanisms. Only when three or less straight lines are obtained we can be reasonably certain that the equation thus found possesses theoretical significance.

We are now in the position to study the result obtained by Sapirstein *et al.*[1]

A single dose of creatinine (2.0 gram in 50 cc of saline) is rapidly (i.e. "instantaneously") injected intravenously. The plasma "disappearance curve" was then determined by taking blood samples at 5-minute intervals in heparinized tubes for 60 minutes. (The plasma creatinine was determined by the method of Bonsnes and Taussky.)

The results were plotted on semi-log paper (time as abscissa) and are shown in Fig. 153.

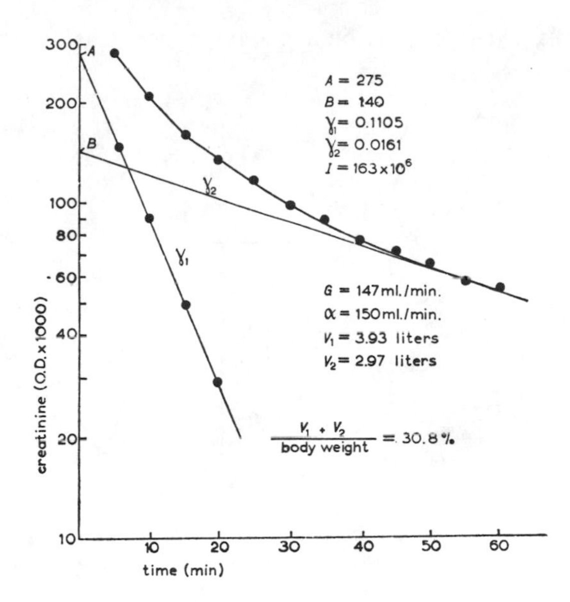

Fig. 153. Disappearance curve for creatinine after single injection in a dog.[1] A, B, γ_1, γ_2, V_1, V_2, C_{cr}, and α have the significance given in the text.

It is seen from this figure that the curve can be resolved into two straight lines by the method described above. This means that the curve can, at least formally, be represented by an equation of the type

$$y = C_1 \mathrm{e}^{-\alpha t} + C_2 \mathrm{e}^{-\beta t}. \tag{1}$$

Our problem will be to justify and derive this equation from first principles. This is important, since it permits us to see the physiological meaning of the constants and to draw important practical conclusions. As must always be done, we shall replace the actual conditions in the body by a model. We shall assume that there are two spaces present, namely the blood-space and the tissue space (this classification is a simplification of that of the original paper) and that free diffusion of creatinine occurs at their boundary. Diffusion means that we assume that Fick's law holds, i.e. that the amount of substance transported in unit time is proportional to the difference in concentration of the substance in the two compartments. We further assume that there is a "leak" in the blood compartment, this leak representing the kidneys and it is assumed that creatinine is lost through this pathway at a rate proportional to its concentration in the blood compartment.

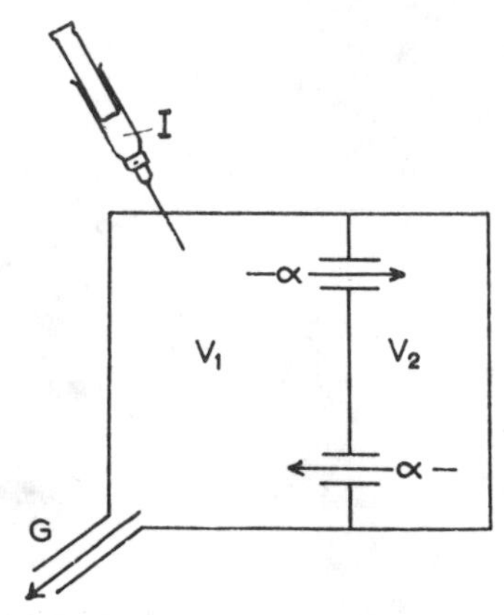

Fig. 154. System for which the equations are derived. The excretory clearance is designated G, and is the virtual (not the actual) volume leaving V_1 for the outside. [From Sapirstein *et al.*[1]].

From the above, it is seen that our model may be represented by Fig. 154.

After these major assumptions we shall specify our further assumptions. It is essential to be aware of the (tacit) assumptions that are introduced if we construct a mathematical model. If the assumptions are too far removed from actual conditions, then, of course, the result becomes of little value.

1) We assume that the creatinine injection in the blood compartment is "instantaneous" (this can be approximated by a very rapid injection) and that the creatinine is instantaneously homogeneously mixed in the blood space. This assumption means that at $t = 0$, *all* the injected creatinine is homogeneously distributed in the blood compartment. This can, of course, never be true, but must be considered as a simplification. In reality, some creatinine will escape to the tissue space by diffusion during the (short) time necessary for homogeneous distribution in the blood space. But if (as we tacitly assume) the rate of escape to the tissue space is slow relative to the mixing process in the blood space, then we may use the above assumption as a mathematical idealization.

2) We assume that creatinine is also homogeneously distributed in the tissue compartment.
3) It is assumed that exogenous creatinine is not metabolized to any significant degree and that it disappears from the body only by way of the kidneys.
4) It is assumed that the volumes of the compartments are constant.

To avoid needless repetition we shall no longer give complete explanations at every step. In particular, the solution of the differential equation we shall now derive is explained in detail in Chapter 9, §10.

Let V_1 and V_2 be the volumes of the blood space and tissue-space, respectively (see Fig. 154).

Let C_1 and C_2 be the concentrations in V_1 and V_2 respectively.

Let m be the amount creatinine present in V_1 at any instant. Then, from the above assumptions, it should be clear that the rate of decrease of this amount is given by

$$\frac{\mathrm{d}m}{\mathrm{d}t} = -GC_1 - \alpha(C_1 - C_2) \tag{8}$$

where G and α are constants of proportionality. (G is called the excretory clearance by the authors.)

Since $C_1 = m/V_1$, we have $m = C_1 V_1$, so that (8) may be written as

$$\frac{\mathrm{d}V_1 C_1}{\mathrm{d}t} = -GC_1 - \alpha(C_1 - C_2)$$

and since by assumption V_1 is constant, it may be brought before the differential sign (show this) to give

$$V_1 \frac{\mathrm{d}C_1}{\mathrm{d}t} = -GC_1 - \alpha(C_1 - C_2). \tag{9}$$

We must try to eliminate C_2 from (9).

For this purpose we observe that the amount injected I, is equal to the amount in V_1, i.e. $V_1 C_1$, plus the amount in V_2, i.e. $V_2 C_2$, plus the amount excreted (at time t), which we shall call L_t, or

$$I = C_1 V_1 + C_2 V_2 + L_t. \tag{10}$$

Such a relation is called a condition of continuity. It is mostly a rather obvious relation, but may be of great value for establishing the differential equation.

From (8) or (9) it may be concluded that

$$\frac{dL_t}{dt} = + GC_1. \tag{11}$$

Hence, the amount excreted by the kidneys at time t is from (11)

$$L_t = G\int_0^t C_1 \, dt. \tag{12}$$

Substituting (12) in (10) and solving for C_2 we have

$$C_2 = \frac{I - G\int_0^t C_1 \, dt - C_1 V_1}{V_2}. \tag{13}$$

Substituting (13) in (9) yields

$$V_1 \frac{dC_1}{dt} = -GC_1 - \alpha C_1 + \frac{\alpha I}{V_2} - \frac{\alpha}{V_2}\int_0^t GC_1 \, dt - \frac{\alpha C_1 V_1}{V_2}. \tag{14}$$

We can easily get rid of the integral by differentiating (14). Then,

$$V_1 \frac{d^2 C_1}{dt^2} = -G\frac{dC_1}{dt} - \alpha \frac{dC_1}{dt} - \frac{\alpha G}{V_2} C_1 - \frac{\alpha V_1}{V_2}\frac{dC_1}{dt} \tag{15}$$

or, rearranging,

$$\frac{d^2 C_1}{dt^2} + \left[\frac{G+\alpha}{V_1} + \frac{\alpha}{V_2}\right]\frac{dC_1}{dt} + \left(\frac{\alpha G}{V_1 V_2}\right) C_1 = 0 \tag{16}$$

which is clearly a homogeneous linear second order differential equation with constant coefficients, whose general solution is

$$y = Ae^{-\lambda_1 t} + Be^{-\lambda_2 t} \tag{17}$$

where λ_1 and λ_2 are the roots of the characteristic (or auxiliary) equation

$$\lambda^2 + \left[\frac{G+\alpha}{V_1} + \frac{\alpha}{V_2}\right]\lambda + \frac{\alpha G}{V_1 V_2} = 0. \tag{18}$$

It is thus seen that the experimental finding of the second-order response (forcing zero) can be explained by theory.

Because of lack of space we cannot discuss the clinical implications and the analysis of the creatinine clearance which results from the above theory. For this the reader is referred to the original paper.

It should be pointed out that in Fig. 153 the linear part of the experimental curve is a bit too short. It would have been better if the experiment had been continued somewhat longer. Then, the slope of the linear part might have been established with greater certainty.

It should also be stressed that the correspondence between the experimental curve and the theoretical curve does *not* guarantee that the theory is correct. Often additional evidence is required. But if the "first principles" of the theory are not too hypothetical, but based upon experimental evidence or plausible "guesses" then this correspondence may strongly suggest the "correctness" of the theory.

Very many studies, both clinical and theoretical, based on similar models and analyses have been published. See, for example, references 2 to 5 for this section and references 1, 2, 6, 7 and 8 for §16.

16. The Form of Physiological Clearance Curves in General: a Different Approach

For metabolic studies in particular, the human subject or animal receives a single injection of a labelled molecule, which is often a radioactive tracer (Chapter 7, §18). The tracer then very quickly becomes uniformly mixed with the whole of the blood, and then the curve for tracer concentration as a function of time provides useful physiological information.

The main problem is very similar to the one in §15. However there was no tracer in that case, and this brings in two essential new points. One is that the tracer must behave chemically and biologically like the "tracee". For example if the tracer is ^{131}I in serum albumin, or ^{47}Ca or ^{45}Ca in blood plasma, tissue and bone, the radioactive isotope must behave like normal iodine or calcium. Then what is measured in order to obtain the tracer concentration is the radioactivity in (for example) a sample of blood, at time t from injection. If m_0 molecules were injected, from equation (4) in Chapter 7, §18a, $m_0 e^{-kt}$ molecules will still be radioactive at time t, irrespective of where they are at that moment. It follows that the number of molecules that were radioactive *when injected* is always e^{kt} times the number observed to be radioactive at time t, and that this correction factor must always be applied when converting an observed count of radioactivity into the corresponding number or concentration of labelled molecules. In what follows it is assumed that this correction has been made. The result then corresponds to a concentration time curve like that for creatinine in the last section.

In this example, the analysis and interpretation was based on a 2-compartment model in which this curve was assumed to be of the form

$$z = a_1 e^{-b_1 t} + a_2 e^{-b_2 t} \tag{1}$$

and if there are 3 compartments there will be a sum of 3 such negative exponentials. Many hundreds of published curves of this kind have in fact been fitted in this way.[1,2]

Many of those curves however also fit either a negative power of the time

$$z = At^{-\alpha} \tag{2a}$$

or a product of negative power and exponential

$$z = At^{-\alpha} e^{-\beta t} \tag{2b}$$

which is a gamma distribution (see Fig. 95 and the associated text), even when they are well fitted by 2 or 3 negative exponentials, over the same range of time. This has led to a controversy in which the advocates of fitting by power function question most of these hundreds of fits by negative exponentials, and the so called "compartmental models" on which they are based.[3–5]

Obviously it is important to consider what the function z in (1) looks like if $y = \log z$ is plotted against $x = \log t$ so we have to consider

$$y = \log (a_1 e^{-b_1 e^x} + a_2 e^{-b_2 e^x}). \tag{3}$$

Suppose $b_1 < b_2$, then the second exponential clearly becomes negligible, in time, compared with the first one – the ratio of the two terms can be written as

$$a_2 e^{-(b_2 - b_1) t} / a_1$$

which tends to zero as t tends to infinity. However if $a_2 \gg a_1$ there will in most cases be a period during which the second (short term) exponential dominates the course of the curve. Then we shall have:

$$y \simeq y_2(x) = \log a_2 - b_2 e^x .$$

Whilst later on, and more exactly, we shall have:

$$y \to y_1(x) = \log a_1 - b_1 e^x .$$

The curves for both $y_1(x)$ and $y_2(x)$ will be like those of Fig. 76, (Chapter 6, §2) but upside down and displaced upwards. This is clearly seen in Fig. 155. This is a plot of $\log z$ vs. $\log t$ for the sum of two exponentials shown

$$z = 0.0244e^{-0.289t} + 0.975e^{-7.2t}.$$

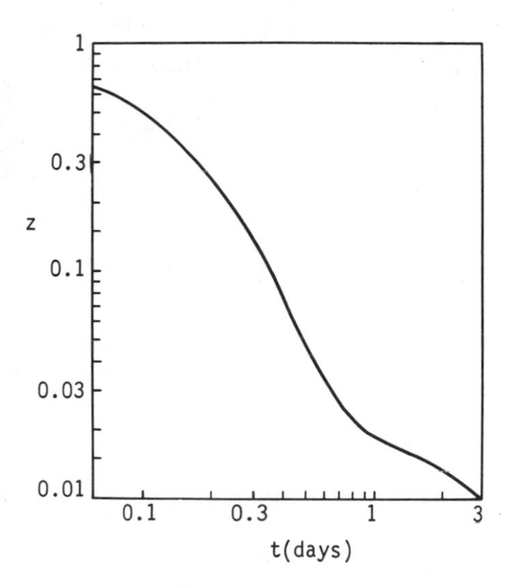

Fig. 155. Replot on log-log paper of a fit by Pollycove and Mortimer [J. Clin. Invest. **40** (1961) 753–782] to radioactive iron ^{59}Fe blood (plasma) activities in a healthy man.

This was fitted to plasma activities of radioactive iron ^{59}Fe up to 3 days from injection.[6] Beyond 3 days the curve continues to bend downwards; this part corresponds to $y_1(x)$ above. Clearly, also, the second derivatives $y_1''(x)$ and $y_2''(x)$ are negative throughout, but whenever there are two clearly separated exponentials, there will be a transition period – in this case between about $\frac{1}{2}$ and 1 day, where the curve is bending back, since the functions are continuous; thus $y''(x)$ is first negative, then positive, and then negative again. Hence by Rolle's theorem (Chapter 3, §8) there must be two points of inflexion.

Similarly, if

$$z = A_1 e^{-b_1 t} + A_2 e^{-b_2 t} + A_3 e^{-b_3 t},$$

and ($\ll$ means "much less than")

$$b_1 \ll b_2 \ll b_3, \quad A_1 \ll A_2 \ll A_3,$$

$y''(x)$ will be in turn −, +, −, +, and −; there will be 4 points of inflexion. These are shown in Fig. 156. It was obtained in a similar way to the previous one for a patient with a blood disease. It shows another surprising feature: $y''(x)$ is almost zero from 1 to 5 days: the curve is almost straight; and in fact

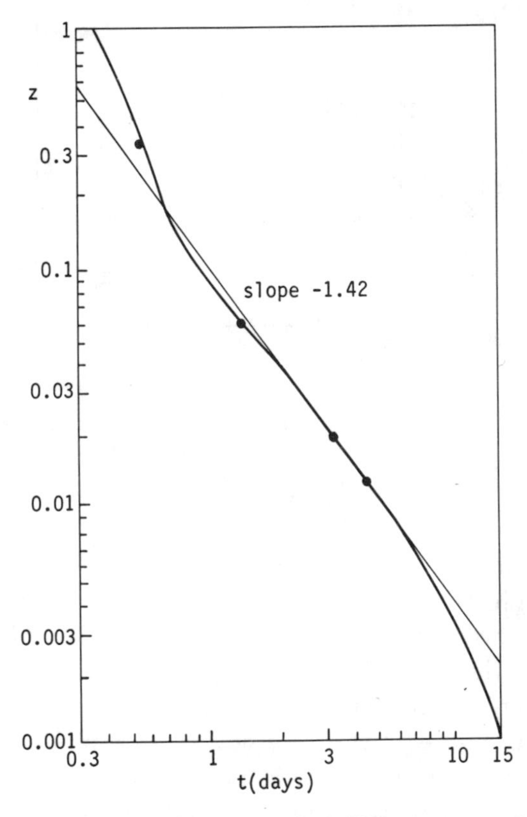

Fig. 156. Replot of $z = 0.981e^{-6.74t} + 0.016e^{-1.034t} + 0.0028e^{-0.21t}$ from the same series as Fig. 155, for a patient with endogenous hemochromatisis; 4 points of inflexion are shown, and the straight line joining the third and fourth.

$$z \simeq At^{-1.42}.$$

From very many numerical studies it has been shown that, if

$$z = At^{-\alpha} \quad \text{or} \quad z = At^{-\alpha}e^{-\beta t} \qquad 0 < t_1 < t < t_2,$$

then z can be fitted by sums of 2 or more negative exponentials: the number of them depends on the relative range of time, t_1 to t_2.

The converse is not true, as Fig. 155 shows. It ought to be possible to find a mathematical rule for testing whether a power function approximation exists. No simple one has been found, but the following result could be useful. It applies to two successive exponentials, so for a sum of 3 it could be used for the first and second, and then the second and third. We will leave the full derivation as an exercise in applying the expansions for logarithms and exponentials in Chapter 6 but we give the main steps. Starting with the expression for z, equation (1), we write

$$B = \tfrac{1}{2}(b_1 + b_2); \quad b = \tfrac{1}{2}(b_2 - b_1)$$

$$A = \tfrac{1}{2}(\log A_1 + \log A_2); \quad a = \tfrac{1}{2}(\log A_2 - \log A_1).$$

Then, writing $y = \log z$ as before,

$$y = -Bt + A + \log(e^{a-bt} + e^{-a+bt}).$$

Now we put $T = a/b$, and $\log t - \log T = u$. This gives

$$y + BTe^u = A + \log 2 + \log\cosh\{a(e^u - 1)\}$$

$$= A + \log 2 + \tfrac{1}{2}a^2(e^u - 1)^2 - \tfrac{1}{12}a^4(e^u - 1)^4 + \ldots \qquad (4)$$

(cf. Chapter 6, §7, equation (16) and problem 4). Write $BT = ba/B = c$. Expanding in powers of u now gives

$$y = A + \log 2 - c - cu + u^2(a^2 - c)/2 + (3a^2 - c)u^3/6 +$$

$$(7a^2 - 2a^4 - c)u^4/24 + (15a^2 - 20a^4 - c)u^5/120 + \ldots. \qquad (5)$$

According to (5) the $\log z$ $\log t$ plot should be nearly a straight line when t is fairly near to T and $a^2 - c$ is small. This has been verified numerically, but this rule appears not to cover all cases, and of course only applies strictly when there are only two terms for z. The straight lines obtained by joining successive points of inflexion appear to be closer still to the exact curve as Figs. 156 and 157 show.

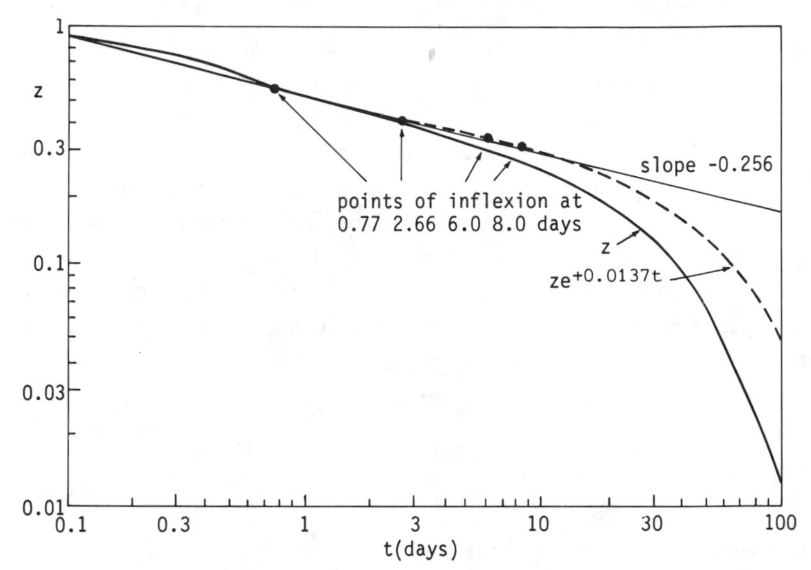

Fig. 157. Replot of $z = 0.47e^{-2.17t} + 0.19e^{-0.359t} + 0.34e^{-0.0327t}$ fitted by Takeda and Reeve, J. Lab. Clin. Med. **61** (1963) 183–202 to plasma activities of ^{131}I labelling albumin in a healthy man, and the corresponding curve for $ze^{\beta t}$ where $\beta = 0.0137$, on log-log paper, with the 4 points of inflexion almost exactly in line.

By differentiating (4) twice we find, writing $\theta = t/T$, that

$$\frac{d^2y}{du^2} = \theta\,\{a^2\theta\,\mathrm{sech}^2(a\theta-a) + a\tanh(a\theta-a) - BT\}$$

$$= \theta f(\theta)\text{ say}\ \dots \tag{6}$$

Since $BT = a(b_2 + b_1)/(b_2 - b_1)$, $BT > a$. So as $\theta \to \infty$, $f(\theta)$ approaches $a - BT$ which is negative; also $f(0) = -BT - a\tanh a$, again negative. The maximum of $\theta f(\theta)$ can be either positive or negative, if it is positive there are two points of inflexion where $f(\theta) = 0$, otherwise there are none.

Finally, if we write $a(\theta - 1) = X$, and expand $f(\theta)$ in powers of X, we find that the points of inflexion are where X satisfies:

$$2aX = c - a^2 + a^2X^2 + 4aX^3/3 - 2a^2X^4/3 - 4aX^5/5\dots. \qquad (7)$$

If $c - a^2$ is fairly small, there is one solution where X is fairly small, and approximately equal to $\frac{1}{2}(c - a^2)/a$. Using this, the exact value and the other solution could be found as in Chapter 4, §9.

The fit to the product of a power function and an exponential

We can write (2) as

$$\log z = \log A - \alpha \log t - \beta t. \qquad (8)$$

Hence:

$$\frac{dz}{dt} = -z(\beta + \alpha/t)$$

which tends to $-\beta z$ as t tends to infinity. This is the limiting end situation, when the tracer is uniformly mixed with the tracee, and the relative rate the tracer leaves the system is a constant β. Such an end exponential is observed in many accurately analysed curves for radioactive iodine ^{131}I labelling human serum albumin,[7,8] for some weeks after a single injection. These often fit excellently to the sum of three negative exponentials, such as the case of Fig. 157. Here the log z log t curve has two points of inflexion. But the corresponding plot of

$$\log z + 0.0137t$$

has 4 points of inflexion almost exactly in line, up to about 20 days. In fact, from ca. 0.1 to 20 days,

$$z = 0.518t^{-0.256}e^{-0.0137t}. \qquad (9)$$

For an observed curve thought to fit (2), β would be estimated from the log z log t plot by measuring the deviations when this is curving downwards from a straight line; from (5) these deviations will be proportional to t.

The same computing procedure has been applied to the curve in Fig. 153 and to other sums of exponentials fitted to this set of clearance curves. These, too, appear to be fitted by (2). The values of G, α, V_1 and V_2 were given by the authors for every curve, but not the exponentials. Recalculating these by computer showed that between ca. 10 minutes and 1 hour from injection, Fig. 153 corresponds to

$$y \triangleq \text{constant} \times t^{-0.646} e^{-0.0052t} . \tag{10}$$

At the time of our first edition, the compartmental model was accepted by almost everyone working on clearance data. We have stated the assumptions in full, and so we leave this section unchanged, so that the reader can see the two approaches side by side. We think compartmental models can be valid and useful where the so called compartments are not geographically separate parts of the body, but, for example, constituents of a fluid. For example a red cell in blood always has contact with the plasma surrounding it.

So far (up to December 1972) some 80 published sets of compartmental data have been reanalysed by one of the authors of this book (M.E. Wise). For most of them this had to be done by investigating $y(x)$ as above for the corresponding sums of negative exponentials which now amount to more than 900. In at least three quarters of these strong indications were found that parts of the clearance curves are very close to negative powers of time and/or to gamma curves.

This means that the curve can be described by fewer parameters than are needed to fit them by sums of negative exponentials. Hence by the economy principle (Occam's razor) we ought to accept the power functions; in fact this is a good illustration of this wider issue. For although over a particular range of time, $At^{-\alpha}$ or $At^{-\alpha}e^{-\beta t}$ can almost always be fitted well by a sum $\Sigma A_i \exp - b_i t$, $n = 2$, 3 or 4, the converse is clearly not true: we have only to imagine A_2 say, as in the sum in Fig. 155, getting larger and larger, so that this exponent comes to dominate more and more of the curve. So if we accept the compartmental model, we have the problem of explaining why the 4 or 6 parameters A_i, b_i should always happen to have a particular relationship between them as indicated by (4) for example.

There is another general issue. Just as in the last section but one, we have to make a clear distinction between an observed curve, a mathematical model that fits it or fits a whole set of such curves, (or a purely empirical description of the curve by mathematical formulae) and a biological or physical (or biophysical) model that is consistent with the mathematical model.

It is surprisingly hard to define precisely the term mathematical model, which we have already used in several different contexts.* If an empirical curve-fitting formula has an impressively low number of parameters then it is often given a higher status (elevation to knighthood) and called a mathematical model, in the hope that there is more "theory" hidden in it than in a many-parameter fit. The sets of exponentials provide a valid *mathematical* model if they fit the curves well and if there are not too many exponentials, as they do for the sets as in Figs. 153, 155, 156 and 157 and many others. So a report of these fits usefully describes the data, even if the corresponding biophysical model is wrong. But so often, alas, only the biophysical parameters are given, hence if an interested reader wants to try out another hypothesis he cannot do so. It seems to us that it is always worth while to look for mathematically simple forms and report them if they are found, even if they cannot be interpreted biologically.

We have still said nothing about our interpretation of all these power laws. Briefly, it is that individual tracer molecules tend to move *through* parts of a physiological system, with preferred directions of motion. Their lifetimes in the system will then have distributions something like the random walk ones of equations (1) and (6), Chapter 7, §15, see also Chapter 9, §21, equation (24). But such tracer particles will be moving through heterogeneous tissue, and back and forth from one part of the body, to another, e.g. between blood and bone if the tracer is radioactive calcium.[9] Wise has shown that under such conditions the powers $-\frac{1}{2}$ and $-\frac{3}{2}$ can become almost any other negative ones.[10, 11] The mathematical statistics is not too advanced, but it is lengthy and rather outside the scope of this book. However the simple direct interpretation in terms of physiological compartments is lost, and nothing as complete has as yet taken its place. But there seems to be a gain in describing the data more simply, and the approach is biologically very reasonable.

Concluding note (added July 1973)

The position has got no clearer recently. Many more clearance curves are being analysed and interpreted as sums of negative ex-

* E.g. in §§ 2, 6, 7, 14 and 15 in this chapter.

ponentials than as negative powers or gamma curves. The fields of application differ widely and the workers in one field have, inevitably, few bio-medical interests in common with those in another. The subjects include cerebral blood flow, flow through the kidneys, various applications in pharmokinetics, and the metabolism of bone seeking radioactive isotopes, such as those of barium, calcium and strontium.

Clearance curves for ^{45}Ca, ^{47}Ca and ^{85}Sr have been extensively studied, especially in order to assess radiation doses from fall-out following nuclear tests or following an accident where radioactive atoms are released. The curves for calcium (possibly strontium also) for individuals nearly fit *two* straight lines in a plot of $\log z$ vs $\log t$, which intersect at about 1 day[3, 11], and the second straight line is always much steeper than the first. A task group concerned with radiation protection have proposed describing the corresponding retention curves, i.e. the proportion still in the body at time t:

$$R(t) = \frac{\int_t^\infty z\,\mathrm{d}t}{\int_0^\infty z\,\mathrm{d}t}\,, \tag{11}$$

with a *mixture* of power or gamma functions as eq. (2) and sums of negative exponentials[12], namely*

$$R = (1-p)\mathrm{e}^{-mt} + p\epsilon^b (t+\epsilon)^{-b}\,\{\beta \mathrm{e}^{-r\lambda t} + (1-\beta)\mathrm{e}^{-\sigma r\lambda t}\}. \tag{12}$$

Clearance curves for kidneys have nearly always been analysed by compartmental models[13, 14], but these may really be complicated functions of time involving negative powers. On the other hand very many clearance curves for drugs do consist mainly of single negative exponentials, and many others have been analysed as in § 15.[15] Our impression up to now is that possibly two thirds or more of these would fit equation (2a) or (2b) better, over the same range of time.

* Here all the constants are positive, and p, r and β are ratios that lie between 0 and 1; $0.1 < b < 0.5$, ϵ is about 1 day, and is introduced mainly to make $R(0)$ finite. The time constant m is between 0.1 and 0.8 per day, and $\sigma\lambda/m < 0.001$. Hence when t is less than a year the terms in β and $1-\beta$ are negligibly different from 1. The authors of this extensive study point out that R does not fit well for t less than 1 day, but they have endeavoured to estimate its order of magnitude for weeks or years after intake of a single dose of radioactive Ca, Sr, Ba, or radium!

The best advice we can give at present is, we believe, to assume nothing in advance and in all cases simply to try plotting $\log z$ against both t and $\log t$; if many points of one plot then lie on or (randomly about) a straight line there ought to be signs of systematic deviations from a straight line in the other plot.

17. The Form of the Physiological CO_2 Dissociation Curve

Introduction

The CO_2 absorption curve of blood depends upon the oxygen saturation of hemoglobin. Similarly, the so-called oxygen dissociation curve of blood depends upon the pH. Let us, to illustrate the problem, fix our attention on the CO_2 absorption curve of (whole) blood.

In the mixed venous blood (coming to the lungs), the oxygen saturation is about 70 per cent, while in the arterial blood (leaving the lungs), this value is close to 100 per cent. It thus follows that during the passage of blood through the lung capillaries, the "point" in the CO_2 concentration – CO_2 pressure diagram traces a path from the venous CO_2 absorption curve (saturation 70 per cent), to the arterial CO_2 absorption curve (saturation 100 per cent).

This path is commonly called the physiological dissociation curve, for which we have already had a clinical application (Chapter 2 §7). Another related application is in measuring arterial compliance.[5]

We shall briefly present a theoretical analysis of the nature of (1) the physiological CO_2 dissociation curve, (2) the physiological oxygen dissociation curve, (3) the simultaneous diffusion of O_2 and CO_2 along the lung capillary, and (4) the pH changes along the lung capillary.[1–4]

The basic equations

Consider a single lung capillary. Variable diameter and variable permeability properties with respect to O_2 and CO_2 along its length are permitted. Let the capillary be subdivided into "volume elements" having equal diffusion capacities with respect to O_2 and CO_2. Let the distance along the capillary be denoted by x.

Consider a "volume element" of thickness Δx, where Δx is sufficiently small in comparison with the total length of the capillary*.

Let the blood flow through the capillary, Q (l/min.), flow from left to right in Fig. 158. Subscripts 1 and 2 in Fig. 158 denote en-

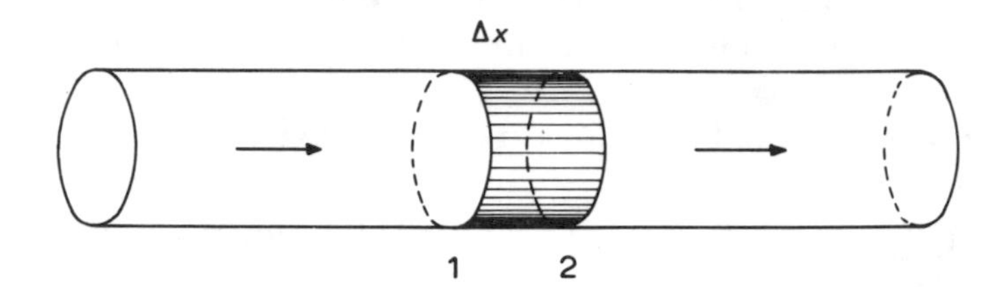

Fig. 158. For explanation see text.

trance and exit respectively of the volume element. Let the total CO_2 concentration (bound and free) in the blood be denoted by $C^{b}_{CO_2}$, where the superscript b denotes whole blood (later we shall have to consider $C^{p}_{CO_2}$, where the superscript p denotes plasma). This concentration is expressed in l CO_2 (S.T.P.D.)/l blood (non-dimensional). The volume of CO_2 entering the volume element per unit time, is clearly $QC^{b}_{CO_2,1}$ where, as stated, the subscript 1 indicates the entrance of the volume element. Similarly, the volume of CO_2 leaving the volume element per unit time, may be written as $QC^{b}_{CO_2,2}$.

Their difference, which can be written as $Q\Delta C^{b}_{CO_2}$ represents the volume of CO_2 leaving the volume element per unit of time by diffusion across the capillary membrane to the lungs (alveolar space). Using the firmly established diffusion "hypothesis", we may clearly write:

$$Q\Delta C^{b}_{CO_2} = -A(P_{CO_2} - P_{A_{CO_2}})\Delta x \tag{1}$$

where the constant A represents the "diffusion capacity of the volume element"** with respect to CO_2. P_{CO_2} denotes the CO_2

* Instead of choosing the distance x along the capillary as the independent variable, it is more logical to select the "diffusion capacity" (for O_2 or CO_2) along the capillary as the independent variable. This has actually been done in the numerical computations, but it is felt that presenting the theory in terms of distance x gives a more vivid picture of the situation.

** The term "diffusion capacity" is not confined to the permeability properties of the capillary membrane with respect to the particular gas, but includes such factors as diffusion limitation in the red cell and reaction rates.

pressure in the blood, while $P_{A_{CO_2}}$ denotes the CO_2 pressure in the alveolar space (lungs). The minus sign in front of A indicates that the quantity on the left is diminished by the diffusion process.

It is assumed here that the $P_{A_{CO_2}}$ is constant, that is, independent of x. In other words, it is assumed that no P_{CO_2} gradient exists in the alveolar space along the capillary. This assumption, although not necessary mathematically, is entirely reasonable, as can be shown by numerical computations based on the physical properties of the system.

It should be noted that no constancy of $P_{A_{CO_2}}$ with respect to time is postulated. The introduction of time as an independent variable leading to partial differential equations should be viewed as an elaboration of the theory. It should be realized, however, that ignoring the time dependency of the $P_{A_{CO_2}}$ (during a single breathing cycle) in no way limits the validity of the present approach. It is clear that in order to obtain a solution, the quantity $C^b_{CO_2}$ on the left in equation (1) must be expressed in terms of P_{CO_2}. This is easily done by using an empirical equation describing the CO_2 dissociation curve (CO_2 absorption curve) of whole blood.

When the CO_2 absorption curve of (whole) blood is plotted on double log paper, a straight line is obtained over the range of P_{CO_2} values from 0 to 100 mm Hg. In our present analysis of the normal case, we are concerned only with P_{CO_2} values between 40 and 46 mm Hg. It is found that within this small range the CO_2 dissociation curve of (whole) blood may be described by the empirical linear equation:

$$C^b_{CO_2} = \alpha + \beta P_{CO_2} - \gamma S \tag{2}$$

where α, β, γ are positive constants and where S denotes the oxygen saturation of hemoglobin (fractional). Numerical values for α, β, and γ will be given later. Substituting equation (2) into (1) gives

$$\frac{Q\Delta(\alpha + \beta P_{CO_2} - \gamma S)}{\Delta x} = -A(P_{CO_2} - P_{A_{CO_2}}). \tag{3}$$

Since Δx is "sufficiently small" we may replace the difference quotient by the derivative, and write, noting that Q, α, β, and γ are all constant,

$$Q\beta \frac{dP_{CO_2}}{dx} - Q\gamma \frac{dS}{dx} = -A(P_{CO_2} - P_{A_{CO_2}}). \qquad (4)$$

Let us here pause again. Before continuing the development of equation (4) where dS/dx is obviously the "bottleneck" we shall turn to the diffusion equation for oxygen, since we require a set of 2 simultaneous differential equations.

Just as in the case of CO_2, we may write down the diffusion equation of O_2 for the volume element,

$$Q\Delta C_{O_2} = B(P_{A_{O_2}} - P_{O_2})\Delta x \qquad (5)$$

where C_{O_2} is the oxygen concentration in (whole) blood (sum of the oxygen bound to hemoglobin and in physical solution). The oxygen concentration is expressed as l O_2 (S.T.P.D.)/l blood (non-dimensional)*, B is the diffusion capacity of the volume element with respect to O_2, $P_{A_{O_2}}$ is the O_2 pressure in the alveolar space, while P_{O_2} is the O_2 pressure in blood. Note that normally oxygen diffuses from the alveolar space into the capillary. Now, the oxygen concentration, being the sum of bound and free O_2, may be written as

$$C_{O_2} = KS + LP_{O_2} \qquad (6)$$

where K is the oxygen capacity of the blood, and L is the solubility coefficient of oxygen for blood (note that the S values lie between 0 and 1). The second term on the right clearly represents Henry's law. Substituting equation (6) into (5) yields, after some rearrangements and replacing the difference quotient by the derivative,

$$QK \frac{dS}{dx} + QL \frac{dP_{O_2}}{dx} = B(P_{A_{O_2}} - P_{O_2}). \qquad (7)$$

When we compare equation (4) and (7) we note that in both equations dS/dx appears. Our next problem is therefore to obtain an expression for this quantity. This requires an expression for the oxygen dissociation curve. Adair's theoretically sound formula (see Chapter 2, §4) is unwieldy and quite unsuitable for the present purpose.

* The superscript b used in the case of CO_2 may be omitted here, since we are not using *plasma* oxygen concentration.

Visser and Maas[6] have presented an empirical equation that adequately describes the oxygen dissociation curve over the whole range,

$$S = \{1 - e^{-\delta P_{O_2}}\}^2 \tag{8}$$

where δ is a positive factor. Equation (8) contains only a single parameter, δ, that actually depends upon the pH. By fitting equation (8) to experimental points obtained at different pH values (7.2, 7.4, 7.6), taken from the Handbook of Respiration,[7] the relationship between the quantity δ and pH could be found. In the normal case under study, the pH values fall within the narrow interval 7.35–7.45 (this is a liberal estimate) where we may use a linear approximation:

$$\delta = \theta\,(\mathrm{pH}) - \sigma\,! \tag{9}$$

Here θ is found to be 0.06778 and σ to be 0.4540. Substituting equation (9) in (8), we obtain

$$S = \{1 - e^{(\sigma - \theta\,\mathrm{pH}) P_{O_2}}\}^2\,. \tag{10}$$

$\mathrm{d}S/\mathrm{d}x$ is now obtained by noting that, since S is a function of 2 variables ($S = f(\mathrm{pH}, P_{O_2})$), we can write,

$$\frac{\mathrm{d}S}{\mathrm{d}x} = \frac{\partial S}{\partial \mathrm{pH}} \cdot \frac{\mathrm{d\,pH}}{\mathrm{d}x} + \frac{\partial S}{\partial P_{O_2}} \cdot \frac{\mathrm{d}P_{O_2}}{\mathrm{d}x}\,. \tag{11}$$

$\partial S/\partial \mathrm{pH}$ and $\partial S/\partial P_{O_2}$ are easily found by differentiating equation (10) with respect to pH and P_{O_2} respectively. Substituting these results in equation (11) we finally obtain, when we introduce, to simplify the notation, $u = \mathrm{pH}$, $z = P_{O_2}$ and $w = \sigma - \theta u$,

$$\frac{\mathrm{d}S}{\mathrm{d}x} = 2(e^{zw} - e^{2zw})\left(\theta z \frac{\mathrm{d}u}{\mathrm{d}x} - w \frac{\mathrm{d}z}{\mathrm{d}x}\right). \tag{12}$$

Substituting equation (12) in (4) and (7), and writing $y = P_{CO_2}$,

$$Q\beta \frac{\mathrm{d}y}{\mathrm{d}x} - 2Qy(e^{zw} - e^{2zw})\left(\theta z \frac{\mathrm{d}u}{\mathrm{d}x} - w \frac{\mathrm{d}z}{\mathrm{d}x}\right) =$$

$$= -A(y - P_{A_{CO_2}}), \tag{13}$$

$$2QK(e^{zw}-e^{2zw})\left(\theta z\frac{du}{dx}-w\frac{dz}{dx}\right)+QL\frac{dz}{dx}=$$

$$=B(P_{A_{CO_2}}-z). \qquad (14)$$

We still need a (third) relationship between u, z, and y. This is obtained as follows.

Visser[8] has shown that the nomogram of Van Slyke and Sendroy[9] for the relationship between CO_2 content of whole blood $C^b_{CO_2}$, CO_2 content of plasma $C^p_{CO_2}$ (superscript p denotes plasma), pH and oxygen saturation, S, may be expressed by the following empirical formula (except for very high values of pH),

$$\frac{C^b_{CO_2}}{C^p_{CO_2}}=1-\frac{2.15\,K}{(2.244-0.422S)(8.740-\text{pH})} \qquad (15)$$

where, as before, K is the oxygen capacity.

Futher, the Henderson–Hasselbach equation[7] relating P_{CO_2}, pH, and $C^p_{CO_2}$ can be written as

$$C^p_{CO_2}=0.00067\,P_{CO_2}\,\{1+10^{(\text{pH}-6.10)}\}. \qquad (16)$$

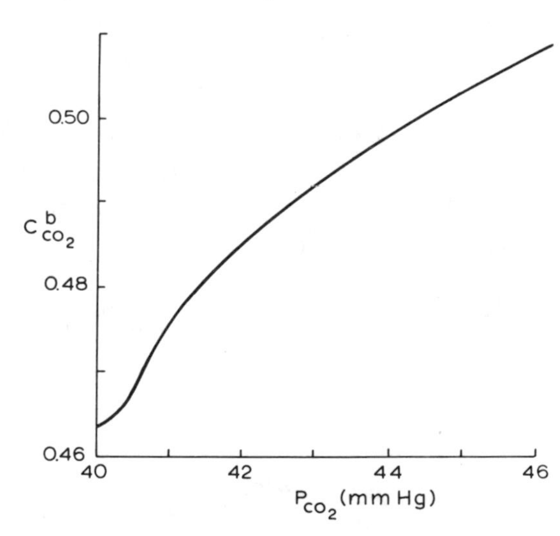

Fig. 159. The physiological CO_2 dissociation curve based on "normal" numerical values of the parameters.

Substituting equations (10) and (16) in (15) finally yields

$$\frac{0.3478+0.00540y-0.0650\{1-e^{(\sigma-\theta u)z}\}^2}{0.00067y\{1+10^{(u-6.10)}\}}$$

$$=1-\frac{2.15K}{[2.244-0.422\{1-e^{(\sigma-\theta u)z}\}^2](8.740-u)}. \qquad (17)$$

Equations (13), (14) and (17) constitute a set of equations that can be solved (at least numerically) for y, u, and z.

In Fig. 159 the physiological dissociation curve as obtained by numerical computation (digital computer) is shown.

18. From the Concluding Note to the First Edition

This note ended a section on "the respiratory system as a feedback regulator", which described a model with very many physiological parameters. We have reluctantly left out the account of this pioneer study[1]; it would now require a large book even to review recent developments[3]. We introduced cybernetics at the end of Chapter 8 in the present edition, and the respiration problem as follows in the last edition:

To quote from Grodins *et al.*[1]:-

"The essence of physiology is regulation. It is this concern with 'purposeful' system responses which distinguishes physiology from biophysics and biochemistry. Thus, physiologists study the regulation of breathing, of cardiac output, of blood pressure, of water balance, of body temperature and of a host of other biological phenomena.

In recent years, stimulated particularly by the practical demands of World War II, a closely related branch of physical science has emerged.

It is concerned with the theory, design, and operation of man-made regulators and servomechanisms. It is usually referred to as control system engineering, but it could very well be called the physiology of physical systems."

This kind of purposeful regulation is sometimes referred to as cybernetics (Wiener, Tsien; cf. Chapter 8, § 19).

We shall give an example of the mathematical analysis of such biological systems by studying the *regulation of respiration*.

Before proceeding we wish to mention the following point. It will be seen that the mathematical analysis leads to such highly complicated non-linear differential equations that it would be impossible to solve them without the aid of so-called electronic computers, a term which includes both the analogue computer and the digital computer.

We shall not explain these machines. Suffice it to say that the necessity of using these machines should not impress the student to the point of paralysis, since once the basic (differential) equations are set up, and the numerical values of the parameters and initial conditions specified, the final steps may be left to those specialized in the handling of electronic computers.

THE RESPIRATORY SYSTEM AS A FEED-BACK REGULATOR

Let us now turn to our problem.

When a subject inhales a CO_2 gas mixture, the CO_2 tension in his body will tend to rise.

Now this initial rise of body CO_2 "acts as a stimulus" for the ventilation so that the ventilation increases.

As a result of this increased ventilation extra quantities of CO_2 are "washed out" through the lungs, resulting in a lowering of the CO_2 tension in the body.

In this way the increase of body CO_2 tension resulting from the inhalation of CO_2 is mitigated.

This constitutes a typical example of a feed-back system. Increase in body CO_2 tension causes an increase in ventilation which in turn causes the body CO_2 tension to fall towards its "normal" level.

The feed-back mechanism (employing so-called negative feed-back) thus seeks to maintain the regulated quantity (CO_2 in this example) at a desired level.

The model then treated relations between the various physiological parameters. The final comment was more general:

Would failure of the theory to predict *all* experimental facts, imply that the theory is false? Certainly not! Bohr's theory of the atom was adequate to explain a great many experimental facts concerning the atom (e.g. it accurately predicted the frequency of the spectral lines of the hydrogen atom and even predicted the existence of lines unknown at that time) but it failed to predict certain other phenomena.

Although then, Bohr's theory failed to predict *all* the phenomena it certainly was not wrong in the sense of being rejectable. On the contrary, it forms the firm foundation of all further developments. Once the fundamental correctness of the theory is verified by experimental verification, the deductions made from the theory can be used to explain the overall response of the system in terms of its elements. It is certainly no exaggeration to say, that only when the 'mathematical' behaviour of the system is fully understood, the physiological behaviour of the system is understood since these are equivalent.

We feel confident in saying that the application of mathematics to physiological problems, made almost unrestrictedly possible through the use of electronic computers, will transform a science which is even to-day still largely an (impressive) collection of empirical data, into the coherent logical structure of a fully developed science.

It should be clearly understood, however, that mathematics does not replace experimentation.

On the contrary, our mathematical assumptions should be based, whenever possible, on firm empirical (physiological) relationships and not necessarily on so-called physical laws (diffusion etc.) alone. This use of physiological relationships in the basic assumptions is illustrated by our analysis.

We made use of the oxygen dissociation curve, of Cunningham's experimental data[2] concerning the dependence of ventilation on alveolar P_{O_2}, and of a host of other physiological, not purely physical, data.

Biology is advanced by the use, not the misuse of mathematics. Extreme theoretical studies, based on very slender experimental evidence, or wholly lacking it, have done much to bring the application of mathematics to biology into discredit in the eyes of the experimental biologist.

The great chemist Bunsen is reported to have said, "Ein Chemiker, der kein Physiker ist, ist gar nichts." (A chemist, who is not a physicist, is almost valueless.) In physiology, someone is reported to have said, "Ein Mathematiker, der kein Physiologe ist, ist gar nichts." Some day this thesis may well prove reversible.

References for Chapter 10

Section 2

1. O. Frank, Z. Biol. **37** (1899) 483.
2. O. Frank, Z. Biol. **85** (1926) 91.
3. E.O. Attinger, Pulsatile blood flow (McGraw-Hill, New York, 1964).
4. D.A. McDonald, Blood flow in arteries (Arnold, London, 1960).
5. K. Wezler and A. Boger, Ergebnisse der Physiol. **41** (1939) 292.
6. P. Lambossy, Helv. Physiol. Acta **8** (1950) 209.
7. R. Macey, Bull. Math. Biophys. **17** (1955) 169*.
8. F.W. Cope, Bull Math. Biophys. **22** (1960) 19.
9. Julia T. Apter, Bull. Math. Biophys. **27** (1965) 27.
10. J.G. Defares, J.J. Osborn and H.H. Hara, Bull. Math. Biophys. **27** (1965) 71.
11. S. Roston, J.J. Osborn and H.H. Hara, Bull. Math. Biophys. **27** (1965) 167.
12. P.J. Lewi, J.J. Osborn and H.H. Hara, Bull. Math. Biophys. **27** (1965) 271.
13. G.S. Malidzek, Jr., Math. Biosci. **7** (1970) 273.

* This gives a complete bibliography. The papers by G. Karreman and by N. Rashevsky in the same journal are especially recommended (1945, 1952, 1953, 1954).

Section 5

1. N.T.J. Bailey, The mathematical approach to biology and medicine (Wiley, New York, 1967) (especially p. 154).
2. J.M. Reiner, Bull Math. Biophys. **22** (1967) 349.
3. J. Kiefer, J. Theor. Biol. **18** (1968) 263.
4. P.E. Hansche, J. Theor. Biol. **24** (1969) 335.
5. P. Holgate and K.H. Lakhanc, Bull. Math. Biophys. **22** (1967) 831 (more advanced).
6. B.V. Bronk, G.J. Dienes and R.A. Johnson, Biophys. J. **10** (1970) 487 (more advanced).

Section 6

1. V. Volterra, Leçons sur la théorie mathématique de la lutte pour la vie (Gauthier-Villars, Paris, 1931).
2. C. Pearce, Biometrics 26 (1970) 387.
3. M.S. Bartlett, Stochastic population models (Methuen, London, 1960).
4. A.J. Lotka, Elements of physical biology (Williams and Wilkins, Baltimore, 1925).
5. V.A. Kostitzin, Mathematical biology (Harrap, London, 1939).
6. N.T.J. Bailey, The elements of stochastic processes (Wiley, New York, 1964).
7. A. Rescigno and I.W. Richardson, Bull. Math. Biophys. **29** (1967) 377.

Section 7

1. A.L. Hodgkin and A.F. Huxley, J. Physiol. **117** (1952) 500.

2. E.R. Lewis, J. Theor. Biol. **10** (1966) 125.
3. W.F. Pickard, J. Theor. Biol. **11** (1966) 30.
4. R.N. Johnson and G.R. Hanna, J. Theor. Biol. 22 (1969) 401, 411.
5. J. Evans and N. Schenk, Biophys. J. **10** (1970) 1090 (advanced).
6. A.C. Scott, Math. Biosci. **13** (1972) 47 (advanced).
7. A.L. Hodgkin, The conduction of the nervous impulse (University Press, Liverpool, 1965).
8. W.F. Ganong, Review of medical physiology (Lange, Los Altos, and Blackwell, Oxford, 1971).

Section 8

1. L. Tasaki and T. Takeuchi, Pflügers Archiv **244** (1941) 696.
2. A. von Muralt, Die Signalübermittelung im Nerven (Birkhäuser, Basel, 1945).
3. L. Tasaki and K. Mizuguchi, J. Neurophysiol. **11** (1948) 295.
4. A.F. Huxley and R. Stämpfli, J. Physiol. **108** (1949) 315.
5. R. Stämpfli and Y. Zottermann, Helv. Phys. Acta **9** (1951) 208.
6. R. Stämpfli, Ergebnisse der Physiol. **47** (1952) 70.
7. I. Tasaki, Conduction of the nerve impulse, in: Handbook of Physiology, Neurophysiology **1** (1959) 75.
8. I. Tasaki, Nerve excitation. A macromolecular approach (C.C. Thomas, Springfield, Ill., 1967).
9. B. Katz, Nerve, muscle and synapse (McGraw-Hill, New York, 1966).
10. N. Rashevsky, Mathematical biophysics. The physio-mathematical foundations of biology, 2 volumes (Dover, New York, 1960).
11. B.A. Curtis, S. Jacobson and E.M. Marcus, An introduction to the neurosciences (W.B. Saunders, London, 1972).

Section 9

1. N.T.J. Bailey, The mathematical theory of epidemics (Griffin, London, 1957).
2. N.T.J. Bailey, see § 5, reference 1.
3. R. Robson, F. Kahrs and J.A. Baker, J. Theor. Biol. **17** (1967) 47.
4. K. Dietz, J. Roy. Statist. Soc. A **130** (1967) 565. Review of recent work with 135 references.

Section 11

1. E.J. Harris, Transport and accumulation in biological systems (Butterworth, London, 1956), Chapter 2.
2. M.H. Jacobs, Ergebnisse der Biol. **12** (1935) 1.
3. L. Garby, Nature, London **173** (1954) 444.
4. T.A. Massaro and I. Fatt, Bull. Math. Biophys. **31** (1969) 327.
5. H. Passon, J. Theor. Biol. **17** (1967) 383 (more advanced).

Section 12

1. S. Hecht, J. Opt. Soc. Amer. **32** (1942) 42.
2. S. Hecht, S. Schlaer and M.H. Pirenne, J. General Physiol. **25** (1942) 819.
3. M.A. Bouman, Doc. Opthalmologica **4** (1950) 23.
4. M.A. Bouman, J. Opt. Soc. Amer. **45** (1955) 36.
5. M.A. Bouman and H.A. v.d. Velden, J. Opt. Soc. Amer. **38** (1948) 570.
6. M.A. Bouman, Physicomathematical aspects of biology, International School of Physics 'Enrico Fermi', Varenna, Italy, course 16 (Academic Press, N.Y., 1962) p. 142.
7. M.H. Pirenne, Vision and the Eye (Science Paperbacks, London, 1967).

Section 13

1. J.N. Franklin, J. Molecular Biol. **41** (1969) 61.
2. Hideyuki Ogawa and Jun-ichi Tomizawa, J. Molecular Biol. **30** (1967) 1.
3. G.E. Gulford and A.L. Koch, J. Theor. Biol. **22** (1969) 271.
4. J.G. Defares, Lancet i for 1971, 135.
5. J. Gani, J. Roy. Statist. Soc. A **134** (1971) 26; section 3 on models in virology.

Section 14

1. P.J. Cook, R. Doll and S.A. Fellingham, Internat. J. Cancer **4** (1969) 93.
2. R. Doll, J. Roy. Statist. Soc. A **134** (1971) 133.
3. M.E. Wise, J. Roy. Statist. Soc. A **134** (1971) 164.
4. A.W. Kimball, Biometrics **27** (1971) 462.
5. A.W. Kimball, Biometrics **25** (1969) 329.
6. C.L. Chiang, Biometrics **26** (1970) 767.
7. H.A. David, Biometrics **26** (1970) 336.
8. D.G. Hoel, Biometrics **28** (1972) 475.
9. J.C. Fisher, Nature, London **181** (1958) 651.
10. P. Armitage and R. Doll, Brit. J. Cancer **8** (1954) 1.
11. P. Armitage and R. Doll, Brit. J. Cancer **11** (1957) 161.
12. M.C. Pike, Biometrics **22** (1966) 142.
13. R. Doll at a meeting of the medical section of the Royal Statistical Society, London (1960).
14. A. Einstein, Ann. der Physik **17** (1905) 549; see also A. Einstein, Investigations on the theory of Brownian movement, Dover Publications (1956) and: E. Schrödinger, Physik. Z. **16** (1915) 289.
15. P.R.J. Burch, An inquiry concerning growth, disease and ageing (Oliver and and Boyd, Edinburgh, 1968). Also unpublished analyses.
16. P.R.J. Burch, in: Radiation induced cancer (International Atomic Energy Agency, Vienna, 1969) p. 29; see also Nature, London **225** (1970) 512.
17. M. Spiegelman and H.M. Marks, Amer. J. Public Health **36** (1946) 26.
18. B. Malzberg, Mental Hygiene **39** (1955) 196.
19. Registrar General, Statistical review for England and Wales for 1961, Part I, medical tables (Her Majesty's Stationary Office, London, 1963).

Section 15

1. L.A. Sapirstein, D.G. Vidt, M.J. Mandel and G. Hanusek, Amer. J. Physiol. **181** (1955) 330.
2. J.P. Aubert and F. Bronner, Biophys. J. **5** (1965) 349.
3. T.L. Schwarz and F.M. Snell, Biophys. J. **8** (1968) 805, 818.
4. J.G. Wagner, J. Theor. Biol. **20** (1968) 173.
5. F. Hagmeijer and P. van Remoortere, J. Theor. Biol. **25** (1969) 236.

Section 16

1. P.E. Bergner and C.C. Lushbaugh, editors, Compartments, pools and spaces in medical physiology (Atomic Energy Commission Symposium series 11, USAEC, 1967).
2. International Atomic Energy Agency, Vienna (1971). Report of symposium at Rotterdam 1970 on 'Dynamic studies with radioisotopes in medicine'.
3. J. Anderson, S.B. Osborn, R.W.S. Tomlinson and I. Weinbren, Phys. Med. Biol. **8** (1963) 287.
4. J. Anderson, S.B. Osborn, R.W.S. Tomlinson and M.E. Wise, Phys. Med. Biol. **14** (1969) 498 and **15** (1970) 567. See also J.S. Beck and A. Rescigno, ibid. **15** (1970) 566.
5. M.E. Wise, Biometrics **27** (1971) 97, 262.
6. Myron Pollycove and R. Mortimer, J. Clin. Invest. **40** (1961) 753.
7. Y. Takeda and E.B. Reeve, J. Lab. Clin. Med. **61** (1963) 183.
8. P.W. Dykes, Clin. Sci. **34** (1968) 161.
9. M.E. Wise, S.B. Osborn, J. Anderson and R.W.S. Tomlinson, Math. Biosci. **2** (1968) 199.
10. M.E. Wise, Statistica Neerlandica **25** (1971) 159.
11. M.E. Wise, Math. Biosci. (1973). In the press.
12. J.H. Marshall, Elizabeth L. Lloyd, J. Rundo, J. Liniecki, G. Marotti, C.W. Mays, H.A. Sissons and W.S. Snyder, Alkaline earth metabolism in adult man. I.C.R.P. publication **20** (Pergamon Press, London, 1973). Also, Health Phys. **22** (1973) 129.
13. S.M. Rosen, N.K. Hollenberg, J.B. Denby, Jr., and J.P. Merrill, Clin. Sci. **9** (1968) 287.
14. J. Ladafoged, Scand. J. Clin. Lab. Invest. **18** (1966) 299.
15. J.M. van Rossum, in: Drug design **1** (Academic Press, N.Y., London, 1971) 470.

Section 17

1. J.G. Defares and B.F. Visser, Ann. New York Acad. Sci. **96** (1962) 939.
2. H. Metzger, Kybernetik **6** (1969) 97.
3. H.T. Milhorn, Jr. and P.E. Pulley, Jr., Biophys. J. **8** (1968) 337.
4. B. Pennock and E.O. Attinger, Biophys. J. **8** (1968) 879.
5. J.G. Defares and M.E. Wise, Bull. Math. Biol. Special issue (1973) 237.
6. B.F. Visser and A.H.J. Maas, Phys. Med. Biol. **3** (1959) 264.

7. J.W. Severinghaus, in: Handbook of Physiology, Respiration **2** (1965) 1475 (Amer. Physiol. Soc., Washington D.C.).
8. B.F. Visser, Phys. Med. Biol. **5** (1960) 155.
9. D.D. van Slyke and J. Sendroy, Jr., J. Biol. Chem. **79** (1928) 781.

Section 18

1. F.S. Grodins, J.S. Gray, K.R. Schroeder, A.L. Norins and R.W. Jones, J. Appl. Physiol. **7** (1954) 283.
2. D.J.C. Cunningham, S. Lahir, B.B. Lloyd and D.G. Shaw, J. Physiol. **148** (1959) 71.
3. See, for example, various articles in "Further reading" at the end of this book, the series "Progress in biocybernetics" edited by Wiener and Schadé (Elsevier, Amsterdam, London, N.Y.), and a recent introduction by A.A. Verveen, Annals of Systems Research **2** (1972) 117. Defares has published a modification of Grodins' model (p. 196 of reference 6 to section 12).

DETERMINANTS

1. Determinants and Linear Algebraic Equations

If we solve the simultaneous algebraic equations

$$\begin{aligned} a_1 x + a_2 y + a_3 &= 0, \\ b_1 x + b_2 y + b_3 &= 0, \end{aligned} \tag{1}$$

we find that we can write the solution in the form

$$\frac{x}{a_2 b_3 - a_3 b_2} = \frac{-y}{a_1 b_3 - a_3 b_1} = \frac{1}{a_1 b_2 - a_2 b_1}. \tag{2}$$

If we now introduce the notation

$$\begin{vmatrix} a_1 & a_2 \\ b_1 & b_2 \end{vmatrix} = a_1 b_2 - a_2 b_1, \tag{3}$$

we see that we can write the solution of the equations (1) in the form

$$\frac{x}{\begin{vmatrix} a_2 & a_3 \\ b_2 & b_3 \end{vmatrix}} = \frac{-y}{\begin{vmatrix} a_1 & a_3 \\ b_1 & b_3 \end{vmatrix}} = \frac{1}{\begin{vmatrix} a_1 & a_2 \\ b_1 & b_2 \end{vmatrix}}. \tag{4}$$

The expression

$$\begin{vmatrix} a_1 & a_2 \\ b_1 & b_2 \end{vmatrix}$$

defined by equation (3) is called a *determinant* with two rows and columns or simply a 2×2 determinant, the numbers a_1, a_2, b_1, b_2 being called its *elements*. If we notice the relation of the elements in the determinants

$$\begin{vmatrix} a_2 & a_3 \\ b_2 & b_3 \end{vmatrix}, \qquad \begin{vmatrix} a_1 & a_3 \\ b_1 & b_3 \end{vmatrix}, \qquad \begin{vmatrix} a_1 & a_2 \\ b_1 & b_2 \end{vmatrix}$$

to the array of coefficients

$$\begin{matrix} a_1 & a_2 & a_3 \\ b_1 & b_2 & b_3 \end{matrix}$$

of the pair of equations (1) we see that the determinant notation expresses the results in a simple way.

Often in mathematics we are presented with the problem of eliminating the two variables x and y from the set of three equations

$$\begin{aligned} a_1x + a_2y + a_3 &= 0, \\ b_1x + b_2y + b_3 &= 0, \\ c_1x + c_2y + c_3 &= 0. \end{aligned} \tag{5}$$

We find the required relation between the coefficients by solving for x and y from two of these equations and substituting that solution in the third equation. For instance if we solve the second and third of these equations for x and y we find that

$$\frac{x}{\begin{vmatrix} b_2 & b_3 \\ c_2 & c_3 \end{vmatrix}} = \frac{-y}{\begin{vmatrix} b_1 & b_3 \\ c_1 & c_3 \end{vmatrix}} = \frac{1}{\begin{vmatrix} b_1 & b_2 \\ c_1 & c_2 \end{vmatrix}}, \tag{6}$$

the solution being finite only if

$$\begin{vmatrix} b_1 & b_2 \\ c_1 & c_2 \end{vmatrix} \neq 0. \tag{7}$$

If we now substitute from equations (6) into the first of the equations (5) we find that

$$a_1 \frac{\begin{vmatrix} b_2 & b_3 \\ c_2 & c_3 \end{vmatrix}}{\begin{vmatrix} b_1 & b_2 \\ c_1 & c_2 \end{vmatrix}} - a_2 \frac{\begin{vmatrix} b_1 & b_3 \\ c_1 & c_3 \end{vmatrix}}{\begin{vmatrix} b_1 & b_2 \\ c_1 & c_2 \end{vmatrix}} + a_3 = 0$$

which is equivalent to the equation

$$a_1 \begin{vmatrix} b_2 & b_3 \\ c_2 & c_3 \end{vmatrix} - a_2 \begin{vmatrix} b_1 & b_3 \\ c_1 & c_3 \end{vmatrix} + a_3 \begin{vmatrix} b_1 & b_2 \\ c_1 & c_2 \end{vmatrix} = 0. \tag{8}$$

If we define a 3×3 determinant by the equation

$$\begin{vmatrix} a_1 & a_2 & a_3 \\ b_1 & b_2 & b_3 \\ c_1 & c_2 & c_3 \end{vmatrix} = a_1 \begin{vmatrix} b_2 & b_3 \\ c_2 & c_3 \end{vmatrix} - a_2 \begin{vmatrix} b_1 & b_3 \\ c_1 & c_3 \end{vmatrix} + a_3 \begin{vmatrix} b_1 & b_2 \\ c_1 & c_2 \end{vmatrix}, \tag{9}$$

we see from equation (8) that, if we eliminate x and y from the three equations (5), we obtain the relation

$$\begin{vmatrix} a_1 & a_2 & a_3 \\ b_1 & b_2 & b_3 \\ c_1 & c_2 & c_3 \end{vmatrix} = 0. \tag{10}$$

If we insert the expressions for the 2×2 determinants occurring on the right-hand side of equation (9) we find that the right-hand side of that equation becomes

$$a_1(b_2 c_3 - b_3 c_2) - a_2(b_1 c_3 - b_3 c_1) + a_3(b_1 c_2 - b_2 c_1)$$

which can be rearranged to give

$$a_1(b_2 c_3 - b_3 c_2) - b_1(a_2 c_3 - a_3 c_2) + c_1(a_2 b_3 - a_3 b_2)$$

showing that

$$\begin{vmatrix} a_1 & a_2 & a_3 \\ b_1 & b_2 & b_3 \\ c_1 & c_2 & c_3 \end{vmatrix} = a_1 \begin{vmatrix} b_2 & b_3 \\ c_2 & c_3 \end{vmatrix} - b_1 \begin{vmatrix} a_2 & a_3 \\ c_2 & c_3 \end{vmatrix} + c_1 \begin{vmatrix} a_2 & a_3 \\ b_2 & b_3 \end{vmatrix}. \tag{11}$$

In the definition (9) we say that we have expanded the determinant on the left in terms of the elements of the first row, whereas in (11) we say that we have expanded in terms of the elements of the first column.

The definitions (9) and (11) suggest how we may define determinants of order n in terms of those of order $n - 1$. We define the determinant

$$\begin{vmatrix} a_1 & a_2 & a_3 \ldots a_{n-1} & a_n \\ b_1 & b_2 & b_3 \ldots b_{n-1} & b_n \\ c_1 & c_2 & c_3 \ldots c_{n-1} & c_n \\ \cdot & \cdot & \cdot & \cdot \\ \cdot & \cdot & \cdot & \cdot \\ \cdot & \cdot & \cdot & \cdot \\ v_1 & v_2 & v_3 \ldots v_{n-1} & v_n \\ w_1 & w_2 & w_3 \ldots w_{n-1} & w_n \end{vmatrix} \tag{12}$$

to be either

$${}_1\begin{vmatrix} b_2 & b_3 \ldots b_n \\ c_2 & c_3 \ldots c_n \\ \cdot & \cdot \\ \cdot & \cdot \\ \cdot & \cdot \\ w_2 & w_3 \ldots w_n \end{vmatrix} - a_2 \begin{vmatrix} b_1 & b_3 \ldots b_n \\ c_1 & c_3 \ldots c_n \\ \cdot & \cdot \\ \cdot & \cdot \\ \cdot & \cdot \\ w_1 & w_3 \ldots w_n \end{vmatrix} + \ldots + (-1)^{n-1} a_n \begin{vmatrix} b_1 & b_2 \ldots b_{n-1} \\ c_1 & c_2 \ldots c_{n-1} \\ \cdot & \cdot \\ \cdot & \cdot \\ \cdot & \cdot \\ w_1 & w_2 \ldots w_{n-1} \end{vmatrix} \tag{13}$$

or

$$a_1 \begin{vmatrix} b_2 & b_3 \dots b_n \\ c_2 & c_3 \dots c_n \\ \cdot & \cdot & \cdot \\ \cdot & \cdot & \cdot \\ \cdot & \cdot & \cdot \\ w_2 & w_3 \dots w_n \end{vmatrix} - b_1 \begin{vmatrix} a_2 & a_3 \dots a_n \\ c_2 & c_3 \dots c_n \\ \cdot & \cdot & \cdot \\ \cdot & \cdot & \cdot \\ \cdot & \cdot & \cdot \\ w_2 & w_3 \dots w_n \end{vmatrix} + \dots + (-1)^{n-1} w_1 \begin{vmatrix} a_2 & a_3 \dots a_n \\ b_2 & b_3 \dots b_n \\ \cdot & \cdot & \cdot \\ \cdot & \cdot & \cdot \\ \cdot & \cdot & \cdot \\ v_2 & v_3 \dots v_n \end{vmatrix} \quad (14$$

the two expansions giving the same result.

For example, we define

$$\begin{vmatrix} a_1 & a_2 & a_3 & a_4 \\ b_1 & b_2 & b_3 & b_4 \\ c_1 & c_2 & c_3 & c_4 \\ d_1 & d_2 & d_3 & d_4 \end{vmatrix} = a_1 \begin{vmatrix} b_2 & b_3 & b_4 \\ c_2 & c_3 & c_4 \\ d_2 & d_3 & d_4 \end{vmatrix} - a_2 \begin{vmatrix} b_1 & b_3 & b_4 \\ c_1 & c_3 & c_4 \\ d_1 & d_3 & d_4 \end{vmatrix} + a_3 \begin{vmatrix} b_1 & b_2 & b_4 \\ c_1 & c_2 & c_4 \\ d_1 & d_2 & d_4 \end{vmatrix} - a_4 \begin{vmatrix} b_1 & b_2 & b_3 \\ c_1 & c_2 & c_3 \\ d_1 & d_2 & d_3 \end{vmatrix}.$$

There are many different ways of defining an nth order determinant, though all the definitions lead to the same numerical value in the end. The definition given here has the merit that it arises naturally in elementary algebra. For the proof of general theorems about determinants it is, however, more convenient to have the following definition: the determinant (12) is that function of the a's, b's, . . ., w's which satisfies the following three conditions:

(1) It is an expression of the form $\Sigma \pm a_r b_s \cdots w_t$, in which the sum is taken over the $n!$ possible ways of assigning to $r, s, \dots, t$ the values $1, 2, \dots, n$ in some order and without repetition;
(2) The sign $+$ occurs before the "diagonal term" $a_1 b_2 \cdots w_n$;
(3) The sign prefixed to any other term is such that the interchange of any two letters* throughout the sum reproduces the same set of terms, but in a different order of occurrence, and with the opposite signs prefixed.

It is easily shown** from this definition that it yields one function of the a's, b's, . . ., w's and only one, and also that from it we can derive the expansions (13) and (14).

Let us return now to our consideration of systems of linear equations.

* By the phrase "interchange a and b" we mean the simultaneous interchanges of a_1 and b_1, a_2 and b_2, . . ., a_n and b_n.

** For a proof see W. L. Ferrar, Algebra (Oxford University Press, 1941), p. 8.

If we solve the three equations

$$\begin{aligned} b_1 x + b_2 y + b_3 z + b_4 &= 0, \\ c_1 x + c_2 y + c_3 z + c_4 &= 0, \\ d_1 x + d_2 y + d_3 z + d_4 &= 0, \end{aligned} \tag{15}$$

for x, y and z by the method of successive eliminations, we can readily show that we may write the results in the determinantal form

$$\frac{x}{\begin{vmatrix} b_2 & b_3 & b_4 \\ c_2 & c_3 & c_4 \\ d_2 & d_3 & d_4 \end{vmatrix}} = \frac{-y}{\begin{vmatrix} b_1 & b_3 & b_4 \\ c_1 & c_3 & c_4 \\ d_1 & d_3 & d_4 \end{vmatrix}} = \frac{z}{\begin{vmatrix} b_1 & b_2 & b_4 \\ c_1 & c_2 & c_4 \\ d_1 & d_2 & d_4 \end{vmatrix}} = \frac{-1}{\begin{vmatrix} b_1 & b_2 & b_3 \\ c_1 & c_2 & c_3 \\ d_1 & d_2 & d_3 \end{vmatrix}} \tag{16}$$

If, in addition to satisfying the three equations (15), the variables x, y, z satisfy the fourth equation

$$a_1 x + a_2 y + a_3 z + a_4 = 0, \tag{17}$$

then the coefficients of the four equations must satisfy a relation which we obtain by substituting the solution (16) into equation (17). In this way we obtain the equation

$$a_1 \begin{vmatrix} b_2 & b_3 & b_4 \\ c_2 & c_3 & c_4 \\ d_2 & d_3 & d_4 \end{vmatrix} - a_2 \begin{vmatrix} b_1 & b_3 & b_4 \\ c_1 & c_3 & c_4 \\ d_1 & d_3 & d_4 \end{vmatrix} + a_3 \begin{vmatrix} b_1 & b_2 & b_4 \\ c_1 & c_2 & c_4 \\ d_1 & d_2 & d_4 \end{vmatrix} - a_4 \begin{vmatrix} b_1 & b_2 & b_3 \\ c_1 & c_2 & c_3 \\ d_1 & d_2 & d_3 \end{vmatrix} = 0$$

which, by the definition (13) gives the result

$$\begin{vmatrix} a_1 & a_2 & a_3 & a_4 \\ b_1 & b_2 & b_3 & b_4 \\ c_1 & c_2 & c_3 & c_4 \\ d_1 & d_2 & d_3 & d_4 \end{vmatrix} = 0 \tag{18}$$

showing that the result of eliminating x, y, and z from equations (15)—(17) is equation (18).

This result can readily be generalized. If the n variables $x, y, \ldots, z$ satisfy simultaneously the $n + 1$ equations

$$\begin{aligned} a_1 x + a_2 y + \ldots + a_n z + a_{n+1} &= 0, \\ b_1 x + b_2 y + \ldots + b_n z + b_{n+1} &= 0, \\ \vdots \quad\quad \vdots \quad\quad\quad\quad \vdots \quad\quad \vdots \quad & \quad \vdots \\ w_1 x + w_2 y + \ldots + w_n z + w_{n+1} &= 0, \end{aligned} \tag{19}$$

then the coefficients satisfy the relation

$$\begin{vmatrix} a_1 & a_2 & a_3 \dots & a_{n+1} \\ b_1 & b_2 & b_3 \dots & b_{n+1} \\ \cdot & \cdot & \cdot & \cdot \\ \cdot & \cdot & \cdot & \cdot \\ \cdot & \cdot & \cdot & \cdot \\ w_1 & w_2 & w_3 & w_{n+1} \end{vmatrix} = 0. \tag{20}$$

2. Rules for the Manipulation of Determinants

In this section we shall describe and illustrate the main properties of determinants, with a view to their evaluation. No proofs will be given. For these the reader is referred to pp. 12—18 of the book by Ferrar cited above.

THEOREM I. *A determinant is unaltered in value if its rows and columns are interchanged.*

This theorem states that

$$\begin{vmatrix} a_1 & a_2 \dots & a_n \\ b_1 & b_2 \dots & b_n \\ \cdot & \cdot & \cdot \\ \cdot & \cdot & \cdot \\ \cdot & \cdot & \cdot \\ w_1 & w_2 \dots & w_n \end{vmatrix} = \begin{vmatrix} a_1 & b_1 \dots & w_1 \\ a_2 & b_2 \dots & w_2 \\ \cdot & \cdot & \cdot \\ \cdot & \cdot & \cdot \\ \cdot & \cdot & \cdot \\ a_n & b_n \dots & w_n \end{vmatrix}. \tag{1}$$

THEOREM II. *The interchange of two columns, or of two rows, in a determinant changes the value of the determinant by a factor* -1.

In the case of interchange of columns, for instance, we have

$$\begin{vmatrix} a_2 & a_1 \dots & a_n \\ b_2 & b_1 \dots & b_n \\ \cdot & \cdot & \cdot \\ \cdot & \cdot & \cdot \\ \cdot & \cdot & \cdot \\ w_2 & w_1 \dots & w_n \end{vmatrix} = - \begin{vmatrix} a_1 & a_2 \dots & a_n \\ b_1 & b_2 \dots & b_n \\ \cdot & \cdot & \cdot \\ \cdot & \cdot & \cdot \\ \cdot & \cdot & \cdot \\ w_1 & w_2 \dots & w_n \end{vmatrix}.$$

It follows as a consequence of this theorem that, if a column (row) is moved past an even number of columns (rows), then the value of the determinant is unaltered. For instance

$$\begin{vmatrix} a_1 & a_2 & a_3 & a_4 \\ b_1 & b_2 & b_3 & b_4 \\ c_1 & c_2 & c_3 & c_4 \\ d_1 & d_2 & d_3 & d_4 \end{vmatrix} = \begin{vmatrix} a_2 & a_3 & a_1 & a_4 \\ b_2 & b_3 & b_1 & b_4 \\ c_2 & c_3 & c_1 & c_4 \\ d_2 & d_3 & d_1 & d_4 \end{vmatrix},$$

the first column having been moved past an even number (two) of columns. Similarly, if a column (row) is moved past an odd number of columns (rows) the value of the determinant is multiplied by -1.

The most important corollary of Theorem II is:

THEOREM III. *If a determinant has two columns (or two rows) identical, its value is zero.*

From the definition of a determinant we have:

THEOREM IV. *If each element of one column (or row) of a determinant is multiplied by a constant factor C, the value of the determinant is altered by a factor C.*

For instance,

$$\begin{vmatrix} Ca_1 & a_2 & a_3 \\ Cb_1 & b_2 & b_3 \\ Cc_1 & c_2 & c_3 \end{vmatrix} = C \begin{vmatrix} a_1 & a_2 & a_3 \\ b_1 & b_2 & b_3 \\ c_1 & c_2 & c_3 \end{vmatrix}.$$

In particular (but this follows directly from the definition), if a determinant has a column (or row) each of whose elements is zero, the value of the determinant is zero.

An important result, which is used a great deal in the numerical evaluation of determinants, is:

THEOREM V. *The value of a determinant is unaltered if to each element of one column (or row) is added a constant multiple of the corresponding element of another column (or row).*

For instance

$$\begin{vmatrix} a_1 + xa_2 & a_2 & a_3 \\ b_1 + xb_2 & b_2 & b_3 \\ c_1 + xc_2 & c_2 & c_3 \end{vmatrix} = \begin{vmatrix} a_1 & a_2 & a_3 \\ b_1 & b_2 & b_3 \\ c_1 & c_2 & c_3 \end{vmatrix},$$

and

$$\begin{vmatrix} a_1 + yb_1 & a_2 + yb_2 & a_3 + yb_3 \\ b_1 & b_2 & b_3 \\ c_1 & c_2 & c_3 \end{vmatrix} = \begin{vmatrix} a_1 & a_2 & a_3 \\ b_1 & b_2 & b_3 \\ c_1 & c_2 & c_3 \end{vmatrix}.$$

By repeated applications of this theorem we see that we may add to each column (row) of a determinant fixed multiples of the *preceding* columns (rows) without altering the value of the determinant. For example

$$\begin{vmatrix} a_1 & a_2 + xa_1 & a_3 + ya_2 + za_1 \\ b_1 & b_2 + xb_1 & b_3 + yb_2 + zb_1 \\ c_1 & c_2 + xc_1 & c_3 + yc_2 + zc_1 \end{vmatrix} = \begin{vmatrix} a_1 & a_2 & a_3 \\ b_1 & b_2 & b_3 \\ c_1 & c_2 & c_3 \end{vmatrix}.$$

There is a similar result with *preceding* replaced by *subsequent.*

To illustrate the use of these results we shall evaluate the determinant

$$\Delta = \begin{vmatrix} 3 & 2 & 1 & -1 \\ 0 & -2 & 3 & 0 \\ 1 & 4 & -1 & 3 \\ -2 & 6 & 0 & 0 \end{vmatrix}.$$

If we move the last column into the first column we shall have to move it past three columns, so the determinant will change its value by a factor -1; hence, we have

$$-\Delta = \begin{vmatrix} -1 & 3 & 2 & 1 \\ 0 & 0 & -2 & 3 \\ 3 & 1 & 4 & -1 \\ 0 & -2 & 6 & 0 \end{vmatrix}.$$

Suppose, now, that we add to the third row of this determinant three times the first row; then we obtain

$$-\Delta = \begin{vmatrix} -1 & 3 & 2 & 1 \\ 0 & 0 & -2 & 3 \\ 0 & 10 & 10 & 2 \\ 0 & -2 & 6 & 0 \end{vmatrix}.$$

Using expansion (14), § 1, we see that the determinant on the right-hand side of this last equation has the value

$$-1 \begin{vmatrix} 0 & -2 & 3 \\ 10 & 10 & 2 \\ -2 & 6 & 0 \end{vmatrix}$$

so that

$$\Delta = \begin{vmatrix} 0 & -2 & 3 \\ 10 & 10 & 2 \\ -2 & 6 & 0 \end{vmatrix}.$$

If, in this last determinant, we add five times the third row to the second row we obtain the equation

$$\Delta = \begin{vmatrix} 0 & -2 & 3 \\ 0 & 40 & 2 \\ -2 & 6 & 0 \end{vmatrix}$$

and, using expansion (14) again, we see that this gives

$$\Delta = -2 \begin{vmatrix} -2 & 3 \\ 40 & 2 \end{vmatrix} = -2(-4 - 120),$$

showing that $\Delta = 248$.

This example is introduced merely to illustrate the way in which the theorems we have quoted work. It is not intended to act as a guide to anyone interested in the systematic computation of determinants. For a description of a systematic method of computing determinants the interested reader is referred to E. T. Whittaker and G. Robinson, The Calculus of Observations (Blackie, Glasgow, 1926) Chapter V.

We have done nothing more here than to indicate how determinants arise in mathematics, and to quote, without proof, their most elementary properties. The reader who finds he needs more than this should consult W. L. Ferrar, Algebra (Oxford University Press, 1941) the first part of which is devoted to a clear account of the more elementary properties of determinants.

PROBLEM

Given the 3 simultaneous equations

$$\begin{aligned} b_1x_1 + b_2x_2 + b_3x_3 &= b_4 \\ c_1x_1 + c_2x_2 + c_3x_3 &= c_4 \\ d_1x_1 + d_2x_2 + d_3x_3 &= d_4 \end{aligned} \tag{2}$$

let

$$\Delta = \begin{vmatrix} b_1 & b_2 & b_3 \\ c_1 & c_2 & c_3 \\ d_1 & d_2 & d_3 \end{vmatrix}, \tag{3}$$

and let Δ_j be the determinant obtained from Δ by replacing its jth column by that on the right of (2); thus b_j is replaced by b_4, c_j by c_4 and d_j by d_4.

Show that the solution of (2) can be written

$$x_j\Delta = \Delta_j \quad \text{for} \quad j = 1, 2 \text{ and } 3. \tag{4}$$

3. Linear Dependence of Functions

We shall now generalize the simple result on the linear dependence of functions stated in § 9 of Chapter 9 (Theorem 42). If the functions $f_1(x), f_2(x), \ldots, f_n(x)$ are linearly dependent, then there exist constants $b_1, b_2, \ldots, b_n$ such that

$$b_1 f_1(x) + b_2 f_2(x) + \ldots + b_n f_n(x) = 0, \tag{1}$$

and we can label the functions in such a way that $b_n \neq 0$, so that the condition for linear dependence can be written in the form

$$f_1(x)c_1 + f_2(x)c_2 + \ldots + f_{n-1}(x)c_{n-1} + f_n(x) = 0, \tag{2}$$

with $c_r = b_r/b_n$. If now we differentiate equation (2) with respect to x successively $n - 1$ times, we obtain the $n - 1$ relations

$$\begin{aligned}
f_1'(x)c_1 + f_2'(x)c_2 + \ldots + f_{n-1}'(x)c_{n-1} + f_n'(x) &= 0,\\
f_1''(x)c_1 + f_2''(x)c_2 + \ldots + f_{n-1}''(x)c_{n-1} + f_n''(x) &= 0,\\
\vdots \qquad\qquad \vdots \qquad\qquad\qquad \vdots \qquad\qquad \vdots \quad &\quad \vdots\\
f_1^{n-1}(x)c_1 + f_2^{n-1}(x)c_2 + \ldots + f_{n-1}^{n-1}(x)c_{n-1} + f_n^{n-1}(x) &= 0,
\end{aligned} \tag{3}$$

where we have written f_s^r to denote $d^r f_s/dx^r$. Now we can regard the n equations (2), (3) as n equations in the $n - 1$ unknowns $c_1, c_2, \ldots, c_{n-1}$. If we eliminate these $n - 1$ unknowns we obtain the condition

$$W(f_1, f_2, \ldots, f_n) = \begin{vmatrix} f_1(x) & f_2(x) \ldots & f_n(x) \\ f_1'(x) & f_2'(x) \ldots & f_n'(x) \\ \vdots & \vdots & \vdots \\ f_1^{n-1}(x) & f_2^{n-1}(x) \ldots & f_n^{n-1}(x) \end{vmatrix} = 0. \tag{4}$$

The function $W(f_1, f_2, \ldots, f_n)$ is called the *Wronskian* of the functions.

For example

$$W(1, \cos x, \cos^2 x) = \begin{vmatrix} 1 & \cos x & \cos^2 x \\ 0 & -\sin x & -2\cos x \sin x \\ 0 & -\cos x & 2(\sin^2 x - \cos^2 x) \end{vmatrix} = -2\sin^3 x \neq 0,$$

showing that 1, $\cos x$, $\cos^2 x$ are linearly independent.

4. A note on Matrices

Sets or arrays of numbers can be considered as a single entity, and mathematical laws for working with such entities have been formulated that are consistent and useful. This involves extending the meanings of addition and multiplication introduced in Chapter 1.

Matrix algebra is thus much more advanced than that of determinants, but they are often included together in textbooks. Many matrix formulae and numerical calculations involve determinants. They enable long lists of equations such as (19) in §1 to be set out concisely, and the underlying ideas on arrays are involved in computer programming.

By way of a very brief introduction, the 3×3 determinant in equation (3) of §2 can be called the *determinant of a* 3×3 *matrix* that consists of the same 9 numbers arranged in rows and columns. The number of rows and columns need not be the same; for example, a 3×1 matrix, with 3 rows and 1 column is called a *column vector*. Equation (2) of §2 in matrix notation is:

$$\begin{pmatrix} b_1 & b_2 & b_3 \\ c_1 & c_2 & c_3 \\ d_1 & d_2 & d_3 \end{pmatrix} \begin{pmatrix} x_1 \\ x_2 \\ x_3 \end{pmatrix} = \begin{pmatrix} b_4 \\ c_4 \\ d_4 \end{pmatrix}.$$

The rule for multiplying matrices is such that the product on the left is itself a 3×1 column vector, and its three elements are the three terms on the left of equation (2), §2.

FURTHER READING

Papers

Studying original papers efficiently is something most of us have never been taught. This can be like swimming in unknown waters which may be either very rough or smooth but with hidden undercurrents, when the swimmer learned only in a calm pool.

We are trying a new kind of list that may help. It consists of the first impressions of one of us (J.G.D.) on about 50 recent articles, mainly in journals that contain many articles in our field. The subject matter is given very briefly, and in particular the estimated level and difficulty of 1) the mathematics, 2) the biology and 3) the physics and chemistry. These are all graded from 1 to 5, and the grades are given in the above order.

The broad division is by the *mathematical* level, and for this levels 1 to 3 fall well within the scope of this book; there may occasionally be a mathematical technique that we have not included. Level 4 indicates more advanced mathematics and level 5 is much more advanced.

Abbreviations

APPN	Acta Physiologica et Pharmacologica Neerlandica
BMB	Bulletin of Mathematical Biophysics
BJ	Biophysical Journal
ISPEF	International School of Physics "Enrico Fermi" Proceedings (Italian Physical Society)
JTB	Journal of Theoretical Biology
K	Kybernetik
MB	Mathematical Biosciences

Group A1

G. Cook and L. Stark. *Derivation of a model for the human eye positioning mechanism.* BMB **29** (1967) 153.
1,3,2. Vision. Control system, first order differential equations, empirical.

J.G. Defares, J.J. Osborn and H.H. Hara. *On the theory of the cardiovascular system.* BMB **27** (1965) 71.
2,2,1. Circulation, Non-linear differential equations, analogue computer. 10, §2.

F. Hagmeijer and P. van Remoortere. *Limitations of the graphical analysis of elution curves.* JTB **25** (1969) 236.
2,3,2. Washout curves, Compartmental analysis, analogue computer. 10, §15.

D.F. Horrobin. *The female sexcycle.* JTB **22** (1969) 80.
1,3,1. Sexcycle. Conceptual, no mathematics.

E.R. Lewis. *An electronic model of neuroelectric point processes.* K **5** (1968) 30.
2,3,4. Neurons. Hodgkin-Huxley equations. 10, §7.

J.M. Reiner. *Molecular biology and the kinetics of growth: I. cell growth.* BMB **29** (1967) 349.
2,1,1. Growth. Differential equations. 10, §5.

S. Roston. *The cardiovascular effects of the carotid sinus mechanism.* BMB **27** (1965) 167.
2,1,1. Circulation. Feedback, first order differential equations. 10, § 2.

D.M. Shaw, A.L. Johnson and R. Short. *Preliminary study of the effects of lithium, imipramine and reserpine on the relative sizes of tryptophan pools and flow constants in rabbits.* MB **15** (1972) 137.
2,2,1. Tracer study. Compartmental models, differential equations, Laplace transforms. 10, §15.

C.P.S. Taylor. *Isometric muscle contraction in the active state: an analog computer study.* BJ **9** (1969) 759.
2,3,2. Muscle. Differential equations.

A.A. Verveen and H.E. Derksen. *Amplitude distribution of axon membrane noise voltage.* APPN **15** (1969) 553.
1,4,4. Axon membranes. Experimental, statistical moments, arc tan plot.

D. Winne. *Formal kinetics of water and solute absorption with regard to intestinal blood flow.* JTB **27** (1970) 1.
2,3,2. Transport. Solution of third degree equation.

Group A2

F.W. Cope. *A theory of ion transport across cell surfaces by a process analogous to electron transport across liquid–solid interfaces.* BMB **27** (1965) 99.
2,4,3. Ion transport. Kinetics obey Elovich rate equation.

N. Rashevsky. *Mathematical biology of automobile driving: III.* BMB **30** (1968) 153.
2,2,2. Car driving. Extreme values.

R.P. Spencer. *Sample theory of transfusions.* BMB **27** (1965) 9.
1,1,1. Blood transfusion. Search for optimal strategy, probability concepts.

T.A. Wilson. *A theoretical prediction of the normal cardiac oxygen consumption.* BJ **7** (1967) 585.
2,3,4. O_2 metabolism. Entropy production, minimum problem.

Group B

E. Ackerman, L.G. Gatewood, J.W. Rosevear and G.D. Molnar. *Model studies of blood-glucose regulation.* BMB **27** (1965) 21.
3,2,1. Glucose levels. Second order linear system.

S. Ainsworth. *A comparative study of some mechanisms describing interaction effects in a dimeric, two substrate enzyme catalyzed reaction.* JTB **19** (1968) 1.
3,4,3. Enzyme reaction. Determinants, matrices, differentiation, points of inflexion.

H. Atlan. *Application of information theory to the study of the stimulating effects of ionizing radiation, thermal energy and other environmental factors.* JTB **21** (1968) 45.
3,3,4. Information theory. Concepts, entropy.

J.J. di Stefano III. *A model of the normal thyroid hormone glandular secretion mechanism.* JTB **22** (1969) 412.
3,3,2. Endocrinology. Differential equations.

K. Horsfield and G. Cumming. *Angles of branching and diameters of branches in the human bronchial tree.* BMB **29** (1967) 245.
3,2,2. Lungs. Optimal design, partial differentiation.

H.T. Malhorn, jr., R. Benton, R. Ross and A.C. Guyton. *A mathematical model of the human respiratory control system.* BJ **5** (1965) 27.
3,3,2. Respiration. Differential equations, solved by digital computer.

T.A. Massaro and I. Fatt. *Oxygen diffusion in large single-celled organisms.* BMB **31** (1969) 327.
3,3,3. Gas transport. Second order differential equations, graphical solutions. 10, §11.

K.H. Norwich and R. Reiter. *Homeostatic control of thyroxin concentration expressed by a set of linear differential equations.* BMB **27** (1965) 133.
3,2,1. Endocrine system. Linear differential equations, mean value theorem.

T.R. Reiff. *Colloid osmotic homeostasis in humans. I. Theoretical aspects and background.* JTB **28** (1970) 1.
3,5,5. Homeostasis. Thermodynamics, partial differentials.

A. Rescigno, R.B. Stein, R.L. Purple and R.E. Poppele. *A neuronal model for the discharge problems produced by cyclic inputs.* BMB **32** (1970) 337.
3,2,2. Neurons. Simple electric current model.

R. Robson, F. Kahrs and J.A. Baker. *Bounds on the mean recurrence time of sub-clinical epidemics.* JTB **17** (1967) 47.
3,4,2. Epidemics. Moments, numerical.

R.N. Stiles and D.M. Alexander. *A viscoelastic-mass model for muscle.* MB **14** (1972) 343.
3,3,2. Muscle. Second order differential equation, experiments.

J.G. Wagner. *Kinetics of pharmocologic response.* JTB **20** (1968) 173.
3,3,2. Pharmacology, one compartment open model system.

Group C

B.V. Bronk, C.J. Dienes and R.A. Johnson. *Co-operative regulation of cellular proliferation by intercellular diffusion* BJ **10** (1970) 487.
4,3,2. Cell growth. Coupled non-linear differential equations. 10, §5.

K. Dähnert. *Die Abhängigkeit des strategischen Verhaltens von Sportlern bei Wettläufen von den Möglichkeiten der Kommunikation und Informationsverarbeitung.* BMB **29** (1967) 625.
4,1,1. Athletics. Optimal strategy, partial derivatives.

H.E. Derksen. *Axon membrane voltage fluctuations.* APPN **13** (1965) 373.
4,4,5. Membranes, noise. Experimental, power spectra, autocorrelation function.

V.L. Engstrom-Heg. *Predation, competition and environmental variables: some mathematical models.* JTB **27** (1970) 175.
4,2,1. Ecology. Lotka-Volterra equations, digital computer. 10, §6.

J.T. Enright. *The spontaneous neuron subject to tonic stimulation.* JTB **16** (1967) 54.
4,3,3. Neurons. Fourier analysis, noise, logfunctions.

W.F. Forbes, D.A. Sprott and M. Feldstein. *A model to account for mortality curves of various species.* JTB **29** (1970) 293.
4,4,1. Mortality. Empirically valid probability function derived from model. 10, §14.

L.L. Gatlin. *The information content of DNA, II.* JTB **18** (1968) 181.
4,4,4. DNA. Entropy, information theory.

A. Helfgott, E.O. Tuck, R.J. Gray and P.S. Hetzel. *A simple mathematical model of muscle induced ejection flows.* MB **13** (1972) 335.
4,3,3. Flow in tubes. Laplace transform solution of linear wave equation.

J. Kiefer. *A model of feedback-controlled cell populations.* JTB **18** (1968) 263.
4,4,2. Cell populations. Differential equations with delay times, computer. 10, §5.

G.S. Malindzak. *Fourier analysis of cardiovascular events.* MB **7** (1970) 273.
4,3,2. Circulation. Fourier series.

P. Nelson. *Comparaison des modèles de neurone de Hodgkin-Huxley et de Nelson.* BMB **28** (1966) 347.
4,3,3. Neurons. Generalized Hodgkin-Huxley model. 10, §7.

J.D. Parrish and S.B. Saila. *Interspecific competition, predation and species diversity.* JTB **27** (1970) 207.
4,2,1. Ecology. Coupled non-linear differential equations, computer. 10, §6.

R. Rosen. *Two factor models, neural nets and biochemical automata.* JTB **15** (1967) 282.
4,5,3. Neural nets. Differential equations, "molecular automata".

A.C. Scott. *Effect of the series inductance of a nerve axon upon its conduction velocity.* MB **11** (1971) 227.
4,3,5. Nerve axons. Hodgkin-Huxley partial differential equations. 10, §7.

R.B. Stein. *A theoretical analysis of neuronal variability.* BJ **5** (1965) 173.
4,3,3. Neurons. Partial differential equations, cumulants.

Group D

R. Bellman and K.J. Aström. *A structural identifiability.* MB **7** (1970) 329.
5,2,2. Tracers. Differential equations, determinants, matrices.

L. Bess. *Network model as a biological pacemaker.* JTB **28** (1970) 59.
5,4,3. Pacemaker. Probability theory, random walk distribution.

C.E. Blumenson. *Random walk and the spread of cancer.* JTB **27** (1970) 273.
5,3,2. Cancer. Random walk, partial differential equation.

B. van Bronk. *On radioactive labelling of proliferating cells: the graph of labelled mitosis.* JTB **22** (1969) 468.
5,4,4. Mitosis. Generating functions, probability, Laplace transforms.

C.C. Heyde and E. Heyde. *A stochastic approach to a one substrate one product enzyme reaction in the initial velocity phase.* JTB **25** (1969) 169.
5,5,5. Enzymes. Probability theory, difference-differential equations.

R.M. May. *Stability in multispecies community models.* MB **12** (1971) 59.
5,3,1. Predator-prey. Eigenvalues, stability of systems. 10, §6.

W. Perl, R.M. Effros and F.P. Chinerd. *Indicator equivalence theorem for input states and regional masses in multi-inlet steady state systems with partially labelled input.* JTB **25** (1969) 297.
5,4,4. Tracers. Convolution integrals, Laplace transforms.

N. Rashevsky. *A suggestion of a new approach to the theory of some biological periodicities.* BMB **30** (1968) 751.
5,3,1. Periodicities. Exploratory, conceptually difficult.

J. Richardson, L.J. Haywood, V.K. Murthy and G. Harvey. *A mathematical model for ECG wave forms and power spectra.* MB **12** (1971) 321.
5,2,2. Electrocardiogram. Fourier transforms, power spectra, mathematical model.

R. Rikmenspoel and M.A. Sleigh. *Bending moments and elastic constants in cilia.* JTB **28** (1970) 81.
5,2,3. Cilia movement. Partial differential equations, experimental.

C.W. Sheppard and M.B. Uffer. *Stochastic model for tracer experiments in the circulation.* JTB **22** (1969) 188.
5,4,4. Tracers. Diffusion equations.

P.J. Staff. *A stochastic development of the reversible Michaelis-Menten mechanism.* JTB **27** (1970) 221.
5,3,5. Enzymes. Difference-differential equations, probability theory.

A.M. Uttley. *The informon: a network for adaptive pattern recognition.* JTB **27** (1970) 31.
5,5,5. Cybernetics. Information theory.

L.G. Vargo. *A note on crime control.* BMB **28** (1966) 375.
5,1,1. Social ecology. Extension of Volterra theory of struggle.

A.A. Verveen. *An introduction to the use of time series in physiology.* ISPEF **43** (1968) 291.
5,4,4. Time series. Fourier series, autocorrelation function. Didactic.

F.E. Warburton. *A model of natural selection based on a theory of guessing games.* JTB **16** (1967) 78.
5,3,2. Evolution. Entropy, information theory. 8, §21.

Periodicals

Of the journals in the above list, BMB probably has the highest proportion of articles within the scope of this book. It was edited by N. Rashevsky, who founded it in 1939, until he died in 1971. The chief editor is now H.D. Landahl, and it has been renamed the *Bulletin of Mathematical Biology*.

MB was started in 1967; its chief editor is R. Bellman. It has a very wide range of mathematical level and difficulty, and of biomedical applications. Articles in JTB tend to be more advanced biologically.

Physics in Medicine and Biology, despite its name, is mainly concerned with physical apparatus so used, but sometimes has biomathematical articles. It has a valuable section of classified abstracts.

Many physiological journals publish mathematical applications, as our reference lists show. These include the American Journal of Physiology, the Journal of Applied Physiology, Circulation Research, Brain Research, and Pflügers Archiv (the European Journal of Physiology). Unfortunately APPN is no longer published: it was edited by J.W. Duyff (see our 2nd preface and foreword).

Clinical journals, such as Clinical Science, and the Journal of Clinical Investigation, publish much quantitative data whose interpretation depends on biomathematical models, which from the nature of these journals are usually given very briefly. Clinical Pharmacology and Therapeutics has also a regular feature "Clinical biostatistics" by A.R. Feinstein, which is stimulating and provocative.

Biometrics and Biometrische Zeitschrift mainly publish biostatistics and methodology, but also biomathematics, and nearly always with good numerical illustrations. Much of the biomathematics used to be in genetics, for historical reasons, and still is, but this emphasis seems to be gradually changing.

Many journals devoted to computers in medicine have started recently, and they naturally contain mathematical and statistical applications. They include Computers and Biomedical Research, Methods of Information in Medicine, Revue d'informatique médicale, and the more technical Computer Programmes in Medicine.

This is a small fraction of journals that at least sometimes contain articles within our field.

Books

This is, inevitably, a small selection from very many books.

Biomathematics

E. Batschelet, *Introduction to mathematics for life scientists.* (Springer, Berlin, 1971). The scope and aims of this book are similar to ours. The mathematics is given more briefly and intuitively, and is throughout interspersed with a large number and variety of biological and medical illustrations.

C.A.B. Smith, *Biomathematics, Vol. 1. Algebra, geometry and calculus* (4th. ed.; Griffin, London and Hafner, New York, 1966). This is revised from the first half of the same author's *Biomathematics* (Griffin, 1954) which is based on the pioneer book by W.M. Feldman (1923 and 1935). It is relatively more mathematical in content than our book and Batschelet's.

N.T.J. Bailey, *The mathematical approach to biology and medicine* (Wiley, New York, London, Sydney, 1967). This consists mainly of review essays on the role of mathematical, numerical and statistical methods, with examples, including the role of operational research and computers.

G. Fuchs, ***Mathematik für Mediziner und Biologen*** (Springer, Berlin, New York, 1969). This is a pocket book based on lecture notes of mathematical methods, mainly differential and integral calculus; also numerical methods.

D.S. Riggs, *The mathematical approach to physiological problems. A critical primer* (M.I.T. Press, paperback, 1970). This is an unusual treatment by a physiologist. Most parts assume some knowledge of physiology; there are also accounts of pitfalls in numerical analyses, including those involving negative exponentials. (Cf. Chapter 10, § 15.)

D.A. Franklin and G.B. Newman, *A guide to medical mathematics* (Blackwell, Oxford, 1973). This as the title implies is mainly mathematics, and with emphasis on numerical applications and computing (of all kinds, i.e. simple plotting, or using tables, desk machines, or electronic computers).

Computing

There are many introductions. One for medical readers is: T.R. Taylor, *The principles of medical computing* (Blackwell, Oxford, 1967).

For lengthy numerical calculations within the scope of our book a digital computer would generally be programmed in Algol or Fortran. (One of us, M.E.W., uses Fortran IV for all his computer work.) This kind of programming has become easier to learn and to carry out as the computers acquire new technical facilities. Also, more and more electronic desk calculating machines are being produced; some of them can be programmed. The developments are so rapid that we feel we should not give detailed advice, but refer instead to computer journals (see e.g., p. 650) and their reviews.

Statistics

The second half of the 1954 edition of C.A.B. Smith's *Biomathematics* listed above provides a good mathematical introduction. This has now become: *Biomathematics, Vol. 2, Numerical methods, matrices, probability, statistics* (4th ed.; Griffin, London, 1969).

P. Armitage, *Statistical methods in medical research* (Blackwell, Oxford, 1971). This introduces clearly every method in common use without advanced mathematics.

P.D. Oldham, *Measurement in medicine. The interpretation of numerical data* (English University Press, London, 1968). He concentrates on "why", Armitage more on "how". Both have very many numerical examples.

At an elementary level, in handling numerical bio-medical data we believe that mathematics is just as important or more so than statistical methods alone. The conclusion reached recently at a conference of experienced medical statisticians in London was even that statistics should not be taught to medical students! (reported in Appl. Statistics **20** (1971) 319).

Mathematics in general

Fortunately the basic mathematics does not change. The emphasis and method of presentation can well do so. Computer programming languages may help to change customs and conventions in notation; there are indications of this in Franklin and Newman's book mentioned above. We would still recommend the various books mentioned in our mathematical sections, most of which are unchanged from the first edition; for many of these books later editions are available, and with only minor revisions. The authors are included in the list on page 667.

There is now more self-examination of scientists than ever before. For the role of the mathematician in general and in our subject in particular, see: R. Bellman, *Some vistas of modern mathematics; dynamic programming, invariant imbedding and the mathematical biosciences* (University of Kentucky Press, 1968).

MORE ADVANCED THEORETICAL AND MATHEMATICAL BIOLOGY AND BOOKS ON SPECIAL TOPICS

See also the lists at the end of chapters 6, 7, 8 and 10. A comprehensive review is being prepared by J.A.J. Metz of the Institute for theoretical biology in Leiden University. We are grateful to him for his help with this list.

All subjects

R. Rosen, editor, *Foundations of mathematical biology*. Vol. I *Subcellular Systems*, II *Cellular Systems*, III *Super-cellular Systems* (Academic Press, New

York, London, 1972, 1973). This is a comprehensive survey of theoretical biology, mainly from the point of view of the "Rashevsky" school (see our periodicals list). The technical level is high, the biology is advanced, the mathematics includes most levels in our book and is sometimes more advanced.

Introductory lectures, with many mathematical applications, are given in: N. Rashevsky, editor, *Physicomathematical aspects of biology* (Academic Press, New York, London, 1962).

Simulation

F. Heinmets, editor, *Concepts and models of biomathematics: simulation techniques and methods* (Marcel Dekker, New York, 1969). The sub-title describes the content better than the main title. A great variety of simulation projects are reviewed.

A controversial work recently in the news describes results of a computer simulation of a bio-mathematical model: D.L. Meadows, *The limits to growth; a report for the Club of Rome project on the predicament of mankind.* (Universe Books, New York, 1972). The model may be too arbitrary, but the work could provide a starting point.

Methodology and formal logic

M. Beckner, *The biological way of thought* (University of California Press, 1968). This was one of the first studies of the methodology of particular biological problems.

L. van Bertalanffy, *General system theory; foundations, development, applications* (G. Braziller, New York, 1968). These articles are especially important historically.

Numerical taxonomy

This is more a branch of applied statistics. We recommend: R.R. Sokel and P.M.A. Sneath, *Numerical taxonomy; the principles and practice of numerical classification* (2nd. ed.; Freeman, San Francisco, 1973).

Populations, ecology

A good book, on demography only, is: N. Keyfitz, *Introduction to the mathematics of population* (Addison Wesley, Reading, Massachusetts, 1968).

Metabolism

Almost every part of theoretical biology can be applied to metabolic problems. The book of the Heligoland symposium, 1967, demonstrates this. A. Locker, editor, *Quantitative biology of metabolism* (Springer, Berlin, Heidelberg, New York, 1968).

Neurophysiology and Neurobiology

J.S. Griffith, *Mathematical neurobiology; an introduction to the mathematics of the nervous system* (Academic Press, New York, London, 1971). There are many recent applications of mathematics of about the same level and content as much of our book.

M.A. Arbib, *Brains, machines and mathematics* (McGraw-Hill, New York, 1961). This small volume is mathematically elementary and conceptually advanced.

K.S. Cole, *Membranes, ions and impulses: a chapter of classical biophysics* (University of California Press, Berkeley, 1968).

R. Fitzhugh, *Mathematical models of excitation and propagation in nerve.* Chapter 1 in: H.P. Schwan, editor, *Biological engineering* (McGraw-Hill, New York, 1969).

H.M. Lieberstein, *Mathematical physiology; blood flow and electrically active cells* (Elsevier, New York, 1973). This is a numerical treatment based on modifications of the Hodgkin-Huxley equations (Chapter 10, §7).

Physiological tracers

C.W. Sheppard, *Basic principles of the tracer method* (Wiley, New York, 1962). This is a good introduction to interpreting tracer data both by compartmental models and by means of skew probability (e.g. random-walk) curves.

The mathematical of compartmental models is treated in several recent books including, in increasing order of mathematical level:
G.L. Atkins, *Multiconpartment models for biological systems* (Methuen

The mathematics of compartmental models is treated in several recent books, including, in increasing order of mathematical level:

G.L. Atkins, *Multicompartment models for biological systems* (Methuen, London, 1969), and
F. Chevallier, *Biodynamique et indicateurs* (Gordon and Breach, London, New York, Paris, 1972).

A. Habermehl, E.H. Graul and H. Wolter, *Computer in der Nuklearmedizin. Die Anwendung des Analogrechners zur Untersuchung biologischer Systeme und Prozesse* (Hüthig, Heidelberg, 1969).

J.A. Jacquez, *Compartmental analysis in biology and medicine. Kinetics of distribution of tracer-labelled materials* (Elsevier, Amsterdam, 1972).

Jacquez also considers non-linear models, circadian rhythms and feedback and control. All four books give many numerical illustrations, mainly in terms of negative exponentials.

For a more critical treatment of the fundamentals, see especially: E.E. Bergner, *The concept of mass, volume and concentration,* in the symposium book edited by himself and C.C. Lushbaugh: *Compartments, pools and spaces in medical physiology* (U.S.A.E.C. Symposium, no. 11, Oak Ridge, 1967). The 20 contributions include a great variety of applications and there are many critical discussions.

ANSWERS TO THE PROBLEMS

CHAPTER 1

p. 7. **3.** $y = (1 + x^2)/(1 - x^2)$.
4. $(x + y)^3 + 2(x^3 + y^3) = 3(x + y)(x^2 + y^2)$.
5. $n^2 = (n - 1)(n + 1) + 1$; 99 980 001.

p. 10. **1.** $(1 + x^2)/2x$.
2. (i) $(x^2 + y^2)(x - y)^{-\frac{4}{3}}$; (ii) $a(a^2 + 2a + 3)^{-\frac{1}{2}}$.
3. 4.

p. 15. **1.** $b = a/(a - 1)$.

p. 22. **1.** $n^2(2n^2 - 1)$. **2.** $(1 - x)^{-2}$. **3.** 16; 0.902.
4. $n/(2n + 1)$; $\frac{1}{2}$. **5.** $\frac{1}{2}$.

p. 28. **1.** $\dfrac{40!}{33!\,4!}$. **2.** 369. **4.** 36. **5.** $(n - 1)!$. **6.** $\frac{1}{2}\{(n - 1)!\}$.

p. 34. **1.** $\frac{28}{31}$. **2.** $-1 + 3x^2 + 10x^3$. **4.** 1.931.
5. $a = 0$, $b = 1$, $c = 1$; 6.

p. 39. **4.** The error in r must be less than $\frac{1}{3}\%$.
5. Possible error in η is $7\frac{1}{2}\%$.

p. 44. **2.** $\dfrac{x}{x^2 + 1} + \dfrac{1}{x - 1} - \dfrac{1}{(x - 1)^2}$. **3.** $\dfrac{3}{2(3 + 2x)} - \dfrac{1}{2(5 - 2x)^2}$.

CHAPTER 2

p. 59. **1.** 21.8, 34.1, 49.8, 60.0, 64.4.
2. (i) 62.5°; 12.25 p.m., 7.35 p.m.;
(ii) 71°, 5 p.m.;
(iii) 9 a.m. to 2 p.m.; 4.1° per hour;
(iv) 6.30 p.m. to 8.30 p.m.; 7.2° per hour.
3. 1.15 secs, 4.85 secs.

p. 72. **1.** $f(0) = 1,\ f(1) = 15,\ f(-1) = 3,\ f(2) = 129,$
$f(-2) = 57,\ f(3) = 547,\ f(-3) = 319.$

2. (i) $y = x + 1$; (ii) $3y = 2x - 1$; (iii) $x + 2y = 2$.

3. $w = 0.8t + 71$; 112.5, 36.3.

4. $R = \dfrac{2240 + v^2}{768}$.

5. $a = 13.92,\ b = 0.18$.

6. $y = ax^2 + b$, with $a = 2.1,\ b = 4.5$.

p. 87. **1.** $b = 30.4,\ k = -0.955$.

2. $y = 0.1905x^{0.62}$, where x = total length, y = pre-ocular length.

3. $y = 72.44(x/100)^{3.54}$, where x=mean cranium length (mm), y = mean face length (mm).

4. $y = 0.1445x^{1.2}$, where x = mean carapace length (mm), y = mean abdomen length (mm).

5. The scatter of the points is appreciable but a reasonable "fit" is given by the relation $y = 1.03x^{0.16}$ where y denotes the index of folding and x the weight of the hemisphere.

6. The population is $180\,000(x - 1918)^{0.39}$; 1962.

p. 96. **1.** (1, 8) with tangent $y = 6x + 2$;
(−5, 8) with tangent $6x + y + 22 = 0$.

4. +3, −1.5.

p. 108. **1.** $2x\sqrt{(1 - x^2)}$.

CHAPTER 3

p. 126. **1.** $\frac{1}{2}$. **2.** 3, −4. **3.** a/p.

p. 129. **1.** p_n, 0. **2.** 1, 1.

p. 132. **1.** (i) $12x(x^2 + x - 1)$;
(ii) $-5/x^3$; (iii) $\frac{1}{2}x^{-\frac{3}{2}}(x - 1)$;
(iv) $2x - 3x^{-2} - 3x^{-4} + 12x^{-5}$; (v) $12(2x + 1)^5$.

2. (i) $\dfrac{1}{2\sqrt{(1 + x)}}$; (ii) $-\dfrac{2}{(1 + x)^3}$; (iii) $\dfrac{1}{3\sqrt[3]{(1+x)^2}}$.

p. 134. **1.** $y' = -Aa \sin(ax + b)$, $y'' = -Aa^2 \cos(ax + b)$.
2. $y' = 4Ax^3 + 2Bx$, $y'' = 12Ax^2 + 2B$, $y''' = 24Ax$, $y^{IV} = 24A$.

CHAPTER 4

p. 153. **1.** (i) $-\dfrac{(1 + x^2)\sin x + 2x\cos x}{(1 + x^2)^2}$;

(ii) $\dfrac{(x^3 + 2)\cos x - 3x^2 \sin x}{(x^3 + 2)^2}$;

(iii) $\tan x + x\sec^2 x$;

(iv) $\dfrac{(1 - x^2)\tan x + x(1 + x + x^2)\sec^2 x}{(1 + x + x^2)^2}$.

2. $y' = -\dfrac{2Ax}{(x^2 + 1)^2} + 2Bx, \quad y'' = \dfrac{2A(3x^2 - 1)}{(x^2 + 1)^3} + 2B.$

p. 156. **1.** $an(ax + b)^{n-1}$. **2.** $n(2ax + b)(ax^2 + bx + c)^{n-1}$.
3. $(a/2x)^{\frac{1}{2}}$. **4.** $-\frac{1}{2}(k - x)^{-\frac{1}{2}}$. **5.** $-8x(a^2 - x^2)^3$.
6. $-\cos^{n-1} x \cdot \sin x$. **7.** $3x^2 \cos(x^3 + 3)$.
8. $a\sec^2(ax + b)$. **9.** $2a/(a - x)^2$.
10. $a(a + x)^{-\frac{1}{2}}(a - x)^{-\frac{3}{2}}$. **11.** $\frac{1}{2}(2a + 3x)(a + x)^{-\frac{1}{2}}$.
12. $mnx^{n-1}\sin^{m-1}(x^n)\cos(x^n)$.

13. $\dfrac{2x\{1 - (n - 1)(1 + x^2)^n\}}{[1 + (1 + x^2)^n]^2}$.

14. $1 - \dfrac{x}{\sqrt{(x^2 - 1)}}$. **15.** $\dfrac{5(x + 1)}{2x\sqrt{x}}\left(\sqrt{x} - \dfrac{1}{\sqrt{x}}\right)^4$.

16. $\frac{20}{3}$ sq. meters/sec. **17.** $\frac{8}{3}$ cm/sec.

p. 160. **1.** (i) $12(9x^2 - 24x + 32)^{-1}$; (ii) $\tan^{-1} x + \dfrac{x}{1 + x^2}$;

(iii) $\dfrac{2x}{\sqrt{(1 - x^4)}}$; (iv) $\dfrac{1}{2\sqrt{(x - x^2)}}$.

4. $x - 6y + 3 = 0$.

p. 164. **2.** $\frac{1}{6}$.

p. 173. **1.** $x = -1$ gives a point of inflexion; $x = 5$ gives a minimum.
2. Maximum at $x = \frac{1}{2}\pi$, minima at $x = \frac{1}{6}\pi, \frac{5}{6}\pi$.
4. Selling price \$2, profit \$625.

p. 178. **1.** There must be a small decrease of $2n$ % in the height.
2. $x = 1$.

p. 184. **1.** $y = 7x - 6$.

p. 188. **1.** 1.41429, 2.15538. **2.** 2.180. **3.** 2.414.

p. 189. **1.** (i) 1; (ii) 3; (iii) $-\frac{1}{3}$; (iv) $\frac{1}{2}\sqrt{3}\pi$.
2. (i) 1; (ii) 0.

CHAPTER 5

p. 199. **3.** (i) $\frac{242}{5}$; (ii) 60; (iii) 1; (iv) $\frac{1}{2}\pi$; (v) 1; (vi) $\frac{1}{4}\pi$.

p. 214. **1.** $x + x^2 + 2x^3$; 4; 44. **2.** $\frac{1}{8}\pi$; $\frac{1}{4}\pi$.

p. 216. **1.** $\frac{1}{16}(8^8 - 5^8)$.

CHAPTER 6

p. 227. **1.** (i) $\dfrac{2x}{1+x^2}$; (ii) $\dfrac{2}{1-x^2}$; (iii) $\dfrac{2x}{x^4-1}$;
(iv) $-\dfrac{n(1+\log x)^{n-1}}{x(\log x)^{n+1}}$; (v) $\sec x$;
(vi) $-\dfrac{2a}{e^{ax}-e^{-ax}}$; (vii) $-\dfrac{4\sin x}{4-\cos^2 x}$.
2. Minimum value $-e^{-1}$ at $x = e^{-1}$.

p. 230. **1.** (i) $(1-x^2)e^{-x}$; (ii) $-2e^{-2x}[\cos(2x+3)+\sin(2x+3)]$;
(iii) $-\cos x\, e^{-\sin x}$; (iv) $(1-2x^2)e^{-x^2}$.
2. Curve has maximum at $(-1, 4e^{-1})$, and minimum at $(1, 0)$.
3. Curve has point of inflexion at $x = 0$ and a maximum at $x = 6$.
4. $b = \sqrt{2}a$, $\beta = \alpha + \frac{5}{4}\pi$;
$d^2y/dx^2 = 2ae^{-x}\sin(x + \alpha + \frac{5}{2}\pi) = -2ae^{-x}\cos(x+\alpha)$.

p. 261. **3.** $x - \frac{1}{2}x^2 + \frac{1}{6}x^3 - \frac{1}{12}x^4$.

CHAPTER 7

(Here C denotes an arbitrary constant.)

p. 279. **1.** (i) $\frac{1}{2}e^{x^2} + C$; (ii) $x + \log(2x+1) + C$;

(iii) $\frac{1}{2}a\log(x^2+c^2) + \frac{b}{c}\tan^{-1}\left(\frac{x}{c}\right) + C$;

(iv) $\frac{1}{4}\log[(1+x^2)/(1-x^2)] + C$.

2. $2\sqrt{(\sin x)} + C$.

p. 286. **1.** (i) $\frac{1}{2}x^2 + \log(x+1) + \log(x+2) + C$;

(ii) $\tan^{-1}x + \dfrac{1}{x+1} + C$;

(iii) $\frac{1}{2}\log(x^2+1) - \dfrac{1}{2(1+x^2)} - \log(x+1) + C$.

3. (i) $\log(x-1) - \frac{2}{3}\log(3x+2) + C$;

(ii) $x^2 + \log x + \frac{1}{2}\log(x^2-1) + C$;

(iii) $x + \log x - x^{-1} + C$;

(iv) $\log x - \log(2x+1) + \frac{1}{2}(2x+1)^{-1} + C$;

(v) $\log\dfrac{x}{x-1} - \dfrac{x+2}{x^2-1} + C$;

(vi) $\log(x+1) - \dfrac{2x+3}{2(x+1)^2} + C$.

p. 294. **1.** (i) $(\frac{1}{2}x+1)\sqrt{x^2+4x+8} + 2\log[x+2+\sqrt{x^2+4x+8}] + C$;

(ii) $(\frac{1}{2}x+1)\sqrt{x^2+4x+3} - \frac{1}{2}\log[x+2+\sqrt{x^2+4x+3}] + C$;

(iii) $(\frac{1}{2}x+1)\sqrt{5-4x-x^2} + \frac{9}{2}\sin^{-1}\left(\frac{x+2}{3}\right) + C$.

2. (i) $\frac{2}{3}\tan^{-1}(\frac{1}{3}\tan\frac{1}{2}x) + C$; (ii) $C - \dfrac{1}{1+\sin x}$;

(iii) $2\tan^{-1}(\tan\frac{1}{2}x) + \dfrac{2}{1+\tan\frac{1}{2}x} + C$;

(iv) $\log(\tanh\frac{1}{2}x) + C$; (v) $\log(1+e^x) + C$;

(vi) $\dfrac{2}{\sqrt{5}}\tan^{-1}\left(\dfrac{1}{\sqrt{5}}\tanh\tfrac{1}{2}x\right) + C;$

(vii) $\dfrac{1}{\sqrt{5}}\log\dfrac{\sqrt{5}+\tanh\frac{1}{2}x}{\sqrt{5}-\tanh\frac{1}{2}x} + C.$

6. (i) $\pi/40$; (ii) $\frac{1}{4}$; (iii) $\frac{9}{4}\pi$.

p. 303. **2.** $x(\log x)^3 - 3x(\log x)^2 + 6x\log x - 6x.$

p. 310. **1.** $\frac{7}{24}$.

3. $$J(2m) = \frac{(2m-1)(2m-3)\cdots 3\cdot 1}{2m(2m-2)\cdots 4\cdot 2}\left(\frac{\pi}{2}\right);$$

$$J(2m-1) = \frac{(2m-2)(2m-4)\cdots 4\cdot 2}{(2m-1)(2m-3)\cdots 3\cdot 1}.$$

4. (i) $\dfrac{63\pi}{512}$; (ii) $\dfrac{35\pi}{256}$; (iii) $\dfrac{128}{215}$; (iv) $\dfrac{1}{40}$;

(v) $\dfrac{5\pi}{2048}$; (vi) $\dfrac{8}{315}$.

p. 322. **1.** $\dfrac{1}{2}\dfrac{3!2!}{6!} = \dfrac{1}{120}$; $\dfrac{1}{2}\dfrac{3!3!}{7!} = \dfrac{1}{280}$.

2. $\dfrac{2!7!}{10!} = \dfrac{1}{360}$; $\dfrac{5!5!}{11!} = \dfrac{1}{2772}$.

CHAPTER 8

p. 381. **4.** $$\frac{\partial^2\psi}{\partial r^2} = n(n-1)r^{n-2}\cos(n\theta+\alpha),$$

$$\frac{\partial^2\psi}{\partial\theta^2} = -n^2r^n\cos(n\theta+\alpha),$$

$$\frac{\partial^2\psi}{\partial r\partial\theta} = -nr^{n-1}\sin(n\theta+\alpha).$$

p. 390. **1.** $\delta\rho = \rho\left\{\dfrac{\delta h}{h} - \dfrac{0.00367\delta t}{1+0.00367t}\right\}$; -0.073.

2. 719 calories.

3. $\frac{1}{2}l + \dfrac{hH}{r+H}\cdot$

4. $p(\mu) \leqq \dfrac{\pi\phi}{216}\cot\frac{1}{2}(\delta + A) + \dfrac{\pi\theta}{216}[\cot\frac{1}{2}A - \cot\frac{1}{2}(\delta+A)];$

$p(\mu) < 0.03\ \%.$

p. 394. **1.** (i) exact: $\frac{1}{4}(x^4 + y^4) + \frac{3}{2}x^2y^2$; (ii) not exact;
(iii) exact: $x - y^2/x$;
(iv) exact: $\frac{1}{3}(x^3 + y^3) - 2x^2y - 2xy^2$;
(v) not exact; (vi) exact: $-\dfrac{1}{x} - \dfrac{y}{x^2}\cdot$

3. $n = 2$.

p. 397. **1.** 57/10. **2.** 10/3.
3. Along $y = 3x - 2$, $\int \Delta f = 527/15$, $\int x^{-2}\Delta f = 32$;
along $y = x^2$, $\int \Delta f = 1964/21$, $\int x^{-2}\Delta f = 32$;
$x^{-2}\Delta f = \mathrm{d}\phi$, where $\phi(x, y) = xy^2 + y/x$, and
$\phi(2, 4) - \phi(1, 1) = 34 - 2 = 32$.

p. 414. **1.** (i) $x = y = \frac{1}{6}$ yield the minimum value 0;
$(0, \frac{1}{3})$ is a saddle point.
(ii) $x = \pm\sqrt{2}$, $y = \mp\sqrt{2}$ yield minimum values -8.
(iii) $x = \pm\frac{1}{3}\pi$, $y = \pm\frac{1}{3}\pi$ yield minimum values $-\frac{1}{8}$;
$x = 0$, $y = 0$ yield the maximum value 1.
2. The maximum value is $ab/27$ given by $x = \frac{1}{3}a$, $y = \frac{1}{3}b$.

p. 422. **2.** $10\pi a^3$.
4. $\frac{3}{4}(a^2 + b^2)^2$.

CHAPTER 9

(Here c, c_1, c_2, . . . denote arbitrary constants.)

p. 447. **1.** $y = cxe^x$.
2. $y^2 + ay = cx$.
3. $(x + 1)(y^2 + 1) = cx$.
4. $\sin y = c\sqrt{(\sec 2x)}$.
5. $x(x^2y + 1) = cy^2$.

p. 451. **1.** $(x+y)^2 = c(x-y)^3$.
2. $2x = (x-y)(\log x + c)$.
3. $(y-x)^2 = cx\mathrm{e}^{-y/x}$.
4. $(x-1)^2 + (y-1)^2 = c \exp\{2\tan^{-1}(y-1)/(x-1)\}$.
5. $(x-y+3)^4 = c\{(x+2)^3 - (y-1)^3\}$.

p. 452. **1.** $\frac{1}{3}x^3 + 2xy + 3y^4 = c$.
2. $2x^4 + 2x^2y + 3y^4 = c$.
3. $3x^2 + 4xy + 3y^2 = c$.
4. $x^2 + xy + y^2 = c$.
5. $x^2 - 6xy - 2y^{-1} = c$.

p. 456. **2.** (i) $3y^2 + 6x^3y + 2x^3 = c$; (ii) $9y^2 + 12x - 4 = c\mathrm{e}^{-3x}$; (iii) $x^2/y^2 - x + y = c$; (iv) $y = x^2 \cot(c - x^2)$.
3. $f(xy) = (xy)^{-3}$; $x^2y^2 = c(y-x)$.

p. 459. **1.** $y = (x+c)(1-x^2)^{-\frac{1}{2}}$.
2. $y = c\mathrm{e}^{\frac{1}{2}x^2} - \mathrm{e}^{ax}$.
3. $y = \frac{1}{6}x^5 + c/x$.
4. $3y(x^2-1) = x(c + 3x - x^3)$.
5. $ay = x\sqrt{(c+x^2)}$.

p. 463. **1.** (i) $y = \mathrm{e}^{\frac{1}{2}x^2}(c+2x)^{-\frac{1}{2}}$; (ii) $y = \mathrm{e}^{\frac{1}{2}x}(x+c)^{-\frac{1}{2}}$; (iii) $y = x(c + 3\sin x)^{-\frac{1}{3}}$.
2. $(x+c)y = x^2 - 2x - c$.
3. $(x^3 + 3x + c)y = x^4 - 2cx + 3$.
4. $y = (x \pm 2)/x^2$ are particular integrals. The general solution corresponding to the $+$ sign is

$$x^2y = 2 + x + 4(c\mathrm{e}^{-4/x} - 1)^{-1}.$$

p. 468. **2.** The particular integrals are e^{-x}, $1 + x^2$ and they are linearly independent since $W(1+x^2, \mathrm{e}^{-x}) = -(1-x)^2\mathrm{e}^{-x} \neq 0$. Hence the general solution is

$$y = A\mathrm{e}^{-x} + B(1+x^2).$$

p. 478. **1.** $y = c_1 \cos nx + c_2 \sin nx + (n^2 - a^2)^{-1} \cos ax$.

2. $y = c_1\mathrm{e}^{3x} + c_2\mathrm{e}^{4x} + \dfrac{12x+7}{144}$.

3. $y = c_1 e^{\frac{2}{3}x} + c_2 e^{-\frac{3}{2}x}$.
4. $y = c_1 e^{20x} + c_2 e^{-3x}$.
5. $y = c_1 e^{-x} + c_2 e^{-4x} + 8x^2 - 20x + 21$.
6. $y = c_1 e^{2x} + (c_2 - x)e^{-x}$.
7. $y = c_1 + c_2 e^{-\frac{1}{2}x} + x^2 - 4x - xe^{-\frac{1}{2}x}$.
8. $y = c_1 e^{-x} + c_2 + 2x - x^2 + \frac{1}{3}x^3$.
9. $y = (c_1 - x)e^{x} + c_2 e^{2x} + x^2 + 3x + \frac{7}{2}$.
10. $y = (c_1 \cos 3x + c_2 \sin 3x)e^{-x} + \frac{1}{37}(\cos 3x + 6 \sin 3x)$.

p. 487. **1.** $y = (c_1 + c_2 x + c_3 x^2 + c_4 x^3)e^{x}$.
2. $y = (c_1 + c_2 x) \cos 2x + (c_3 + c_4 x) \sin 2x$.
3. $y = (c_1 + c_2 x)e^{2x} + c_3 e^{-x}$.
4. $y = c_1 e^{x} + c_2 e^{-x} + c_3 e^{-2x}$.
5. $y = c_1 \cos x + c_2 \sin x + c_3 e^{2x} + \frac{3}{4} - \frac{1}{2}x - \frac{1}{2}x^2$.
6. $y = c_1 + (c_2 - x + \frac{1}{2}x^2)e^{x}$.
7. $y = c_1 e^{x} + (c_2 - \frac{1}{4}x) \cos x + (c_3 - \frac{1}{4}x) \sin x$.
8. $y = c_1 e^{x} + c_2 e^{-x} + c_3 \cos x + (c_4 - \frac{1}{4}x) \sin x$.
9. $y = (c_1 + \frac{1}{8} \sin 2x - \frac{1}{8} \cos 2x)e^{-x} + c_2 e^{-3x}$.
10. $y = \left(c_1 + \dfrac{x^2}{4n^2}\right) \cos nx + \left(c_2 - \dfrac{x}{4n^3} + \dfrac{x^3}{6n}\right) \sin nx$.

p. 490. **2.** $y = \frac{2}{3}e^{-\frac{1}{2}t} \cos (\frac{1}{2}\sqrt{3}t) + \frac{1}{3}e^{t} - 1$.
3. $y = \frac{1}{2}t - \frac{1}{4}e^{-t} \sin 2t$.
4. $y = -\frac{1}{4}e^{-2t} - \frac{1}{2}te^{-t}$.
5. $y = 2e^{-t} - e^{-\frac{1}{2}t}\left\{\cos (\frac{1}{2}\sqrt{7}t) + \dfrac{1}{\sqrt{7}} \sin (\frac{1}{2}\sqrt{7}t)\right\}$.

p. 495. **1.** $y = c_1 \cos (\log x + c_2) + \log x$.
2. $y = e^{-x}(c_1 x^2 + c_2/x)$.
3. $y = x \log x + c_1 \left\{\frac{1}{2}x \log \dfrac{x+1}{x-1} - 1\right\} + c_2 x$.
4. $y = (x + 1)^{-1}\{x^2 + c_1 + c_2 \log x\}$.

p. 502. **2.** $x = -b \sin kt,\ y = b \cos kt + \dfrac{a}{k} \sin kt$.

3. $x = \frac{V}{2c}\left\{\sin(ct) + \frac{1}{\sqrt{3}}\sin(\sqrt{3}ct)\right\},$

$$y = \frac{V}{2c}\left\{\sin(ct) - \frac{1}{\sqrt{3}}\sin(\sqrt{3}ct)\right\}.$$

4. $x_0 = e^{-t},\ x_1 = te^{-t},\ x_2 = \frac{1}{2}t^2e^{-t},\ x_3 = \frac{1}{6}t^3e^{-t}.$

5. $x = \frac{1}{2}(3e^{2t} + 2e^{3t} + e^{6t}),\ y = -e^{3t} + e^{6t},$
$z = \frac{1}{2}(-3e^{2t} + 2e^{3t} + e^{6t}).$

p. 510. **1.** The equation transforms to

$$\frac{\partial z}{\partial \eta} = \eta - \frac{\xi^2}{\eta}$$

with solution

$$z = \tfrac{1}{2}\eta^2 - \xi^2 \log \eta + f(\xi)$$

where $f(\xi)$ is an arbitrary function of ξ. The required solution is therefore

$$z = 2x^2y^2 - (x^2 + y^2)^2 \log(2xy) + f(x^2 + y^2).$$

3. The equation transforms to

$$\frac{\partial z}{\partial \xi} = \tfrac{1}{2}\xi \sin \xi$$

with solution

$$z = \tfrac{1}{2}(\sin \xi - \xi \cos \xi) + f(\eta),$$

when $f(\eta)$ is an arbitrary function of η. The required solution is therefore

$$z = \tfrac{1}{2}\{\sin(xy) - xy\cos(xy)\} + f\left(\frac{x}{y}\right).$$

4. The equation transforms to

$$\frac{\partial z}{\partial \xi} = 4\xi\eta^2 \log \xi$$

with solution

$$z = 2\xi^2\eta^2 \log \xi - \xi^2\eta^2 + f(\eta)$$

where $f(\eta)$ is an arbitrary function of η. The required solution is therefore

$$z = 2x^2 \log (x + y) - x^2 + f\left(\frac{x}{x + y}\right).$$

5. The equation transforms to

$$\frac{\partial^2 u}{\partial \xi \partial \eta} = 0$$

with solution

$$u = f(\xi) + g(\eta)$$

where f, g denote arbitrary functions. Hence the solution of the original equation is

$$u = f\left(\frac{x^2}{2a} - ct\right) + g\left(\frac{x^2}{2a} + ct\right).$$

If we take

$$f\left(\frac{x^2}{2a} - ct\right) = \tfrac{1}{2} \cos\left(\frac{x^2 \omega}{2ac} - \omega t\right),$$

$$g\left(\frac{x^2}{2a} + ct\right) = \tfrac{1}{2} \cos\left(\frac{x^2 \omega}{2ac} + \omega t\right),$$

we get the solution given.

ADDENDA AND CORRIGENDA

Pages 431, 432 and 437. In Fig. 123 the values of H have been obtained in terms of logarithms to the base 2. The corresponding unit of information is called the *bit*. This is shortened from *binary digit*, since a 0 or 1, each occurring with equal probability $\frac{1}{2}$, then carries 1 unit of information.

It follows that the ratio of 1 Hartley (p. 437) to 1 bit is:

$$\log_2 p / \log_{10} p = \log_2 10 = 1/\log_{10} 2 = 3.32.$$

NAME INDEX

See also the reference lists ending Chapters 6, 7, 8 and 10, and "further reading".

SUBJECT INDEX